Contemporary Research Topics in
NUCLEAR PHYSICS

Contemporary Research Topics in NUCLEAR PHYSICS

Edited by

Da Hsuan Feng and Michel Vallières

Drexel University
Philadelphia, Pennsylvania

and

Michael W. Guidry and Lee L. Riedinger

University of Tennessee
Knoxville, Tennessee

PLENUM PRESS • NEW YORK AND LONDON

Library of Congress Cataloging in Publication Data

Main entry under title:

Contemporary research topics in nuclear physics.

"Proceedings of a workshop held at Drexel University from September 1 to September 3, 1980, under the joint auspices of Drexel University, the University of Tennessee, and Vanderbilt University"—Pref.
Bibliography: p.
Includes index.
1. Nuclear physics—Congresses. I. Feng, Da Hsuan, 1945- II. Drexel University. III. University of Tennessee (Knoxville campus). IV. Vanderbilt University.
QC770.CC6 539.7 82-3677

ISBN-13:978-1-4684-1136-2 e-ISBN-13:978-1-4684-1134-8

DOI: 10.1007/978-1-4684-1134-8

Proceedings of a workshop on Nuclear Physics, held September 1–3, 1980, at Drexel University, Philadelphia, Pennsylvania

Softcover reprint of the hardcover 1st edition 1982
A Division of Plenum Publishing Corporation
233 Spring Street, New York, N.Y. 10013

PREFACE

This volume contains the proceedings of a workshop held at Drexel University from September 1 to September 3, 1980, under the joint auspices of Drexel University, The University of Tennessee and Vanderbilt University.

The workshop dealt with subjects of topical importance to the nuclear physics community: high spin phenomena, heavy ion reactions, transfer reactions, microscopic theories of nuclear structure and the interacting boson model, and miscellaneous topics. This proceedings contains all of the invited papers plus short manuscripts expanding on the materials of the invited papers.

A total of about 85 participants came to the workshop. The format of the conference was kept informal on purpose, so as to facilitate the discussions. Unfortunately, these discussions, at times intense, could not be included in this volume due to the lack of secretarial help during the meeting.

A great deal of current information was exchanged during the conference. However, the full impact of a conference can only be realized when the proceedings have been published and read by participants as well as other colleagues in this field of physics who were not in attendance. We sincerely hope that these proceedings will be useful in this regard.

The Organizing Committee is particularly grateful to the Conference participants for their patience and in helping us prepare the proceedings. The generous financial help of Drexel University, University of Tennessee and Vanderbilt University was instrumental in making this conference a reality, and is greatly appreciated. Last but not least, the professional help of Mr. Palliparambil Joseph Antony deserves to be especially mentioned, without which completion of this volume would have been impossible.

The Editors

CONTENTS

Note: The name in capitals indicates the author who presented the talk.

QUASIPARTICLE MOTION IN ROTATING NUCLEI

S. Frauendorf

Central Institute for Nuclear Research
Rossendorf, 8051 Dresden, PF 19
German Democratic Republic

I. ABSTRACT

The rotational bands of deformed nuclei with high spin can be interpreted as configurations of independent quasiparticles in a deformed potential rotating with a constant frequency and with a pairfield present. This generalization of the shell model to rotating potentials is called the Cranked Shell Model (CSM). The classification with respect to the symmetry quantum number signature is introduced. Several CSM-predictions, such as the additivity of quasiparticle energies and angular momenta, the occurrence of characteristic frequencies around which band crossings cluster and the interference between quasiparticle orbits are discussed and compared with experimental data on high spin states in the light rare earth region.

II. INTRODUCTION

The ascent of heavy ion accelerators has created the possibility to study nuclei that carry as much angular momentum as they can accomodate without fissioning. The region of a few MeV excitation above the yrast line is being intensively studied. In this yrast region the rotating nucleus is cold, i.e. its level density is comparable with that near the ground state. Moreover, it takes the rotating nucleus at least 10^{-12}s to de-excite into the ground state. This time is very long in comparison with the typical nuclear time of 10^{-22}s and the states in the yrast region may be considered as stationary ones. This led to the concept of "Yrast-spectroscopy" which extends the spectroscopy of low lying states to the whole yrast region[1]. Recent experimental yrast-spectroscopy of discrete γ-lines has reached angular momentum of $I \simeq 30\,\hbar^2$. The region of higher angular momenta is being explored by studying the γ-continua[3].

Since the nucleus is cold in the yrast region the mean free path of the nucleons is large enough that the independent particle model is expected to work. The majority of the excited states above the yrast line should correspond to a rearrangement of a few nucleonic orbitals. Thus it is obvious to base the analysis of the yrast region on an appropriate version of the shell model. One may distinguish two important cases:

i. The nucleus rotates about its symmetry axis. The rotation does not directly modify the nucleonic orbits (only via changes of the potential) and one may employ the standard shell model with a spherical or deformed potential. The experimental data are consistent with such an approach[1,4].

ii. The nucleus rotates about an axis perpendicular to its symmetry axis. The rotation of the deformed potential modifies the nucleonic orbits which depend on the angular frequency. The rotation is collective and the states group into rotational bands.

My talk will only deal with case (ii). It is based mainly on the work by R. Bengtsson and S. Frauendorf[5,6] (cf. also ref.1).

III. THE CRANKED SHELL MODEL

We assume that the deformed shell-model potential rotates with the constant angular frequency ω about an axis perpendicular to its symmetry axis. In the frame of reference attached to the potential the shell model Hamiltonian reads

$$h' = h - \omega j_x \tag{1}$$

where h is a standard deformed shell model Hamiltonian and z and x are chosen to be the symmetry and rotation axis, respectively. We suggest calling the Hamiltonian the "Cranked Shell Model" (CSM) in order to relate it to the Cranking Model. The latter notation (also Cranking HFB) is usually used in the context of a self-consistent treatment of the independent particle Hamiltonian (1), providing total excitation energies with respect to the ground state. The simpler CSM continues the tradition of shell models assuming that h does not depend on the configuration; h may depend on ω, but so far the consequences of this possibility have not been studied in a detail. As any shell model, CSM aims only at the calculation of excitation energies with respect to a reference configuration that depends on the frequency ω (a band which is yrast of near yrast).

TABLE I

TYPE OF SHELL MODEL	INDEPENDENT PARTICLE HAMILTONIAN	SYMMETRIES	CONSERVED QUANTITIES
Spherical	h_{sph} = kinetic energy + spherical potential	$\mathcal{R}_x$, $\mathcal{R}_y$, $\mathcal{R}_z$ $\mathcal{P}$, $\mathcal{T}$, $\mathcal{R}_G$	j, K, π, N
Deformed	$h_{def} = h_{sph} + V_{def}$ V_{def}: deformation potential (axial, reflection symmetric)	$\mathcal{R}_z$, $\mathcal{R}_x(\pi)$, $\mathcal{P}$, $\mathcal{T}$, $\mathcal{R}_G$	K, α, π, N
Quasi-Particle	$h_{qp} = h_{def} - \lambda + \Delta$ Δ: monopole pair-field λ: chemical potential	$\mathcal{R}_z$, $\mathcal{R}_x(\pi)$ $\mathcal{P}$, $\mathcal{T}$	K, α, π
Cranked	$h'_{sp} = h_{def} - \omega j_x$ ωj_x: cranking potential	$\mathcal{R}_x(\pi)$, $\mathcal{P}$, $\mathcal{R}_G$	α, π, N
Cranked Quasi-Particle	$h'_{qp} = h_{qp} - \omega j_x$	$\mathcal{R}_x(\pi)$,	α, π

In order to explore the basic structure of the spectra, the residual interaction between particles is neglected. Then, the spectrum may be generated by adding the single particle energies given by the CSM-Hamiltonian[1]. Obviously this additivity holds only if states with the same frequency ω are combined, which may correspond to rather different angular momenta I. Hence, to reveal the underlying single particle structure the experimental spectra are characterized by the angular frequency, which is essentially equal to one half of the frequency of the E2 γ-radiation.

The detailed structure of the CSM Hamiltonian[1] is shown in Table I, which contains a hierarchy of nuclear shell modles in the chronological order of their appearance. In the region $I \sim 30\,\hbar$ the yrast spectra indicate significant pair correlations and the analysis must be based on the cranked quasiparticle Hamiltonian h_{qp}. We employ for h_{def}. Nilsson's modified harmonic oscillator potential; other versions of the deformed shell model potential may be used equally well. The deformation perameters ε_2, ε_4 and the chemical potential λ are kept equal to the ground state values, and the pair field Δ is chosen to be 0.8 of the even-odd mass difference in order to account for blocking in many quasiparticle configurations.

Table 1 demonstrates how new components of the potential break the symmetries. The CSM-Hamiltonian has only two distinct symmetries. Space inversion implies the parity π. The invariance of h' with respect to a reflection through the x-y plane implies the second symmetry quantum number called the signature[1]. We define it by the rotation $_x(\pi)$ about the x-axis by an angle π:

$$_x(\pi)\psi_\alpha = e^{-i\pi\alpha}\psi_\alpha, \qquad \alpha = \begin{matrix} 0, & 1 \\ -1/2, & +1/2 \end{matrix} \quad \text{for} \quad \begin{matrix} \text{even} \\ \text{odd} \end{matrix} \; N \tag{2}$$

In the case of even particle number the signature just determines whether the wavefunction is even or odd with respect to $_x(\pi)$:

$$\psi_\alpha = \psi_{\pm}(\phi) \qquad \text{for} \quad \alpha = \begin{matrix} 0 \\ 1 \end{matrix} \tag{3}$$

where ϕ is the polar angle in the y-z plane and

$$_x(\pi)\psi_{\pm}(\phi) = \psi_{\pm}(\phi - \pi) = \pm\,\psi_{\pm}(\phi) \tag{4}$$

In the case of odd particle number the spin of the unpaired particle

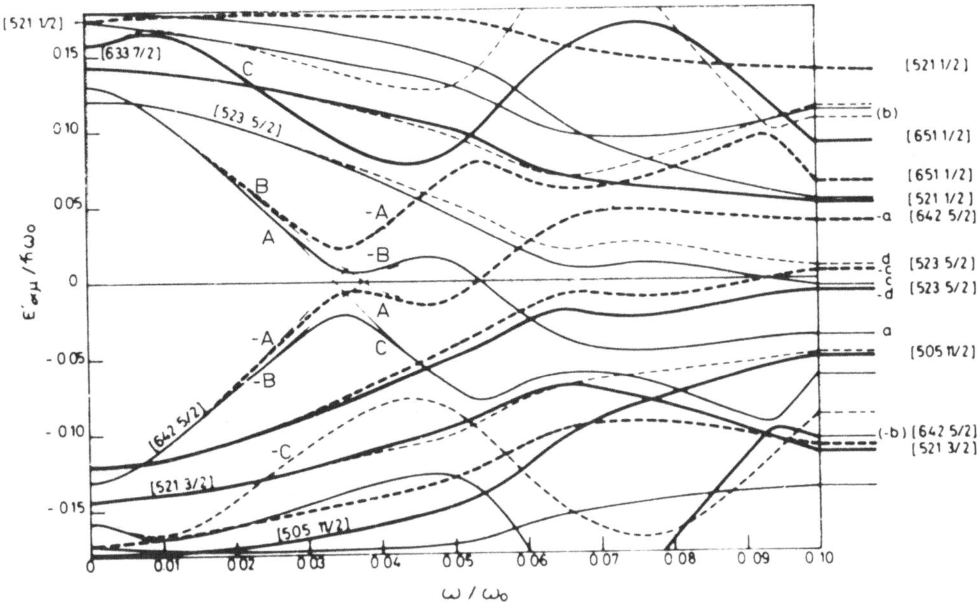

Fig.1 Quasineutron energies for N $\sim$ 96 as functions of the rotational frequency. The energy unit is the $h\omega_0$ = 41 MeV $A^{-1/3}$. The figure is calculated for Δ = 0.12 $\hbar\omega_0$ and ε_2 = 0.26. For $\omega > 0.05\ \omega_0$, Δ decreases linearly, reaching 0 at $\omega = 0.1\ \omega_0$.

contributes an additional phase factor $e^{i\phi/2}$, i.e.

$$\psi_\alpha(\phi) = e^{i\phi/2}\psi_{\pm}(\phi)\ ,\ \text{for}\ \alpha = \begin{cases} 1/2 & = 1/2 + 0 \\ -1/2 & = 1/2 - 1 \end{cases} \tag{5}$$

Fig.1 shows an example of cranked quasineutron levels as functions of the frequency ω. The states are classified with respect to α and π. The Nilsson quantum numbers $[Nn_z\Lambda K]$ serve as additional labels for the trajectories. This classification is only relevant for $\omega = 0$.

In order to construct the quasiparticle configurations we refer to the double dimensional representation of Fig.1. Traditionally one associates quasiparticles only with the states $e' > 0$. This provides a complete description. However, at crossings between trajections $e'(\omega) > 0$ and $e'(\omega) < 0$ the quasiparticle vacuum changes dramatically and represents an inconvenient reference. We believe that explicit inclusion of the $e' < 0$ trajectories leads a more transparent and flexible description. The corresponding occupation number representation deviates somewhat from the usual occupation number representation for Fermions. That is, for each level there is a conjugate partner which lies symmetric to the zero line and has the opposite signature and the same parity, as seen in Fig.1. The number of possible configurations is restricted by the rule that if a level is occupied, its conjugate partner must be free. Furthermore, only the excitation energy of one of the conjugate partners interchanging occupation must be taken into account. The equivalence of this occupation number representation with the more familiar representation in terms of quasiparticle operators is discussed in ref.5.

The quasiparticle levels form trajectories that reflect the change of the quasiparticle motion with the frequency ω. Each configuration defined by a certain distribution of quasiparticles over the trajectories corresponds to a rotational band. The smooth variation of the trajectories with ω implies a similar intrinsic structure of adjacent band members. Rapid variations at the quasi-crossings between trajectories (like the AB-crossing at $\omega = 0.034\ \omega$ in Fig.1 are interpreted as band crossings (cf. Section VI.)

The lowest configuration is the vacuum, which has all levels of negative energy occupied. It represents states of the yrast line (the convex envelope) that may serve as a reference from which to count the excitation energy. The excitation spectrum is obtained by exciting one, two, ... quasiparticles from the reference configuration obeying the rules mentioned above.

In the region of the first crossings between positive energy and negative energy trajectories the ground state configuration (g-conf.) is a more instructive reference, because it represents a purely collective rotation to which the excited quasiparticles add extra energy and angular momentum. Below the first crossing (-AB in Fig.1) the g-configuration coincides with the vacuum, while above it corresponds to the configuration -A, -B occupied (-A, -B free), which preserves the character of the wavefunction. The mixing region may be bridged by the interpolation (thin lines in Fig.1, suggested in ref.6).

Instructive information about the geometry of the quasiparticle orbits is provided by the expectation values of the angular momentum components. Besides the familiar z-component, denoted by K, the rotation induces a finite x-component which is called the aligned angular momentum i. It is given by the negative slope of the trajectories,

$$i = \langle j_x \rangle = -\langle \frac{\partial h'}{\partial \omega} \rangle = -\frac{de'}{d\omega} \tag{6}$$

IV. GEOMETRY OF THE QUASIPARTICLE ORBITS

The quasiparticle motion is governed by the competition between the quadrupole force caused by the deformation of the potential and the Coriolis force. The quantal orbitals react to these forces like gyroscopes. The competition results in the different coupling schemes, illustrated in Fig.2 by means of the precession cones described by the angular momentum vector of the quasiparticle. The figure also shows a fingerprint of the quasiparticle trajectories corresponding to the coupling scheme.

At low frequency the torque od the deformed potential causes $\vec{j}$ to process about the z-axis, i.e. $\langle d_x \rangle \equiv i$ is about zero, whereas $\langle j_z \rangle \equiv K$ is nearly constant. The trajectories are horizontal and degenerate with respect to the signature α. Since the deformed field tries to align the quadrupole moment of the orbital with its own the coupling scheme is called deformation aligned (DAL).

At large frequency the Coriolis force prevails, which causes $\vec{j}$ to precess about the x-axis, i.e., i is constant and $K \not\approx 0$. The quantization of the orbitals corresponds to the quantization of j_x. Therefore, the trajectories have a constant slope and are ordered

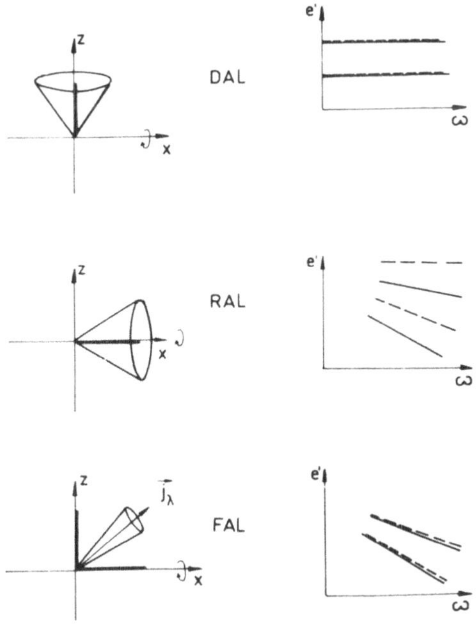

Fig.2 The three coupling schemes and their corresponding fingerprints in the quasiparticle energy diagram.

with respect to the eigenvalue of j_x (= α + even number). This coupling scheme is called rotation aligned (RAL).

The pairfield supports the rotation alignment of the orbitals lying close to the Fermi-surface, because the mixing of particle and hole states (whose quadrupole moments have the opposite sign) effectively reduces the quadrupole moment of the quasiparticle and thus its coupling to the deformed potential. As discussed below, the pairfield may cause a new coupling scheme that corresponds to a precession of $\vec{j}$ about an axis $\vec{j}_\lambda$ defined by the intersection of the precession cone (at ω = 0) of the Fermi-level with the xy-plane ($x > 0$). As seen in Fig.2, both K and i are finite and approximately constant. The fingerprint corresponds to constantly sloping trajectories which are degenerate with respect to α. We suggest calling the precession axis the Fermi-axis and the coupling scheme Fermi-aligned (FAL).

The different coupling shchmes are clearly developed for the high-j intruder states, which determine the high spin behavior. Following Bohr and Mottelson[7], let us consider the classical orbits of the angular momentum vector $\vec{j}$. Since the intruder states are almost pure with respect to j the motion is restricted to the sphere

$$j_x^2 + j_y^2 + j_z^2 = j^2 \tag{7}$$

Without pairing the equienergy surfaces are well approximated by the parabolic cylinders

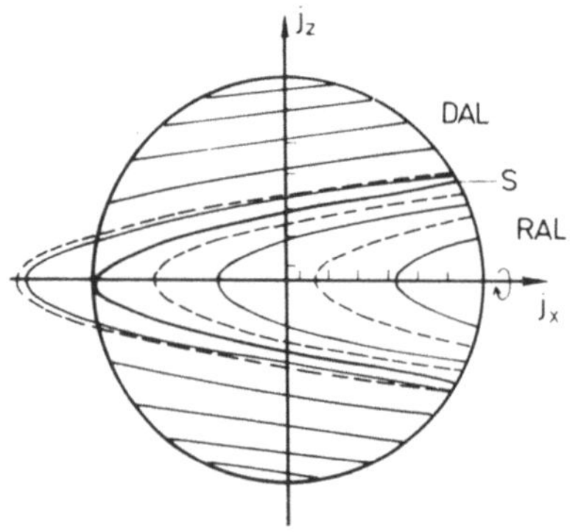

Fig.3 Classical orbits of an $i_{13/2}$ particle in angular momentum space. The orbits lie on the surface of a sphere and the figure is understood as projections onto the xz-plane. The deformation of the potential is ε_2 = 0.26 and the rotational frequency ω = 0.025 $_o$

$$e' = \kappa j_z^2 - \omega j_x \tag{8}$$

The orbits lie at the intersection between cylinder and sphere. The curvature of the parabola is κ/ω and its vertex lies at $j_x = e'/\omega$. Hence, with increasing energy the cylinder moves from the right to the left through the sphere as illustrated by Fig.3.

The equienergy lines in Fig.3 are obtained by putting a spline through the exact $i_{13/2}$-energies of the Nilsson-Hamiltonian h_{def}. The position of the vertices corresponds to the eigenvalues e_i' ($\omega = 0.025\ \omega_o$) of the cranked Nilsson-Hamiltonian h_{sp}', shown in Fig.4 There are two differend kinds of orbits that are separated by the separatrix S whose vertex touches the sphere. The DAL orbits revolve the z-axis and the RAL orbits the x-axis. As in any symmetric one-dimensional potential, where even and odd states follow each other, consecutive RAL orbits correspond to opposite signature. For each energy there exist two degenerate DAL orbitals ($K \lessgtr 0$). Combining them into an even and an odd superposition, one obtains the characteristic signature doublets. Tunneling through the classically forbidden region causes some splitting of the doublets near the separatrix. Fig.4 illustrates how the separatrix devides the spectrum into the two coupling schemes easily recognized by their fingerprints. Fig.7 demonstrates how with increasing ω the RAL region occupies a larger and larger part of the phase space.

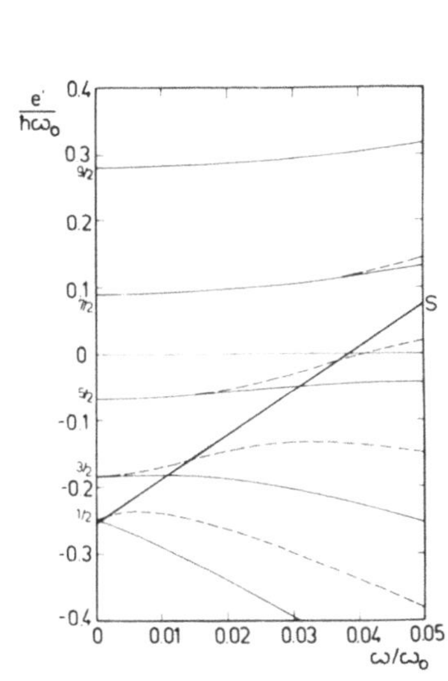

Fig.4 Energies of an $i_{13/2}$-particle in a potential rotating with the frequency ω. The thick line is the separatrix S and the numbers at the levels are K.

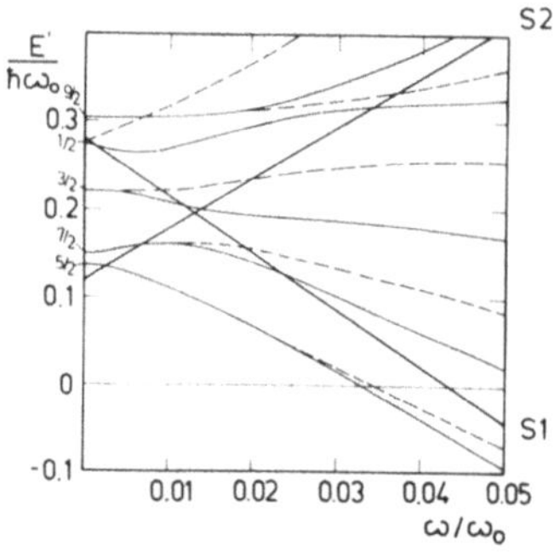

Fig.5 Similar to Fig.4 for the e_+ solutions of an $i_{13/2}$ quasiparticle. The gap parameter is $\Delta = 0.12\ \hbar\omega_o$ and λ is the zero line in Fig.4

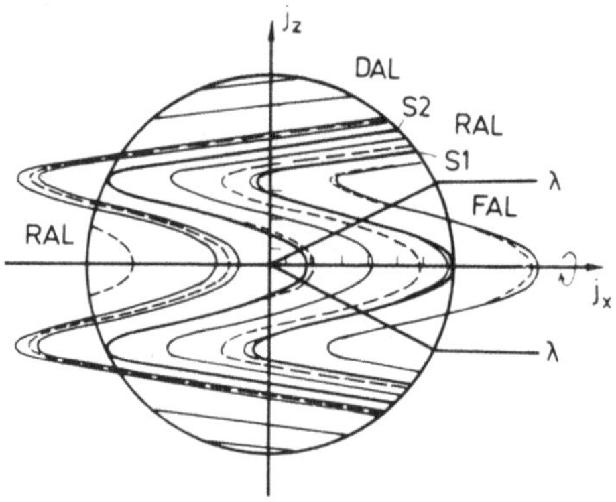

Fig.6 Similar to Fig.3 for the $i_{13/2}$ quasiparticle solutions presented in Fig.5

The CSM-Hamiltonian with (h'_{qp}) pairing is analyzed in the familiar basis of the quasiparticle states at $\omega = 0$, which consists of the two sets of positive and negative energy solutions e'_+ ($\omega = 0$) and e'_- ($\omega = 0$), respectively (cf. Fig.6). The couplings among the positive solutions dominate the quasiparticle motion. Fig.5 shows the eigenvalues $e'_+(\omega)$ obtained by diagonalizing $h'_{qp}(\omega)$ in the basis of only the states $e'_+(\omega = 0)$ and neglecting the couplings to the states $e'_-(\omega = 0)$. As seen, the $e'_+(\omega)$ curves approximate rather well the exact quasiparticle energies displayed in Fig.1. Using this approximation the Hamiltonian for the positive energy quasiparticles reads (approximately)

$$h'_+ = \sqrt{(\kappa j_z^2 - \lambda)^2 + \Delta^2} - \omega j_x \tag{9}$$

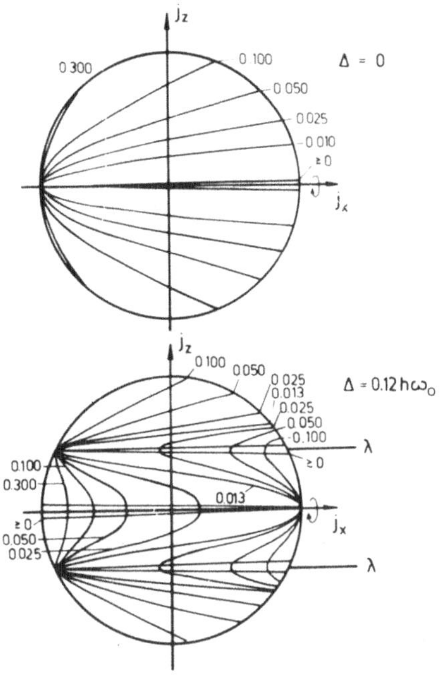

Fig.7 Sequence of separatrices. The separatrices are labeled by ω/ω_0.

The angular momentum sphere now intersects a 4th order energy cylinder. Fig.6 shows the orbits, which again are based on the exact $i_{13/2}$-Nilsson energies. The location is fixed by the cranked quasiparticle energies shown in Fig.5.

The topology is more complex. As a new type, the FAL orbits revolve the Fermi-axis marked by λ. Like for the DAL states there are always two degenerate FAL orbitals ($K \lessgtr 0$) that may be split by tunneling into a signature doublet. The FAL states are separated by the separatrix S_1 from an RAL region, which is divided by a second separatrix S_2 from two distinct regions, one containing the DAL particle-like the other RAL hole-like orbits. Fig.7 demonstrates that an FAL region is only developed for low ω and λ lying about in the middle of the shell. The two separatrices divide the spectrum into the three discussed regions which are easily recognized by their fingerprints in Fig.5. The crossings of S_1 and S_2 corresponds to the single separatrix $\omega = 0.013\ \omega_0$ in Fig.7. Below the crossing the topology is changed: S_1 divides the FAL regions from the DAL hole-like orbitals and above S_2 like the RAL hole-like and DAL particle-like states.

Perfect FAL coupling, i.e., a very narrow precession cone around the Fermi-axis $\vec{j}_\lambda$, corresponds to the classical

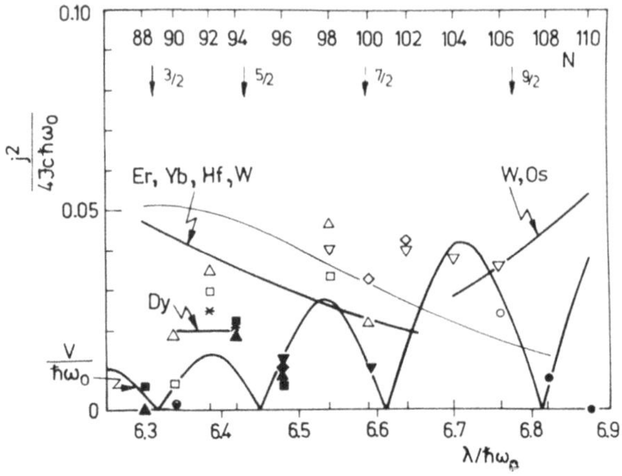

Fig.8 Interaction Matrix element $|V|$ at the -AB crossing between the $i_{13/2}$-FAL orbitals with the lowest $|e'|$. The experimental values are obtained from the analysis of the g-s crossing, where the various symbols denote isotopes: *Dy, □ Er, △ Yb, ◇ Hf, ▽ W, ○ Os. The matrix elements are compared with the back-bending limit (cf. Ref.6) denoted by the heavy sloping lines. The arrows denote the position of the $i_{13/2}$ states at $\omega = 0$.

orbits arising when the energy cylinder coming from the right only slightly penetrates into the angular momentum sphere. This can be achieved by increasing j and scaling the parameters Δ and κ such that only the scale of Fig.5 is changed. Since the phase space of the lowest FAL orbital remains about 1 $\hbar$, its fraction of the surface of the sphere and consequently the opening angle of the precession cone shrink.

The full cranked-quasiparticle Hamiltonian contains the couplings between the states $e'_+(\omega)$ and $e'_-(\omega)$, which causes the repulsion between the crossing trajectories seen in Fig.1. The interaction matrix elements oscillate as functions of the chemical potential[6,8]. Fig.8 illustrates this point by means of the crossing between the lowest e'_+ and the highest e'_--trajectories (-AB in Fig.1).

The oscillations are caused by interferences similar to the ones observed in the classical two slit experiment of optics. The classical action W is the generalization of the optical path. Using j_z as the coordinate and the polar angle χ of the angular momentum projection onto the xy-plane as the conjugate momentum, the action reads

$$W = \int \chi \, dj_z \tag{10}$$

The lowest e'_+ level corresponds to the FAL orbit situated furthest right. The orbit of highest e'_- level is related to it by

$$\chi_- (j_z) = \chi_+ (j_z) + \pi \tag{11}$$

i.e., it corresponds to turning the e'_+ orbit around the z-axis by an angle of π. This implies for the actions

$$W_- (j_z) = W_+ (j_z) + \pi j_z \tag{12}$$

The coupling is accomplished by pair-transfer due to the parifield and dominates therefore at the Fermi-axis $\vec{j}_\lambda$. Assuming a δ-function, one finds that the matrix element becomes proportional to a trigonometric function of the difference between the actions (optical paths) and therefore oscillates with a half-period of

$$\Delta j_z/\hbar = \Delta K = 1 \tag{13}$$

what is indeed born out by Fig.8. Tunneling and the finite width of the coupling as a function of j_z cause deviations from Eq.13, which dominate at the small $j_{\lambda z}$.

Our discussion of the quasiparticle motion in the presence of pairing has started from the separate solutions e'_+ and e'_- and considered selected interaction matrix elements afterwards. This is a reasonable approximation up to the first crossings between the e'_+ and e'_--trajectories. At larger ω one must treat all couplings on the same footing. Actually, this is the region where the pairfield becomes unimportant, i.e., the quasiparticle spectrum approaches the single particle spectrum (except for the trivial doubling of states obtained by reflecting the single particle spectrum through the Fermi surface $e'(\omega) = \lambda$). For example, the FAL level called A in Fig.1 is crossed by more and more trajectories and gradually changes into the separatrix S in Fig.4 (reflected through the zero line). Hence, with increasing ω the lowest level A does not become the fully aligned RAL level as expected from the analysis of the e'_+-solutions (cf. Fig.5 and 7), but looses its identity as a state instead.

V. ANALYSIS OF THE EXPERIMENTAL SPECTRUM

The spectra are grouped into rotational bands which are interpreted as quasiparticle configurations. The total signature of a configuration restricts the angular momentum to the values

$$I = \alpha + \text{even number}. \tag{14}$$

This relation determines the experimental signature of the band. (We apply the term "band" for a $\Delta I = 2$ sequence, what is at variance with the conventional definition).

The (average) angular velocity with which the potential rotates about the x-axis can be obtained by approximating the classical relation

$$\omega = \frac{dE}{dI_x} \tag{15}$$

by the quotient of finite differences

$$\hbar\omega(I) = \frac{E(I+1) - E(I-1)}{I_x(I+1) - I_x(I-1)} \tag{16}$$

calculated for adjacent levels within the band.

Semi-classically, the component I_s is determined as

$$I_x(I) = \sqrt{(I+1/2)^2 - K^2} \tag{17}$$

K being the angular momentum component along the symmetry axis.

For $K = 0$, Eq.16 expresses the familiar result of classical electrodynamics that a charged body which has reflection symmetry radiates electromagnetic waves with twice the frequency of its rotation. For $K \neq 0$, Eq.16 corresponds to the frequency ratio

$$\omega = \frac{\omega_{RAD}}{2} \sqrt{1 - K^2/I^2} \tag{18}$$

containing an additional geometry factor due to the fact that the radiation is generated by rotation about the angular momentum vector I which is tilted by $\sin\phi = \sqrt{1 - K^2/I^2}$ with respect to the xy-plane

The analysis by means of Eqs.16 and 17 provides the projection of the angular velocity vector onto the xy-plane, which indeed is the quantity controlling the quasiparticle motion.

Classically, energy in the rotating and laboratory systems are related by the Legendre transformation

$$E' = E - \omega I_x \tag{19}$$

which is approximated by

$$E'(I) = \frac{1}{2}[E(I+1) + E(I-1)] - \omega(I) I_x(I) \cdot \tag{20}$$

Employing the terminology of classical mechanics, E' may be called a Routhian.

As discussed above, the CSM aims only at excitation energies with respect to a reference band. For moderate angular momentum the g-band is a reasonable reference. Thus we define the energy e' and the aligned angular momentum i of the excited quasiparticles as

$$\begin{aligned} e'(\omega) &= E'(\omega) - E'_g(\omega), \\ i(\omega) &= I_x(\omega) - I_{xg}(\omega) \end{aligned} \tag{21}$$

where the reference quantities E'_g and I_{xg} are calculated from the experimental g-band levels. The experimental g-band is usually not well known in the crossing region with other bands. Therefore, the reference is generated by extrapolating the g-band with the help of the Harris (VMI) expressions

$$\begin{aligned} E'_g(\omega) &= -\frac{\omega^2}{2}\mathcal{J}_o - \frac{\omega^4}{4}\mathcal{J}_1 + \frac{1}{8\mathcal{J}_o} \\ I_{xg}(\omega) &= \omega_o \mathcal{J}_o + \omega^3 \mathcal{J}_1 \end{aligned} \tag{22}$$

The reference parameters are most easily obtained by plotting the moment of inertia as a function of ω^2 (backbending plot). For the ground-state band this becomes the linear function (cf. Eq.22)

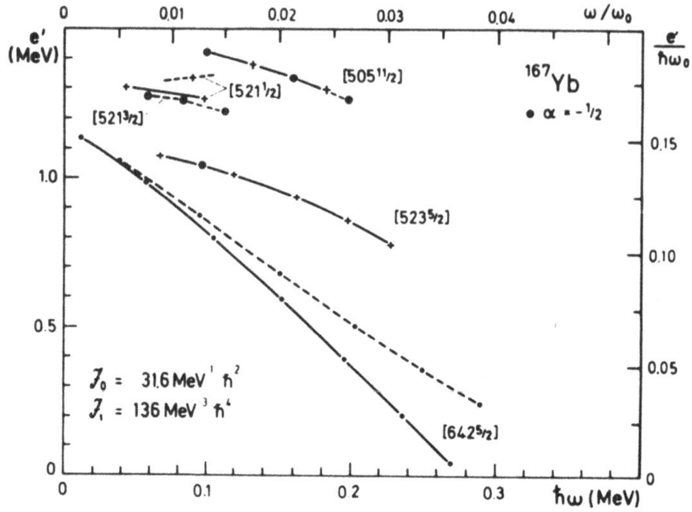

Fig.9 Experimental Routhians of ^{167}Yb from the data of ref.9. The reference parameters are the mean values of ^{166}Yb and ^{168}Yb.

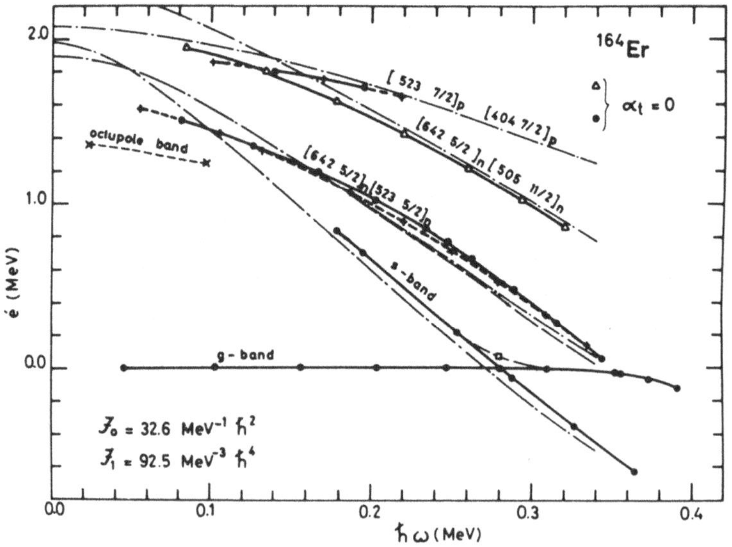

Fig.10 Experimental Routhians of ^{164}Er, from the data of ref.10. The yrast line follows the g-band until the dashed (open circle) line connects it with the S-bands.

$$\mathscr{J}^{(1)}_{g\text{-band}}(\omega^2) \equiv I_{xg}/\omega = \mathscr{J}_o + \omega^2 \mathscr{J}_1 \tag{23}$$

The intercept with the ordinate and the slope determine $\mathscr{J}_o$ and $\mathscr{J}_1$, respectively.

This determination of the reference is reliable for well deformed nuclei. It is used to obtain Fig. 8-10 and Table II. However, for less well deformed species the functions $i(\omega)$ determined in this way have a tendency to fall off at high ω. Such a decrease is not expected for low lying excitations (c.f. Fig. 1 and 11 and Section IV). Moreover, the downbending occurs for all bands that are expected to have constant i. Therefore, it seems obvious to interpret the discrepancy as a change of $I_{xg}(\omega)$ due to the presence of excited quasiparticles. Improved reference parameters $\mathscr{J}_o$ and $\mathscr{J}_1$ may be determined from the aligned few-quasiparticle bands. This is possible if a constant aligned angular momentum i may be ascribed to the band. Then Eqs. 21 and 22 become the three parameter expression

$$I_x(\omega) = i + \omega \mathscr{J}_o + \omega^3 \mathscr{J}_1 \tag{24}$$

The parameters $\mathscr{J}_o$ and $\mathscr{J}_1$ may be obtained by plotting the moment of inertia

TABLE II

Neutron orbitals active in ^{167}Yb. The aligned angular momenta are taken at $\hbar\omega = 0.23$ MeV. The data is taken from Fig. 9, the parameters of the calculation are the same as in Fig. 1, except that λ corresponds to N $\approx$ 97.

ORBITAL	α^π	i_{cal}	i_{exp}	$\|K\|$
[642 5/2]	$1/2^+$	4.9	5.1	5/2
[642 5/2]	$-1/2^+$	4.4	3.1	5/2
[523 5/2]	$1/2^-$	2.1	3.1	5/2
[523 5/2]	$-1/2^-$	1.8	-	5/2

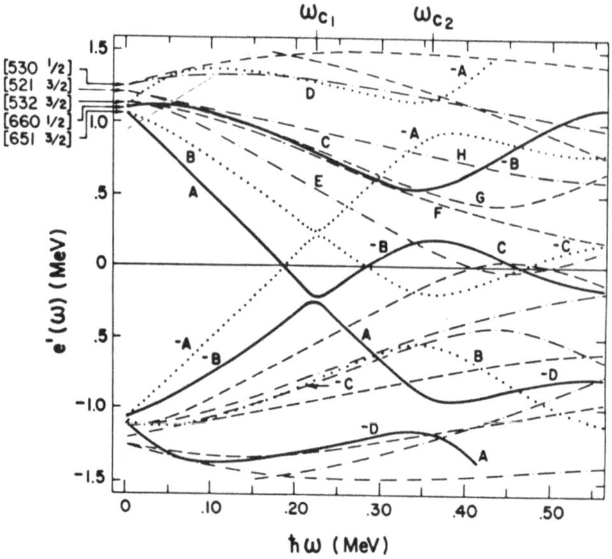

Fig.11 Quasi-neutron energies for N $\approx$ 90 (cf. Fig.1). The figure is calculated for ε_2 = 0.20, ε_4 = -0.02 and Δ = 1.02 MeV. The frequencies denoted by ω_{c1} and ω_{c2} are refered to in the text by ω_1 and ω_2.

$$\mathcal{J}^{(2)}_{band} \equiv dI_x/d\omega = \mathcal{J}_o + 3\omega^2 \mathcal{J}_1 \qquad (25)$$

versus ω^2, which may be obtained from the differences of the transition energies

$$\mathcal{J}^{(2)}_{band}(\omega^2) = \frac{I_x(I) - I_x(I-2)}{\omega(I) - \omega(I-2)} \,, \qquad (26)$$

$$\omega^2 = \frac{1}{4}(\omega(I) + \omega(I-2))^2 \,.$$

This method is used to obtain Fig.12 and Tables III - V.

VI. QUASIPARTICLE EXCITATION ENERGIES

Fig.9 shows the experimental Routhians of the N = 97 nuclid ^{167}Yb obtained from the data of ref.9. The excitations are of the one-quasineutron type and, hence, represent the experimental counterpart of the quasi-particle trajectories in Fig.1. The two lowest $\pi = +$ trajectories form the first FAL-signature doublet, the splitting of which grows with ω. The next trajectories have $\pi = -$, a smaller slope and no apparent signature splitting. They may be interpreted as perturbed DAL orbitals. The quantitative agreement between calculated and experimental quasiparticle trajectories is rather good.

A crucial test of the independent particle approach is provided by the spectra of even-even nuclei, because the lowest (non-collective) excitations are two-quasiparticles states. Neglecting the residual interaction, the excitation energy e' must be the sum of the energies of the excited quasiparticles. The same holds for other additive quantities as i or α. From Fig.1 the following two-quasineutron excitations are expected: an $\alpha^{\pi} = 0^{+}$ band [AB] called the s-band, and somewhat higher two $\alpha^{\pi} = 0^{-}$ and two 1^{-} bands [642 5/2 $\alpha = \pm 1/2$, 523 5/2 $\alpha = \pm 1/2$] lying close in energy.

As a notation for the configuration we indicate in square brackets all levels becoming occupied by quasiparticles excited from the g-configuration. For example, the two-quasiparticle configuration (s-conf.) with AB filled and -A,-B free is denoted by [AB].

Fig.10 shows the experimental Routhians for ^{164}Er obtained from the data of ref.10. The s-band and two bands of the $\pi = -$ quadruplet are observed. Moreover, all other observed non-collective bands may be interpreted as two-quasiparticle excitations. The excitation energy and the aligned angulare momentum of the bands agree well with the sums of the respective quasiparticle quantities. Fig.10 clearly demonstrates the predictive power of the CSM. Other two- and three-quasiparticle spectra in the rare-earth region have been analyzed with similar success (cf. refs. 5, 11, 12).

VII. THE QUASIPARTICLE ANGULAR MOMENTA

For FAL-states the projections i, and K, of the angular momentum are approximately conserved (cf. Section IV). Only the absolute value $|K|$ may be ascribed to states with good signature α, because they contain +K and -K with equal probability (cf. Section III and IV. Table II displays the angular momentum of the lowest orbitals in ^{167}Yb. The calculations reasonably reproduce the experimental values.

The two $i_{13/2}$-orbitals [642 5/2] differ noticeably in their i-values, indicating deviations from the FAL coupling limit. Further evidence comes from the sum $i^2 + K^2 = (4.1)^2 + (2.5)^2 = 23$, which should approach $j^2 = (6.5)^2 = 42$ in the FAL limit. Thus, even the angular momentum of $j = 13/2$ is not large enough to realize the perfect FAL coupling (cf. Section IV).

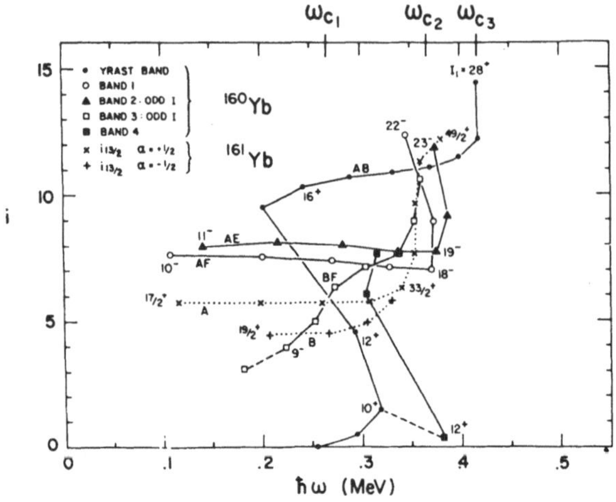

Fig. 12 Experimental aligned angular momentum of ^{160}Yb from the data of ref.11. The letter combinations suggest the quasiparticle configurations of the band. The reference parameters are $\mathcal{J}_0$ = 16 MeV^{-1} and $\mathcal{J}_1$ = 90 MeV^{-3}. The frequencies ω_{c1}, ω_{c2}, and ω_{c3} are refered to in the text by ω_1, ω_2 and ω_3.

TABLE III

Neutron orbitals active in ^{160,161}Yb. The third column refers to the labeling of the trajectories in Fig.1, i_{cal}[ħ] corresponds to the same calculation. The experimental values of i are taken from the one quasineutron bands in ^{161}Yb (ref.12). The $i_{13/2}$ bands are shown in Fig.2.

ORBITAL		LABEL	i_{cal}	i_{exp}
$i_{13/2}$	$1/2^+$	A	5.8	6
$i_{13/2}$	$-1/2^+$	B	4.1	4
$i_{13/2}$	$1/2^+$	C	2.5	-
$i_{13/2}$	$-1/2$	D	1.5	-
$h_{9/2}$	$1/2^-$	E	3	2
$h_{9/2}$	$-1/2^-$	F	2	-

TABLE IV

Few-quasineutron configurations in $^{160,161}Yb$. Column 2 gives the number of excited quasiparticles, 3 denotes the Configuration (both in terms of the high spin orbitals and the letters used in Fig.11): col.5 gives the sum of i_{cal} of Table III: col.6 gives the sum of i_{exp} of Table III: the values in col.7 are taken from Fig.12 (ref.11), except for No.8 and 9, which come from more recent measurements at Oak Ridge[15].

No.		CONFIGURATION			i_{cal}	$\sum i_{exp}$	i_{exp}
1	2	$(i_{13/2})^2$	AB	0^+	9.9	10	10
2	2	$(i_{13/2})^2$	AC	1^+	8.3		-
3	2	$(i_{13/2})^2$	BC	0^+	6.6		7.5
4	2	$h_{9/2}\ i_{13/2}$	AE	1^-	8.8	8	8
5	2	$h_{9/2}\ i_{13/2}$	AF	0^-	7.8		7
6	3	$h_{9/2}\ (i_{13/2})^2$	ABE	$1/2^-$	12.9	12	11.5
7	3	$h_{9/2}(i_{13/2})^2$	ABF	$-1/2^-$	11.9		11.5
8	4	$h_{9/2}(i_{13/2})^3$	ABCE	1^-	15.5		14.5
9	4	$h_{9/2}(i_{13/2})^3$	ABCF	0^-	14.5		13.5

A second example is provided by the Copenhagen data[11,12] on $^{160,161}Yb$ in Fig.12 The corresponding quasineutron diagram, Fig.11, shows the fingerprint of RAL states. Thus, the aligned angular momentum i of the quasiparticles may be used to classify the configurations. Table III lists the active orbitals. The i-values indicate deviations from the RAL limit. Table IV demonstrates the additivity of i and α for several few-quasineutron configurations in ^{160}Yb.

The considered examples show the occurrence of the coupling schemes discussed in Section IV although there are noticeable deviations from the limiting cases. Nevertheless, the CSM reproduces the experimental quasiparticle angular momenta rather well. It is essential that the additivity of i expected for independent quasiparticles is confirmed by the data.

VIII. BAND CROSSINGS

The independent particle picture implies that each crossing between two quasiparticle trajectories gives rise to a family of band crossings at one at the same frequency ω_ν. The crossing trajectories may be combined with an arbitrary distribution of "spectator" quasiparticles over the remaining levels. Crossings between trajectories with the same α and π, one coming from the region e' > 0 the other from e' < 0, are of special importance because they correspond to a large gain of i and are therefore energetically favored. For example, the band crossing pattern due to the $i_{13/2}$ neutron system is governed by the three crossings seen in Fig.1 and 11: ω_1(-AB), ω_2(-BC) and ω_3(-AD).

No selection rule prevents the mixing between the configurations, which is manifested by the repulsion of the trajectories of the same α and π (and consequently of the configurations) near the crossing point. The mixing causes irregularities of the bands at the crossing. The experiments clearly reveal these structural changes as a backbending sequence of γ-lines.

The crossing between the g- and the s-band in even rare-earth nuclei is the best studied case. It belongs to the family of band crossings implied by the crossings of the trajectories -AB in Fig.1 and 11. For ^{164}Er shown in Fig.10, the crossing is observed at the expected frequency of $\hbar\omega_c \approx 0.28$ MeV. The yrast line follows the g-band up to its crossing with the s-band, then it is continued by the latter. Since the aligned angular momentum of the s-band amounts to about 8 $\hbar$, the yrast sequence corresponds to the backbending functions I(ω) or i(ω), (compare to the Yb nuclei in Fig.12) and the double valued function e'(ω) shown in Fig.10.

The description of the crossing region by the CSM is not correct. Angular momentum conservation implies that bands cross and mix at fixed angular momentum and not at fixed frequency. Bengtsson and Frauendorf[6] suggested a way to circumvent this difficulty. Near the crossing the two quasiparticle trajectories may be decomposed into unperturbed states with a (constant) interaction matrix element (cf. Section V). The unperturbed trajectories (corresponding to the thin lines in Fig.1) are used to construct the unmixed bands whereas the interaction matrix element is identified with the coupling between states of the same I belonging to these bands.

The rapidity of the backbending is controlled by the interaction matrix element V, which determines the width of the hybridization region. As discussed in Section IV, this matrix element oscillates as a function of the neutron number due to interference between the FAL orbitals. Fig.9 expresses the rapidity of backbending by means of $|V|$, which is clearly correlated with the oscillations predicted by the CSM.

From the calculations in Fig.11 one expects three families of band crossings corresponding to the following lowest trajectory crossings: $-AB(\omega_1)$, $-BC(\omega_2)$ and $-AD(\omega_3)$. The occurrence of these three basic frequencies is a general feature of high-spin intruder orbitals if a strong pairfield is present (compare Fig.1). In the considered case ω_2 and ω_3 are almost equal. Fig.12 shows the expected bunching of band crossings (up- or back-bends) into two groups. Table V lists the observed crossings, indicating the "spectator" quasiparticles. The gain of aligned angular momentum

TABLE V

Band crossings in ^{160}Yb and ^{161}Yb. The data corresponds to Table III and come from refs.11, 12, and 15.

No.	CROSSING	α^{π}	SPECTATOR	ω_{cal}	Δi_{cal}	ω_{exp}	Δi_{exp}
1	$-AB\ \omega_1$	0^+	no	0.23	10	0.28	10
2		$1/2^+$	E			0.24	9.5
3	$-BC\ \omega_2$	$1/2^+$	A	0.35	6.6	0.35	6.5
4		1^-	AE			0.35	6.5
5		0^-	AF			0.37	7
6		0^+	no			0.34	7.4
7	$-AD\ \omega_3$	$-1/2^+$	B	0.36	7.3	0.36	>4

Δi, is equal to the sum of $|i|$ of the two crossing trajectories and may be obtained from Table III.

For the most cases the CSM-pattern seems to be confirmed by the observed band crossings. The spectator quasiparticles do not modify the crossing. An exception is the $1/2^-$-band in ^{161}Yb, for which the presence of the quasiparticle E reduces $\hbar\omega_1$ by 40 keV.

A quite stringent test of the nature of a family of band crossings is delivered by the blocking pattern extensively discussed in connection with the first backbend. In the context of CSM, it is just the consequence of the quasiparticle Fermi-statistics. The crossing ω_ν may only show up in configurations with one of the crossing trajectories occupied, the other one free; if both are occupied or free, the crossing is "blocked", i.e. no structural change of the wave function appears.

The blocking pattern is revealed by Fig.12 and Table V. The -AB crossing manifests itself as the crossing of the g-band with the s-band [AB], (No.1, Table V), which is the familiar first backbend of the yrast sequence. The s-band behaves smoothly at ω_2 because the crossings -BC and -AD are blocked. The presence of the spectator in level E does not change the pattern (the $1/2^-$-band, No.2). If the level A is occupied and B is free (and thus -B is occupied) the -AB crossing is blocked, whereas the -BC crossing is active. Indeed, in the $1/2^+$-band [A] the upbend is delayed until ω_2 (No.3). The addition of spectators in the levels E or F leading to the 1^- and 0^- side bands leaves the blocking pattern unchanged (No.4, 5, respectively). Finally, the -AD crossing is active and -AB blocked in the $-1/2^+$band [B], (No.7).

Two further side bands, 0^- and 1^-, in ^{160}Yb show an upbend at ω_2 corresponding to the -AB or the -BC crossings. In the g-band all three crossings are active. Indeed, after its crossing with the the s-band the continuation of the g-band crosses with the [BC]-band at the expected frequency ω_2 (No.6). In the configuration [AC] all three crossings are blocked. This interesting 1^+ band has not been observed so far.

The $h_{11/2}$ intruder states in the proton system bahave similar to the $i_{13/2}$ neutron states. Thus, the next basic frequency corresponds to the -AB crossing in the proton system. It is observed at $\hbar\omega_c \approx 0.42$ MeV in ^{160}Yb (cf. Fig.12 and refs.11 and 13) and ^{158}Er (ref.2).

As a whole, one may state that the data from the lower half of the rare earth region agrees rather well with the CSM predictions assuming a shape and a parifield similar to the ground state values[5]. Significant neutron pair-correlations up to $\hbar\omega = 0.35$ MeV are necessary to understand the observed aligned angular momenta

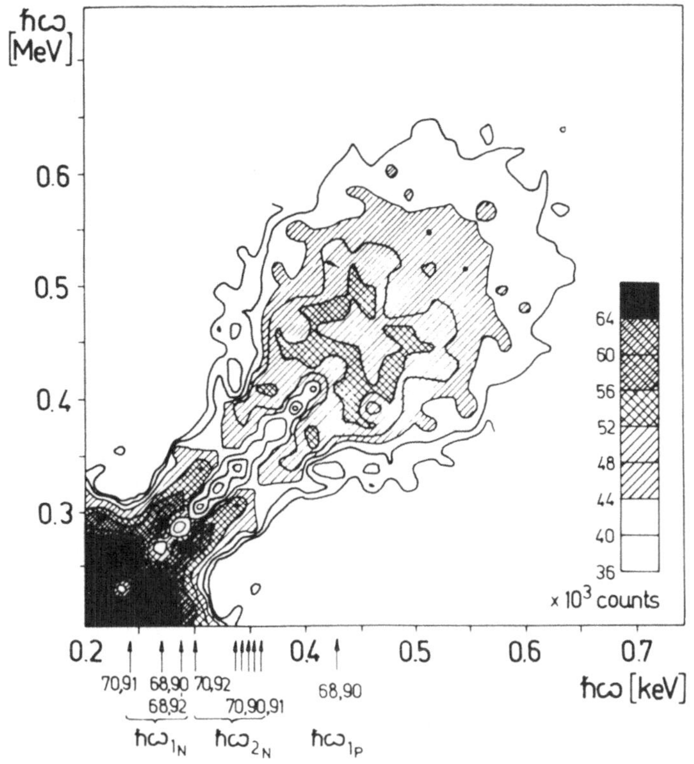

Fig.13 Coincidence spectrum obtained in the reaction $^{124}Sn + ^{60}Ar \rightarrow ^{164}Er^*$ with 185 MeV projectile energy[3]. The arrows indicate crossing frequencies obtained from the discrete spectra, where the numbers corre spond to Z, N.

and the band-crossing pattern.

Recent progress in continuum yrast spectroscopy has demonstrated the possibility to discern the basic frequencies of band crossings from the γ-continua[3]. Fig.13 shows a γ-γ coincidence spectrum. The scale $\hbar\omega$ is half of the transition energy. The figure displays the probability for two transitions $\hbar\omega_1$ and $\hbar\omega_2$ to belong to the same cascade along a band. At a band crossing the cascade follows the lower branches. The distance $\Delta\omega$ from the central valley is equal to the difference between the frequencies of the transitions. The first ridge corresponds to consecutive transitions differing by two units of angular momentum. Its distance from the diagonal is equal to

$$\Delta\omega = \omega(I) - \omega(I-2) \approx \frac{1}{2}\frac{d\omega}{dI} . \tag{22}$$

Thus, if there is an upbend of the function $I(\omega)$, the valley is pinched. A vertical upbend or a backbend corresponds to a bridge. The spectrum in Fig.13 is generated by the dominating evaporation residues $Z = 68$ and $N = 91, 92$. The figure also shows the values of the experimental basic frequencies obtained from the discrete spectra. Part of the information had to be taken from the $Z = 70$ isotones discussed above. Due to the smaller deformation the frequencies are somewhat reduced in Yb compared to Er. The first two bridges presumably correspond to $\hbar\omega$ in (-AB neutron) in the odd and even mass isotopes, respectively. The following two bridges should belong to $\hbar\omega_{2N}$ (-BC, -AD neutron). The valley is closed by a faint bridge at the frequency $\hbar\omega_{1P}$ (-AB $h_{11/2}$proton). The disappearance of the valley at high frequency might be due to a clustering of crossings. There are many candidates: -BC $h_{9/2}$ proton, -EG $h_{9/2}$ neutron (cf. Fig.11) but also crossings due to the $j_{15/2}$ neutron or $h_{9/2}$ and $i_{13/2}$ proton orbitals, which may come into play at these frequencies. It might also be that there are experimental problems at the highest frequencies, which are not yet understood. Nevertheless, the example demonstrates the possibility of discerning information about the quasiparticle motion up to an angular momentum of about 40 $\hbar$.

IX. CONCLUSIONS AND PERSPECTIVES

The examples presented demonstrate that the complex yrast spectra may be disentangled into configurations of quasiparticles moving in the deformed rotating potential. The quasiparticle diagrams calculated from the CSM-Hamiltonian form a sound basis for such an analysis.

Now that we understand the structure of the spectra it is time to extend and refine the investigations. The study of the excitation energies must be complemented by calculations of the transition probabilities. The first attempt into this direction[14] turned out to be quite promising. Another important aspect is the residual interaction between the quasiparticles. The spread of the basic frequencies in Fig.13 might be attributed to it. In particular, the configuration mixing leading to collective excitation should be studied. The collective γ- and octupole bands are observed up to the region where they cross with the two quasiparticle bands. The study of the quadrupole mode is closely related to corrections to the CSM due to the angular momentum conservation. The latter has the same order of magnitude and also a similar form as rotationally induced time odd components of the average field.

So far, the assumption of an ω-independent rotating field has not led to any obvious contradiction with the data. However, major changes are expected, like the break down of the pairfield or significant changes of the nuclear shape, which remain a challenge for further spectroscopic work.

REFERENCES

1. A. Bohr and B. Mottelson, Proc. Int. Conf. on Nucl. Structure Tokyo, 1977; J. Phys. Soc. Japan 44, suppl. p.157 (1978).
2. I.Y. Lee, M.M. Aleonard, M.A. Deleplanque, Y. El Masri, J.O. Newton, R.S. Simon, R.M. Diamond, and F.S. Stephens, Phys. Rev. Lett. 38, 1454 (1977).
3. O. Andersen J.D. Garrett, G.B. Hagemann, B. Herskind, O.L Hillis and L.L. Riedinger, Phys. Rev. Lett. 43, 687 (1979); J.D. Garrett, B. Herskind, Talks at the "Study Weekend on Nuclei far from Stability", Daresbury, Sept. 1979.
4. R.M. Diamond and F.S. Stephens, Annual Review of Nuclear and Particle Science, Vol.30, 85 (1980).
5. R. Bengtsson and S. Frauendorf, Nucl. Phys. A327, 139)1979).
6. R. Bengtsson and S. Frauendorf, Nucl. Phys. A314, 27 (1979).
7. B. Mottelson, Proc. Sym. on High-Spin Phenomena in Nuclei, Argonne, 1979, Report ANL/PHY-79-4.
8. R. Bengtsson, I. Hamamoto and B. Mottelson, Phys. Lett. 73B, 134 (1978).
9. Th. Lindbald, Nucl. Phys. A238, 287 (1975).
10. O.C. Kistner, A.W. Sunyar and E. der Mateosian, Phys. Rev. 17, 1417 (1978); N.R. Johnson, D. Cline, S.W. Yates, F.S. Stephens, L.L. Riedinger and R.M. Ronningen, Phys. Rev. Lett. 40, 151 (1978).
11. L.L. Riedinger, O. Anderson, S. Frauendorf, J.D. Garrett, J.J. Gaardhøje, G.B. Hagemann, B. Herskind, Y.W. Makovetzky, J.C. Waddington, M. Guttormsen and P.O. Tjorn, Phys. Rev. Lett. 44, 568 (1980).
12. J.J. Gardehøje, L.L. Riedinger et al., in preparation.
13. F.A. Beck, E. Bozek, T. Byrski, C. Gehringer, J.C. Merdinger, Y. Schutz, J. Styczen and V.P. Virein, Phys. Rev. Lett. 42, 492 (1979).
14. I. Hamamoto and H. Sagawa, Nucl. Phys. A327, 99 (1979).
15. L.L. Riedinger, D.R. Haenni, S.A. Hjorth, N.R. Johnson, I.Y. Lee, W.K. Luk and R.L. Robinson, to be published.

NUCLEAR SPECTROSCOPY AT VERY HIGH AND VERY LOW ROTATIONAL FREQUENCIES

J.D. Garrett

The Niels Bohr Institute
University of Copenhagen
DK-2100 Copenhagen Ø, Denmark

I ABSTRACT

The present contribution discusses the possibility of extending the study of the spectroscopy of multiple quasi-particle bands both to very high and to very low rotational frequencies using experimental techniques to isolate specific features of rotational transitions in the continuum. At high rotational frequencies the correlation between the γ-ray transition energies gives information on transitions which are specific to a particular γ-ray cascade. From such measurements it is possible to ascertain, on average, both the rotational behavior and an estimate of the in-band moment of inertia, $\mathcal{J}_c$, as a function of the rotational frequency. The existing data for rotational nuclei show a rotational pattern which, even though it is broken at certain specific frequencies, does extend to high frequencies. A decrease also is observed in the values of $\mathcal{J}_c$ at the highest rotational frequencies. Such observations are consistent with at least a significant fraction of the cascades at very high frequencies proceeding along rotational-like bands, which at specific $\hbar\omega$'s cross with a group of other bands involving the additional alignment of a particular high-j, low-K pair of nucleons.

It also is interesting to follow a particular aligned band to low rotational frequency where the total angular momentum, I, becomes less than the aligned angular momentum of the unpaired particles. The ^{161}Dy(^{3}He, α)^{160}Dy reaction has been used to pick-up an $i_{13/2}$ neutron from a target nucleus, whose ground state has an unpaired $i_{13/2}$ neutron. From among the spectrum of $(\nu i_{13/2})^2$ states candidates have been selected from the 4^+, 6^+, and 8 members of the aligned "S band" using the relative strength of the (^{3}He, α) population and the γ-decay of these states, which were known previously. The excitation energies of systematics of these states indicate that at low frequency the angular momentum of the two non-paired particles do not remain fully aligned, but instead recouple as predicted qualitatively by the particle plus rotor model.

II. INTRODUCTION

One of the most active subfields of nuclear-structure physics of the last decade has been the study of high-spin states[1]. Much of the effort has been devoted to establishing and understanding the systematics of the yrast band, which is composed of, at small rotational frequencies, the ground-state band, and at large rotational frequencies, a band (the S band) based on a configuration in which a pair of quasiparticles with high j are aligned along the nuclear rotational axis[2]. Recently detailed level schemes have become available for nonyrast bands (see e.g. refs. 3 and 4) showing

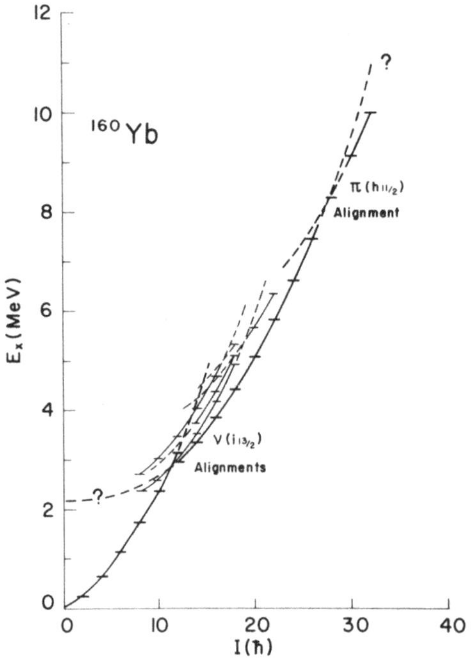

Fig.1 Plot of excitation energy versus angular momentum for the known[4,8] even-spin bands in ^{160}Yb. The questions which the present report proposes to address, i.e. what happens at high angular momentum and what happens to bands based on aligned configurations when the angular momentum is less than the alignment, are indicated by the question marks.

a variety of rotational and nonrotational (i.e. "back-bends" and "up-bends") features. The features of the yrast and near-yrast bands have been explained in terms of the crossings of rotational bands based on multiple quasiparticle configurations with verying alignments (see e.g. refs.4-6). "Yrast spectroscopy" resulting from the interpretation of such data was the subject of the preceding paper by Frauendorf[7].

The subject of the present report is the extension of such spectroscopic studies to higher rotational frequencies and further away from the yrast line. This is illustrated in fig.1, where the known even spin bands based on ground-state and various two- and four-quasiparticle configurations are shown as a function of excitation energy, E_x, and spin, I, for ^{160}Yb, one of the best studied cases[4,8]. The concentration of the γ-ray intensity in the yrast and specific near-yrast bands is diluted rapidly with increasing angular momentum. This can be seen from typical γ-ray spectra,

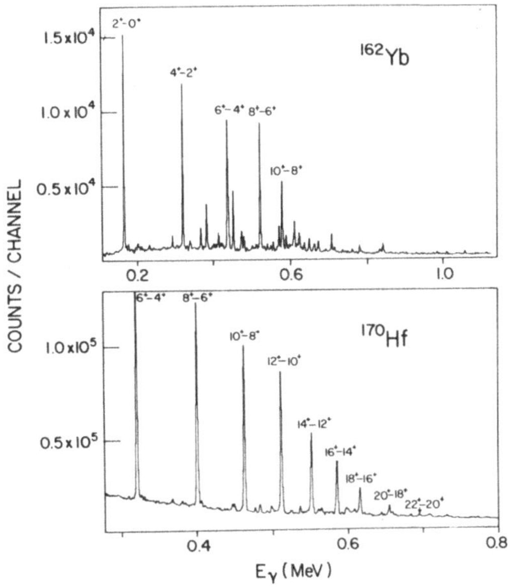

Fig.2 Comparison of "gated" γ-ray spectra for ^{170}Hf, a case in which the ground-state and S bands are strongly interacting, and ^{162}Yb, a case in which these bands are weakly interacting. For ^{170}Hf the spectra shown in the sum of spectra gated on the 6→4, 8→6 and 10→8 transitions, and ^{162}Yb is the sum of 2→0, 4→0 and 6→4 gated spectra.

e.g. those shown in Fig.2. Even with the use of anticompton suppression, it is not possible to resolve the continuum of γ-rays[9], thereby extending spectroscopic studies to much higher spins and away from the yrast line.

The experimental γ-ray spectroscopist is faced with a problem similar to that of the relativistic heavy-ion physicist. In the collision of a relativistic heavy ion with a nucleus, a final state with perhaps fifty particles may be formed. To obtain detailed information on the physics of such collisions it is necessary to either collect data for a large fraction of the projectiles using a multidetector, large solid angle device or to find the appropriate correlation between certain parameters which is specific to a particular physical quantity. For the γ-decay of a rare-earth system with maximum angular momentum, over thirty γ-rays may be involved. One technique for the spectroscopic studies of such systems is the large solid angle arrays of detectors, "crystal balls" or "spin spectrometers", now being constructed at Oak Ridge and at Heidelberg. We, however, feel that an equally attractive and complementary approach is the study of the correlation among the energies of the γ-rays[10] which allows specific features to be isolated from the γ-ray continuum. This technique and some examples of the type of spectroscopic information available from such studies is the subject of Section III.

Among the states more distant from the yrast line, the low-spin states of an aligned configuration are of particular interest. For these states the total angular momentum can be smaller than the possible fully-aligned angular momentum; therefore, the details of the interplay between rotational and single-particle motion can be studied[11]. Such states are neither populated in the γ-ray cascades following (heavy-ion, xn) reactions, nor specifically selected in (light-ion, xn) reactions. Candidates for such states, however, can be selected from the spectrum of $(\nu(i_{13/2}))^2$ states populated in pickup and stripping reactions on target nuclei which have $\nu(i_{13/2})$ ground-state configurations. An example of such a study for ^{160}Dy is discussed in Section IV, and in ref.12.

III. GAMMA RAY ENERGY CORRELATIONS

The energy of the spectrum of excited states of a rotating, axially-symmetric, quantum-mechanical system is

$$\begin{aligned} E(I) &= \frac{\hbar^2}{2\mathcal{J}_c} R(R+1) + E_j \\ &\approx \frac{\hbar^2}{2\mathcal{J}_c} (I - j_a)^2 + E_j \end{aligned} \qquad (1)$$

$\mathcal{I}_c$ is the moment of inertia of the system, E_j is the excitation energy due to the internal excitation, and R, the rotational angular momentum, is the difference between the total angular momentum, I, and the angular momentum of the unpaired nucleons, j_a. The transition energy, E_γ, between $\Delta I=2$ states in such a rotational band is

$$E_\gamma = 2 \left.\frac{dE}{dI}\right|_{j_a} = \frac{\hbar^2}{2\mathcal{I}_c} 4(I - j_a) \tag{2}$$

and the difference in transition energies for consecutive transitions in the cascade is

$$\Delta E_\gamma = 4 \left.\frac{d^2E}{dI^2}\right|_{j_a} = 8 \frac{\hbar^2}{2\mathcal{I}_c} \tag{3}$$

Consider the E_γ-E_γ pattern of possible coincidences for the yrast transitions in ^{170}Hf, a "good" rotor (see lower portion of Fig.2). If a γ-ray from a specific cascade following the yrast line is observed in a detector, it is possible to record coincidence events in a second detector for the remaining transitions in the cascade, but not for the transition detected in the original detector. A regular pattern of coincidences, shown in Fig.3, emerges. The coincidence points are separated from neighboring points in the E_γ-E_γ plot by $8 \frac{\hbar^2}{2\mathcal{I}_i}$. Coincidences corresponding to recording equal-energy γ-rays in both detectors, however, are missing in the

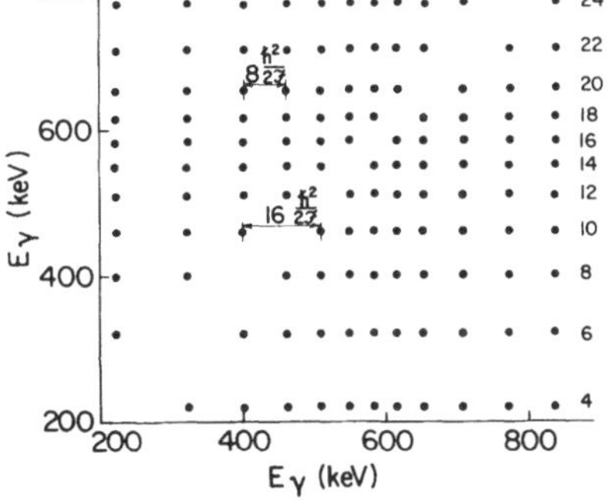

Fig.3 Transition-energy coincidence pattern for the yrast band in ^{170}Hf.

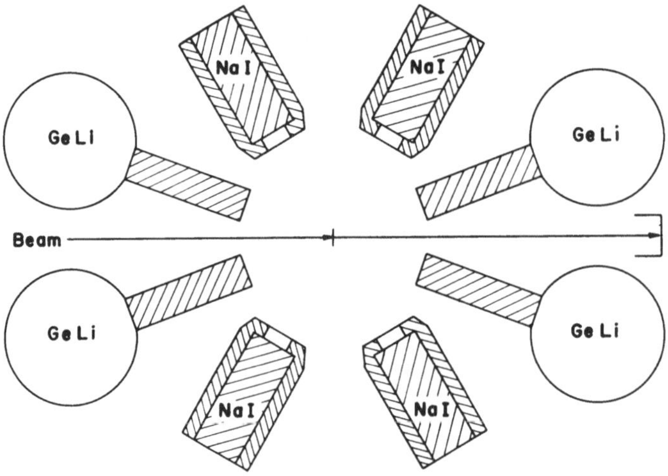

Fig.4 Typical experimental arrangement for transition-energy correlation experiments.

case of rotors, because only a single γ-ray of a particular energy is present in a specific cascade. Therefore, the transition energy difference between the coincidence points across an equal-energy, central "valley", is $16\,\frac{\hbar^2}{2\mathcal{I}_c}$, and the points forming the "ridge" on either side of the central "valley" correspond to coincidences between consecutive transitions in the cascade.

It should be possible to obtain a pattern similar to that shown in Fig.3 from γ-γ coincidence data[13] for ^{170}Hf, since the yrast transitions dominate the E_γ spectra (see Fig.2). A typical experimental setup for γ-γ coincidences using four Ge(Li) and four NaI detectors is shown in Fig.4. The data, recorded for an experiment in which E_γ-E_γ coincidences between Ge(Li) detectors are desired, are the events in which γ-rays are detected in at least two Ge(Li)'s and at least three of the total number of detectors. (Sodium-iodide detectors with their poorer resolution, but larger photo-efficiency, also can be used as the primary detectors for E_γ correlation studies, particularly for light and medium mass systems[10,14], where the moments of inertia are smaller and, therefore, the separation in E_γ for consecutive transitions is greater (see ref.10)). By requiring a coincidence with a third detector, most of the low-multiplicity events resulting from, e.g., Coulomb excitation and the decay of radioactive daughters are suppressed. The gains of the n detectors which are to be used in the correlation experiments are carefully matched so that the data from the $\binom{n}{2}$ possible coincident pairs can be added. Finally, the correlation between the photopeak events in the detectors, corresponding to the γ-ray transition energies, must be isolated from the "background" of photo-Compton and Compton-Compton coincidences. For coincidences between two large Ge(Li) detectors with a 20% photo-efficiency, only about 4% of the events correspond to the interesting photo-photo coincidences. (Photo-efficiencies as large as 50%, corresponding to photo-photo coincidences of about 25%, can be obtained for well collimated NaI detectors). Therefore, $\tilde{N}_{ij}$, the average number of uncorrelated events with coincident energies E_i, E_j is estimated, assuming that all events are uncorrelated, to be[10]

$$\tilde{N}_{ij} = \frac{\sum_j N_{ij} \sum_i N_{ij}}{\sum_{ij} N_{ij}} \tag{4}$$

N_{ij} is the number of coincident events with energies E_i, E_j. The difference between the number of coincident events and the average number of uncorrelated events in a particular channel,

$$\Delta N_{ij} = N_{ij} - \tilde{N}_{ij} \tag{5}$$

represents the excess or deficiency of correlated events for trans-

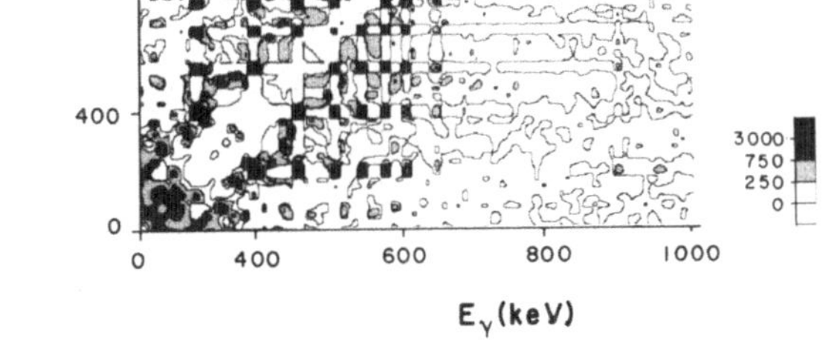

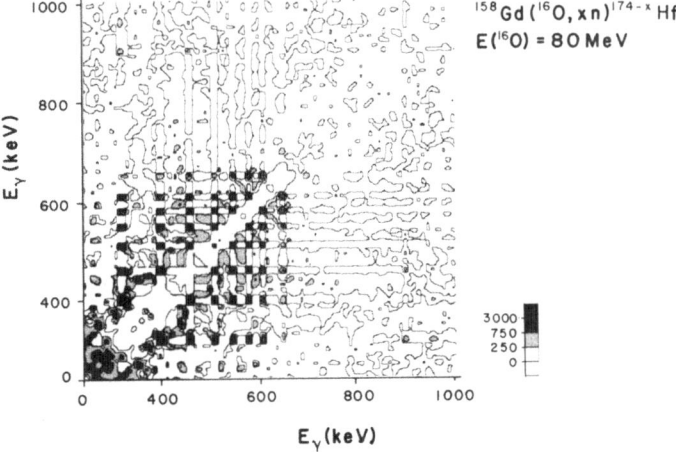

Fig.5 Transition-energy correlation spectrum for the reaction $^{158}Gd(^{16}O,xn)^{174-x}Hf$ at $E(^{16}O)$=80 MeV[13]. The dominant reaction product is ^{170}Hf.

sition energies E_i and E_j. (For a more detailed discussion of the technique used to isolate the correlated events from the average number of uncorrelated events, as well as several improvements using an iterative procedure, the reader is referred to references 15 and 16).

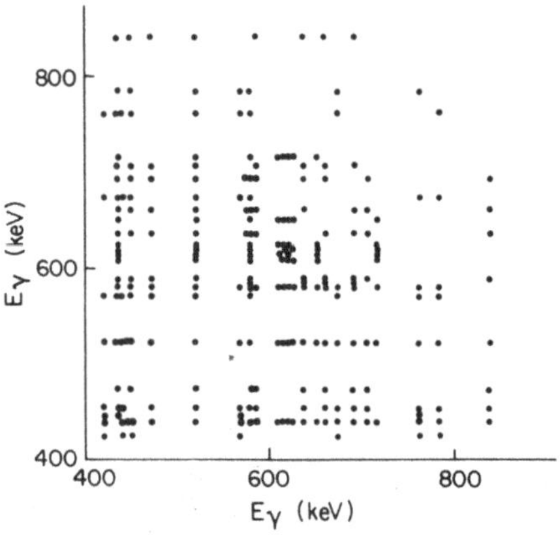

Fig.6 Transition-energy coincidence pattern for the ^{162}Yb ground-state band at the three strongest two-quasiparticle bands feeding the ^{162}Yb ground-state band in the data obtained from the $^{149}Sm(^{16}O, 3n)$ reaction[17,18].

A correlation spectrum, i.e. the distribution of the excess or deficiency of correlated events, N_{ij}, as a function of E_i and E_j, is shown in Fig.5 for the reaction $^{158}Gd(^{16}O, xn)^{174-x}Hf$ at $E(^{16}O)=80$ Mev[13]. The dominant reaction is the 4n channel with ^{170}Hf as the main product. The expected features are observed (compare with Fig.3). The correlation spectrum is dominated by the coincidences between the yrast states in ^{170}Hf (see Fig.2). A valley is clearly visible along the equal transition energy diagonal for $E_\gamma \lesssim 800$ keV. The width of this valley defined by consecutive transitions between known yrast transitions at low E_γ and by a smear of correlated transitions from the continuum at large E_γ, decreases with increasing E_γ up to $E_\gamma \approx 550$ keV. At the transition energies where the central valley is narrowest, the yrast band corresponds to a mixture of the ground-state band and the aligned S band. The interaction strength between the ground-state band and the aligned S band is very strong in ^{170}Hf, which has 98 neutrons. At higher energies the moment of inertia, corresponding to the width of the valley, is characteristic of the moment of inertia of the aligned S band.

For contrast, transition-energy correlations have been studied[17,18] for ^{162}Yb, a case where there is a weak interaction between the ground-state and S bands. The γ-ray energy spectra for ^{162}Yb are much more complex than that for ^{170}Hf (see Fig.2). The yrast band is not dominant above the $10^+ \rightarrow 8^+$ transition and very strong tansition in non-yrast bands are observed at low transition energies. The correlation pattern for cascades proceeding along the the three bands in ^{162}Yb which are known to be populated strongest by the ^{149}Sm(^{16}O, 3n) reaction for E(^{16}O)=80 MeV is shown in Fig.6; this pattern has been constructed using our known discrete-line information. The complete correlation spectrum from the same data is shown in Fig.7. Nonrotational behavior is expected, and indeed is observed for the γ-ray transition energies corresponding to the rotational frequencies ($\hbar\omega \sim E_\gamma/2$) at which the band crossings are known[17,18]. In particular the very strong feature observed in the valley at $E_\gamma \sim 615$ keV can be correlated with "upbends" in several bands which involve the unpairing of particular pair of $i_{13/2}$ neutrons (the BC pair in the nomenclature used by Frauendorf in the preceding paper[7]). In ^{162}Yb such crossings are predicted in cranked shell-model calculations[5-7] at $E_\gamma \sim 650$ keV. Furthermore, the total strength of the 615-keV feature in the valley (Fig.7) is more than two times greater than the strength expected from the coincidences between known states in the γ-decay scheme (Fig.6)[15]. Therefore, cascades along several additional bands, which individually are weakly fed in ^{149}Sm(^{16}O, 3n), but which collectively contain considerable strength, also must have a nonrotational behavior at $E_\gamma \sim 615$ keV. It is assumed that such a behavior is the result of

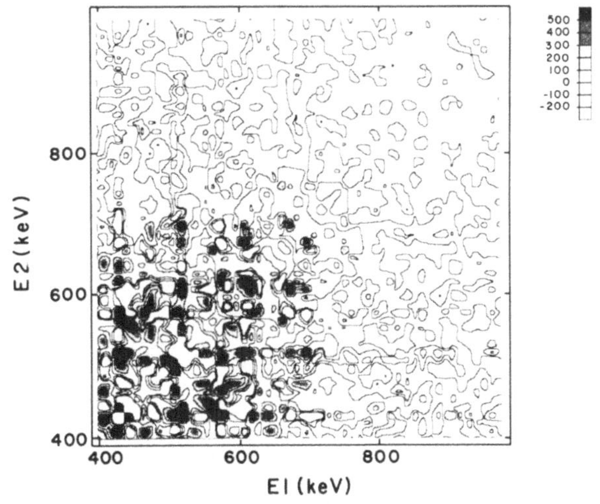

Fig.7 Transition-energy correlation spectrum for the reaction ^{149}Sm(^{16}O,xn)$^{165-x}$Yb at E(^{16}O) =80 MeV. The dominant reaction product is ^{162}Yb.

multiple band crossings[19], i.e. the crossing of a group of bands in which the BC orbits are unaligned with a similar group in which the orbit BC is aligned. (For example, bands based on aligned quasiparticles XY cross with bands based on aligned quasiparticles XYBC. XY designates arbitrary multi-quasiparticle configurations which do not block the alignment of the BC pair). Therefore, the observed enhancements of the nonrotational features ("bridges") in the valley of the correlation spectra indicate that the scheme of near-yrast bands pictured in Fig.1 continues to excitation energies more distant from the yrast line. That is, in the near-yrast region, at intermediate rotational frequencies, the cascades follow several bands corresponding to rather large alignments, which cross with bands of smaller alignment at specific frequencies that are characteristic of the pair of aligning quasiparticles. Except for the few bands, which lie nearest to the yrast line, most are too weakly populated for the γ-ray transitions to be resolved from the continuum. The additional strength, however, is accounted for by the increased level density for quasiparticle configurations more distant from the yrast line. The population of such bands, however, must decrease more rapidly than the increase of the level densities

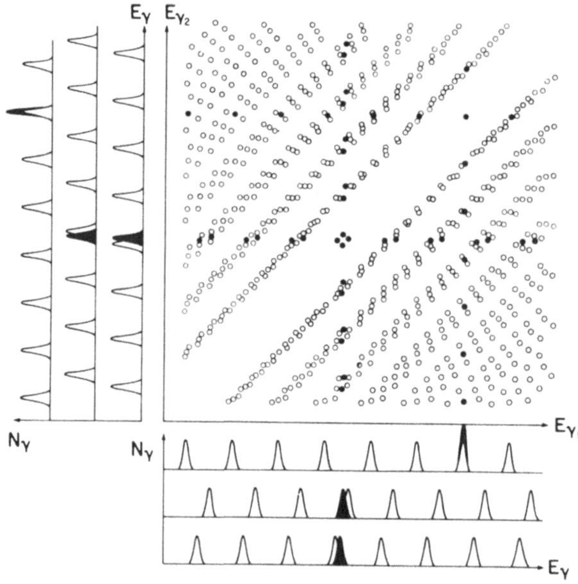

Fig.8 Coincidence pattern for twelve rotational bands with moments of inertia which differ by 20%. In the three bands shown at the bottom and left hand side of the figure a nonrotational transition has been added. Coincidences with the nonrotational transitions are shown by the solid points.

in the excitation region more distant from the yrast line.

With the considerable success in applying the γ-ray energy correlation techniques to the low and intermediate region of nuclear spins, it seems natural to extend such measurements to the highest spins which the nucleus can sustain[20]. For $I \gtrsim 35\hbar$ no discrete states are observed; therefore, it is necessary to ascertain the effect of varying the moments of inertia on the features of the E_γ-correlation spectra. The effect of varying the moment-of inertia by $\pm 20\%$, illustrated in Fig.8, is to increasingly smear the "ridges" parallel to the central "valley" as a function of the distance from the "valley". Even with moments-of-inertia differing by 20%, the "ridges" bounding the "valley", which correspond to consecutive transitions in a cascade, remain sufficiently well defined so that there is hope that information on the moments-of-inertia can be obtained as a function of transition energy (or rotational frequency) from the γ-ray continuum. This statement, of course, assumes that the continuum transitions correspond to correlated decays along rotational bands. The effects of decay along a few bands which have nonrotational features at a specific rotational frequency, i.e. "backbends" or "upbends", also are shown by the solid dots in Fig.8. Not only are features observed in the central valley, "bridges", but additional coincidences are observed with the other transitions in the cascade. If there are many cascades with two or more transitions of about the same energy, then the enhanced coincidences with the other transitions in the cascade should show

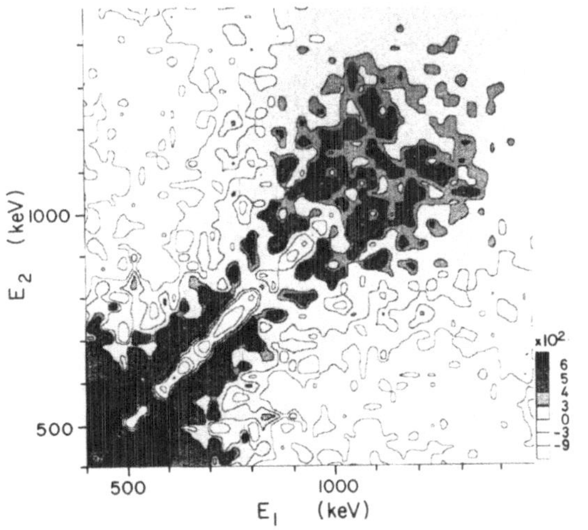

Fig.9 Transition-energy correlation spectra for the reaction $^{124}Sn(^{40}Ar,xn)^{164-x}Er$ at $E(^{40}Ar)=185$ MeV[22]. The dominant reaction products are $^{158-160}Er$.

up as vertical and horizontal lines or "stripes" intersecting at the central valley at the "upbend" energy, as observed for the solid points in Fig.8.

In order to study rare-earth nuclei at the highest angular momentum it is necessary to use heavier beams[19]. Therefore, we have used the ^{40}Ar beam from the 88" cyclotron at the Lawrence Berkely Laboratory to populate the residues of ^{164}Er using the ^{124}Sn(^{40}Ar, xn) reaction at E(^{40}Ar)=185 MeV[21]. Dominant residues at this incident energy are $^{158-160}$Er. The E_γ-E_γ correlation spectrum is shown in Fig.9. The "valley" is visible up to a transition energy of about 1 MeV above which a large number of cascades apparently have transitions with nearly identical energies. The "valley"

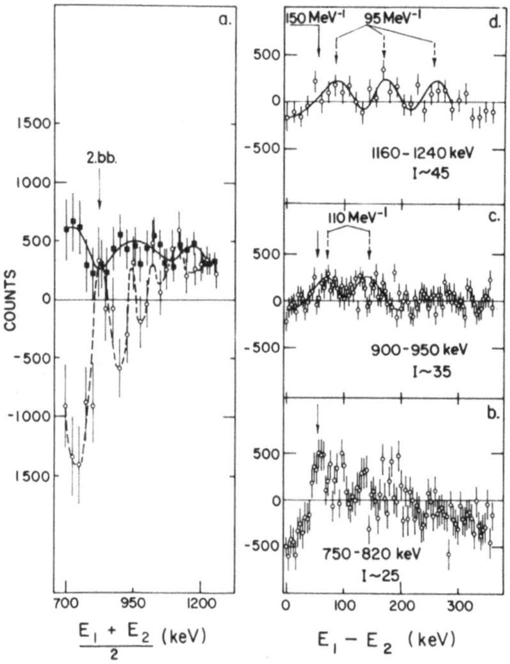

Fig.10 Projections along the "valley" and along the "ridge" bounding the "valley" from the ^{40}Ar+^{124}Sn correlation spectra shown in Fig.9 are indicated by the open and closed points respectively in the left-hand side of the figure. Projections across the vaelly at the right-hand side. The transition energy corresponding to the second backbend in ^{158}Er is indicated as is the expected positions of the first ridges for $2\mathcal{J}/\hbar^2$=95, 110, and 150 MeV^{-1}.

seems to reappear at a transition energy of about 1125 keV and to continue upward to about 1250 keV.

In order to determine if the features of the ridge structure are statistically significant, projections have been made along the central "valley" and the "ridge" bounding the "valley", as well as across the central valley at three transition energies where the "valley" is visible. These projections are shown in Fig.10. The position of the first ridge corresponding to $2\mathcal{J}/\hbar^2=150\ \text{MeV}^{-1}$ is shown in Fig.10. This value of the moment-of-inertia was obtained for the same E systems from the transition energy of the edge of "quadrupole bump" in the spectrum of $\langle M_\gamma\rangle$ as a function of E_γ, ref.9. For the projection across the "valley" at E_γ=750-820 keV the "ridge" structures of the correlation plots are not in serious disagreement with a moment-of-inertia of about 150 MeV^{-1}. At higher transition energies, however, the "valley" is observed to become considerably wider indicating a significant reduction in the moment-of-inertia. Finally at the highest transition energy where a "valley" can be defined, a moment-of-inertia $\approx$95 MeV^{-1} is indicated. The values of the moments-of-inertia obtained from the width of the "valley" in the correlation plots are compared in Fig.11 with moments-of-inertia for ^{158}Er obtained from traditional γ-γ discrete-line spectroscopy[22] and at higher rotational frequencies values for $^{156-160}$Er from the edge of the "quadrupole bump" in the spectrum of $\langle M_\gamma\rangle$ as a function of E_γ[9]. At intermediate rotational frequencies the moments-of-inertia obtained from the width of the "valley" in

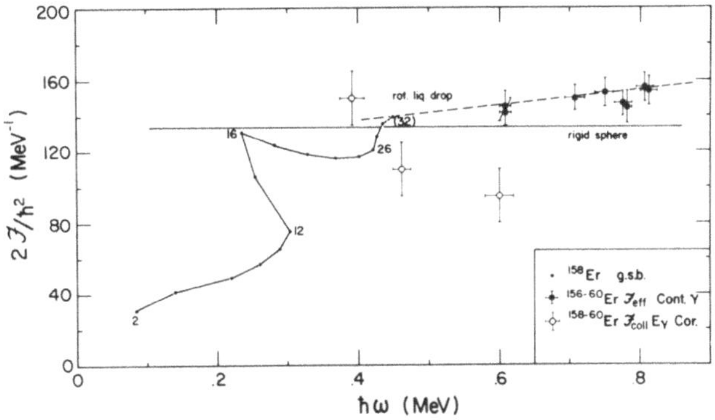

Fig.11 Moments of inertia as a function of $\hbar\omega$ for ^{158}Er from discrete line γ-γ spectroscopy[22], for $^{156-160}$Er from the edge of the "quadrupole bump" in the $\langle M_\gamma\rangle$ versus E_γ spectrum[9], and from the width of the valley in che γ-ray correlation data shown in Figs.9 and 10. For reference, the moments of inertia of a rotating rigid sphere and that expected from a rotating liquid drop[20] are shown.

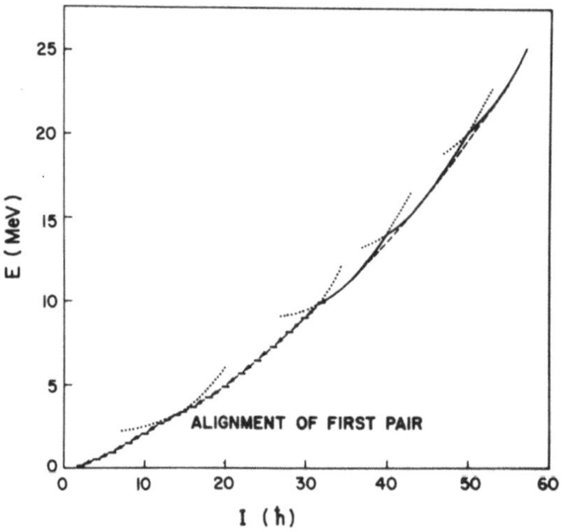

Fig.12 Typical decay sequence expected for γ-ray de-excitation of a rotational-like system with large angular momentum. For lower angular momentum transitions between the yrast states in ^{158}Er, ref.23, are shown. At larger angular momentum only one of the many possible yrast-like sequences is shown. For such a decay sequence along crossing bands the average moment of inertia of the envelope of transition, $\mathcal{J}_e$, (shown by dotted lines) will differ from the moment of inertia within a particular b and, $\mathcal{J}_c$.

the correlation data are in reasonable agreement with that expected for a rigid rotor. However, at the higher frequencies the values obtained from E_γ-E_γ correlations are as much as 30% lower than those obtained from the $\langle M_\gamma \rangle$ vs. E_γ data which are in agreement with the values expected for a rotating liquid drop[20]. The solution to this paradox is that the moment-of-inertia, $\mathcal{J}_c$, obtained from the width of the valley in the γ-ray transition energy correlation data not only is specific to a particular cascade, but is determined from consecutive transitions within such a cascade. The moments-of-inertia within a particular rotational band, $\mathcal{J}_c$, may be very different (see Fig.12) from the effective moments-of-inertia, $\mathcal{J}_e$, obtained from associating the highest γ-ray energy in a cascade with the highest angular momentum of that cascade[9,23], i.e.,

$$E_\gamma = 4 \frac{\hbar^2}{2 \mathcal{J}_e} I \tag{6}$$

The effective moment-of-inertia, $\mathcal{I}_e$, is characteristic of the envelope of bands, and therefore, is larger than the value in a particular band because of the larger curvature of a particular band (see Fig.12). The larger curvature or smaller moment-of-inertia in a particular band is a consequence of the offset in angular momentum, i.e. the alignment j_a, in the parabolic expression, eq.(1), relating E and I. Equating the transition energy from eqs.(2) and (6), a relation between $\mathcal{I}_c$ and $\mathcal{I}_e$ is obtained:

$$\mathcal{I}_c = (1 - \frac{j_a}{I})\, \mathcal{I}_e \qquad (7)$$

indicating that the fractional change between $\mathcal{I}_c$ and $\mathcal{I}_e$ is the percentage of the total angular momentum which corresponds to particle alignment. For the $^{158-160}Er$ systems the data, therefore, would indicate that about 1/3 of the angular momentum at $I \approx 45$ corresponds to particle alignment.

The increasing alignment observed for increasing angular momentum is consistent with the "bridge" observed at high transition energies in the "valley" of the $^{40}Ar+^{124}Sn$ E_γ correlation data (see Fig.9). Such features must represent multiple-band crossings based on the alignments of specific pairs of nucleons with an associated increase in the aligned angular momentum. In particular, the "bridge" at $E_\gamma \approx 830$ keV is not nearly the same transition energy as the known second "backbends" in the yrast bands of ^{158}Er (ref.22) and ^{160}Yb (ref.8, 24). At this crossing in ^{160}Yb the nucleus gains about 6.5 units of alignment . The other intense "bridges" which we observed at E_γ=950, 1020, and 1120 keV, presumably correspond to the alignment of other high-j, low-K orbits. Even more pronounced such structures corresponding to two or more γ-rays of identical energy in the same cascade have been observed in other studies of the E_γ correlations[25]. When the iterative procedure described in refs.15, and 16 is used to isolate the correlated events, such features are enhanced, and enhanced coincidences are observed with the lower lying transitions in a particular nucleus ("stripes").

Finally, I would like to discuss a new technical development which shows great promise for the future study of γ-ray transition-energy correlations. It is obvious that the increased photo efficiency of Compton suppressed Ge(Li)'s would be advantageous for E_γ-E_γ correlations if such a system could be constructed with sufficient solid angle. This system has the additional advantage that the uncorrelated photo-Compton and Compton-Compton spectrum can be measured simultaneously. After scaling to the appropriate fraction of such events expected to contaminate the measured suppressed coincidences, this experimental background can be subtracted from the measured suppressed spectrum. Such a procedure circumvents the problems associated with statistical methods for the removal of the background of uncorrelated events. A system containing five Comtpon-suppressed Ge-(Li) detectors has been constructed and tested at the Niels Bohr Institute Tandem Accelerator Laboratory at Risø, in collaboration with a group from the Universities of Liverpool and Manchester[26]. Preliminary measurements made only last month with this apparatus were very promising.

IV. LOW-SPIN MEMBERS OF THE S BAND

The extension of rotational bands which are based on aligned configurations (e.g. those shown in Fig.12) to low angular momentum is an interesting topic. For values of $I<j_a$ the rotational angular momentum, R, would be antiparallel with j_a (see right side of Fig.13), and the rotational energy would increase with decreasing spin (see eq.1). Therefore, as the value of I approaches j_a, either the individual particles recouple to reduce j_a or the alignment remains constant and the energy of the band members reaches a minimum for $I=j_a$ and then increases for a further reduction in I. Particle-rotor model calculations, which include the cross terms in I and j_a (i.e. the "recoil" terms) in the Hamiltonian, predict[11] that it is energetically more economical for the particles to recouple their angular momenta at low spin.

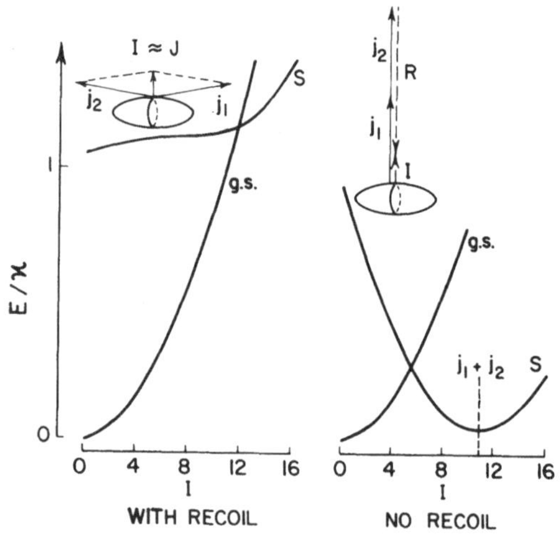

Fig.13 Particle-rotor model predictions for the low-spin portion of the S band with (left side) and without (right side) "recoil" terms included in the Hamiltonian. The corresponding angular momentum coupling schemes are indicated. For comparison the ground-state band also is shown.

The low-spin members of highly aligned bands are difficult to establish. The members of these bands are increasingly excited with respect to the yrast states for lower angular momentum (see Fig.,12). Therefore, such states are not populated in the near yrast γ-ray cascades following (H.I.,xn) reactions. The low-spin members of aligned abnds are not excluded from the γ-ray cascaded following (L.I., xn) reactions; however, such reactions are not specific to these states. Furthermore, these states are often sufficiently high in excitation so that transitions to low-lying states by high-energy γ-rays dominate, even though such transitions are not favored intrinsically. Candidates for members of the S band, i.e. the highly-aligned, two quasiparticle configuration wich lies lowest in excitation, however, have been suggested for ^{156}Dy (ref.27) and ^{172}Yb (ref.28) from the systematics of rotational bands established using (L.I.,xn) reactions and for ^{164}Yb from possible transitions observed in the E_γ correlated spectra of transition feeding various low-lying states . In such studies, however, it was not possible to ascertain the configuration of the proposed members of the aligned band.

In the rare earth region the dominant configuration of the S band is aligned $i_{13/2}$ neutrons[2]. It, therefore, should be possible to suggest candidates for the low-spin members of this configuration from the spectrum of $(\nu(i_{13/2}))^2$ states populated in ℓ =6 pickup or stripping reactions on target nuclei with $\nu(i_{13/2})$ ground-state configurations. Three such nuclei are stable, ^{161}Dy, ^{167}Er, and ^{179}Hf.

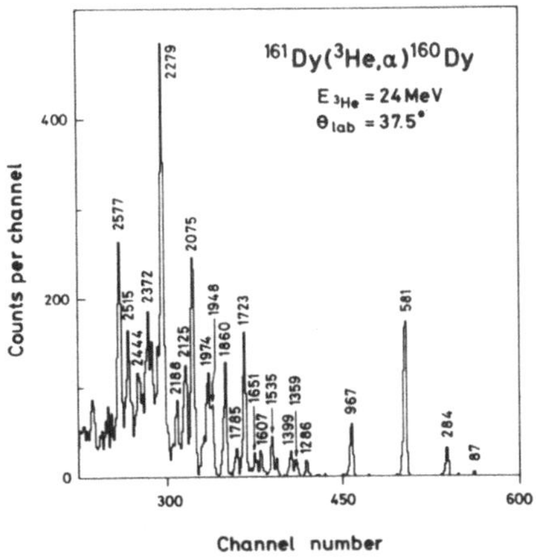

Fig.14 Alpha-particle spectrum for the reaction ^{161}Dy(^{3}He,α) populating states in ^{160}Dy. The spectrum was obtained at 37.5° in the laboratory and with 24 MeV incident ^{3}He projectiles.

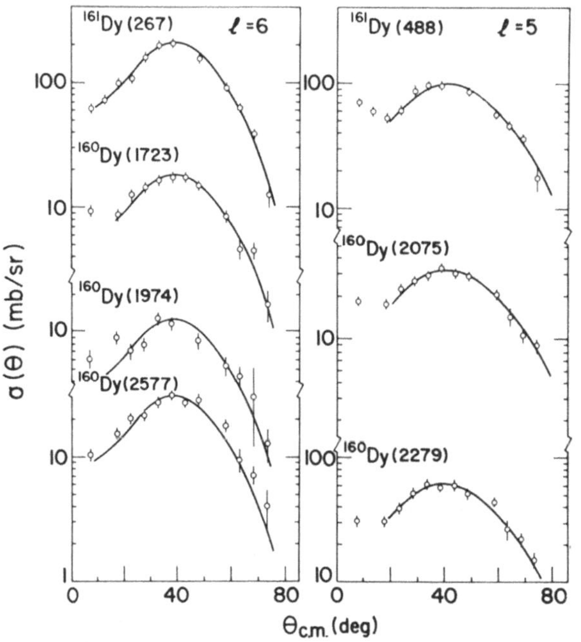

Fig.15 Angular Distributions of selected ℓ_n=5 and 6 transitions from the reactions ^{161,16}Dy(^{3}He,α)^{160,161}Dy. The empirical shapes for ℓ_n=5 and 6 established from the transitions to ^{161}Dy(488) and ^{161}Dy9267), respectively, are shown with the other transitions.

The ℓ_n=5 and 6 reactions are selectively populated in (^{3}He, α) re-(α, ^{3}He) reactions[29]. A spectrum of the ^{161}Dy(^{3}He, α) reaction populating ^{160}Dy (ref.12) is shown in Fig.14. It is possible to distinguish between ℓ_n=5 and 6 (^{3}He, α) pickup from the angular distribution shapes established[12] from the (^{3}He, α) transitions on ^{162}Dy populating known[30] $h_{11/2}$ and $i_{13/2}$ states in ^{161}Dy (see Fig. Fig.15).

The low-lying spectra of $(\nu(i_{13/2}))^2$ states populated in the ^{161}Dy(^{3}He, α) reaction are quite simple. The 4^+, 6^+ and 8^+ states of the ground-state band (4_g^+, 6_g^+ and 8_g^+) are populated with relative strengths of 0.20, 1.0 and 0.33 (see Table 1) in agreement with the angular momentum coupling for $i_{13/2}$ pickup from $I^\pi=5/2^+$ target $(\alpha|\langle 5/2, 5/2, 13/2, -5/2|I, K=0\rangle|^2)$. The remaining ℓ_n=6 transitions below E =2.5 MeV are to states at E_x=1723 and 1974 keV. For states with low K (i.e. aligned states), which are predicted to be populated in ℓ_n=6 pickup, the angular-momentum coupling coefficients are large for $I^\pi=6^+$, 7^+, and 8^+. The 1723 keV level agrees

TABLE I

Summary Of $\ell_n = 6$ $^{161}Dy(^3He, \alpha)^{160}Dy$ Transitions

E_x (keV)	I^π	σ_{ex}/σ_{DW} [a]	Suggested config.
283	4^+	0.09	4^+_g
583	6^+	0.45	6^+_g
967	8^+	0.15	8^+_g
1607 [b]	(4^+) [b]	$\lesssim 0.06$ [b]	4^+_g
1723	6^+	0.33	6^+_s
1974	$8^+(4-9)^+$	0.20	8^+_s
2577	$4^+ - 9^+$	0.73	c)

For primary data, see ref.12

a) σ_{DW} calculated using code DWUCK and ave. optical model parameters from ref.29.

b) Weak transitions - not established by angular distribution slope - see text.

The next strong very $\ell_n = 6$ transition is to 7^+ member of odd spin $(\nu(i13/2))^2$ configuration.

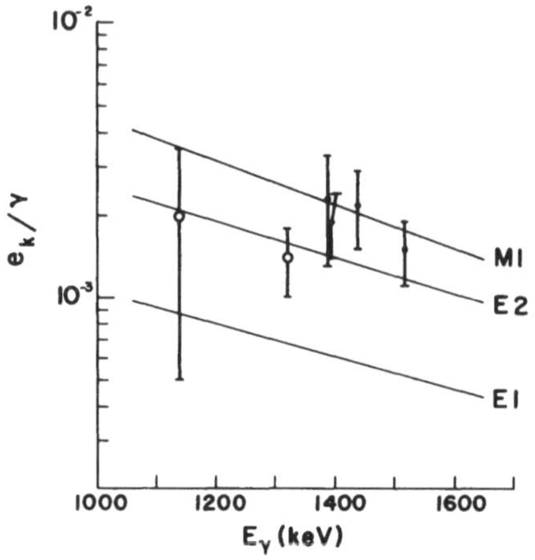

Fig.16 Internal conversion coefficients from ref.31, for the transitions in ^{160}Dy which are proposed as decay between the low-spin portion of the S band and the ground-state band (see Fig.17).

in E_x with a state at 1722.5 keV, established[31] from γ-emission following the β-decay of ^{160}Ho. The large population of this level in the (^{3}He, α) reaction together with the E2 and/or M1 and E2 or E1 transitions from the 1722.5 keV level to 4_g^+ and 6_g^+ respectively (see Figs.16 and 17) suggest $I^\pi=6^+$ for the 1722.5 keV level. The 1974 keV level was not previously known in ^{160}Dy; however, two possible candidates (both of E2 and/or M1 multipolarity - see Fig.16 and 17 - exist for the γ-decay of this state to 6_g^+.

The relative $\ell_n=6$ pickup strengths and the selectivity of the pickup reaction, together with the γ-decay data[31], suggest a special configuration of $(\nu(i_{13/2}))^2$ states at intermediate excitation energies near to that predicted for the S band from a fit of empirical data[32], as well as from cranked shell-model[6,7] and particle-rotor model[11] calculations. Such states should be populated in the ^{161}Dy(^{3}He, α) reaction. Therefore, the 1723 and 1974 keV levels of ^{160}Dy are suggested as candidates for the 6^+ and 8^+ states (6_s^+ and 8_s^+) of the extension of the S band to low spin.

If the identification of the 1723 and 1974 keV levels with 6_s^+ and 8_s^+ is correct, a 4_s^+ state should be weakly populated by a $\ell_n=6$ transition at a somewhat lower excitation energy than the 6_s^+ state. A candidate for this state is the level 1605.5 keV which is known[31] to decay by E2 and/or M1 and by E2 transitions to the 2_g^+ and 4_g^+ levels (see Fig.16 and 17). This state would correspond to the level at 1607 keV which is weakly populated in the (^{3}He, α)

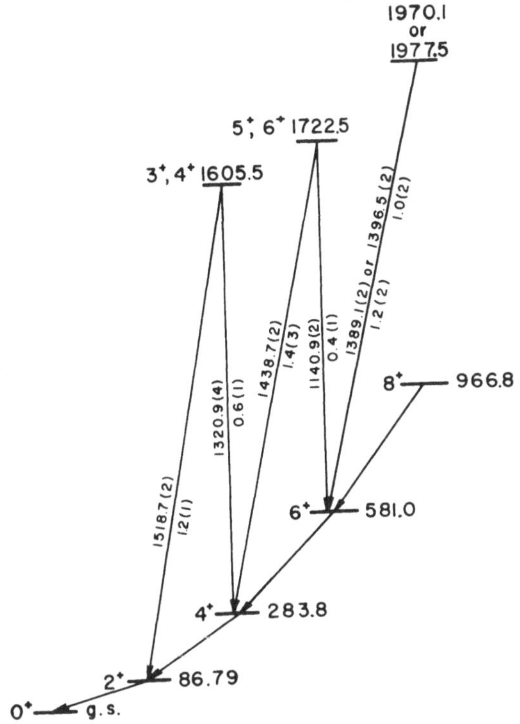

Fig.17 Summary of proposed γ-ray transitions between the members of low-spin portions of the S band and the ground-state band in ^{160}Dy. The intensities shown are from γ-decay following the β-decay of ^{160}Ho (ref.31).

reaction and therefore, cannot be definitely established as $\ell_n=6$ pickup.

The strengths of the suggested $I_s \to (I-2)_g$ transitions for all three S-band candidates, shown in Fig.17, are nearly equal, and the strengths of the suggested $I_s \to I_g$ transitions from 4^+_s and 6^+_s are from 2-4 times less than the strength of the corresponding $I_s \to (I-2)_g$ transition. For the intra S-band transition to compete with $I_s \to (I-2)_g$ transitions, the intraband transition must be enhanced by $\gtrsim 10$ and $\gtrsim 10^3$ compared to the intraband transitions for the decay of 6^+ and 8^+, respectively. No intraband transitions have been observed between the proposed members of the S band.

The excitation energy versus I(I+1) of the suggested low-spin members of the S band in ^{160}Dy is compared in Fig.18 with similar values for the members of the ^{160}Dy yrast band. For reference a dashed line with slope corresponding to the spherical rigid body

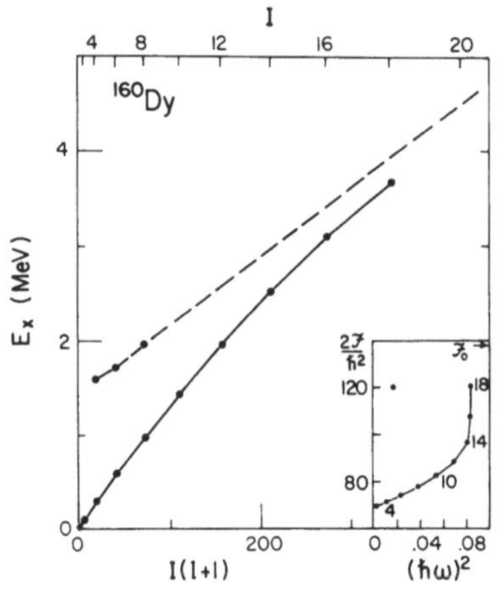

Fig.18 Plot of E_x versus I(I+1) for the yrast band and the proposed low-spin portion of the S band in ^{160}Dy. A dashed line with slope corresponding to the spherical rigid-body moment of inertia is extended form the 8_s^+ to higher spins. For reference, a "bankbending" plot for the yrast band in ^{160}Dy is shown in the inset. The moment of inertia corresponding to he $8_s^+ \rightarrow 6_s^+$ transition also is shown in the "back-bending" plot as well as, $\mathcal{J}_o$, the spherical rigid-body moment of inertia.

moment of inertia ($2\mathcal{J}/\hbar^2$=137 MeV^{-1}) is extended to higher spins from the candidate for 8_s^+. A band with a moment of inertia of about 15% larger than the spherical rigid-body value would cross the yrast line at the 18^+ state, which is in the "upbend" region (see the inset in Fig.18). The excitation energy systematics of the suggested candidates for the 4_s^+, 6_s^+ and 8_s^+ are reasonable. It, however, would seem to be important to extend the yrast band to higher spins to establish the high-frequency portion of the S band. The ^{150}Nd(^{14}C, 4n) reaction or Coulomb excitation appears to be the most promising reactions for such extensions of the yrast band in ^{160}Dy.

The experimental alignments, i, and Routhians[6,7], e', corresponding to the assumed S-band members are compared in Fig.19 with such quantities extracted for the yrast band of ^{160}Dy and the two signatures, $\alpha=\pm 1/2$ of the lowest $i_{13/2}$ band in ^{159,161}Dy. The values shown for the S band were calculated assuming K=0; however, $K \leq 2$ would not make large differences in the extracted i and e' values. The two unpaired neutrons, which combine to form the S band, give an alignment similar in magnitude[11] to a single unpaired

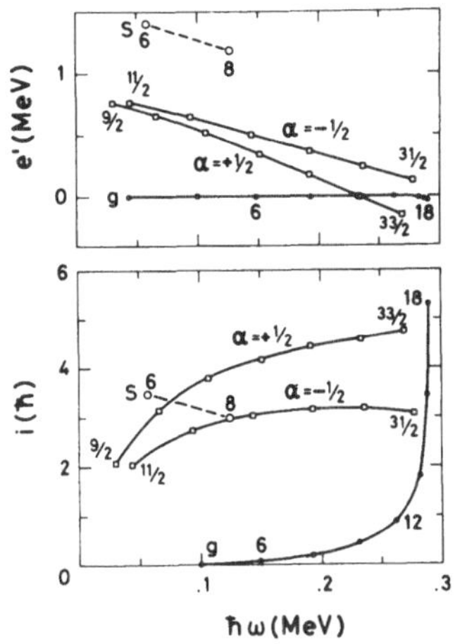

Fig.19 Alignment, i, and experimental Routhians, plotted as a function of the rotational frequency for yrast (solid points) and S (open points bands in ^{160}Dy and the $\alpha = \pm 1/2$ portions of the lowest $i_{13/2}$ bands in ^{159,161}Dy (squares). The prescription for obtaining values of these parameters from the excitation energies and spins is given in ref.6.

neutron at the same rotational frequency. These data, therefore, indicate that for the low-spin members of the S band the alignment along the nuclear symmetry axis is reduced by an adjustment of the angle between the quasiparticle spins. Such a recoupling of the angular momentum vectors of the quasiparticles is predicted for the low-spin members of the S band by the particle-rotor model when the "recoil terms" are treated correctly[11]. It is noted, however, that such calculations for ^{160}Dy predict[11] the low-spin members of the S band at an excitation energy that is somewhat higher than that of the suggested candidates. An average alignment of 8.5 for the S band between $I^{\pi}=8^{+}$ and the crossing with the g.s. band is obtained from the experimental Routhians shown in $(i = -\frac{de'}{d\omega})$. (The crossing frequency is taken either from the rotational frequency of the "upbend" in ^{160}Dy or from the frequency where the sum of the e' values for $\alpha = \pm 1/2$ bands in the odd Dy nuclei equals zero). This value is in reasonable agreement with i = 8.9 predicted from cranked shell-model calculations[6,7]. From the experimental two quasiparticle Routhians shown for the S band in Fig.19, the pairing energy, Δ, is estimated to be in the range of of 0.75 to 0.79 MeV depending on the procedure used to extrapolate the values of e' to $\hbar\omega = 0$. This is somewhat below the odd-even mass difference (1.08 MeV) calculated using neutron binding energies.

Presumably, this is the effect of a reduced pairing in the odd-A nuclei due to "blocking" by the unpaired neutron. Indeed, realistic calculations for ^{160}Dy place the two quasi-particle band heads (i.e. 2Δ) at about 1.5 MeV[33].

Information about the magnitudes of the K quantum number (i.e., the projection of the spin of the "aligned" $i_{13/2}$ neutrons on the symmetry axis) is available both from the γ-ray branching ratio for $I_s \rightarrow I_g$ and $I_s \rightarrow (I-2)_g$ and from the relative $\ell_n=6$ (^3He, α) transition strengths to the S band members ($\propto |\langle 5/2, 5/2, 13/2, -5/2 | I_s K\rangle|^2$). The pickup cross sections[12] together with the γ-decay data[31] are consistent with average K=0, 1, and 2 mixture in the low-spin part of the S band in an approximate ratio of 0.35:0.15:0.50 respectively[12].

IV. CONCLUDING REMARKS

The purpose of this article is threefold: First, to introduce some of the techniques, which are specific in isolating particular features from the continuum of states near the yrast line both at high and low spins. Second, to give a few examples illustrating the "flavor" of the physics obtainable using such techniques. And final finally, to attempt to convey some of my enthusiasm for such studies

Several techniques for studying the γ-ray continuum exploiting correlations between the transition energy and other parameters e.g., $\langle M_\gamma \rangle$, total cascade energy, angular correlations, etc., have been omitted from this report. For a review of these topics, as well as references to the original literature, the reader is referred to reference 19.

The work described in this report has been made in collaboratio with colleagues and guests at the Niels Bohr Institute as well as with coworkers at the Lawrence Berkeley Laboratory, where the ^{40}Ar data were obtained. In particular I wish to acknowledge B. Herskind G.H. Hagemann, O. Andersen, L.L. Riedinger, F.S. Stephens, M.A. Deleplanque, C. Ellegaard, J.J. Gaardøje, Jin Gen-ming, J.C. Lisle, and P. Twin. The theoretical ideas have been developed after developed after extended conversations with Aage Bohr, B. Mottelson, S. Frauendorf, G. Leander, I. Hamamoto and H. Sagawa.

REFERENCES

1. See e.g. R.M. Lieder and H. Ryde, in Advances in Nuclear Physics, ed. M. Baranger and E. Vogt (Plenum, New York, 1978), vol.10, p.1 and refs. therein.

2. F.S. Stephens and R.S. Simon, Nucl. Phys. A183, 257 (1972).

3. O.C. Kistner, A.W. Sunyar, and E. der Mateosian, Phys. Rev. C17, 1417 (1978).
4. L.L. Riedinger, O. Andersen, S. Frauendorf, J.D. Garrett, J.J. Gaardhøje, G.B. Hagemann, M. Guttormsen, B. Herskind, Y.V. Makovetzky, P.O. Tjom, and J.C. Waddington, Phys. Rev. Lett. 44, 568 (1980).
5. Aage Bohr and B. Mottelson, in Proceedings of the International Conference on Nuclear Structure, Tokyo, 1977; J. Phys. Soc. Japan Japan 44 (1978), suppl. p.157.
6. R. Bengtsson and S. Frauendorf, Nucl. Phys. A314, 27 (1979) and A327, 139 (1979).
7. S. Frauendorf, paper in this volume.
8. L.L. Riedinger, in Nobel Symposium on Nuclei at Very High Spin, June 1980, to be published in Phys. Scripta.
9. D.L. Hillis, J.D. Garrett, O. Christensen, B. Fernandez, G.B. Hagemann, B. Herskind, B.B. Back, and F. Folkmann, Nucl. Phys. A325, 216 (1979).
10. O. Andersen, J.D. Garrett, G.B. Hagemann, B. Herskind, D.L. Hillis, and L.L. Riedinger, Phys. Rev. Lett. 43, 687 (1 687 (1979).
11. J. Almberger, I. Hamamoto, G. Leander, and J.O. Rasmussen, Phys. Lett. 90B, 1 (1980).
12. Jin Gen-ming, J.D. Garrett, G. Løvhøiden, T.F. Thorsteinsen, J.C. Waddington, and J. Rekstad, NBI preprint 1980; and in Proceedings of the International Conference on Band Structure and Nuclear Dynamics, New Orleans, 1980, Vol.I, contributed paper.
13. J.C. Lisle, O. Andersen, J.D. Garrett, G.B. Hagemann, and B. Herskind (to be published).
14. Th. Lindblad, A. Johnson, S.A. Hjorth, C.G. Linden, O. Andersen, M.A. Deleplanque, J.D. Garrett, B. Herskind, and F.S. Stephens, Phys. Scripta (to be published).
15. O. Andersen, Ph.D. Thesis, University of Copenhagen (1980).
16. B. Herskind, in Proceedings of the International Conference on Nuclear Behavior at High Angular Momentum, Strasbourgh April 1980.
17. See e.g., L.L. Riedinger, in Proceedings of the International Conference on Band Structure and Nuclear Dynamics, New Orleans, 1980; Nucl. Phys. A347, 171 (1980).
18. L.L. Riedinger, O. Andersen, S. Frauendorf, J.D. Garrett, J.J. Gaardhøje, G.B. Hagemann, B. Herskind, Y.V. Makovetzky, M. Guttormsen, and P.O. Tjøm (to be published).
19. J.D. Garrett and B. Herskind, in Proceedings: Study Weekend on Nuclei Far from Stability, 22-23 Sept., 1979, Ed. W. Gelletly (Daresbury Laboratory, Warrington, U.K., 1980) p.30.
20. S. Cohen, F. Plasil, and W. Swiatecki, Ann. Phys. 82, 557 (1974).

21. M.A. Deleplanque, F.S. Stephens, O. Andersen, J.D. Garrett B. Herskind, R.M. Diamond, C. Ellegaard, D.B. Fossan, D.L. Hillis, H. Kluge, M. Neiman, C.P. Roulet, S. Shih, and R.S. Simon, Phys. Rev. Lett. 45, 172 (1980).
22. J.Y. Lee, M.M. Aleonard, M.A. Deleplanque, Y. El Masri, J.O. Newton, R.S. Simon, R.M. Diamond, and F.S. Stephens, Phys. Rev. Lett. 38, 1454 (1977).
23. M.V. Banaschik, R.S. Simon, P. Colombani, D.P. Soroka, F.S. Stephens, and R.M. Diamond, Phys. Rev. Lett. 34, 892 (1975).
24. F.A. Beck, E. Bozek, T. Byrski, C. Gehringer, J.C. Merdinger, Y. Schutz, J. Styczen, and J.P. Vivien, Phys. Rev. Lett. 42, 492 (1979).
25. See e.g., M.A. Deleplanque, in Proceedings of the Nobel Symposium on Nuclei at Very High Spin, June 1980 (to be published in Physics Scripta); and F.S. Stephens in the Proceedings of the International Conference on Nuclear Physics, Berkeley, 1980 (to be published in Nuclear Physics).
26. P. Twin, P. Noland, J. Sharpey-Schaffer, P. Butler and J.C. Lisle,
27. Y. El Masri, R. Janssens, C. Michel, P. Monseu, J. Steyaert, and J. Vervier, Z. Phys. A274, 113 (1975).
28. P.M. Walker, L.R. Faber, W.H. Bentley, R.M. Ronningen and R.B. Firestone, Nucl. Phys. A343, 45 (1980).
29. See e.g. P.O. Tjøm and B. Elbek, in Advances in Nuclear Physics, ed. M. Barranger and E. Vogt (Plenum, New York, 1978), Vol.3, p.259.
30. T. Grotdal et al., Phys. Norvegica 8, 23 (1975).
31. E.P. Grigor'cu, K.Y. Gromov, Zh. T. Zhelev, T.A. Islamov, V.G. Kalinnikov, U.K. Nayarov, and L.L. Sabirov, Izv. Akad. Nauk. SSSR Fiz. 33, 635 (1969); A.A. Aleksandrov, V.L. Buttsev Ts. Vylov, E.P. Grigorev, K.Y. Gromov, V.G. Kalinninkov, and N.A. Lebedev, ibid. 38, 2096 (1974).
32. J.O. Rasmussen, M.W. Guidry, T.E. Ward, C. Castaneda, L.K. Peker, E. Leber, and J.H. Hamilton, Nucl. Phys. A332, 82 (1979).
33. H. Sagawa (private communication).

DYNAMIC DEFORMATION THEORY: RECENT RESULTS FOR SPECTRA AND FOR CROSS-SECTIONS

Krishna Kumar

Department of Physics
University of Bergen
Bergen, Norway

and

Department of Physics*
Vanderbilt University
Nashville, TN 37235

and

Department of Physics†
Tennessee Technological University
Cookeville, TN 38501

I. ABSTRACT

A global, microscopic (dynamic deformation) theory of nuclear collective motion has been developed in recent years which allows us to calculate the structure wave functions of nuclei as light as ^{12}C, as heavy as ^{240}Pu, as spherical as ^{208}Pb, as deformed as the fission isomer of ^{240}Pu, as transitional as ^{72}Ge, and with as much shape co-existence as in ^{16}O.

Several components of the theory have been improved upon during the past year: treatment of pairing in the limit of vanishing gap; convergence of the shell-correction energy; solution of the collective Schrodinger equation.

---

*Supported in part by the U.S. Department of Energy

†Present address.

Recent results are for: giant quadrupole and giant monopole resonances in ^{208}Pb; core-particle coupling in the odd-A nucleus ^{191}Pt; coupled-channel calculations of neutron (elastic, inelastic, total) cross-sections of Sm isotopes; coupled-channel calculations of proton inelastic cross-sections of Ge isotopes.

II. INTRODUCTION

The main objective of the dynamic deformation theory (DDT) is to provide nuclear wave functions which can be employed not only for nuclear structure studies but also for nuclear reaction studies. The reason behind this objective is quite obvious, but may be well worth repeating. The reaction cross-sections, especially the angular distributions, probe the structure wave functions in much greater detail than the energy levels and the electromagnetic moments.

The reaction part of the DDT is discussed in this volume by Jan Vaagen[1]. I shall discuss mainly the structure part. Before discussing recent results, I shall give a brief comparison of DDT with two of the other theories discussed at this workshop.

Recently, much progress has been made in several basically different structure theories. One of them is the nuclear field theory (NFT) developed by Bohr and Mottelson[2], Bes at al.[3], Wu and Feng[4], and others. Another one is the interacting Boson Model (IBM) developed by Arima and Iachello[5], Talmi[6], Scholten[7], and others.

The DDT differs from both NFT and IBM in the following essential aspects:

1. While only a small configuration space is employed in the NFT and IBM, a greatly extended (660 nucleon) space[8] is employed in the DDT. Consequently, it is not necessary to employ an effective charge. It is not necessary to vary model parameters from one nuclear region to another.

2. While only the spherical single-particle basis is employed in the NFT and IBM, spherical as well as deformed bases are employed in the DDT. A second quantization leading to a second Schrodinger equation[8,9] in the collective coordinates is employed to determine the probability for different types of nuclear shapes: spherical-deformed, axially symmetric-asymmetric, prolate-oblate. The shape probability distribution is calculated not only for the ground state but also for each of the calculated nuclear states. This allows for the possibility of the co-existence of very different shapes in the same nucleus. For instance, the ground state of ^{16}O is nearly spherical but the first excited state is well-deformed[10].

A fission isomeric state of ^{240}Pu has a deformation which is about three times as large as the ground state deformation[10].

3. More of the main effects of nucleon-nucleon interaction are included in the single-particle basis in the DDT compared to the NFT and IBM. Hence, the residual interaction is much weaker. It can be given a simpler form and can be treated in a simpler way. We employ the usual pairing interaction, but improve upon the BCS method by including the particle-hole channel on the same footing as the particle-particle channel[8].

A brief review of the DDT is given in Section III. Calculated results and their comparison with experiment are discussed in Sect. IV, while the conclusions are given in Sect. V.

III. BRIEF REVIEW OF DYNAMIC DEFORMATION THEORY

Details of the theory can be found in the references quoted above. The main steps of the theory and the calculation are indicated below.

We start from the general, microscopic Hamiltonian

$$H = T + V_{int} \tag{1}$$

where V_{int} is the nucleon-nucleon interaction. Eq.1 can be written as

$$\begin{aligned} H &= (T + U) + (V_{int} - U) \\ &= H_{sp} + V_{res} \end{aligned} \tag{2}$$

where U is the average single-particle potential, H_{sp} is the single-particle Hamiltonian, and V_{res} is the <u>residual</u> interaction. In the present version of DDT, U is given by a modified oscillator potential including spin-orbit and l^2 - terms (as in the Nilsson model), while V_{res} is approximated by the pairing interaction.

Our modified oscillator potential depends on three oscillator frequencies $\hbar\omega_x$, $\hbar\omega_y$, $\hbar\omega_z$. These are equivalent to $\hbar\omega$, β, γ where $\hbar\omega$ represents the overall scaling factor for the energies, while (β, γ) represent nuclear deformation. In DDT, (β, γ) are dynamic variables or integration variables. Integrations over β, γ are performed via a numerical approximation method which utilizes the

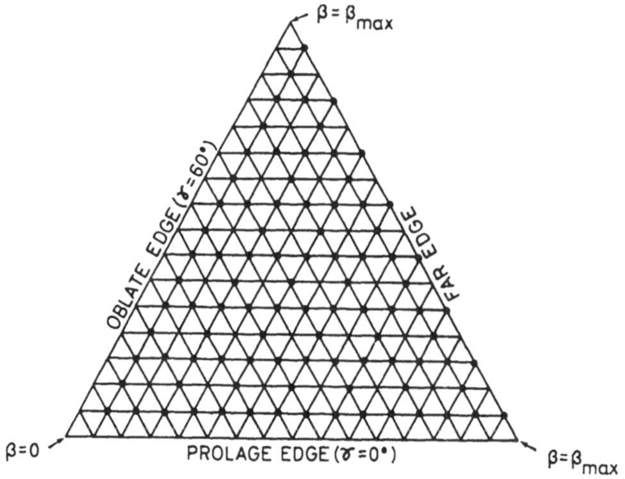

Fig.1 The β-γ triangle. This triangle provides the link between the shell model (one single-particle basis for each mesh point) and the collective model (a β-γ-dependent wave function for each nuclear state). See refs.11, 12 for details.

two-dimensional β-γ mesh of Fig.1. For each point of this mesh, H_{sp} is diagonalized in a spherical basis consisting of N = 0-8 shells. Complete mixing of type ΔN=2; Δℓ=2; Δj=1,2; Δm=2 is included. The configuration space can accommodate 330 neutrons and 330 protons. A scaling method has been found[8] such that this calculation is done once, and the various matrix elements (6 million of them) are stored on tapes and used[9,10] again and again for different nuclei ranging from ^{12}C to ^{240}Pu.

The second step of the calculation concerns the inclusion of the effects of V_{res} (the pairing interaction, in our case). A modified BCS method is employed. Our modification concerns the inclusion of the P-H channel on the same footing as the P-P channel of the pairing interaction. This leads to a modification of the expressions for the quasi-particle energies (E_p of Fig.2) and of the occupation probabilities (V_p^2 of Fig.2). These modifications are particularly important for those regions of the β-γ mesh where the energy gap becomes small.

The third step of the calculation concerns the potential energy of deformation

$$V(\beta, \gamma) = \langle 0(\beta, \gamma)|H|0(\beta, \gamma)\rangle \qquad (3)$$

PAIRING EQUATIONS

P-P CHANNEL: $\underbrace{C^{+}\;C^{+}}\;\underbrace{C\;C}$

P-H CHANNEL: $C^{+}\;C^{+}\;C\;C$ (contractions $C^{+}_{2}C_{3}$ and $C^{+}_{1}C_{4}$)

$$E_p = \frac{1}{2}G + \frac{|\eta_p - \lambda - G/2|}{\sqrt{1 - \Delta^2/E_p^2}}$$

$$2v_p^2 = 1 - \frac{\eta_p - \lambda - G/2}{E_p - 1/2\ G}$$

$$\sum_p{}' 1/E_p = 2/G \quad , \quad \sum_p{}' 2v_p^2 = N.$$

LIMIT OF VANISHING ENERGY GAP ($\Delta \to 0$)

$$E_> = \frac{\eta_> - \eta_< + G}{\omega + 1} \quad , \quad \omega = \sqrt{\frac{\Omega_<}{\Omega_>}}$$

$$E_< = \omega\ E_>$$

$$\lambda = \frac{\Omega\ \eta_> + \eta_< - G}{\omega + 1}$$

Fig.2 Pairing equations and the limit of vanishing energy gap. The particle-hole channel neglected in the usual BCS method is included here (see Refs.8, and 12 for details).

where $|0(\beta, \gamma)>$ is the lowest-energy solution of the microscopic Hamiltonian, H, of Eq.2. In order to improve upon the inaccuracies of the single-particle levels (especially those away from the Fermi surface), the Eq.3 is replaced by that of the shell-correction method of Strutinsky:

$$V(\beta, \gamma) = V_{DM} + \delta U + \delta V_P \tag{4}$$

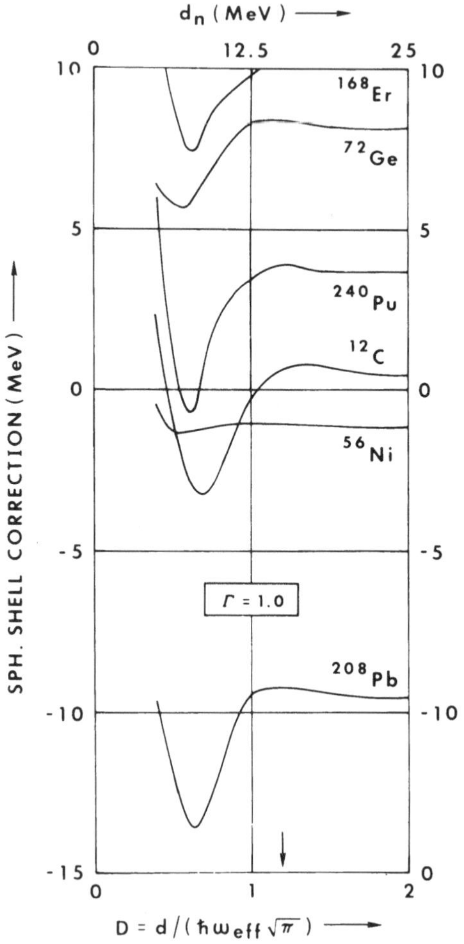

Fig.3 Spherical shell-correction energy vs. configuration-space coefficient D. The Strutinsky width parameter is kept fixed at $\Gamma = \gamma/(\hbar\omega_{eff}) = 1.0$. The quantity $\hbar\omega_{eff}$ is the energy interval between neighboring magic numbers, while d_n is for neutrons in ^{208}Pb.

where V_{DM} is the droplet model energy (depending on surface energy, Coulomb energy, ...), δU is the shell-correction, and δV_P is the pairing-correction. The quantity V_{DM} represents the deformation energy due to the uniform parts of the single-particle levels, while δU represents the contribution due to the non-uniform parts.

The calculation of δU is the most crucial part of the DDT. This term is quite sensitive to two quantities: D, the energy-spacing which determines the number of single-particle levels included in the calculation of δU, and Γ which is the width of the Gaussians which are employed to "smooth" the single-particle levels (one Gaussian peaked around each level). Fig.3 shows the convergence of δU vs. D (for a given Γ) for six different nuclei. Fig.4 shows the

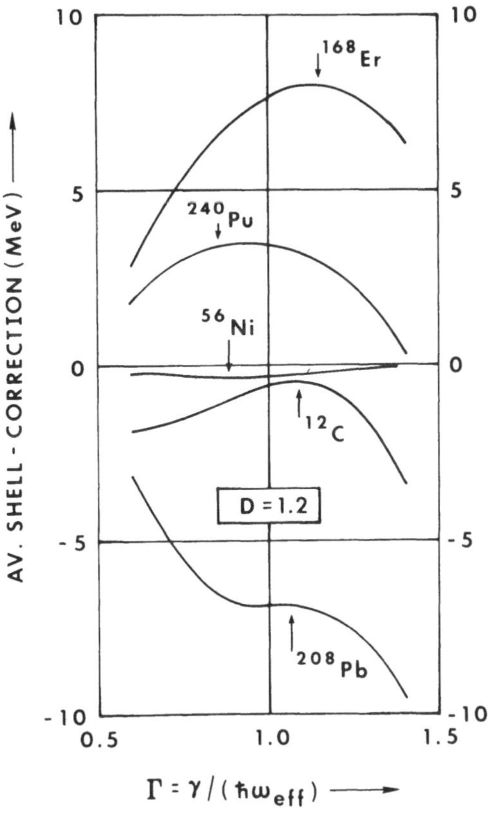

Fig.4 Average shell-correction energy vs. Γ. The configuration-space coefficient is kept fixed at D = 1.2 (space above and below Fermi level = 4.3 $\hbar\omega_{eff}$). The averaging over different shapes is performed via the method of Ref.9.

convergence of δU vs. Γ (for a given D) for five of the same nuclei. The convergence is admittedly not satisfactory. As a temporary measure, the following procedure was employed in most of the calculations reported below. All the levels of the 660 - nucleon space were included in the calculation ($D \sim 4\hbar\omega$). The value of Γ was fixed via the relation

$$\Gamma = 0.3 + 0.2\, A^{1/3} \tag{5}$$

The reasoning leading to Eq.5 has been discussed in ref.9, where other parameters of the theory are also given. All parameters have prescribed Z-A - dependences, and no "free" parameter is employed to fit the nuclear spectra.

Returning to our discussion of the calculation of the shell-correction δU, we have recently found a general method of removing at least part of the uncertainty of the convergence of δU vs. D and Γ. This comes from realizing that (a) of the calculation is more convergent as well as accurate if the difference δU is calculated directly (rather than as a difference of two large numbers), and (b) the calculation can be expressed in terms of two universal functions.

When the quantized levels are smoothed out, one needs to recalculate the Fermi energy λ in order to make certain that both level distributions contain exactly the same number of nucleons. The corresponding condition can be expressed as

$$\sum_i \delta n(\chi_i) = 0 \tag{6}$$

where

$$\chi_i = (\eta_i - \lambda)/\gamma \tag{7}$$

η_i being the energy of level 'i', and γ being the Gaussian width ($\gamma = \Gamma\hbar\omega$). The function δn represents the difference between the occupation probabilities for the two kinds of level distributions. Once the Fermi energy λ has been found, the shell-correction δU can be evaluated via the relation

$$\delta U = \gamma \sum_i \delta u(\chi_i) \tag{8}$$

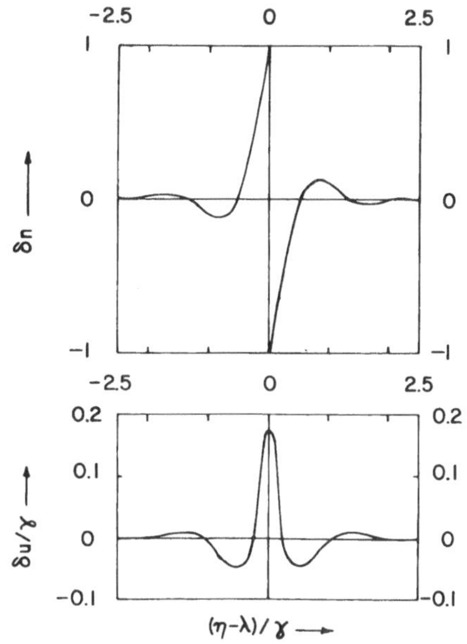

Fig.5 Universal curves for particle number and shell-correction functions. The uniform Fermi energy for a nucleus is determined by solving the equation $\Sigma_i\ \delta n(\chi_i) = 0$, and then the shell correction is given by $\delta U = \gamma\ \Sigma_i\ \delta U(\chi_i)$, where "i" represents single-particle levels. Different nuclei differ mainly via the differences in the level density distribution near the Fermi surface.

Although the functions δn, δu vary with 'i' in a complicated way, their properties can be studied in terms of some simple, universal functions. Taking χ to be a continuous variable, we can calculate δn and δu for any χ value. Variations of δn and δu with $\chi = (\eta - \lambda)/\gamma$ are shown in Fig.5. The two curves are universal curves since they are valid for any nucleus. (One nucleus differs from another only in the level distribution, or the distribution of the discrete χ values.) Furthermore, they show that one must include only the levels within the space $-1 \leq \chi \leq +1$, since the levels outside this space contribute only a long range tail to δU. This long range tail must not be included since that is already included in the droplet energy V_{DM}. Thus, one must set D equal to γ.

The above study does not tell us what the value of γ should be. It would still have to be determined by a condition on the convergence of δU vs. γ. But at least the uncertainty about the

TWO-GAUSSIAN METHOD OF SOLVING
BOHR'S COLLECTIVE HAMILTONIAN

$$H = V(\beta,\gamma) + \sum_{k=1}^{3} \frac{\hbar^2 I_k^2}{2\mathcal{J}_k(\beta,\gamma)}$$

$$- \frac{\hbar^2}{2F}\left(\frac{\partial}{\partial\beta} F_{\beta\beta} \frac{\partial}{\partial\beta} + \frac{2}{\beta}\frac{\partial}{\gamma\gamma} F_{\gamma\beta} \frac{\partial}{\partial\beta} + \frac{1}{\beta^2}\frac{\partial}{\partial\gamma} F_{\gamma\gamma} \frac{\partial}{\partial\gamma}\right)$$

$$H|IM\rangle = E_I|IM\rangle$$

$$|IM\rangle = \sum_K A_{IK}(\beta,\gamma)\, \Phi^I_{MK}(\phi, \theta, \Psi)$$

$$A_{IK} = \sum_p A_p P_{pIK}$$

$$P_{pIK} = D^v g^v_{IK}(\gamma)\, (D^3 \cos 3\gamma)^s$$

$$\times\ (D^2 - D_i^2)^r \exp\left[-\ \frac{1}{2}(D^2 - D_i^2)^2\right]$$

$D = \beta/b$

$p \equiv (v, r, s, i)$

i	Di	GAUSSIAN
1	0.0	SPHERICAL
2	0.4	DEFORMED

Fig.6 Two-Gaussian method of solving Bohr's Collective Hamiltonian. See Refs. 11 and 12 for the details of the earlier versions of this method.

configuration space parameter D can be eliminated in future calculations.

The next step of the calculation concerns the restoration of the rotational symmetries lost in the deformed single-particle basis. This is achieved in the DDT by employing nine-dimensional constraints via the constrained Hamiltonian

$$H' = H - \frac{\hbar}{i} \sum_{\mu} \dot{\alpha}_{\mu} \frac{\partial}{\partial \alpha_{\mu}} \tag{9}$$

where $\dot{\alpha}_{\mu}$ are nine Lagrange multipliers whose values are determined by the constraints

$$\sum_{\nu} B_{\mu\nu} \dot{\alpha}_{\nu} = \langle 0' | \frac{\hbar}{i} \frac{\partial}{\partial \alpha_{\mu}} | 0' \rangle \tag{10}$$

This procedure leads to the collective Hamiltonian

$$H_C = V + \hat{T}_C = V - \frac{\hbar^2}{2} B^{-1/2} \sum_{\mu\nu} \frac{\partial}{\partial \alpha_{\mu}} B^{1/2} B^{-1}_{\mu\nu} \frac{\partial}{\partial \alpha_{\nu}} \tag{11}$$

where α_{μ} are nine collective variables (five quadrupole, four pairing), $B_{\mu\nu}$ is a 9x9 mass-matrix (which itself depends on the collective variables), and B is the determinant of the matrix $B_{\mu\nu}$.

The pairing variables are further constrained by the usual Bogolyubov conditions. Hence, although their time dependences make important contributions to the mass-matrix, the pairing variables do not appear explicitly in the final collective Hamiltonian. This Hamiltonian is expressed in terms of five quadrupole variables in the intrinsic system: two shape variables β, γ and three Euler's angles ϕ, θ, ψ (or equivalently, three angular momentum operators I_1, I_2, I_3). Such a Hamiltonian and its method of solution are indicated in Fig.6. The β-γ mesh shown in Fig.1 is employed to solve this Hamiltonian in such a way that the five dimensional coupling of rotations and β-γ vibrations is solved exactly.

IV. DISCUSSION OF RESULTS AND COMPARISON WITH EXPERIMENT

Some examples of the types of nuclear structure results obtained via the DDT are given in Figs.7-18. Figs.7, 8, and 9 concern the medium-mass transitional nuclei ^{70}Ge, ^{72}Ge, and ^{74}Ge. In Fig.9, numbers along the vertical arrows give the B(E2) values in $e^2 \cdot b^2$, while the numbers along the curved arrows give the electric quadrupole moments in e•b. Note that in this early version of the DDT,

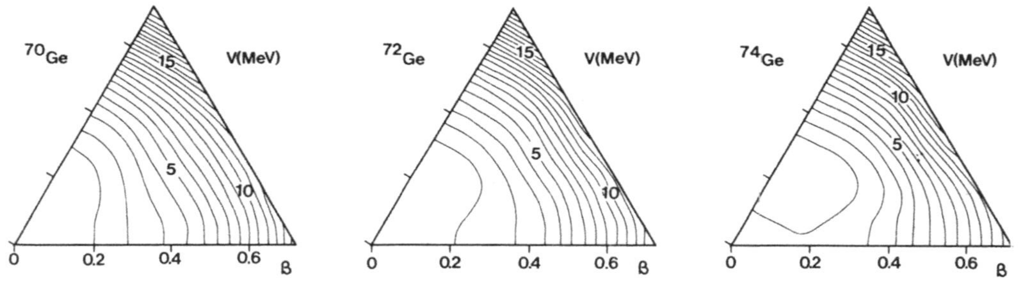

Fig. 7 Contour plots of potential energy for Ge nuclei. The single-particle basis for each (β, γ) mesh point is used to compute 22 shape-dependent functions for each nucleus. A modified oscillator potential is used for the single-particle basis. A modified BCS method is used for the pairing correlations. The shell-correction method is used for the potential energy, and the cranking method for the six inertial functions (three moments of inertia, three mass parameters for $\beta\beta$, $\beta\gamma$, $\gamma\gamma$ vibrations).

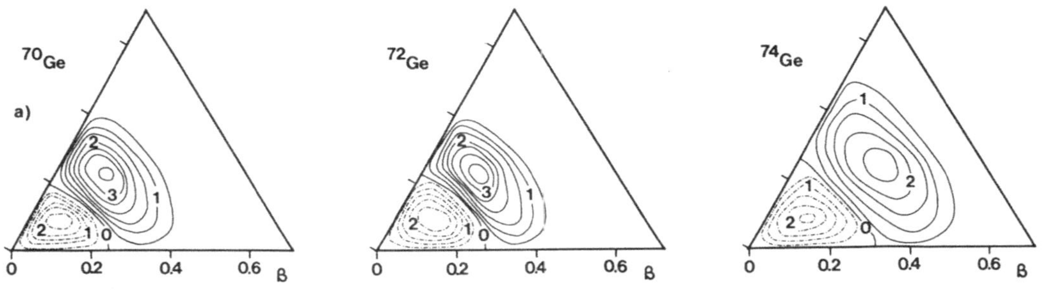

Fig. 8 Contour plots of the collective wave functions for the $0'^{+}$ states of Ge nuclei. These $0'^{+}$ states have β-vibrational character (the wave function changes sign near the ground state shape), but they also represent shape c0-existence since their rms deformation is 25-40% larger compared to the ground states (see the numbers given in paratheses along the respective theoretical states in Fig. 9.

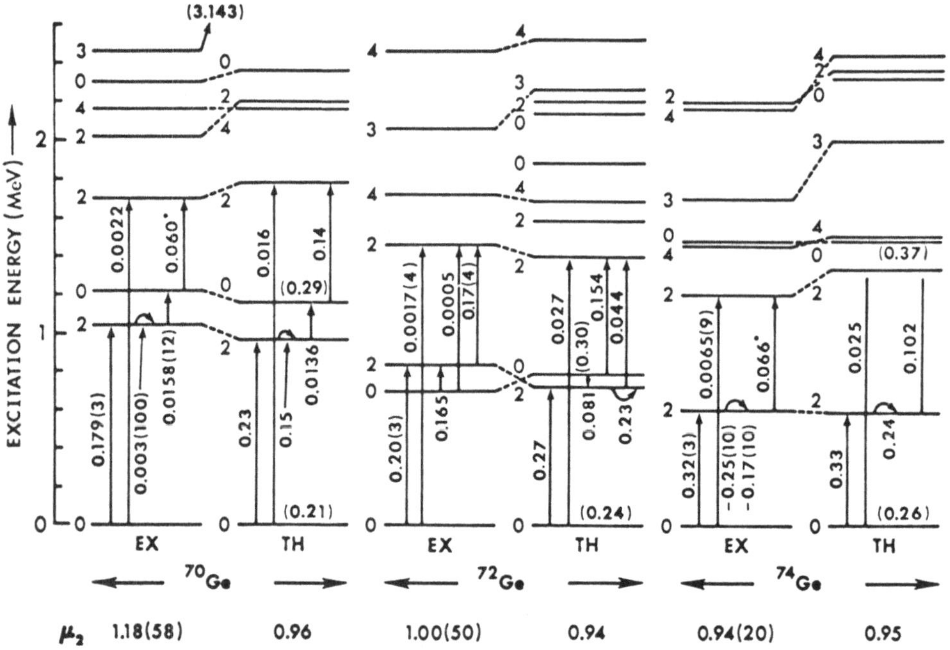

Fig.9 Theoretical and experimental spectra for Ge nuclei. One parameter, the width parameter for the shell-correction energy, was varied to fit the energy of the first 2^+ state (Ref.13).

one free parameter (namely the shell-correction width parameter Γ) was employed to fit the energy of the first 2+ state (see Ref.13). However, no free parameter was employed for the results reported in the following figures.

Figs.10-13 give some of our results of a global study of 25 even-even nuclei ranging from ^{12}C to ^{240}Pu (taken from Ref.9). These nuclei include spherical nuclei (^{16}O, ^{42}Ca, ^{90}Zr and ^{204}Pb), deformed nuclei (^{12}C, ^{24}Mg, ^{104}Ru, $^{164,166,168}Er$, ^{238}U and ^{240}Pu), and transitional nuclie (^{56}Fe, ^{72}Ge, ^{114}Cd, ^{130}Te, ^{142}Ce, $^{152,154}Gd$, $^{184,186,188}Hg$, ^{192}Os, ^{194}Pt and ^{222}Rn). Although a number of important discrepancies remain, the general trends of the low-energy spectra of such a wide range of nuclei are reproduced for the first time within a theory with no free parameters,

Figs.14-16 give some details of the DDT results for two of the most famous cases of shape co-existence: ^{16}O and ^{240}Pu (see Ref.10 for further details). The DDT gives the lowering of the deformed $0'^+$ state (0_β in DDT) in ^{16}O without any fitting parameters. In fact, the calculated 0^+ is even lower than the observed $0'^+$ state

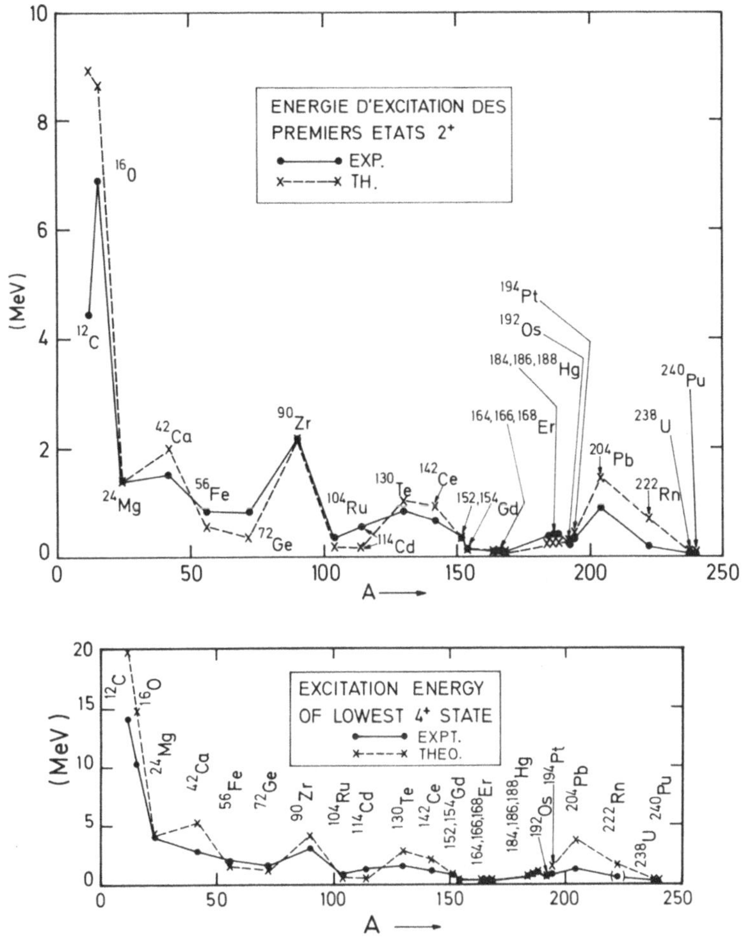

Fig.10 Energies of the lowest 2^+, 4^+ states of 25 even-even nuclei with A = 12 - 240. No parameter has been varied to fit the data. values for different nuclei have been connected by straight lines only to guide the eye -- they do not represent the local Z-A dependence (Ref.9).

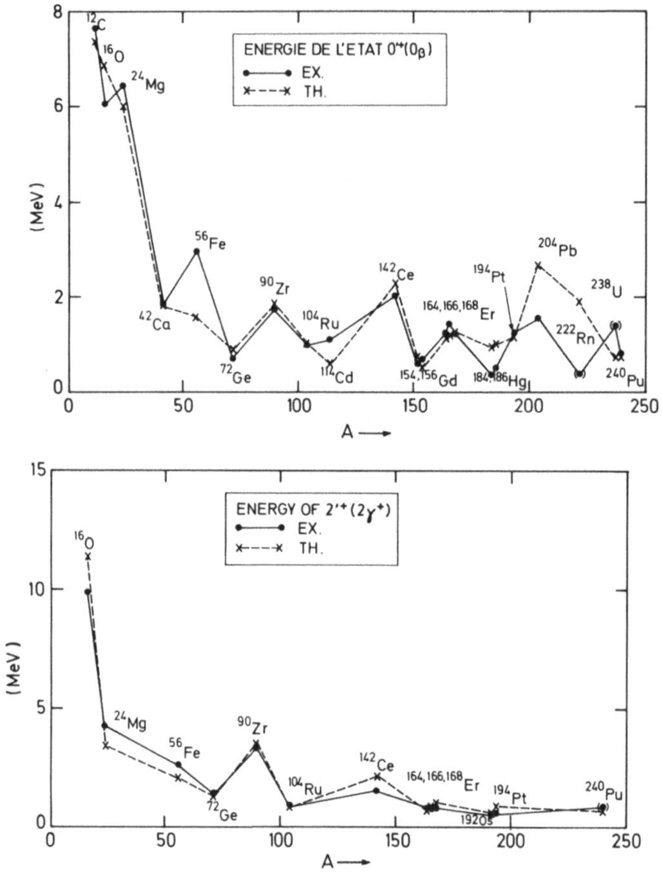

Fig.11 Energies of $0'^+$, $2'^+$ states of selected even-even nuclei. See Fig.10 caption.

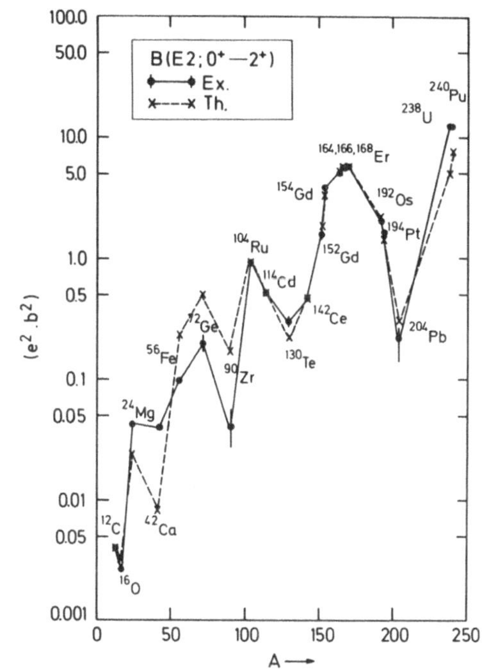

Fig.12 B(E2; $0^+ \rightarrow 2^+$) values for selected even-even nuclei. No effective charge has been used for the theoretical values. See Fig.10 caption.

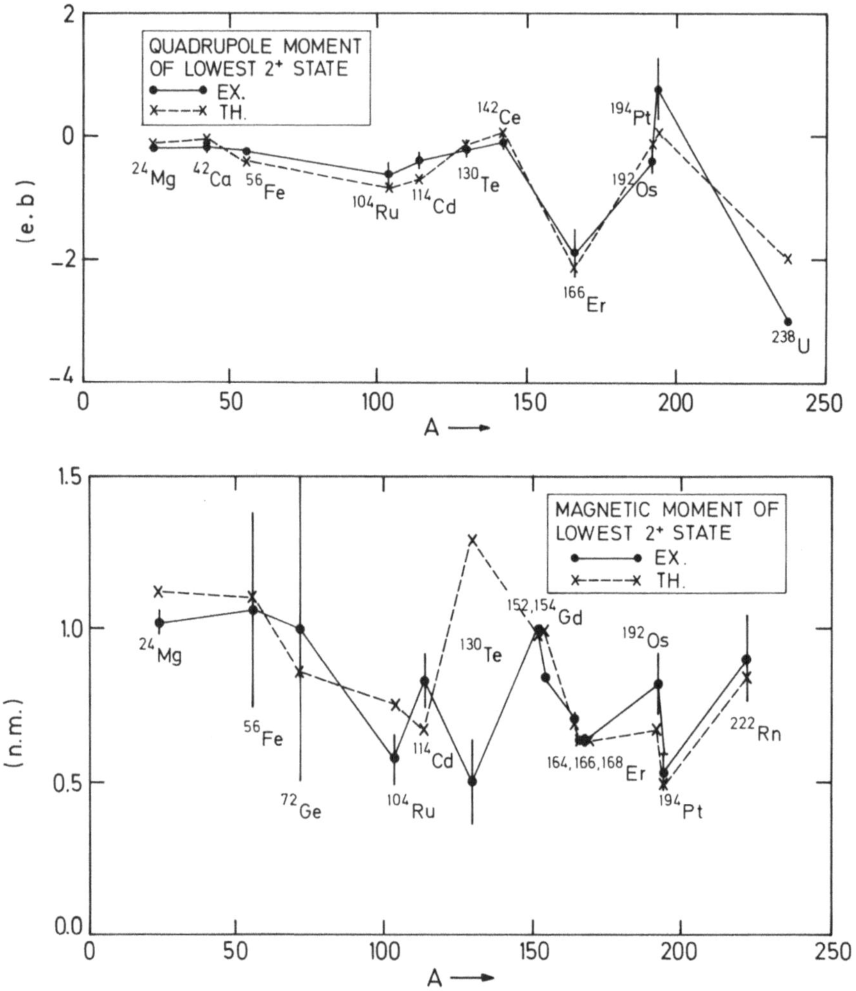

Fig.13 Quadrupole and magnetic moments of the first 2^+ states of selected even-even nuclei. See Fig.12 caption.

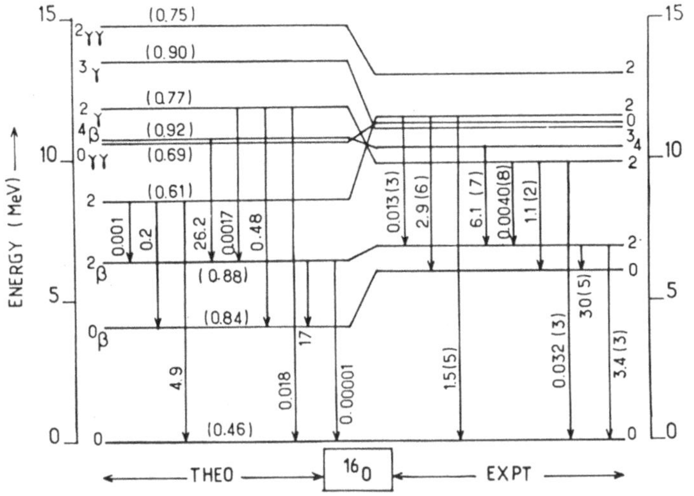

Fig.14 Theoretical and experiemntal spectra for ^{16}O. The calculated values of β_{rms} are given in parantheses along the theoretical states. They support the idea of shape co-existence since the average deformation in the $0'^{+}$ state is about twice as large as that in the ground state (Ref.10).

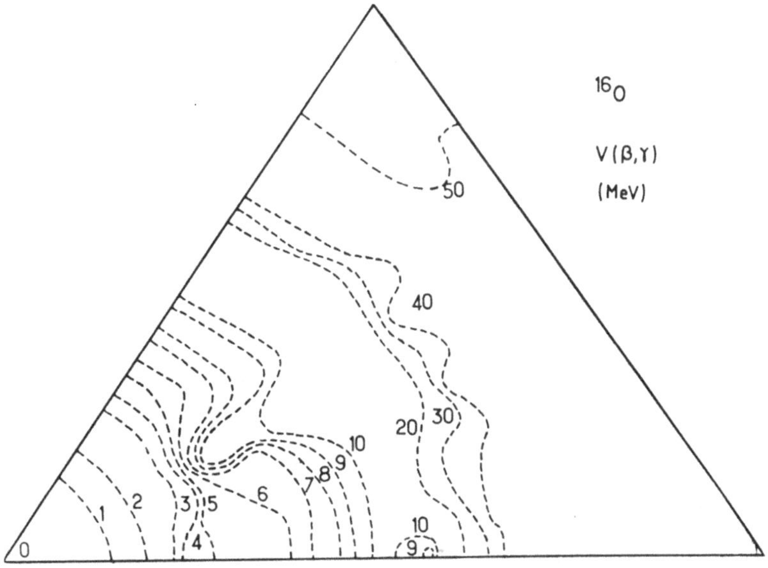

Fig.15 Contour plot of the potential energy of ^{16}O. The second minimum occurs at $\beta = 0.45$, $\gamma = 0°$ which is 8.8 MeV above the first (spherical) minimum and 1.6 MeV below the barrier at (0.4, 0°). Dynamics lowers the $0'^{+}$ state from 8.8 to 4.1 MeV (see Fig.14).

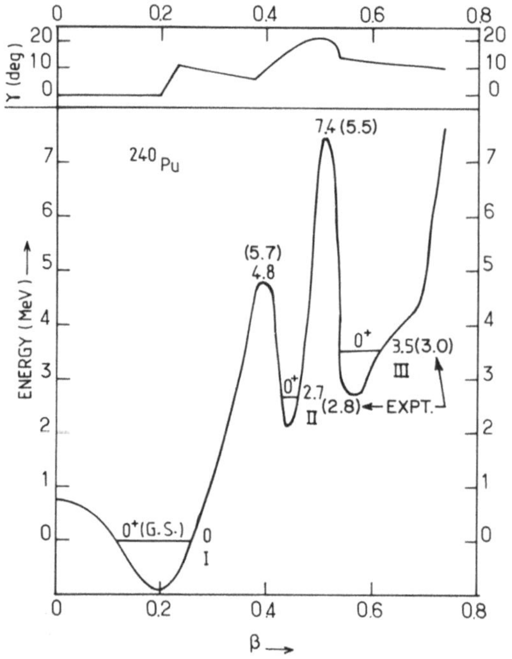

Fig.16 Effective potential energy of ^{240}Pu. The lower part of the figure gives $V_{eff} = V + \langle T_{coll} \rangle_\beta / \langle \ \rangle_\beta$ along the path of steepest descent from one stationary point to another one. The upper part of the figure gives the γ-value along this path (see Ref.10 for details).

(Fig.14). The calculated B(E2) values (given in single-particle-units (s.p.u.) along the vertical arrows in Fig.14) show that the first excited 2^+ state (2_β) has hardly any B(E2) value for decay to the ground state, while its B(E2) value to the $0'^+$ state (0_β) is 17 s.p.u. This is not quite as large as the experimental value (30 s.p.u.), but it is large enough to support the idea of shape co-existence. In order to understand the nature of the calculated $0'^+$ state, we look at the contour plot of $V(\beta, \gamma)$ in Fig.15. This potential has two minima: a spherical one at V = 0, and a deformed on at V = 8.8 MeV (β = 0.45, γ = 0°). The two minima are separated by a barrier of 7.2 MeV at β = 0.4, γ = 0°. The energy difference of 8.8 MeV is reduced to 4 MeV ($E_{0'} - E_0$, see Fig.14) by the strong β-γ - dependence of the mass-matrix. The mass parameter is almost zero at $\beta = 0$ and rises sharply with β. Therefore, the collective kinetic energy is much lower in the $0'^+$ state compared to the ground state.

Another kind of shape co-existence is exhibited by the calculated potential energy function of ^{240}Pu (see Fig.16). The function plotted is V_{eff}, which includes the average kinetic energy, along the fission path in β. The value of γ which optimizes this path at each β value is shown in the top part of Fig.16. The function V_{eff} of ^{240}Pu has three minima, one at $\beta = 0.2$ is for the ground state, while the other two at $\beta = 0.45$ and at $\beta = 0.6$ are for two fission isomeric states. Many other theories are able to explain one of these fission isomers, but to the best of our knowledge none of them explains the second fission isomer. The present version of DDT is not suitable for deformations greater than $\beta \sim 0.6$ and hence, the third fission barrier is not reproduced. The heights of the first two fission barriers agree with the empirical values (given in the parentheses) to the same degree of accuracy as the conventional theories of nuclear fission. But the present theory goes far beyond such theories in reproducing the observed energies of $0'^+$ states associated with the two fission isomers (see Ref.10 for details and for the references).

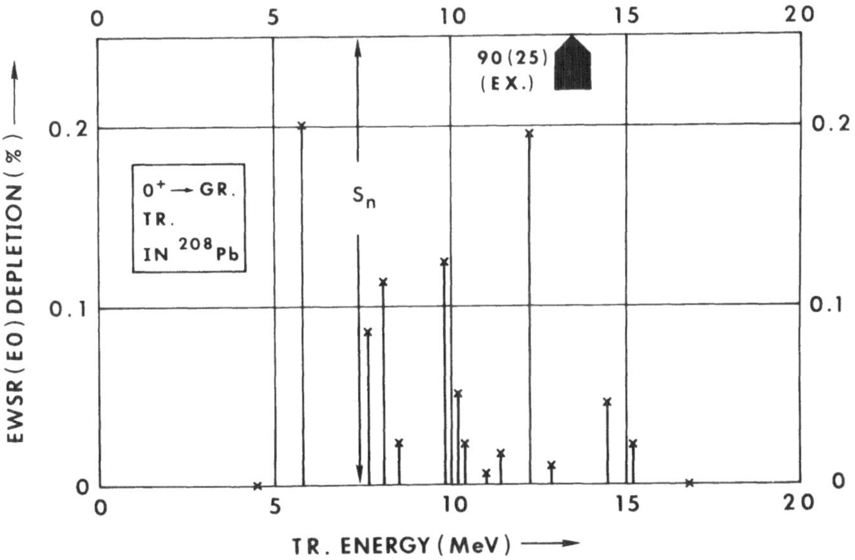

Fig.17 Electrical monopole strengths in ^{208}Pb. The present calculation fails to reproduce the giant monopole (the experimental values are from Ref.14) because only the seniority zero states with +++ symmetry (see Ref.2) were included. S_n represents neutron separation energy

The present type of microscopic theory of collective motion is normally employed only for the low-energy (∿1-2 MeV in heavy nuclei) states. Our recent studies[§] show that it is not unreasonable to extend such calculations to much higher energies (∿10-15 MeV in heavy nuclei). The conventional argument does have important validity since many states (for instance, negative parity states, states of non-zero seniority) do appear at energies above the pair-breakup energy. But our studies suggest that such states can co-exist with the positive parity, zero-seniority states without mixing with them.

Fig.17 shows that the current version of the DDT does not reproduce the observed giant monopole strength at E ∿ 15 MeV in ^{208}Pb. This is understandable since the giant monopole states are associated with seniority two (one broken pair) states which are not included in the theory.

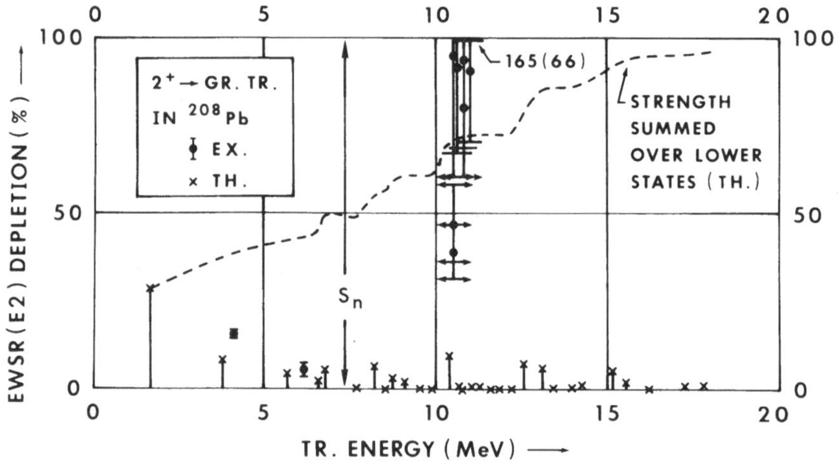

Fig.18 Electrical quadrupole strengths in ^{208}Pb. The calculated strengths distribution is incorrect: The first 2^+ state is too low and has too much strength (see Ref.15 for the experimental low-energy states). The giant quadrupole state comes at the right energy but has too little strength (see Ref.14 for the experimental data). However, the calculated summed strength (dashed curve) does represent most of the sum rule strength.

[§]These studies were suggested by Dr. D.H. Feng. The author is grateful to him for his suggestion and encouragement. These studies have turned out to be crucial for future improvements of the theory.

On the other hand, the DDT does reproduce a giant quadrupole strength at the expected energy of $\sim$10 MeV in ^{208}Pb. The calculated sum rule strength, summed up to $\sim$18 MeV, does exhaust almost 100% of the E2 sum rule (see Fig.18). However, the calculated strength distribution is not correct at all. The major discrepancy in our view, is not so much the lack of concentration of the E2 strength in a single state at $\sim$10 MeV, as is the strength of the first 2^+ state which is too strong as well as too low. This discrepancy suggested strongly that our calculation of the shell correction was not accurate enough. Inclusion of too many configurations in this calculation led to a nucleus which did not have enough magicity (or rigidity against deformation). A re-examination of this problem has led to important suggestions about improving the calculation (see the discussion given above in Section II in connection with Fig.5).

The DDT wave functions have recently been employed as input to several reaction theories which employ the ideas of nuclear deformations and rotations. Such theories have to be extended so as

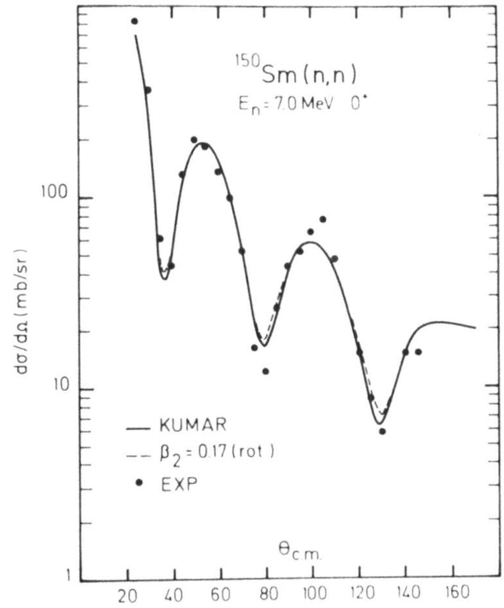

Fig.19 Angular distribution of elastic neutron cross-sections for ^{150}Sm. The coupled channel method of Tamura has been generalized so as to include the γ degree of freedom, K-mixing, and integration over different shapes via DDT structure wave functions (Refs.9, 16).

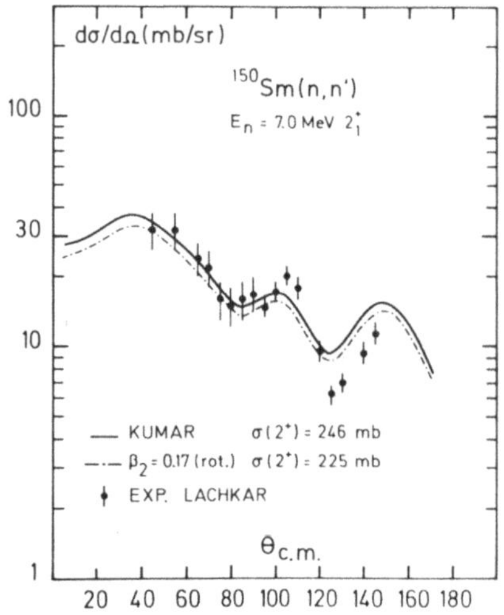

Fig.20 Same as Fig.19 but for inelastic cross-sections leading to the first 2^+ state of ^{150}Sm.

to include the γ degree of freedom, and the various amplitudes have to be calculated for about 100 (β, γ) shapes rather than just one shape. However, this leads to an enormous generalization of reaction theories whereby one takes into account the possibilities of different shapes in different reaction channels and for the initial, final nuclei. The increase in computation time is moderate (at least for the calculations based on the optical potentials) since the basic shape-dependent amplitudes can be calculated once and then multiplied by appropriate scaling factors for different nuclei.

Figs.19-21 give some examples of this application of the DDT wave functions. Fig.19 is for the angular distribution of the cross-section for the elastic scattering of 7 MeV neutrons from ^{150}Sm. The coupled channel method of Tamura, extended along the lines discussed above, has been employed. The DDT curve (called 'KUMAR' in the figure, which was prepared by C. Lagrange, see Ref.9, 16) reproduces the shape and magnitude of the angular distribution extremely well. But so does the conventional theory which assumes that ^{150}Sm is a good rotor with a fixed deformation whose value is treated as a fitting parameter. However, the DDT eliminates the need for such a fitting parameter. Fig.20 gives the same kind of comparison for the inelastic scattering leading to the first 2^+ state of ^{150}Sm. Experiments for inelastic scattering to higher states would be extremely valuable in showing greater differences between the DDT and the conventional theories.

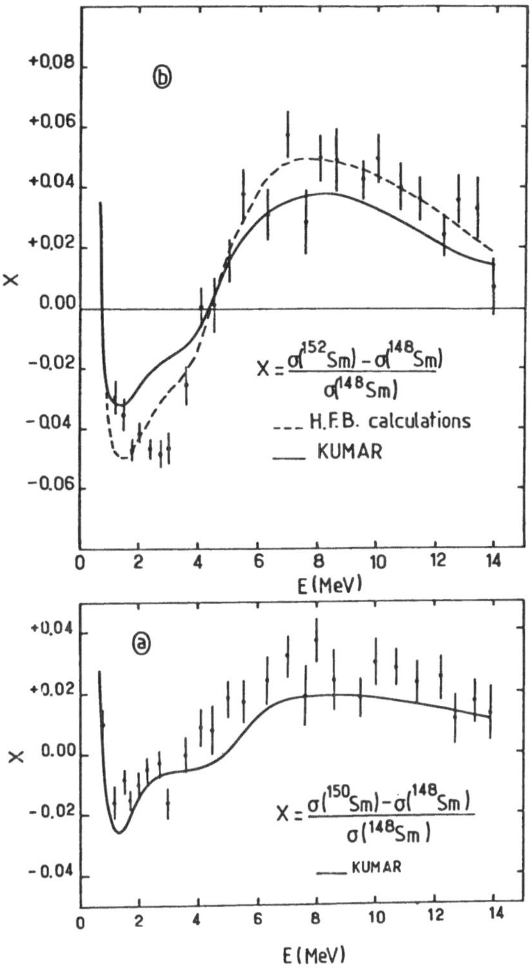

Fig.21 Relative neutron cross-sections for Sm nuclei (Ref.17).

Fig.21 gives some remarkable patterns observed in the relative (total) neutron cross-sections for Sm nuclei. Because of the extremely good resolution obtained at Bruyeres-le-Chatel, it is possible to follow some fine details of the observed cross-sections. As compared to ^{148}Sm, the cross-sections for ^{150,152,154}Sm (only two of them are shown in Fig.12) show significant reductions for

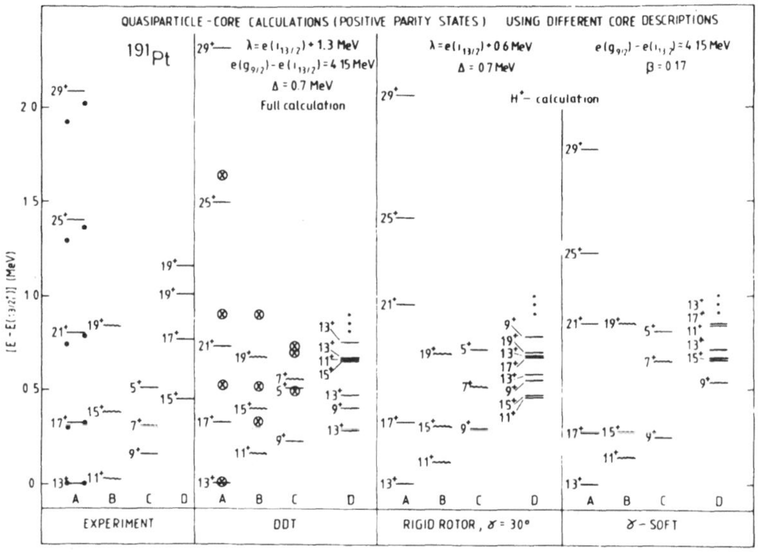

Fig. 22 Core-particle-hole coupling in odd-A ^{191}Pt. The DDT wave functions for $^{190,192}Pt$ have been employed (Ref. 18).

$E_n \sim 2$ MeV. These can be related in a general way to the shape effects (the well-known spherical-deformed shape transition at $N \sim 90$). The DDT-based calculations reproduce the general trends of these fine details of neutron cross-sections. The HFB-based calculations (performed for only one case out of the three mentioned above) do even better. But they also require much more computation time: the structure part of the calculation takes about 10 hours per nucleus (as compared to about 10 minutes for the DDT).

The DDT wavefunctions have recently been employed for yet another application: core-particle-hole coupling in odd-A ^{191}Pt (Fig.22). A simultaneous coupling of a particle to the A-1 (e-e) core and a hole to the A+1 (e-e) core is employed. The two systems are connected by the pairing interaction. A large number of core states($\sim$13 for each core) are employed. The E2 matrix elements connecting the core states are taken from the DDT (instead of assuming rotational cores, or vibrational cores). In these preliminary calculations, performed by A. Vagnes et al.[18] at Bergen, only one particle orbital ($i_{13/2}$) was employed.

V. CONCLUDING REMARKS

A brief review has been given of recent studies of nuclear spectra and cross-sections on the basis of a microscopic theory of collective nuclear motion, called the Dynamic Deformation Theory.

A crucial part of this theory is a scaling method which allows one to work in a very large (660 nucleon, 165 deformed orbitals) configuration space and still keep the computation times quite low (∿10 minutes per nucleus for nuclei ranging from ^{12}C to ^{240}Pu).

It has been demonstrated that it is not necessary to limit such a theory to the low-energy states of heavy, non-closed-shell nuclei. The basic assumption of the theory, that states of different seniority (number of unpaired nucleons distributed over deformed orbitals) do not mix with each other, appears to be valid for practically all nuclei.

It has been shown that the DDT wavefunctions are useful not only for the understanding of nuclear spectra-energy levels, B(E2) values, quadrupole moments, magnetic moments, ... - but also for calculating fine details of fission isomers and reaction cross-sections.

The above discussion has concerned mainly the positive parity states of even-even nuclei. But the theory can be extended to odd-A, odd-odd nuclei, and to the seniority non-zero states of e-e nuclei (needed for backbending, giant monopoles, ...). The main new element of the calculation would be the blocking of one or more orbitals.

As regards experimental data, we need the moments and the B(E2) values for high-spins and for high-energy states in order to test our theories in greater detail.

RERERENCES

1. J.S. Vaagen Contribution to this volume.
2. A. Bohr and B.R. Mottelson, Nuclear Structure, Vol.2 (Benjamin, New York, 1975).
3. D.R. Bes, G.G. Dussel, R.A. Broglia, R.A. Liotta, and B.R. Mottelson, Phys. Lett. 52B, 253 (1974).
4. C.L. Wu and D.H. Feng, Phys. Lett. 96B, 243 (1980); Ann. of Physics (in press); C.L. Wu, M.W. Guidry, J.Q. Chen and D.H. Feng, contribution to this volume.
5. A. Arima and F. Iachello, Ann. Phys. 99, 253 (1976).
6. I. Talmi, Contribution to this volume.
7. O. Scholten, Contribution to this volume.
8. K. Kumar, B. Remaud, P. Aguer, J.S. Vaagen, A.C. Rester, R. Foucher, and J.H. Hamilton, Phys. Rev. C16, 1235 (1977).
9. K. Kumar, in Structure of Medium-Heavy Nuclei 1979, Proc. Conf. at Rhodos, May 1979 (The Institute of Physics, London, 1980, Conf. Ser. No.49) p.169.
10. K. Kumar, in Future Directions in Studies of Nuclei Far From Stability, Proc. Conf. at Nashville, September 1979, ed. by

J.H. Hamilton et al. (North-Holland, Amsterdam, 1980) p. 265.
11. K. Kumar and M. Baranger, Nucl. Phys. A92, 608 (1967).
12. K. Kumar, Nuclear Models, Lectures at University of Bergen 1979-80 (and references quoted there).
13. K. Kumar, J. Phys. G: Nucl. Phys. 4, 849 (1978).
14. F.E. Bertrand, Ann. Rev. Nucl. Sci. 26, 457 (1976); also in Common Problems in Low- and Medium-Energy Nuclear Physics, ed. by B. Castel, B. Goulard, and F.C. Khanna (Plenum, New York,1979) p. 417; F.E. Bertrand, G.R. Satchler, D.J. Horen, and A. van der Woude, Phys. Lett. 80B, 198 (1979).
15. M.B. Lewis, Nucl. Data Sheets B5, 243 (1971).
16. C. Lagrange, private communication, April 1971.
17. C. Lagrange, M. Girod, B. Grammaticos, and K. Kumar, Proc. Int. Conf. on Nuclear Physics, Berkeley, CA, August 1980, in press.
18. A. Vagnes, K. Kumar, and J.S. Vaagen, Proc. Int. Conf. on Nuclear Physics, Berkeley, CA, August 1980, in press.

HIGH SPIN PHENOMENA

J.H. Hamilton and C.F. Maguire

Physics Department
Vanderbilt University
Nashville, TN 37235

I. INTRODUCTION

In this paper, we wish to discuss various aspects of high-spin phenomena. In particular, evidence for bands of states built on two quasiparticle states which are rotation aligned with respect to the rotating core is found in new regions at both ends of the periodic table for ^{68}Ge, $^{76-80}Kr$, and ^{248}Cm. The use of massive transfer reactions to populate high spin states is considered in the final section.

II. HIGH SPIN STATES

The structure of nuclei at higher and higher spins is one of the forefronts in nuclear research today as can be seen by the number of recent conferences devoted entirely to that subject, for examples the 1979 Argonne and 1980 Strasbourg conferences. Studies of high spin states as a major field of research over the last decade received much of its impetus from the discovery of the sudden changes in the effective moments of inertia, $\mathcal{I}$, of deformed rare earth nuclei[1]. Experimentally it was found in certain nuclei that, as one excited higher spin states in well-known ground state rotational bands, GRB, then in contrast to the rotational model where the level energies to first order increase as $[\hbar^2/2\mathcal{I}]I(I+1)$, around spin 12-14 $\hbar$ the energies suddenly increase less rapidly. Thus, in a plot of $\mathcal{I}$ as a function of nuclear rotation frequency $\hbar\omega$, there was a large increase in $\mathcal{I}$ with a sudden decrease in $\hbar\omega$, the so-called "backbending" of the moment of inertia. Bands in some deformed nuclei were found to exhibit backbending and some did not. Then come forward bending of the moments of inertia in

Se and Hg nuclei[2,3]. A rotation aligned (RAL) model[4] was developed to explain the backbending phenomena and indeed its detailed application gives explanations of the variety of different backbending phenomena[5-9]. The forward bending can be understood in terms of the the coexistence of different nuclear shapes with overlapping bands of levels built on these different shapes[3] to confirm the long-standing predictions of Hill and Wheeler[10].

In an RAL model, in deformed nuclei where Fermi energy for protons and neutrons is close the the $\Omega=1/2$, large $j(i_{13/2}, h_{11/2}, g_{9/2})$ levels strong Coriolis coupling can give rise to aligned ($I = I_o, I_o+2, \ldots$) bands based on $(j)^2$ two-particle configurations and to partially aligned rotational bands based on odd-parity two-particle configurations $(i_{13/2}, f_{5/2}, p_{3/2})$, $(h_{11/2}\ d_{5/2}, g_{7/2})$ or $(g_{9/2}; f_{4/2}, p_{3/2})$ with $I = 5^-, 7^-, 9^-, \ldots ; 6^-, 8^-, 10^-, \ldots$ (see for example refs.4 and 5). It is the crossing of the ground band by such RAL bands built on $\nu(i_{13/2})^2$ configuration in the rare earths[6], on the $(h_{11/2})^2$ one in Ba and Ce nuclei[7] and the $\nu(h_{11/2})^2$ one in Ru-Cd nuclei[8] that produces backbending of $\mathcal{I}$.

In nuclei with N>>Z, the condition for a strong Coriolis coupling exists only for proton or neutron configurations but not both for the same j orbital. However, in the A=70 region both N and Z are between 30 and 40, so one may have both proton and neutron aligned bands built on the $g_{9/2}$ orbitals, if there is sufficient deformation for the RAL model to be applied in this lighter mass region. As we shall see, new results have extended the RAL model into this still lighter mass region with the new feature of seeing both $\pi(g_{9/2})^2$ and $\nu(g_{9/2})^2$ bands as well as π and ν negative parity bands . Then there is the question of what hapens in very heavy nuclei in the actinide region where still higher j orbitals like $\pi i_{13/2}$ and $\nu j_{15/2}$ can be active.

Studies of high spin phenomena began to grow in two broad areas. One was to study what happens as we continue to higher energies and higher spins above the then known limits, which were for example about 20^+ in rare earth nuclei but only $\lesssim 8^+$ in most A=70 nuclei. The second is what happens at very high spins 40-50$\hbar$ where we may not be able to see discrete states but where studies of continuum gamma rays may give us insights.

Here we will describe some recent studies that involved the Vanderbilt group in the first area of extending our knowledge of discrete states above the then known limits in three regions, around A=70, the near spherical to deformed region of A = 154-158 and in the heaviest nucleus studied to high spin, ^{248}Cm. These data demonstrate the importance of both collective and single particle motions at higher spins over a wide range of the particle table.

III. A=70 REGION

Evidence for the coexistence of bands built on a different shapes in this region as first reported in ^{72}Se (ref.2) now comes from a veriety of theoretical and experimental directions (see for examples ref.11 and 12). Here we will concentrate on the studies of additional new multiple collective band structures observed in the higher spin states which have come from in-beam gamma ray spectroscopy.

The levels of ^{68}Ge, shown in Fig.1 (ref.9) provide a "textbook" example of rotation-aligned bands including for the first time in one nucleus positive and negative parity bands for both protons and

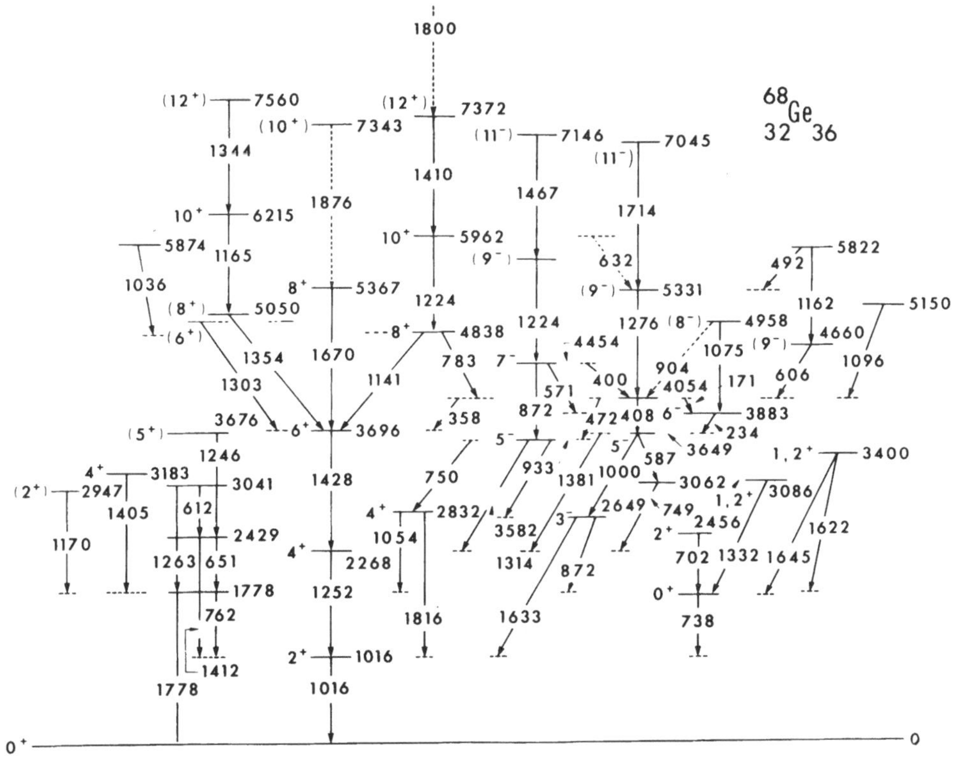

Fig.1 The decay scheme of ^{68}Ge. Above the 6^+ yrast state there are 3 positive parity bands, two based on the rotation alignment of protons and neutrons in the $g_{9/2}$ single particle orbital and one a continuation of the ground band.

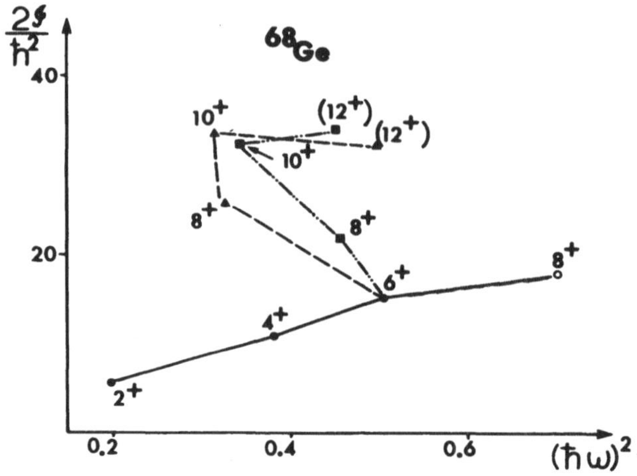

Fig.2 The moment of inertia plot for the positive parity states in ^{68}Ge. The two bands with lower 8^+ energies show strong backbending while the upper band is interpreted as an extension of the ground band.

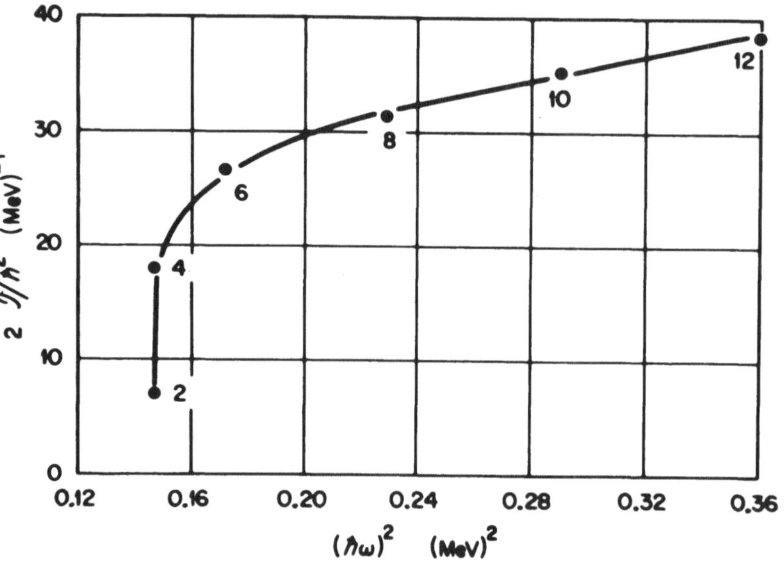

Fig.3 The moment of inertial plot for the yrast band of ^{72}Se showing the strong forward bending above the 4^+ state.

neutrons built on the same orbitals. The first striking feature is three 8^+ levels at similar energies with bands built on all three levels. Then note two negative parity bands with 5^- band heads but quite different level spacings and another beginning at 6^-. Here the moment of inertia of the yrast band, Fig.2, backbends in contrast to the forward bends in nearby $^{72,74}Se$, (Fig.3) where a shape coexistence picture has been invoked[2,3]. The negative parity bands have different spacings from the single very rotational-like band seen from 3^- to 13^- in ^{74}Se and ^{78}Kr (refs.13-15), see Fig.4.

The 5366.8 and 7243 keV levels in Fig.2 are easily interpreted as the continuation of the ground band. The backbendings of $\mathcal{J}$ at the two lowest 8^+ levels clearly indicate significant changes in their structures from the ground band. The similarity to the backbending of $\mathcal{J}$ in Pd, Ba, and rare-earth nuclei suggests the crossing of RAL bands here too.

To test this RAL interpretation, a two-quasiparticle-plus-rotor calculation was performed as in Flaum and Cline[5] with the dependence of $\mathcal{J}$ on the total angular moment based on the VMI model. The calculations describe the band structures of the level scheme very well, independent of any fine parameter adjustments,

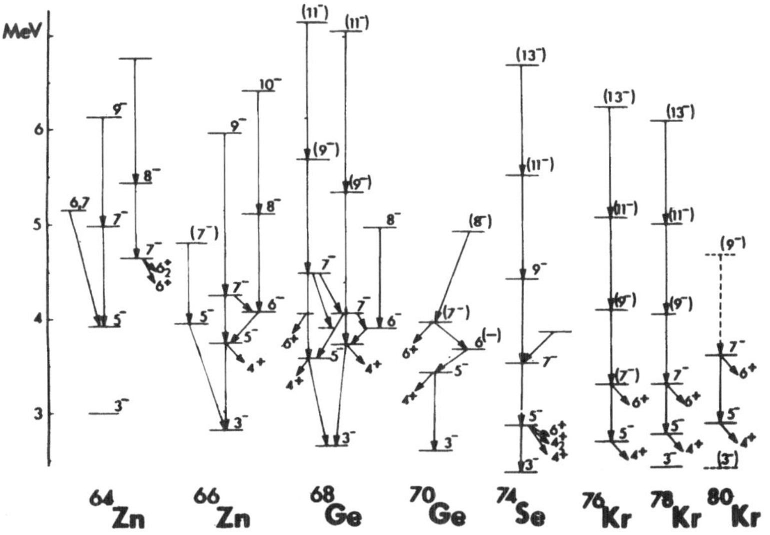

Fig.4 Systematics of negative parity bands in some medium mass even-even nuclei.

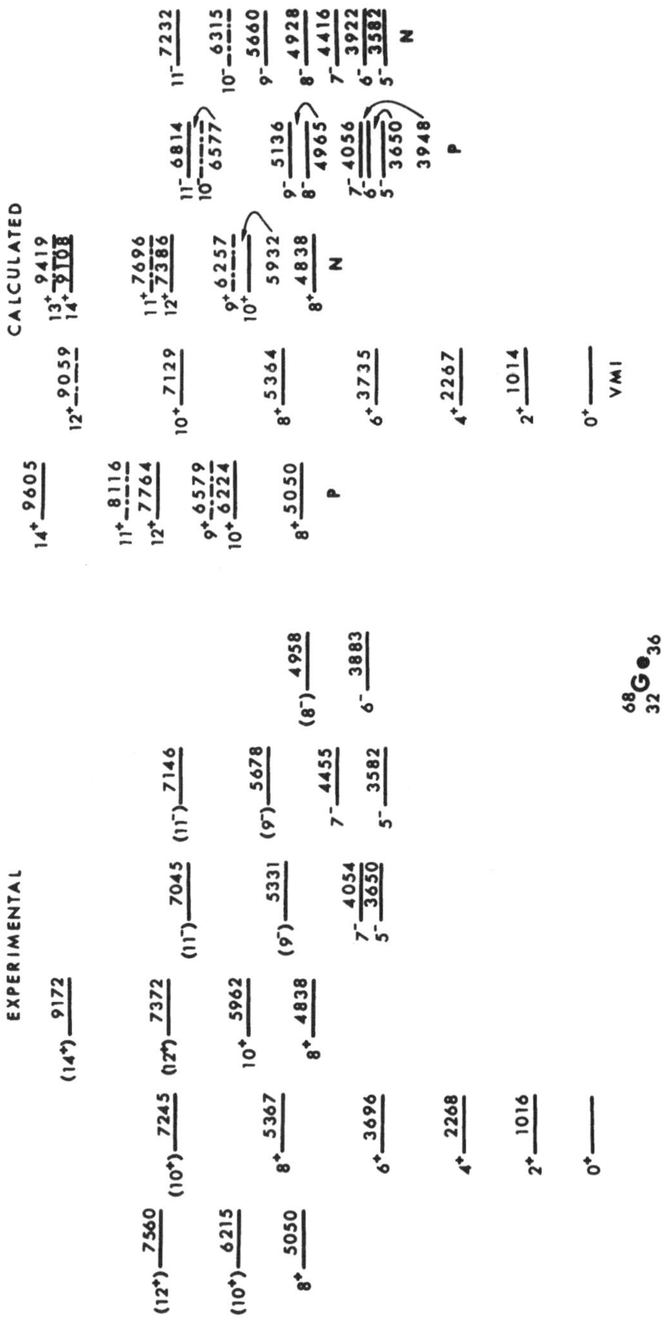

Fig.5 Experimental levels in ^{68}Ge compared to rotation alignment calculations including both proton (π) and neutron (ν) levels.

see Fig.5. For both the two-quasi-neutron and two-quasi-proton negative parity states, one quasiparticle occupies a $g_{9/2}$ orbital. The negative parity orbitals near the neutron Fermi surface are the higher Ω states from $f_{5/2}$ and $p_{3/2}$ orbitals, with the single $p_{1/2}$ orbital close by. The resulting band has a nearly equal-energy-spaced $\Delta I=1$ level sequence built on a 5^- state of mixed parentage. The proton Fermi surface, on the other hand, lies close to the $\Omega = 1/2$, $f_{5/2}$ orbital, and the $\Omega = 3/2$, $p_{3/2}$ orbital with the $p_{1/2}$ orbital far away. Here the 5^- state is only partially aligned while the 7^- state and the ones above it are much more fully aligned, so the 5^- and 6^- states are compressed in energy. Thus the different energy spacings of the two odd-spin negative parity states, which are quite different from negative parity bands in their heavier neighbors, are nicely explained in the RAL calculations. The 6^- and 8^- states probably are neutron-proton mixed states. Some mixing is likely for all the lowest 5^-, 6^-, 7^-, and 8^- states based on their branchings.

So, in contrast to $^{72,74}Se$ where the even-parity yrast bands exhibit "forward bending" of $\mathscr{I}$ with no forking, tripple forking is observed at 8^+ in ^{68}Ge with backbending of $\mathscr{I}$ at the two lowest 8^+ states to indicate a variety of nuclear motions are occurring in this region. The bands built on the two lowest 8^+ states and the new new odd-parity bands in ^{68}Ge can be explained very nicely in the RAL model to extend this model down to a still lighter mass region. A more detailed analysis of the $^{72,74}Se$ data (ref.16), suggests that their yrast cascades may involve three different structures with a shift from near-spherical to deformed around 4^+ and a shift to an RAL band above 8^+ and other bands at high spin are simply not observed because of weak feeding from above in those reactions. Nevertheless it is still a challenge to understand why different motions win out in carrying angular momentum more efficiently in nuclei as close as ^{68}Ge and ^{72}Se.

In the nuclei above $^{72,74}Se$, Hellmeister, et al.[14] have interpreted the yrast cascade in ^{78}Kr in terms of the interacting boson approximation, IBA[17], and found evidence above 12^+ for the dropoff in collective B(E2) strength predicted by the IBA as a result of the finite number of bosons present. However, from our systematic study of the nuclei in this region we suggest that the 12^+ and higher members of the yrast levels in ^{78}Kr are more likely members of a band built on a $(g_{9/2})^2$ single particle configuration. Indeed in ^{80}Kr (refs.18, 19) there is branching in the ground state band at 8^+. Our two-quasiparticle-plus-rotor calculations indicate the second band beginning at 8^+ is an RAL band[20]. Since the ground bands become more compressed as one goes from ^{80}Kr to ^{76}Kr, the crossing by an RAL band could occur at higher spin as N decreases.

One of the more important characteristics of RAL bands is the magnitude of the alignment I_o -- the projection of the intrinsic

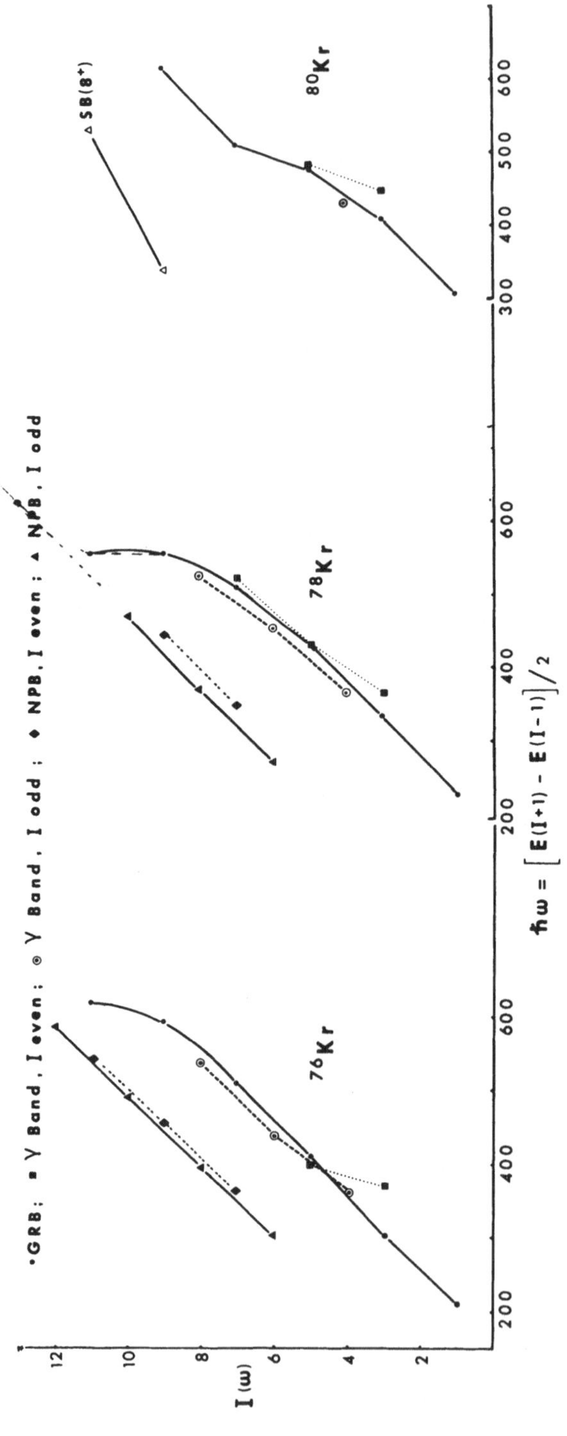

Fig.6 Bohr-Mottelson plots for the aligned angular momentum in the yrast bands in ^{76,78,80}Kr.

angular momentum of both unpaired aligned particles on the axis of rotation. Bohr and Mottelson[21] proposed a way to extract the information on I_0 from an analysis of experimental data on the function $I=I(\omega)$, where the angular frequency $\hbar\omega=\partial E(I)/\partial I \approx E_\gamma(I+1\rightarrow I-1)/2$. The reason for their proposal can be illustrated in the approximation where only the first term is important in the expansion,

$$E_{rot} = \sum_i A_i [R(R+1)]^i \quad (1)$$

Here $R = I-I_0$, and I is the total angular momentum of the rotational state. Then $E_{rot} = A_1 R(R+1)$. In such an approximation, the function $I=I(\omega)$ for a given rotational band is $I = (1/4)A_1 E_\gamma$+constant, and is described by a straight line with a slope proportional to the moment of inertia $\mathcal{I}=\hbar^2/2A_1$. For the GRB and super band, SB, which crosses the GRB these lines would be displaced by a quantity $i_\alpha(\omega)-I_{SB}(\omega)-I_{GRB}(\omega)$, (where α refers to other variables). Bohr and Mottelson[21] call this quantity $i_\alpha(\omega)$ the "aligned angular momentum" and propose to consider it as a measure of the alignment of the SB. In such a point of view, the maximal alignment of two unpaired $i_{13/2}$ neutrons along the rotational axis yields $i_\alpha(\omega)_{max} \approx 12$, whereas when the particles are coupled to the deformed field $i_\alpha(\omega)=0$. In Fig.6, we plot $I(\omega)$ versus $\hbar\omega \approx E_\gamma(I+1\rightarrow I-1)/2$ for $^{76-80}$Kr. A marked shift in $I(\omega)$ from its near-linear variation at lower spins is seen in ^{78}Kr about 10^+ and is apparently just beginning in ^{76}Kr at 12^+. Such shifts in $I(\omega)$ are characteristic of the crossing of an RAL and ground band as seen in ^{68}Ge and may be expected throughout this region. Thus, we interpret the 14^+ and 16^+ levels observed by Hellmeister et al.[14] as members of an RAL band, and so they do not provide evidence for the IBA predicted drop in collectivity.

A further, interesting point seen in Fig.6 is the fact that $i(\omega)$ is only about half that for full alignment and the same as that of the negative parity bands which involve only one $g_{9/2}$ particle. A possible explanation of this will be considered in the next section. In ^{78}Kr there is only one well developed negative parity band from 3^- or 5^- to 13^-(refs.14,15). It looks very much like the single negative parity one in ^{74}Se (ref.13). It is still not completely resolved as to whether the negative parity bands, which are seen from 3^- to 13^- in these nuclei, are really an octupole band explanable in IBA from a coupling of the octupole and quadrupole phonons as suggested by Hellmeister, et al.[14] for ^{78}Kr or whether it is the result of the crossing of the 3^- octupole state by a rotation-aligned band beginning at 5^-. This latter explanation best fits in ^{68}Ge where the 3^- and 5^- states are more widely separated and 2-quasi-particle calculations reproduce the two 5^- bands and their quite different energy spacings. Also, in ^{80}Kr the negative parity bands are nicely reproduce by our RAL calculations. Nevertheless, IBA calculations reproduced the properties of the negative parity bands in ^{78}Kr and ^{74}Se nicely. It is possible that different nuclear motions win out in carrying angular momentum more effectively in the negative parity bands in this region -- rotation-aligned structures in some nuclei and octupole bands in others. This is an interesting question that will provide further tests of both the IBA and RAL models.

IV. ALIGNED ANGULAR MOMENTUM

Because of the large importance of the effect of Coriolis alignment for the understanding of the nature of high spin collective states, Peker et al.[22] have analyzed in more detail the proposed Bohr and Mottelson[21] interpretation of the quantity $i_\alpha(\omega) = I_{SB}(\omega) - I_{GRB}(\omega)$ as defined above. They[22] point out that the function $i_\alpha(\omega)$ depends not only on the properties of the SB (including alignment) reflected in the function $I_{SB}(\omega)$, but by definition depends on the properties of the GRB reflected in the function $I_{GRB}(\omega)$ and as well on the strength of coupling of these two bands, $\approx \hbar/\Delta E(I)$. Here let us look briefly at their arguments on the possible effects of these last two factors of the magnitude of $i_\alpha(\omega)$.

By considering the next simplified approximation of Eq.1, one can easily see that the GRB and the SB have a very different dependence on I:

$$E_{rot} = A_1[R(R+1)] - A_2[R(R+1)]^2 \tag{2}$$

where A_2 characterises the softness of a nucleus. From Eq.2 it follows that the two leading terms of $E_\gamma(I)$ are

$$E_\gamma(I) = 4A_1R - 8A_2R^3 + \ldots$$

Usually $A_2/A_1 \approx 10^{-2} - 10^{-3}$ and at low R the influence of the first term is dominant. But with the increasing of R, the influence of the second term on E_γ increases very fast. For the rotational states with the same I, $R_{GRB} >> R_{SB}$ since $R_{GRB}=I$ and $R_{SB} = I-I_o$ ($I_o \approx 12$ for $i = 13/2$). Therefore, the energies of the GRB at a given I are much more strongly affected by the second and higher order terms in Eqs.1 and 2 than are the eneries in the SB at the same I. Thus, we can expect that in the GRB as I increase, E_γ increases not as fast as in the SB, and the $I=I_{GRB}(\omega)$ curve up bends faster than the $I=I_{SB}(\omega)$ curve.

Use of a more realistic expansion $E_{rot} = \Sigma_i \alpha_i \omega^{2i}$ (refs.23,24) leads to the same conclusion. Therefore $i_\alpha(\omega) = I_{SB}(\omega) - I_{GRB}(\omega)$,

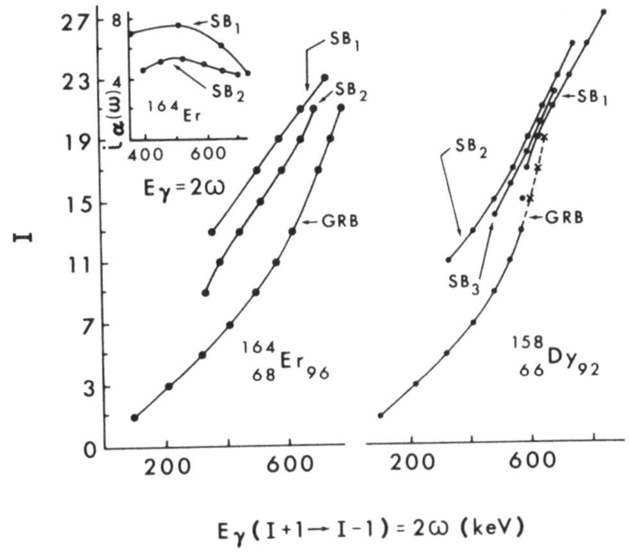

Fig.7 The functions of I=I(ω) for GRB and observed aligned bands SB_1, SB_2 and SB_3 in $^{164}_{68}Er_{96}$ and $^{158}_{66}Dy_{92}$. For ^{164}Er we also present data on $i_\alpha(\omega)$ for the two even spin aligned bands SB_1 and SB_2. Note that $i_\alpha(\omega)_{SB_1} > i_\alpha(\omega)_{SB_2}$ for ^{164}Er and $i_\alpha(\omega)_{SB_1} < i_\alpha(\omega)_{SB_2}$ for ^{158}Dy. It is possible that SB_2 and/or SB_3 shown may have negative parity.

which reflects the difference in the behavior of the SB and the GRB, at large I is expected to decrease with increasing I. At a certain value of I, the two curves have to cross each other and at this moment $i_\alpha(\omega)=0$. So a decreasing of $i_\alpha(\omega)$ with increasing I is expected independent of the magnitude of the real alignment I_o of the SB. This effect can be illustrated by an analysis of the experimental data[24-27] for the GRB and the SB in ^{164}Er and ^{158}Dy (Fig.7). For soft nuclei with N=90, the difference in the behavior of the GRB and the SB with increase of I is particularly strong.

Fig.8 presents the data on I=I(ω) for the GRB and the SB in $^{154}_{64}Gd_{90}$, $^{158}_{68}Er_{90}$ and $^{160}_{70}Yb_{90}$ (refs.24, 28, 29 respectively). It is easy to see that the influence of different softness [A_1(keV) of 19.4, 26.0 and 33.7 and A_2(eV) of 66, 91, 190, respectively] is very strong in the curves of the GRB and small in the SB curves. It means that alignment in the SB of these nuclei is nearly the same, whereas the large difference in $i_\alpha(\omega)$ is almost completely related to different properties of the GRB.

Now consider the effects of the mixing of the GRB and SB on the magnitude of $i_\alpha(\omega)$. Without going into detail, let us note

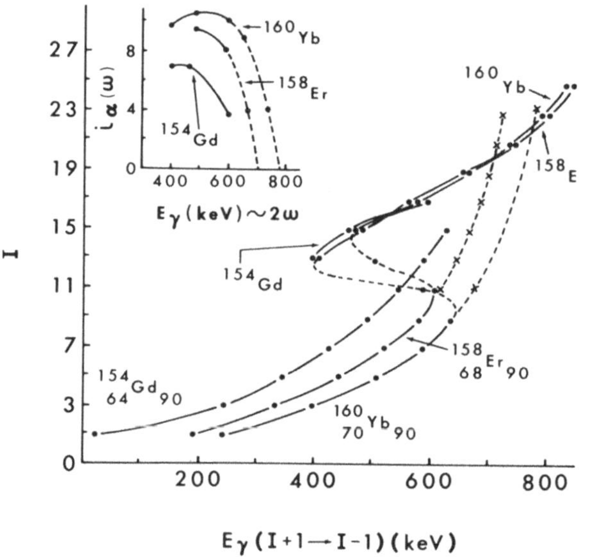

Fig.8 The functions $I=I(\omega)$ and $i_\alpha(\omega)$ for N=90 nuclei. For ^{158}Er and ^{160}Yb and I>10, E_γ in GRB were estimated on the basis of the expansion $E_{lev} = \alpha\omega^2+\beta\omega^4+\gamma\omega^6+\delta\omega^8$.

the following. From the rotational alignment model[4], the configuration $(j \geq 9/2)^2$ can produce many alignment bands with even and odd spins with different degrees of rotational alignment I_o. The band with maximal alignment I_o^{max}, SB_1, and with strongest Coriolis effect has the largest moment of inertia $\mathcal{I}_{max}$ and is the lowest. Therefore, this band is responsible for the observed first backbending in the yrast band. The bands with smaller degrees of alignment, SB_2, SB_3, have by definition smaller I_o and larger energies. Because of weaker Coriolis coupling, they also have a smaller mo moment of inertia. But this effect on $\mathcal{I}$ is true only if we do not take into account the mixing of the SB_1 and GRB. If such mixing is strong enough, the observed $i_\alpha(\omega)$ for SB_1 can be smaller than $i_\alpha(\omega)$ for SB_2 (with even I) which are higher and therefore have larger $\Delta E(I) = E_{SB_2}(I)-E_{GRB}(I)$ and therefore weaker mixing.

The same is true for $i_\alpha(\omega)$ for SB_3 (with odd I) and for the aligned negative parity bands (NPB) which cannot be mixed with the GRB (with even I, even parity levels). From these considerations, it follows that in ^{164}Er, where the mixing of the GRB and SB_1 is weak, we can expect that $i_\alpha(SB_1) > i_\alpha(SB_2$, SB_3 or NPB), whereas in ^{158}Dy with strong mixing of the GRB and SB_1 $i_\alpha(SB_1)$ has to be smaller than $i_\alpha(SB_2$, SB_3 or NPB) [$i_\alpha(SB_1) < i_\alpha(SB_2$, SB_3, NPB)]. As seen in Fig.7, the observed properties of the side bands in these nuclei

(^{164}Er, ref.27, and ^{158}Dy, ref.27) completely support these predictions and therefore demonstrate the influence of the mixing of the GRB and SB_1 on $i_\alpha(\omega)$.

Thus, we see that the magnitude of the "aligned angular momentum", $i_\alpha(\omega)$, introduced by Bohr and Mottelson[21], reflects not only the real alignment in SB_1 but also the properties of the GRB and the strength of the mixing of the GRB and SB_1. Thus, i_α is not a measure of the real alignment of the SB, and one must take these effects into consideration in comparing experimental and theoretical calculations of i_α.

V. HIGH SPIN STATES IN ^{248}Cm

The question of what happens to the collective rotation and single particle configurations as the nuclear angular velocity increases further is of much current interest[30-32]. Important to the extension of our understanding is to observe the behavior of more purely rotational-like states to higher angular momenta and of rotation-aligned configurations in very high j orbitals like $\nu_{15/2}$ and $\pi_{13/2}$ as can occur in actinide nuclei. Even-even actinide nuclei are well suited for studying the rotational characteristics of nuclei to high spin since their known spectra like that of ^{238}U deviate little from the rotational model to as high as 18^+ (ref.33). To provide a firm extension of our understanding of nuclei at very high spins, 30-60h, one would like to study discrete rotational-like yrast bands to as high a spin as possible.

With one of the lowest first excited 2^+ energies and largest collectivity, ^{248}Cm should offer one of the best opportunities to study a rotational band to high spin. Traditional techniques based on heavy ion reactions are of little use for such studies since the main cross-section for decay of the compound system is for fission rather than particle evaporation. Recently a Vanderbilt-GSI collaboration completed the first Coulomb excitation to high spin of a transuranic element.

The ground state rotational band of ^{248}Cm has been studied up to spin 28^+ and tentatively to 30^+ by multiple Coulomb excitation using ^{208}Pb ions of 5.3 MeV/amu from Unilac at GSI[34]. The gamma-rays emitted from the recoiling Curium nuclei were corrected for the strong Doppler broadening by detecting both the scattered Pb and the recoiling Cm particles in two gas detector. A fast "on-line" Doppler correction was achieved by utilizing a specially shaped micro-strip delay line as the cathodes of the gas detectors. A corrected gamma-ray spectrum is shown in Fig.9. A precise measurement of the transition energies was achieved by averaging the energies observed in two gamma detectors, one situated to detect gamma-rays emitted in the direction of the recoiling Cm nucleus and the

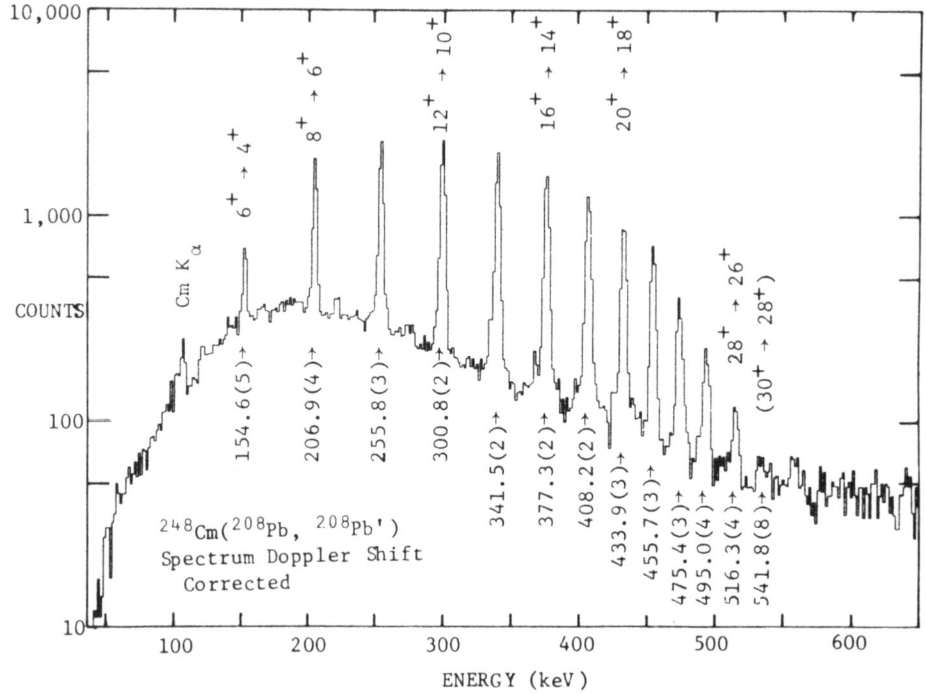

Fig.9 A Doppler-shift corrected spectrum showing the gamma rays observed for small impact parameters in ^{248}Cm. The insert compares the high and low impact parameter data for the region containing the proposed $30^+ \rightarrow 28^+$ transition.

other detector at 180° to the Cm recoil direction. In this newly developed technique, errors associated with uncertainties in the recoil velocity and in the alignment are such that total errors in the transition energies are, for most of the transitions, less than 0.2 keV.

The energy levels of ^{248}Cm determined in this way show an anomalous tendency at high spin to be too high compared to a rotational model. This trend corresponds to a forward bending of at the highest spins and is illustrated in a more dramatic way in Fig.10 as a sharp upbend in the plot of Δ^2E vs. I. These data are in sharp contrast to the backbending of $\mathscr{I}$ seen from ^{68}Ge on up through the deformed rare earths. Since this conference, calculations have been carried out in Cranked Hartree-Fock-Bogoliubov approach. These data can nicely reproduce the anomalous behavior at the highest spins in terms of the alignment of the high j $\pi i_{13/2}$

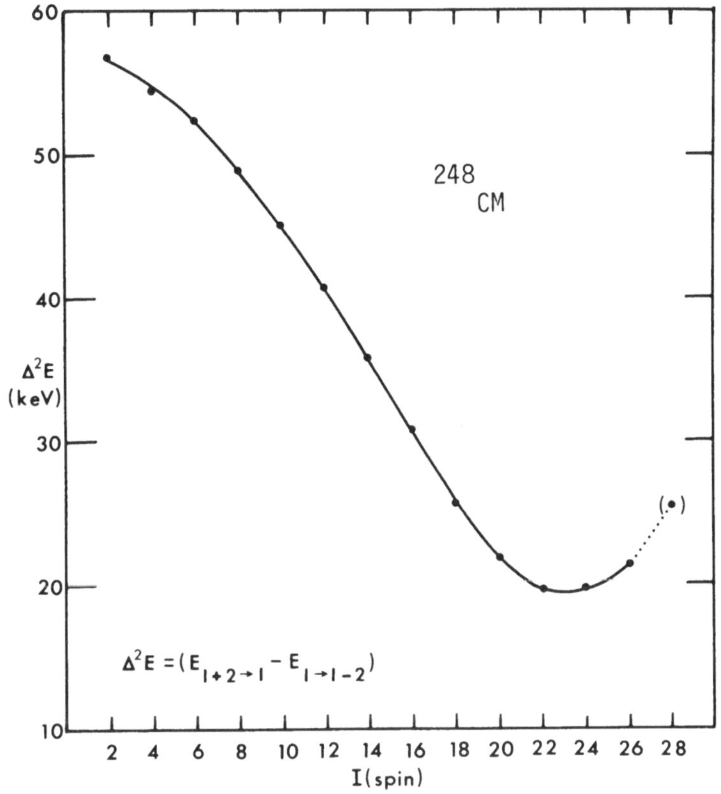

Fig.10 Δ^2E plot for the gamma rays belonging to ^{248}Cm showing the anomalous behavior at high spin.

and $\nu j_{15/2}$ orbitals along the rotation axis[35]. Thus, these data extend on the other end of the periodic table our knowledge of the interplay of collective and single particle motions now seen at higher spins than in lighter nuclei.

VI. MASSIVE TRANSFER REACTIONS

In this last section, we would like to describe the results of two recent Vanderbilt-ONRL-Washington U., St. Louis experiments that involve a new process called massive transfer reactions[36]. The results suggest that this may be a very promising, new technique to study the begavior of discrete energy states to still higher spins than can be reached in heavy-ion, xn reactions with multiplicity filters.

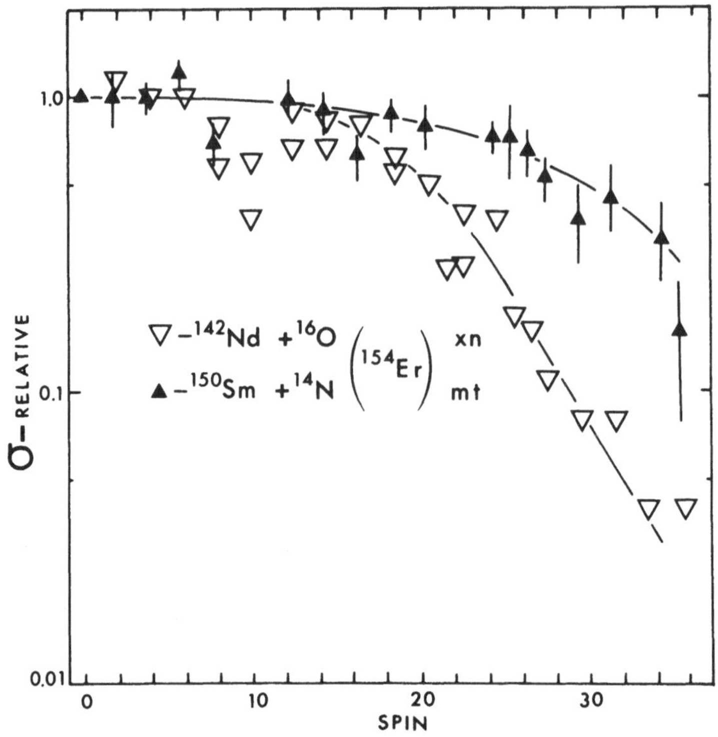

Fig.11 A comparison of relative yields of yrast transitions in ^{154}Er showing the enhancement of the high spin states in a "massive transfer" experiment as compared to the more traditional (HI, xn) reaction.

Based on the first studies of Yamada, et al.[36] of the gamma rays emitted in coincidence with the forward peaked fast protons, electrons and tritons in reactions like ^{154}Sm(^{14}N, pxn)$^{167-x}$Er, Yamada[37] perceived that such reactions could provide a new way to study discrete energy states to higher spin. In a study of the same reaction at 165 MeV, coincidences were reached between fast protons detected in two counter telescopes at 22° to the beam and two large NaI detectors operated in coincidence with an eight NaI detector multiplicity filter.[38] This experiment clearly indicated that the massive transfer takes place in a rather narrow angular momentum window at near the critical angular momenta for fusion. A high γ-ray multiplicity of 30 was measured.

In a second experiment, a ring proton detector was used in the forward direction to catch nearly 70% of the strongly forward peaked protons, deutrons and tritons in coincidence with two large NaI detectors and the same multiplicity filter[39]. The reactions $^{154}Sm(^{14}N, p9n)^{158}Er$ and $^{150}Sm(^{14}N, p9n)^{154}Er$ were studied with 165 MeV ^{14}N ions. In both the well deformed and near-spherical residual nuclei, ^{158}Er and ^{154}Er, one sees the γ-rays out into the known limits of discrete states[40,41]. The relative intensities of the transitions along the known yrast states in ^{154}Er up to spin 35 are shown in Fig.11 and compared with the relative intensities seen in the xn reactions. The massive transfer reaction has a factor of 10 greater relative intensity at the highest spins compared to the relative intensities of these same transitions in the xn reaction. These data and the multiplicity data clearly show that this is a powerful way of populating high spin states. We are planning experiments to use this reaction to study discrete high spin states to still higher spins.

Already with the large NaI detector we looked for gamma-rays from states above 32^+ in ^{158}Er by different techniques, including the E_γ-E_γ correlation technique as recently intorduced by a Copenhagen group . These data suggest that we may be seeing new behavior above the known limits of 32^+ in ^{158}Er. However, the statistics are not sufficient to establish such yet. Nevertheless, here again the results look encouraging.

In summary, the massive transfer appears to be a very promising, new way to extend the limits of our knowledge of discrete states in both deformed and near-spherical nuclei.

ACKNOWLEDGEMENT

We would like to thank our collaborators at Vanderbilt, Oak Ridge National Laboratory, and Gesellschaft fur Schwerionenforschung, Darmstadt, for permission to use the results of various experiments in this paper. This work was supported in part by a grant from the U.S. Department of Energy, Contract No. DE-AS05-76ER05034.

REFERENCES

1. A. Johnson, H. Ryde and J. Sztarkier, Phys. Lett. 34B, 605 (1979 605 (1971).
2. J.H. Hamilton et al., Phys. Rev. Lett. 32, 239 (1974); 36, 340 (1976).
3. J.H. Hamilton, et al., Phys. Rev. Lett. 32, 239 (1974); Selected Topics in Nuclear Structure, ed. V.G. Soloviev (Dubna Press, 1976), Vol. II, p.303.
4. F.S. Stephens, Rev. Mod. Phys. 47, 43 (1975).
5. C. Flaum and D. Cline, Phys. Rev. 14, 1224 (1976).
6. A. Faessler, M. Ploszajczak and K.R. Sandhya Devi, Intern. Symp. on High spin States and Nuclear Structure, ed. L. Funke,

Akademie der Wissenschaften der DDR, ZFK-336 68 (1977); A. Faessler, et al., Nucl. Phys. A286, 101 (1977).
7. C. Flaum, D. Cline, A.W. Sunyar and O.C. Kistner, Phys. Rev. Lett. 33 973 (1974); C.Flaum et al., Nucl. Phys. A264, 291 (1976).
8. J.A. Grau, L.E. Samuelson, F.A. Rickey, P.C. Simms and G.J. G.J. Smith, Phys. Rev. C14, 2297 (1976).
9. A.P. de Lima, J.H. Hamilton, A.V. Ramayya, B. van Nooijen, R.M. Roonningen, H. Kawakami, R.B. Piercey, E. de Lima, R.L. Robinson, H.J. Kim, W.K. Tuttle, L.K. Peker, F.A. Rickey and R. Popli, Phys. Lett. 83B, 43 (1979).
10. D.L. Hill and J.A. Wheeler, Phys. Rev. 89, 1102 (1952).
11. J.H. Hamilton, R.L. Robinson, A.V. Ramayya, Physica Energ. Fort. Phys. Nucl. 3, 355 (1979); Nuclear Interactions, ed. B.A. Robson (Springer-Verlag, 1979), p.253.
12. K. Kumar, Future Directions in Studies of Nuclei Far From Stability, eds. J.H. Hamilton, E.H. Spejewski, C.R. Bingham and E.F. Zganjar, (North Holland Publ. Co., Amsterdam, 1980), p.265.
13. R.B. Piercey et al., Phys. Rev. Lett. 37, 496 (1976).
14. H.P. Hellmeister, J. Keinonen, K.P. Lieb, R. Rascher, R R. Ballini, J. Delaunay, and H. Dunont, Phys. Lett. 85B, 34 (1979).
15. R.L. Robinson, H.J. Kim, R.O. Sater, W.T. Milner, R.B. Piercey, J.H. Hamilton, A.V. Ramayya, J.C. Wells, Jr., and A.J. Coffrey, Phys. Rev. 21, 603 (1980).
16. R.B. Piercey and J.H. Hamilton, in Selected Topics in Nuclear Structure, XVIII Winter School on Physics (Bielsko-Biala, Poland, 1980). to be published.
17. A. Arima and F. Iachello, Ann. Phys. 99, 253 (1976).
18. D.L. Sastry et al., Proc. Int. Conf. on Nuclear Structure, Contributed Papers, (Tokyo, 1977), p.301.
19. L. Funke et al., in Future Directions in Studies of Nuclei Far From Stability, eds. J.H. Hamilton, E.H. Spejewski C.R. Bingham and E.F. Zhanjar (North Holland Publ. Co., Amsterdam, 1980), pg.231.
20. A.V. Ramayya, J.H. Hamilton, R. Soundranayagam and S. Rama-avatharam, private communication.
21. A. Bohr and B.R. Mottelson, Supp. J. Phys. Soc. Japan 44, 157 (1978).
22. L.K. Peker, J.O. Rasmussen and J.H. Hamilton, Phys. Lett. 91B, 365 (1979).
23. A. Bohr and B.R. Mottelson, Nuclear Structure, (Benjamin, New York, 1975), Vol. 2.
24. O. Saethre et al., Nucl. Phys. A207, 486 (1973).
25. N.R. Johnson et al., Phys. Rev. Lett. 40, 151 (1978).
26. O.C. Kistner, A.W. Sunyar and E. Der Mateosian, Phys. Rev. 17C, 1417 (1978).

27. A.W. Sunyar, Proc. Symp. on High Spin Phenomena in Nuclei, ANL/PHY-79-4 (1979) p.77.
28. I.Y. Lee et al., Phys. Rev. Lett. 38, 1454 (1977).
29. F.A. Beck et al., Phys. Rev. Lett. 42, 493 (1979).
30. A. Faessler, M. Ploszajczak, K.R. Sandhya Devi, Nucl. Phys. A301, 529 (1978); A. Faessler, Invited talk, Int. Conf. on Nucl. Behavior at High Angular Momentum, Strasbourg, France (1980).
31. U. Mosel, Invited talk, Int. Conf. on Nucl. Behavior at High Angular Momentum, Strasbourg, France, (1980).
32. R. Bengtsson and S. Frauendorf, Nucl. Phys. A327, 39 (1979); ibid., A314, 27 (1979); S. Frauendorf, Invited lecture, Nobel Symposium (1980).
33. E. Grosse, J. deBoer, R.M. Diamond, F.S. Stephens, P. Tjøm, Phys. Rev. Lett. 35, 565 (1976).
34. R.B. Piercey, H. Emling, P. Fuchs, E. Grosse, J.H. Hamilton, H. Ower, A.V. Ramayya, D. Schwalm, N. Trautmann and H.J H.J. Wollersheim, GSI Annual Report 30-3, July 1980, p.47.
35. R.B. Piercey, J.H. Hamilton, A.V. Ramayya, D. Schwalm, E. Grosse, H. Emling, P. Fuchs, H.J. Wollersheim, N. Trautmann, A. Faessler and M. Ploszajczak, to be submitted to Phys. Rev. to be submitted to Phys. Rev. Lett.
36. H. Yamada, et al., Phys. Rev. Lett. 43, 605 (1979).
37. H. Yamada, Private communication.
38. H. Yamada, et al., Bull. Am. Phys. Soc. 25, 596 (1980).
39. H. Yamada, D.C. Hensley, S. Hjorth, A.V. Ramayya, C.F. Maguire, M. Herath-Banda, J.H. Hamilton, R.L. Robbinson and M.L. Halbert, private communication.
40. O. Anderson, et al., Phys. Rev. Lett. 43, 687 (1979).

INTERPRETATION OF THE 21-ns ISOMER IN ^{190}Hg AS $(\nu i_{13/2})^2$ FROM A g-FACTOR MEASUREMENT

S.A. Hjorth*, I.Y. Lee, J.R. Beene, C. Roulet†, D.R. Haenni, Noah R. Johnson, Felix E. Obenshain and G.R. Young

Oak Ridge National Laboratory
Oak Ridge, Tennessee 37830

ABSTRACT

A 21-ns isomer in ^{190}Hg has earlier been interpreted as the 10^+ member of the $(h_{11/2})^{2n}_{10}$ configuration. The g-factor of the isomer is measured by means of the spin rotation method and is found to be -0.21 ± 0.02. This determines the configuration of the isomer as $(i_{13/2})^{2n}$. The well-known properties of levels within a $(j)^{2n}$ configuration then suggests that the Isomerism is due to a $(i_{13/2})^{2n}_{12+}$ level that exists very close to the 10^+ state, with the 12^+ to 10^+ transition so low in energy that it has not yet been observed.

The $i_{13/2}$ neutron orbital plays an important role in the descri-scription of high-spin states in the A = 150-208 mass region. For example, the first backbend found in many deformed nuclei is explained as an alignment of a pair of $i_{13/2}$ neutrons[1]. Also, in light lead isotopes the level obtained by coupling two neutron holes to the maximum allowable angular momentum of 12 units occurs as a low-lying isomer[2,3]. In contrast, published level schemes of $^{190-196}Hg$ and ^{194}Pt exhibit a low-lying isomeric 10^+ level while the 12^+ level is shown at somewhat higher excitation[4-7]. Such a level sequence does not arise naturally in the $(\nu i_{13/2})^2$ configuration, and the first 10^+ level in $^{190-196}Hg$ and in ^{194}Pt has therefore been explained as a $(\pi h_{11/2})^{-2}_{10+}$ state[4,6,8].

---

*On leave from the Research Institute of Physics, Stockholm, Sweden.

†Present Address: Etudes et Productions, Schlumberger, Clamar, France.

A measurement of the nuclear g-factor should easily distinguish between these possibilities in a model independent way. This is true because the g-factor for a $\pi h_{11/2}$ excitation is close to unity, whereas the g-factor of an $i_{13/2}$ neutron is about -0.2. Since the half-lives of the isomers in $^{190,192}Hg$ are reported[5] as 24 3 and 16 ± 3 ns, respectively, a measurement using the time differential spin rotation method is feasible.

In the present experiment 94-MeV ^{14}N ions from the Oak Ridge Isochronous Cyclotron were used to produce ^{190}Hg nuclei in a 0.114-mm-thick foil of ^{181}Ta. Tantalum is an ideal target material for this kind of study since it forms a body centered cubic lattice and has small paramagnetism. Therefore, the spin alignment produced in the reaction has a good chance to survive during the isomeric decay. In order to avoid spin-relaxation due to internal stresses produced by the rolling procedure, the target was annealed for 30 minutes at 1100°C prior to bombardment.

The γ rays emitted in the reaction were detected in three coaxial Ge(Li) detectors located at 0° and at $\pm135°$ relative to the beam. Pulse-heights from the three detectors and the times relative to the cyclotron beam bursts were stored in list mode on magnetic tape. The time of arrival of a beam burst was obtained from a device[9] containing a thin plastic scintillator inserted in the path of the beam. In this way data were recorded in the same detector geometry for externally applied magnetic fields of 0.0 and ±2.0 T at the target position. The time dependence of the intensity of the γ-ray transitions depopulating the isomer was obtained in an off-line analysis by putting gates on photopeaks and on suitably chosen background intervals between the peaks. To improve statistical accuracy, time spectra for the 2^+ to 0^+, 4^+ to 2^+ and 8^+ to 6^+ transitions were combined. The 10^+ to 8^+ transition is too low in energy to give good time resolution, and the 6^+ to 4^+ transition is partially obscured by delayed γ rays from the Ge(n, n') reaction; consequently, neither has been included in the analysis.

In the measurement with no magnetic field the time distributions observed in all three detectors are consistent with a purely exponential decay. In the measurements with field up and field down, all time distributions deviated from the exponential curve in a way characteristic of a slow rotation of an angular distribution pattern. All curves were fitted with an expression of the form

$$I_\gamma(t, \theta) = N\left\{1 + A_2P_2[\cos(\theta-\omega_L t-\delta)]\right\} \exp(-\lambda t)$$

with $t_{1/2} = \ln 2/\lambda = 21\pm2$ ns and with nearly the same values of the angular distribution coefficient A_2, Larmor frequency $_L$ and beam bending angle δ. The beam bending angle is calculated to be 8.5°

by using measurements of the magnetic field strength along the beam path. Relative normalizations, N, were determined by fixing δ to this value and by fixing ω_L at the weighted mean of values obtained in free fits to individual spectra. These normalizations differed slightly from those obtained with zero field, as would be expected due to beam bending effects, slow drifts in the electronics, etc. The normalized decay spectra were used to form the ratio

$$R(t) = \frac{I_\gamma(t, 135°) - I_\gamma(t, -135°)}{I_\gamma(t, 135°) + I_\gamma(t, -135°)}$$

upon which final fits to ω_L were performed for the two cases $B = \pm 2.0$ T. As shown in Fig.1, good fits corresponding to $g = -0.21 \pm 0.02$ with $A_2 = 0.18$ for $B = +2.0$ T and $A_2 = 0.16$ for $B = -2.0$ T were obtained.

The only seniority 2 excitations that can be responsible for a low-lying 10^+ level in ^{190}Hg arise from the $(\pi h_{11/2})^{-2}$ and $(\nu i_{13/2})^2$ configurations. The widely differing values of the gyromagnetic ratios for these two configurations (cf. the discussion above) let us uniquely establish the configuration of the isomer as $(\nu i_{13/2})^2$. In fact, when an effective magnetic moment operator thought[10] to be appropriate for deformed nuclei, ($g_s = -2.4 \pm 0.4$ and $g_\ell = -0.03 \pm 0.04$) is used, the g-factor for the $(\nu i_{13/2})^2$ state is calculated to be -0.21 ± 0.07. This certainly agrees with our experimental value of -0.21 ± 0.02. The experimental value of the ^{190}Hg g-factor would imply a $\nu i_{13/2}$ moment more negative (and hence nearer the Schmidt value) than the effective $\nu i_{13/2}$ moment measured[11] in ^{207}Pb ($g = -0.14$). Such small differences could easily occur due to variations in core polarization and configuration mixing.

In the $(\nu i_{13/2})^2$ configuration the spacing between the 12^+ and 10^+ levels is considerably smaller than the spacing between any pair of lower spin $(\nu i_{13/2})^2$ states, while in the established level s scheme of ^{190}Hg the 12^+ to 10^+ spacing is more three times the 8^+ to 10^+ spacing. Our result is therefore most easily understood if a 12^+ state exists very close to the 10^+ state, with the 12^+ to 10^+ transition so low in energy that it has not yet been observed. The 21-ns isomer should then be identified with this 12^+ state rather than with the 10^+ state. Further, both the energies and spins of the levels in the yrast band above spin 10^+ should be adjusted to accommodate this unobserved low-energy transition.

Using the value $B(E2; 12^+ \rightarrow 10^+) = 2921\ e^2 fm^4$ measured[12] in ^{198}Hg and the 21-ns half-life determined in our work, we estimate the energy of the 12^+ to 10^+ transition in ^{190}Hg to be a 12 keV.

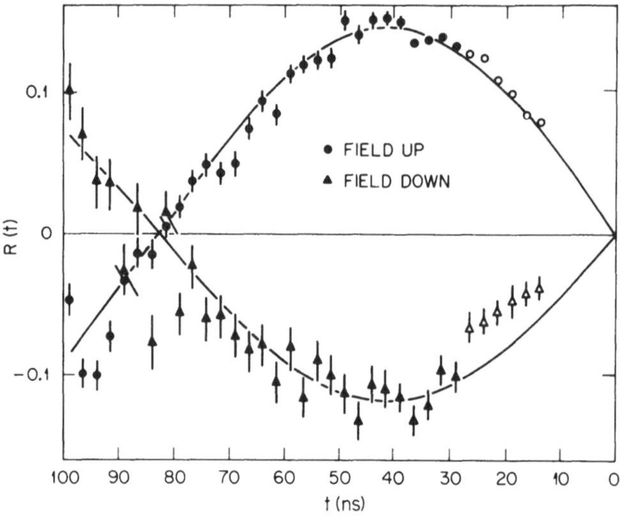

Fig.1 Spin rotation ratio R(t) in ^{190}Hg as a function of time relative to the center of the prompt peak. The figure displays the data for field up and field down as solid round and triangular symbols, respectively. The curves are fits to the data as described in the text. The data for t $\lesssim$ 30 ns are effected by prompt events and are excluded from the fits. This is indicated by open symbols in the figure.

Such a low energy transition could easily have escaped detection in previous work. Also, using the B(E2; 10^+ 8^+) value from ^{198}Hg, we estimate the half-life of the 10^+ state in ^{190}Hg to be shorter than 2 ns, supporting our suggestion that the 10^+ state is not the isomer in ^{190}Hg. Furthermore, in nearby isotopes of platinum and lead the $(\nu i_{13/2})^2$ isomerism is well established with the 12^+ to 10^+ spacings ranging between 3.1 and 60 keV. It is likely that the same type of isomer exists in the mercury isotopes.

ACKNOWLEDGEMENT

The authors wish to thank J.B. McGrory for helpful discussions on this work. This research was supported by the U.S. Department of Energy, partially under Contract No. W-7405-eng-26 with Union Carbide Corporation.

REFERENCES

1. F.S. Stephens, Rev. Mod. Phys. 47, 43 (1975).
2. M. Pautrat, G. Albouy, J.C. David, J.M. LaGrange, N. Poffe, C. Roulet, H. Sergolle, J. Vanhorenbeek and H. Abou-Leila, Nucl. Phys. A201, 449 (1973).
3. I. Bergström, J. Blomqvist, B. Fant, A. Filevich, G. Linde'n, K.-G. Rensfelt, J. Sztarker and K. Widström, Physics Scripta 3, 11 (1971).
4. D. Proetel, R.M. Diamond and F.S. Stephens, Nucl. Phys. A231, 301 (1974).
5. R.M. Lieder, H. Beuscher, W.F. Davidson, A. Neskakis and C. Mayer Böricke, Nucl. Phys. A248, 317 (1975).
6. A.W. Sunyar, G. Scharff-Goldhaber and M. McKeown, Phys. Rev. Lett. 21, 237 (1968).
7. S.A. Hjorth, A. Johnson, Th. Lindblad, L. Funke, P. Kemnitz and G. Winter, Nucl. Phys. A262, 328 (1976).
8. T. Inamura, Y. Tendow, S. Nagamiya and A. Hashizume, J. Phys. Soc. Japan 32, 1163 (1972).
9. C. Roulet, G. Albouy, G. Auger, M.P. Bourgarel, J.C. David, J.M. Lagrange, B. Monsanglant, M. Pautrat, H. Richel, H. Sergolle and J. Vanhorenbeeck, Nucl. Instr. and Meth. 125, 29 (1975).
10. H. Morinaga and T. Yamazaki, In-Beam Gamma-Ray Spectroscopy, (North-Hollandy Amsterdam, New York, Oxford, 1976).
11. D. Riegel, N. Brauer, B. Focke, G. Goldmann, J. Hadijuana, M.V. Hartrot and K. Nishiyama, Phys. Lett. 44B, 456 (1973).
12. C. Gunter, H. Hubel, A. Kleinrahm, D. Mertin, B. Richter, W.D. Schneider and R. Tischler, Phys. Rev. C15, 1298 (1977).

HIGH SPIN STUDIES BY MULTIPLE COULOMB EXCITATION

Eckart Grosse

GSI, D6100 Darmstadt
Federal Republic of Germany

I. ABSTRACT

It is shown that multiple Coulomb excitation is an excellent means to study high spin states in heavy deformed nuclei. Using ^{208}Pb as a projectile and a particle-gamma-coincidence technique the ground state bands in rare earth nuclei can be excited to spins around 20 $\hbar$, and in actinide nuclei spin 30 can be reached. Since Coulomb excitation not only gives the energies and the corresponding rotational frequencies of the transitions in the ground state bands of the target nucleus but also delivers the electric quadrupole transition strengths, a sensitive test of theoretical predictions concerning high spin states becomes possible. The relevance of such data for the understanding of the rotation alignment and the related band crossing phenomena and for the question of the high spin cut off of nuclear collectivity will be discussed.

II. INTRODUCTION

Coulomb excitation, i.e. the excitation of a target nucleus in the electric field of a projectile nucleus passing by just out of reach of the nuclear forces, has been an important tool in nuclear spectroscopy already since about three decades[1].

However, high spin states in heavy nuclei have up to now[2] been studied mainly through compound reactions of the type (HI, xnγ). Using the newly available beams of very heavy ions, states with spins above 20^+ can now also be investigated by multiple Coulomb excitation. Ions of ^{208}Pb are especially well suited because of the low probability to excite this projectile nucleus itself. By

Coulomb excitation one can study the more neutron rich stable nuclei as compared to the neutron deficient nuclei populated by (HI, xn)-reactions; additionally it allows to investigate high spin states in the actinides which are not accessible as compound nuclei because of high fission probabilities. Whereas with the (HI, xn) reaction the gamma transitions observed predominantly belong to the yrast band, Coulomb excitation populates mainly the ground state band. This band is phenomenologically defined as the sequence of states connected to the ground state by a series of strongly collective E2-transitions. In and above the region of the crossing of the two rotational bands, i.e. the ground state rotational band and the Coriolis aligned s-band[3], the ground state band may no longer form the yrast sequence and the two excitation mechanisms may populate different states, when the mixing between the two bands and conquently the B(E2)-values between them are small.

An advantage of Coulomb excitation is the possibility of extracting E2-transition probabilities from an analysis of the experimentally observed gamma ray yields. These B(E2)-values are an important information complementary to the transition energies which deliver apparent moments of inertia in the bands. The B(E2)-values are directly related to the intrinsic quadrupole moment and the deformation. They are sensitive to possible shape changes due to centrifugal effects as well as changes in the collectivity in a band caused by single particle effects. In the region of the band crossing[3] the measurement of B(E2)-values allows the investigation of the admixture of the aligned two quasiparticle configuration into the ground band. Moreover, E2-transition rates between very high spin states are sensitive to the single particle aspect of collective states in a different way: In a microscopic description such states are represented by a superposition of many multi-nucleon wave functions and the limited number of configurations that are allowed for high spin states by the Pauli principle should lead to a reduction of the collective E2-strength. In the Interacting Boson Approximation[4] (IBA) to complex microscopic multi-nucleon shell model calculations, a complete breakdown of the collectivity is predicted above the limiting spin $I_{lim} = 2n$. This is the maximum spin which can be formed from n (L = 2)-bosons and their number n is half the number of nucleons outside the next closed shell. This cutoff and the reduction of the collectivity already at lower spins as the limiting spin, which are in contrast to the predictions of the geometric model[5] for collective states, can be tested by determining B(E2)-values at high spins.

III. EXPERIMENTS

The main experimental problem occuring in the multiple Coulomb excitation experiments with very heavy projectiles is the large

Doppler broadening of the deexcitation gamma rays. This broadening is due to the large recoil velocities induced by the heavy ions and the large spread in magnitude and direction of these velocities. In general the high spin states of interest have a too short lifetime to allow an observation at rest after stopping the recoiling excited nuclei in a thick target backing. We therefore developed an experimental method[7] which allows a correction of the Doppler shift of the gamma radiation. It consists in detecting the deexcitation gamma rays in coincidence with the Coulomb excited nuclei recoiling out of the thin target. For the particle detection position sensitive avalanche detectors are used, whose cathodes are subdivided in narrow stripes corresponding to curves of equal scattering angles. These stripes are connected to form a meander which acts as a delay line readout of the scattering angle information. From this information a nearly complete correction of the Doppler shift of the gamma rays could be achieved resulting in a linewidth of $\sim$3 keV at 400 keV. In the deformed rare earth nuclei gamma transitions in the ground state band could thus be observed[6] up to a spin of about 20 $\hbar$; in the actinide nuclei studied so far[7,8,9] the ground state band was identified up to spin 28 or 30.

Additionally our set-up allowed the determination of gamma ray yeilds $Y_I(\theta)$ for the decay of the members of the ground state band with spin I as a function of the c.m. scattering angle θ covering a range of about $30° \leq \theta \leq 150°$. These data could be used to extract individual B(E2)-values; in the ground state band this analysis was rather straightforward since the yield ratio $R_I(\theta) = Y_I(\theta)/Y_{I-2}(\theta)$ depends mainly on B(E2, I-2 $\rightarrow$ I) for certain θ. The relation between these two quantities, i.e. the sensitivity of the experimental ratios R_I to the respective B(E2)-value, was determined by semiclassical Coulomb excitation calculations. These calculations show, that the sensitivity is large at small scattering angles, e.g. those below an angle θ_I at which $R_I(\theta)$ reaches half its maximum value. It also turns out that for $\theta \leq \theta_I$ the sensitivity to other B(E2)-values in the band as well as to side band excitations is rather small. Because θ_I increases from small angles up to $\sim$180° with increasing spin, a stepwise determination of individual B(E2)-values in a model independent way was possible, especially since our data covered a wide range of scattering angles. The actual B(E2)-value determination was performed by an iterative fit procedure[7] taking into account the full sensitivity matrix.

IV. THE STABLE Dy-ISOTOPES

The excitation energy data for all the stable Dy-isotopes as obtained from our Coulomb excitation experiments are displayed in Fig.1 as a spin I vs. rotational frequency ω plot. For the neutron deficient rare earth nuclei irregularities similar to the ones in our experimental I vs. ω curves have been attributed to the crossing

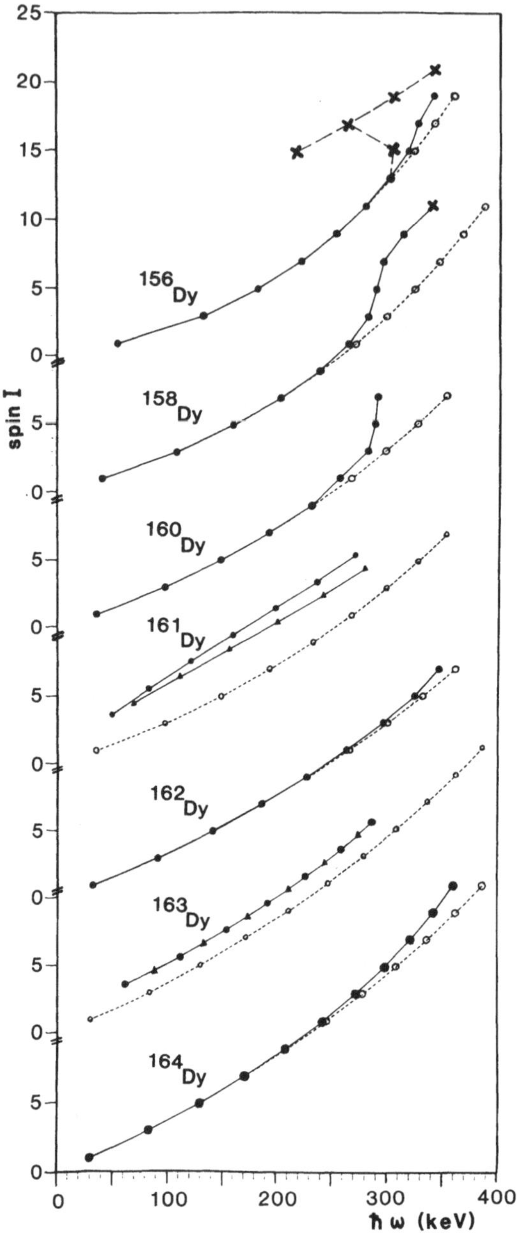

Fig.1 The dependence of the spin I on the rotational frequency $\omega = \Delta E/\Delta I_x$ (cf. ref. 10) as determined by Coulomb excitation of the stable Dy nuclei (full dots and triangles). Some (HI, xn) data[13] are also included as crosses. The dotted lines indicate a VMI fit to the data for $I \leq 6$; they were taken as a reference for the calculation of the aligned angular momentum.

of the ground state rotational band with a band built on two $i_{13/2}$ neutrons aligned to the rotation axis. Calculations in the framework of the cranked shell model (CSM) show[10], that the crossing of the rotation aligned s-band with the ground state rotational band occurs in the light rare earth nuclei at rather the same rotational frequency, i.e. around 290 keV, nearly independent of N and Z. The aligned angular momentum in the s-band, however, as well as the amount of interaction between the two bands are predicted to depend characteristically on the neutron number. Whereas the aligned angular momentum is decreasing when the $i_{13/2}$ subshell is filled, the strength of the interaction is predicted to show a periodic dependence of the neutron number for a series of isotopes, its value depends on the position of the fermi surface relative to the last $i_{13/2}$ Nilsson orbit. Results obtained by Cranked Hartree-Fock-Bogolyubov (CHFB) calculations[9,11] are very similar to the ones resulting from the more phenomenological CSM calculations.

Experimentally the aligned angular momentum is best deduced from a comparison to a reference configuration[10] which can be obtained from a Variable Moment of Inertia (VMI) model fit to the low spin data. Our data obtained from Coulomb excitation of the Dy-isotopes are in qualitative agreement to the theoretical predictions. In ^{156}Dy the interaction is weak and the Coulomb excitation follows the continuation of the ground state rotational band at $\hbar\omega$ = 290 keV. In this nucleus the crossing with the s-band at this frequency is known from (HI, xn)-experiments; some of these data are indicated as crosses in Fig.1. An irregularity occuring in the ground state band at a higher frequency is probably due to a crossing with a higher aligned band. In ^{158}Dy and ^{160}Dy the Coulomb excitation follows the aligned configuration above 290 keV; the interaction between the two configurations is strong. Because of the large energy difference and the reduction of the B(E2)-value due to interference effects resulting from the strong mixing the excitation into the continuation of the ground state rotational band is expected to be reduced. A Coulomb excitation run with high statistical accuracy showed no transitions in ^{158}Dy, which could be identified as an obvious continuation of the ground state rotational band to higher spins. The Coulomb excitation data for ^{162}Dy resemble those for ^{156}Dy, the interaction is predicted to be weak and might be too weak to allow the excitation of the aligned configuration; our data is not in contradiction to this prediction. Because of the periodicity of the interaction strength V a new maximum in V is expected to occur around ^{164}Dy, but because of the rather large filling of the $i_{13/2}$ subshell the s-configuration can have a small aligned an angular momentum only; the data for ^{164}Dy are in qualitative agreement to such a prediction.

Our data on the odd stable Dy-isotopes are further illuminating the picture of the Coriolis force induced alignment of single particle orbits along the axis of the rotating core. The $5/2^+$ [642]

orbit, which forms the ground state of ^{161}Dy, is a member of the $\nu i_{13/2}$ subshell, which because of its high angular momentum - shows large Coriolis effects. When plotting the spin in the ground state band of ^{161}Dy vs. the rotational frequency and comparing it to ^{162}Dy (respectively the VMI-fit to the ground state band of that nucleus), an aligned angular momentum of around $4\hbar$ is observed. The aligned spin is slightly larger for the states with signature $\alpha = +1/2$, i.e. with spins $5/2^+$, $9/2^+$, $13/2^+$ etc. Such a signature splitting has been attributed to Coriolis mixing of bands with different Ω and is closely related to the rotation alignment, which also mixes Ω. No signature splitting is observed in the ground band of ^{163}Dy which has the single particle configuration $5/2^-$ [523]. Obviously Coriolis effects are weak in this band which is based on an intrinsic configuration involving lower j-values.

As described above, the yields observed in Coulomb excitation experiments allow a determination of B(E2)-values, at least within the ground state rotational band. Such an analysis was carried through for the data on ^{156}Dy, ^{158}Dy and ^{164}Dy. In ^{164}Dy our data show now significant deviation from the rigid rotor prediction up to to spin 20, below spin 14 they agree within errors to B(E2)-values determined by other methods[14]. In ^{160}Dy and ^{162}Dy a similar agreement to the rigid rotor model had been found previously[14] for the transitions up to 12^+ respectively 14^+. The situation is different for ^{156}Dy, where the B(E2)-values[13] up to spin 10 indicate a centrifugal stretching effect, i.e. an increase of the intrinsic quadrupole moment by 20% when going from the 0^+ to the 10^+ state. At higher spins the situation is complicated by the crossing with the s-band; we are presently extracting more detailed information about the electric quadrupole transitions in both bands from our Coulomb excitation data and a recoil distance experiment following the ^{25}Mg(^{136}Xe, 5n) reaction. The most detailed data exist for ^{158}Dy, where Coulomb excitation and the ^{26}Mg(^{136}Xe, 4n) reaction populate the same levels; the B(E2)-values from the analysis of the Coulomb excitation yeilds agree to the results of our recoil distance experiment[15]. These are plotted in Fig.2 normalized to the rigid rotor B(E2)-values.

The B(E2)-values in this nucleus increase slightly with spin up to the 6^+ state - indicating centrifugal stretching - and then drop below the rigid rotor value with increasing spin. Between spin 14 and 20, where two aligned $i_{13/2}$ neutrons determine the structure of the ground state band the B(E2)-values are rather constant, whereas at the highest spins they decrease again. This observed reduction of the B(E2)-values excludes that the large apparent moment of inertia observed in the ground state band above spin 14 is due to an increased intrinsic deformation. Within a simple two-band mixing model, the B(E2)-values below and above the bandcrossing, which is the cause for the upbending in I vs. ω at spin 16, are determined by the intrinsic quadrupole moment Q_o of the

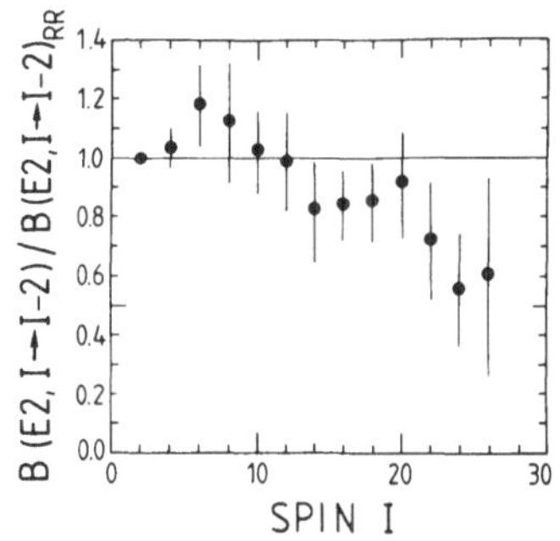

Fig.2 B(E2)-values between the members of the ground state band of ^{158}Dy as deduced from the lifetimes measured following the $^{26}Mg(^{136}Xe,4n)$ reaction and normalized to the rigid rotor prediction using an intrinsic quadrupole moment of Q_o=6.9 eb (ref.14). The results of the analysis of our Coulomb excitation yield data agree quantitatively for I $\leq$ 18 to the values shown here.

ground state band and Q_s of the s-band, respectively. At the bandcrossing there should be a smooth transition between the two. These predictions are at least in qualitative agreement with the observed B(E2)-values, if the intrinsic quadrupole moment of the s-band is assumed to be $Q_s \approx 6$ eb as compared to $Q_o \approx 7$ eb. It is intriguing to compare Q_s with the intrinsic quadrupole moment of ^{156}Dy, which has a Q_o of 6.1 eb. This comparison gives further support to the idea that the s-band in ^{158}Dy has the configuration of two aligned neutrons outside the core ^{156}Dy. However, the observed centrifugal stretching at low spin and its apparent blocking already far below the bandcrossing indicate that a mixing calculation based on two bands of fixed intrinsic properties can describe the observed spin dependence of the B(E2)-values only quantitatively. In the spirit of this discussion the B(E2)-values at the very high spins can be related to one or more further bandcrossings, which have been predicted[10] and some of which have been observed[16] in other nuclei. Thus, the general trend of the observed B(E2)-values might indicate the beginning of a transition away from a prolate nucleus rotating around an axis perpendicular to the symmetry axis into the direction of a shell model configuration consisting of many nucleon pairs aligned along the total angular momentum. This latter configuration can be viewed at as a slightly oblate nucleus "rotating" around its symmetry axis.

V. ACTINIDE NUCLEI

The excitation energy data from our Coulomb excitation experiments[7,8,9] on some actinide nuclei are shown as spin vs. rotational

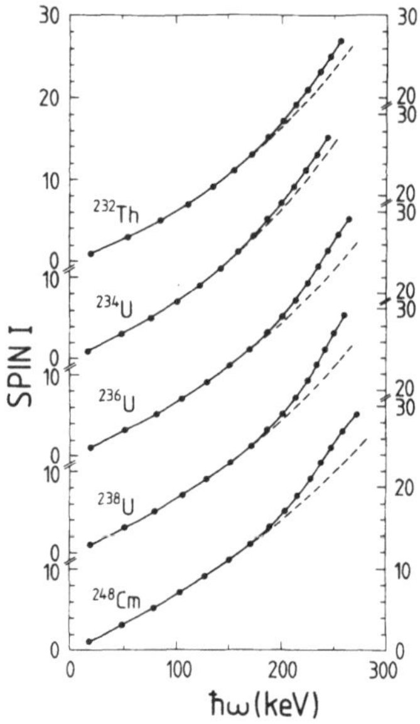

Fig.3 Plot of the angular momentum I(ω) for the ground state bands of ^{232}Th, ^{234}U, ^{236}U, ^{238}U and ^{248}Cm. When using a VMI fit to the low spin states (I $\leq$ 8) as a reference (dashed line) an aligned angular momentum of 2$\hbar$ (for ^{232}Th) to 5$\hbar$ (for ^{248}Cm) can be deduced for the highest spin states observed in the ground state band.

frequency plot in Fig.3. In all the nuclei studied - ^{232}Th, ^{234}U, ^{236}U, ^{238}U and ^{248}Cm - there appears an irregularity above a rotational frequency of 200 keV. This irregularity is similar to the one observed e.g. in the heavy Dy-isotopes and not as pronounced as the one occuring in many neutron deficient rare earth nuclei. In the CHFB calculations[9] as well as in the cranked shell model[17] two close lying s-bands are predicted, one based on the $j_{15/2}$ neutron orbit, the other one on $i_{13/2}$ protons. The interaction between the proton s-band and the ground state band is predicted[17] to be strong for Th, U and Cm. The same should be true for the interaction of the neutron s-bands in ^{232}Th and ^{234}U (N = 142), whereas the neutron s-bands in ^{236}U, ^{238}U and in ^{248}Cm should interact only weakly with the ground state band. The band-crossings due to the two s-bands. are calculated to appear very close to each other at about 240 keV. The experimental findings are in qualitative agreement to the calculations concerning the frequency of the bandcrossing. From a comparison to the VMI fit to the 2^+ to 8^+ levels one obtains an aligned angular momentum of 4 - 5 $\hbar$ for ^{236}U, ^{238}U and ^{248}Cm at $\hbar\omega_c$ = 240 keV. In ^{232}Th and ^{234}U one gets only half as much aligned angular momentum; it is not clear how that observation is related to the prediction[17] which has been made for these two nuclei saying that both the proton and the neutron s-band are strongly mixed with the ground state band. The occurance of the two s-bands complicates the analysis of these actinide data, since it is not obvious if the mixing is in all cases strong enough to make sure that the Coulomb excited ground state band is identical to the yrast sequence. Additionally the rather large filling of the neutron $j_{15/2}$ as well as the proton $i_{13/2}$ shells reduces the observable effect since there are only the high Ω values available causing only a small angular momentum to be gained by rotation alignment[3]. In the case of ^{248}Cm, where a sufficiently high spin was reached to observe the leveling off of the aligned angular momentum a reasonable description of the data was reached by the CHFB calculations[9], showing that the alignment of neutron and proton orbitals is important.

Whereas the rather distinct upbending above spin 20 is obviously caused by a Coriolis force induced breaking and alignment of one or two pairs of nucleons, the smooth deviation of the excitation energies from the rigid rotor at lower spin is not necessarily caused by Coriolis effects alone. At low spins the excitation energies are we well described by the VMI model, but his model is only a phenomenological description of different physical effects; it can fit changes

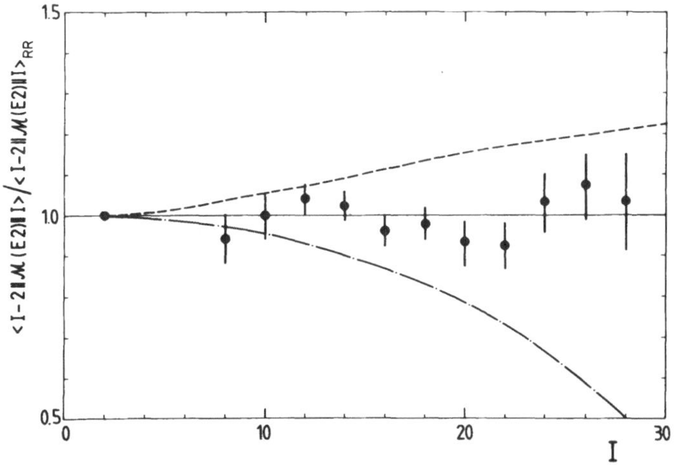

Fig.4 Experimental E2-transition matrix elements for the 8^+ to 28^+ states in the ground state band of ^{238}U, normalized to the rigid rotor values as calculated from Q_o = 11.1 eb. (from ref.21). For comparison the results of calculations in the Rotation-Vibration-Model (dashed line) and in the Interacting-Boson-Approximation (SU(3)-Limit, dash-dotted line) are given.

in the excitation energies caused by the reduction of the pairing[18] due to Coriolis effects as well as changes arising from centrifugal stretching. However, out of these two only the centrifugal stretching as expected to result in a deformation change and consequently in a change of the B(E2)-values which are thus important in order to distinguish between the two processes. The E2-matrix elements for ^{238}U obtained from the analysis of our Coulomb excitation data[7] are shown in Fig.4 in comparison to the rigid rotor values, the Rotation-Vibration-Model[19] (RVM) and the results from the IBA. In the RVM the classical idea of centrifugal stretching of the nucleus is quantum mechanically treated as a coupling of the rotation to the dynamic distrotions of the nuclear surface, which, for an axially symmetric nucleus, are the β- and γ-vibrations. In the basic form of the RVM such calculations have been performed[19] for ^{238}U and the excitation energies in the ground state band are reasonably well reproduced at least up to spin 16. From the comparison of the calculated to the experimental E2-matrix elements it is obvious that the E2-transition strength is considerably overestimated above spin 12. Since this RVM calculation[19] also overestimated the E2-transition rates to the 2^+ members of the β- and γ-band one can assume that in these calculations the rotation-vibration interaction is too strong. On the other hand, our B(E2)-values are consistent with the assumption of a vanishing rotation-vibration interaction. We thus conclude that in ^{238}U at least a considerable part of the increase of the slope in I(ω) vs. ω cannot be attributed to centrifugal stretching effects.

Alternatively to this comparison to the geometric model it is interesting to compare the experimental results to calculations in the algebraic boson expansion approach as well, e.g. the Interacting Boson Approximation (IBA)[4]. As full IBA-calculations (off the limits) are not yet available for ^{238}U, the SU(3)-limit (rotational limit), which is expected to be closest to the results of a full calculation, will be used. In this limit the IBA predicts excitation energies E(I), which are given by the I(I+1)-rule; the deviations of E(I) from experiment at lower spins might be diminished by going away from the SU(3)-limit. The deviations at high spins, however, believed to be due to rotation alignment, cannot be reproduced as alignment effects are not included in the model which only allows for the coupling of nucleon pairs to bosons of L = 0 and L = 2. For the B(E2)-values the IBM predicts a strong reduction which approaches vanishing B(E2) above $I = 2n = I_{lim}$, where n is the number of bosons to be formed from the valence nucleons outside the next closed shell (for ^{238}U this is ^{208}Pb and consequently $n = 15$, $I_{lim} = 30$). Moreover, in the rotational limit the predicted E2-matrix elements decrease smoothly with spin reaching 75% of the rigid rotor value at around spin 22 (Fig.4). The observed E2-transition matrix elements do not show the predicted smooth decrease (not even below the onset of the rotation alignment) nor do they show any effect of the cut-off of the ground state rotational band for

$I \lesssim 30$ expected also in the case the ground state band changes its character around $I \approx 26$ due to the crossing of an aligned band. Although part of the discrepancies at lower spins might be removed in a full calculation, our data still seem to indicate that the configuration space of the IBA is not sufficient to describe high spin rotational states. A similar conclusion can be drawn from our data on the other actinide nuclei, which also show no significant deviation from the rigid rotor prediction[20], and which very clearly show no indication of cutoff of the collective bands at I_{lim}.

SUMMARY

Multiple Coulomb excitation using the very heavy projectiles like ^{208}Pb is a powerful experimental method for the investigation of high spin phenomena in heavy deformed nuclei as it allows the determination of the frequency dependence of the spin I and of the electric quadrupole transition strength. The fact, that the E2-matrix elements for the well deformed heavy Dy-isotopes and for the actinide nuclei studied do not significantly exceed the rigid rotor values excludes all models, which attribute the observed increase in apparent moment of inertia to a deformation change. A description of the smooth increase at low spin as well as the more sudden upbending occurring at higher rotational frequencie by Coriolis effects alone is consistent with our data for the frequency dependence of I and of the E2-transition strengths. In the actinide nuclei studied the E2-transitions between high spin states are not hindered as predicted by the IBA. A high spin cutoff in collectivity is not observed; in ^{232}Th and ^{234}U levels beyond the angular momentum limit as predicted by the IBA have been populated by Coulomb excitation.

ACKNOWLEDGEMENTS

The Coulomb excitation experiments reported here have been performed in a close and stimulating collaboration between the members of the nuclear spectroscopy ground at GSI, A. Balanda and R. Lulessa (Visitors from the Jagiellonian University at Cracow), H. Emling and F. Folkmann (Now at University of Arhus, Denmark), D. Schwalm, R.S. Simon and H.J. Wollersheim, and with outside users from the University of Frankfurt (Th. W. Elze, J. Idzko, H. Ower and K. Stelzer), the University of Mainz (N. Kaddrell and N. Trautmann), the University at Munich (D. Evers) and the Vanderbilt University at Nashville (J.H. Hamilton, A.V. Ramayya and R.B. Piercey). Thanks are also due to the UNILAC crew who put much effort in the production of the ^{208}Pb beams and to H. Folger for the preparation of numerous targets.

REFERENCES

1. Coulomb excitation, K. Alder and A.Winther, eds., New York 1966.
2. R.M. Lieder and H. Ryde, Adv. in Nucl. Phys., M. Baranger and E. Vogt, eds., New York, 10 (1978) 1.
3. F.S. Stephens, Rev. Mod. Phys. 47 (1975) 43.
4. A. Arima and F. Iachello, Ann. Phys. (N.Y.) 99 (1976) 253, ibid. III (1978) 201.
5. A. Bohr and B.R. Mottelson, Nuclear Structure, Vol. 2, Reading (Mass., USA) 1975.
6. E. Grosse, Symp, on High Spin Phenomena in Nuclei, (ANL-PHY-79-4), Argonne (III., USA), 1979.
7. E. Grosse et al., Proc. of the Nobel Symp. on Nuclei at Very High Spin, Orenas (Sweden) 1980, to be published in Physica Scripta.
8. H. Ower et al., Proc. of the Conf. on Nuclear Behaviour at High Angular Momentum Strasbourg (France), 1980, to be pub published in Journal de Physique.
9. R.B. Piercey et al., to be published in Phys. Rev. Lett.
10. R. Bengtsson and S. Frauendorf, Nucl. Phys. A314 (1979) 27 and ibid. A327 (1979) 139.
11. A. Faessler et al., to be published in Progress of Particle and Nuclear Physics.
12. P. Thieberger et al., Phys. Rev. Lett. 28 (1972) 972.
13. D. Ward et al., Nucl. Phys. A332 (1979) 433.
14. F. Kearns et al., Nucl. Phys. A278 (1977) 109.
Ph. Hubert et al., Phys. Rev. C17 (1978) 622
R.O. Sayer et al., Phys. Rev. C17 (1978) 1026
R.M. Ronningen et al., Phys. Rev. C16 (1977) 2208
15. H. Emling et al., Phys. Lett. 98B (1981) 169.
16. L.L. Riedinger et al., Proc. of the Nobel Symp. on Nuclei at Very High Spin, Orenas (Sweden) 1980. to be published in Physica Scripta and Phys. Rev. Lett. 44 (1980) 568.
17. R. Bengtsson, Proc. of the Conf. on Nuclear Behaviour at High Angular Momentum, Strasbourg (France) 1980 to be published in Journal de Physique.
S. Frauendorf, Proc. of the Nobel Symp. on Nuclei at Very High Spin, Orenas (Sweden) 1980, to be published in Physica Scripta.
18. B.R. Mottelson and J.G. Valatain, Phys. Lett. 5 (1960) 511.
19. V. Oberacker and G. Soff, Z. Naturforsch. 32A (1977) 1465.
20. H. Ower, Ph.D. thesis, Frankfurt, 1980.
H. Ower et al., to be published.
Ch. Michel et al., to be published.
21. C.E. Bemis, Jr., F.K. Gowan, J.L.C. Ford, Jr. W.T. Milner, P.H. Stelson, and R.L. Robinson, Phys. Rev. C Vol. 8, 4, 1466 (1973);
J.G. Alessi, J.X. Saladin, C. Baktash and T. Humanic, to be published.

ON THE TRANSITION FROM THE COHERENT TO THE STATISTICAL PHASE IN DEEP INELASTIC COLLISIONS

U. Smilansky

Department of Physics
Weizmann Institute of Science
Rehovot, Israel

S. Mukamel

Department of Physical Chemistry
Weizmann Institute of Science
Rehovot, Israel

D.H.E. Gross and K. Mohring

Bereich Kern und Strahlenphysik
Hahn Meitner Institut fur Kernforschung
D-1000, Berlin 39, West Germany

I ABSTRACT

We propose a theory in which the transition from the "approach" to the "contact" phase in DIC is treated in a unified and consistent way. We consider the coherent excitation of giant resonances as direct excitation and transfers as the dominant mechanism in the "approach" phase, and show how they are gradually replaced by the "statistical" particle transfers. The induced force on the relative motion which is due to the two types of interactions is calculated in a consistent way.

The partition of particles between the two collision partners is governed by equations which look initially like von Neumann's equations, but smoothly turn over to stochastic rate equations in the contact phase of the DIC.

II. INTRODUCTION

The cross-sections for reactions between complex nuclei display a continuous transition from the elastic (EL) and the quasi-elastic (QE) to the deeply inelastic (DI) processes. The experimental efforts which were originally directed to the DI components, recently started to shift towards the understanding of the transition from the DI to the QE and EL collisions. This was motivated by the realization that interesting entrance channel effects are apparent in the cross-sections, and that the distinction between the QE and the DI components is often arbitrary[1-3].

Most of the microscopic theories[4-5] which were put forward in order to explain the DI phenomenon concentrate on the "contact" phase of the collision, that is, the study of the binuclear complex after it was heated up to an excitation energy of 20-50 MeV. During the contact phase, the transfer of energy and angular momentum from the relative motion to the internal degrees of freedom, as well as the exchange of nucleons, can be viewed as a stochastic process. (Hence the name "statistical" phase). The heating up of the internal nuclear system occurs in the "approach" phase of the collision. The detailed description of this phase, as well as its matching to the contact phase is lacking in most theoretical studies. An exception is the approach developed by the Copenhangen group[9], where one follows the excitation of intrinsic giant resonance (GR) modes starting from two reactants in their ground states. This model lacks, however, a consistent treatment of mass transfer, which is introduced ad hoc in terms of a phenomenological diffusion model. Their treatment of fluctuations is also insufficient.

In this lecture we shall outline and discuss a new theoretical framework[17] which may fill the above-mentioned gap in our understanding of collisions between heavy ions. That is, the theory provides a consistent description of a dissipative reation in which the initial stages involve the coherent excitation of simple internal modes, followed by interactions with more complicated modes whose large phase-space and complexity naturally brings in the statistical elements.

III. THE MEAN TRAJECTORY

In a previous paper[6] we have shown how the semi-classical limit of Feynman's influence functional method allows a natural definition of a mean classical trajectory for the relative motion of the collision partners. Our previous discussion which was confined to rather simple dissipative mechanisms was recently extended[7] to the most general cases. The equations of motion for the classical trajectory can be summarized in the following way. Let $H(\vec{P}, \vec{R}, \pi, \xi)$ be the hamiltonian for the system in which the relative motion

$(\vec{P}, \vec{R})$ is coupled to the internal degrees of freedom (π, ξ). Before the collision $(t \to -\infty)$, the internal system is characterized by a given density matrix ρ_0. (For discussion of DIC ρ_0 reduces to $|0\rangle \langle 0|$ where $|0\rangle$ is the product wave function of the two collision partners in their ground state). The equations of motion read

$$\frac{d}{dt}\vec{R}(t) = \mathrm{Tr}\,\{\rho(t)\,\vec{\nabla}_P H(\vec{P}(t), \vec{R}(t), \pi, \xi)\} \tag{1}$$

$$\frac{d}{dt}\vec{P}(t) = -\mathrm{Tr}\,\{\rho(t)\,\vec{\nabla}_R H(\vec{P}(t), \vec{R}(t), \pi, \xi)\} \tag{2}$$

$$i\hbar \frac{d}{dt}\rho(t) = [H(\vec{P}(t), \vec{R}(t), \pi, \xi), \rho(t)] \equiv L(t)\rho(t) \tag{3}$$

The trace in (1) and (2) is taken over the internal system, where-as in (3) $H(\vec{P}(t), \vec{R}(t), \pi, \xi)$ is considered as an explicitly time dependent hamiltonian which drives the internal system. Its time dependence is dictated by the mean trajectory $(\vec{P}(t), \vec{R}(t))$. The simultaneous solution of equations (1-3) provides the informa tion about the mean trajectory, from which the inclusive cross-section can be calculated (see ref.6 for details). These equations of motion are essentially a semi-classical version of the "self-consistent mean field" approximation which is often used in various problems in statistical mechanics . It is known that in order to obtain this approximation one must neglect some correlations between the $(\vec{R}, \vec{P})$ and the (ξ, π) degrees of freedom. In the more recent formulation of the theory[18], we replace equations (1)-(2) by a generalized Fourier-Planck equation for the description of the relative motion, thus taking the fluctuation in the R, P variables into account. The self-consistency of (1)-(3) ensures, however, that those correlations which are required for the calculation of the inclusive cross-sections are maintained. Thus, the internal density matrix $\rho(t)$ is an auxiliary quantity which is needed for the purpose of calculating the mean trajectory. Still, $\rho(t)$ is related to the classical trajectory in a consistent way. Therefore it is expected to give a fair description of the distribution of observables in the internal system.

IV. THE INTERNAL SYSTEM AND THE INTERACTION HAMILTONIAN

The collisions of interest here are peripherial. Therefore we are allowed to expand the intrinsic states in the following basis. We partition the total number of protons and neutrons between two shell-model potential wells. We denote the partitions by $n \equiv (n_p, n_n)$, where n_p (n_n) stands for the excess of protons (neutrons) relative to the projectile in the entrance channel. A basis shell-model state $|n\nu\rangle_s$ is constructed as a product of two Slater determinants, and ν stands for all the quantum numbers required to specify the product state. We neglect continuum states. The vectors $|n\nu\rangle_s$ form an orthogonal basis as long as the two wells do not overlap considerably. The residual interaction has two components. The first, W, acts only within a given partition and is independent of the relative position of the fragments. Its effect is to mix the pure shell-model states. The other component $V(\vec{R}, \xi)$ is due to the interaction between nucleons in different wells, and it is responsible for inelastic and transfer processes. We approximate $V(\vec{R}, \xi)$ by a one-body force .

The entrance partition (n = 0) plays a special role. It was shown by Broglia et al.[9] that the coherent excitation of giant resonances is an efficient mechanism for the initial heating of the interaction W within the 1p-1h subspace in the n = 0 partition. We denote by $|I\rangle$, $|J\rangle$ etc. the coherent superpositions of 1p-1h states which have the collective characteristics of giant resonances. These states are only approximate eigen-states because of their coupling to the more complicated $|0\nu\rangle_s$ states via W. We diagonalize the intrinsic hamiltonian in each partition, and we get a new basis $|n\nu\rangle$. The hamiltonian for the system is

$$H = h_o + \frac{p^2}{2M} + V^{int}(\vec{R}) \tag{4}$$

$$h_o = \sum_I |I\rangle e_I \langle I| + \sum_{n\nu} |n\nu\rangle e_{n\nu} \langle n\nu| \tag{5}$$

$$\begin{aligned} V^{int}(\vec{R}) = & \sum_{IJ} |I\rangle V_{IJ}(\vec{R})\langle J| + \sum_{\substack{n\\m}} |n\nu\rangle V_{n\nu,m\mu}(\vec{R})\langle m\mu| \\ & + \sum_{I,N}\{|I\rangle V_{I,N\nu}(\vec{R})\langle n\nu| + \\ & + |n\nu\rangle V_{n\nu,I}(\vec{R})\langle I|\} + W \end{aligned} \tag{6}$$

$$W = \sum_{I\nu} \{|I\rangle W_{I,0\nu}\langle 0\nu| + |0\nu\rangle W_{0\nu,I}\langle I|\} \tag{7}$$

We denote the g.s. in the entrance channel by $|0\rangle$ and it is included in the I summation. For the sake of simplifying the presentation we do not discuss here the exchange of angular momentum.

V. THE REDUCED EQUATIONS OF MOTION

A detailed solution of eq.(3) is neither possible nor interesting. We shall therefore devise reduced equations of motion (REM) for a limited set of dynamical variables. An optimum choice of the reduction scheme will be the one that will provide all the relevant information needed for the particular application, and at the same time will result in the simplest possible form of the equations. The latter requirement is often related to the time scale on which the dynamical variables are changing. If one can choose a complete set of slow variables, the Markovian limit holds and the REM are considerably simplified.

The information that the REM should supply in the present model is (a) the mass and charge distributions, and (b) the induced force (r.h.s. of eq.(2)) on the relative motion due to internal excitations and transfer processes. Our set should therefore include at least the following dynamical variables,

$$a(n) \quad = \quad \sum_{\nu} |n\nu\rangle\langle n\nu| \tag{8}$$

and

$$F(n,\ m) \quad = \quad \sum_{\nu\mu} |n\nu\rangle V'_{n\nu,m\mu} \langle m\mu| \quad , \qquad (n \neq m) \tag{9}$$

where

$$V'_{n\nu,m\mu} \quad = \quad \langle n\nu| \frac{\partial V^{int}}{\partial R} |m\mu\rangle \tag{10}$$

An improved description can be obtained if the various partitions (especially those which are close to the entrance channel) would be sub-divided into energy bins. In order to clarify the presentation we shall not discuss this issue any further. The formalism which is presented here can be easily extended to deal with the generalized concept of a portion.

The expectation values of a(n) will be denoted by $P_n(t)$ (the population of the n'th partition). The expectation value of F(n, m) measures the induced force due to transitions from the n'th to the m'th partition and will be denoted by $f_{nm}(t)$ (forces).

The phenomenological analysis of DI experiments indeed shows that particle exchange is a slow mode once the contact phase is reached. But during the approach phase this is not the case and one should enlarge the set (8)-(9) in order to meet the requirements which underlie the Markovian limit. We therefore introduce the operators A(n,m),

$$A(n, m) = \sum_{\nu\mu} |n\nu\rangle V^{int}_{n\nu,m\mu}(R(t)) \langle m\mu| \qquad (n \neq m) \tag{11}$$

Their expectation values $S_{nm}(t)$ (the coherences) are related to the time derivatives of the $P_n(t)$, since they arise from the commutator of a(n) with the hamiltonian. They keep track of the mean relative phases between states in the n'th and the m'th partition, and hence their name-coherences.

Similar operators are defined also in relation to modes discussed in the preceding section. Thus, e.g.

$$a(I) = |I\rangle \langle I| \tag{8a}$$

$$A(I, J) = |I\rangle V_{IJ} \langle J| \tag{11a}$$

$$A(n, J) = \sum_{\mu} |n\nu\rangle V^{int}_{n\nu,J} \langle J| \tag{11b}$$

etc.

It is therefore obvious that a(I) and A(I, J) contain the complete quantal information on the density matrix in the subspace of the special states $|I\rangle$. In other words, our reduction refers only to the complicated subspace spanned by the $|n\nu\rangle$ states. Because of this reason, we shall write the equations of motion in terms of the elements of the statistical matrix ρ_{IJ} and not in terms of P_I and S_{IJ}.

The definitions (8)-(11) raise the important question - which intrinsic states in a given partition should be included in the sums. One can take two extreme views. Either that only a few simple particle-hole configurations are relevant, or that the complete spectrum of intrinsic states should be considered. This question is related to the estimate of the nuclear relaxation time relative to the collision time. This quantity is of great significance for the understanding of nuclear structure at high excitations, but its estimate is very difficult. As will be shown later, the theory proposed here depends critically on the choice of relevant states and it is expected that the comparison of the experimental data with cross-sections calculated under different assumptions could shed light on the question of internal nuclear relaxation. For the purpose of this presentation, we shall assume that the complete spectrum participates (short relaxation time). This view is partially supported by the analysis of ref.(12), and it allows us to make use of the methods of ref.(13) in order to evaluate the parameters which appear in the theory.

Independently of the choice of the type of states considered, we should also limit the summations (8, 9, 11) to states whose excitation energy is not in excess of the total available energy. The available excitation energy is a function of the partition and the distance between the collision partners. Hence it depends on the time. In order to simplify the formalism, we take the cut-off energy c(n) as the available energy at the Coulomb barrier in each partition.

$$c(n) = E_{tot} - \frac{e^2(Z_1Z_2)_n}{R_{Coul}(n)} + Q(n) - \frac{\hbar^2 \ell^2}{2MR^2_{Coul}(n)} \tag{12}$$

Here, Q(n) is the ground-ground Q-value relative to the entrance channel, and ℓ is the angular momentum of the relative motion.

The problem of the cut-off energy does not occur when we consider the generalization partitions which refer also to energy. In this case ć(n) lets the limit for the number of bins in each channel.

The derivation of the REM follows the ideas developed in ref.(10-11). It involves a few mathematical steps, which will be deferred to Appendix A. The physical arguments for the validity of the approximations will be discussed in the sequel. We shall present here a slightly simplified varsion of the theory, and write down the REM in terms of the populations $P_n(t)$ and the coherence S_{nm} only. This will be supplemented by a prescription which relates the forces f_{nm} to the coherences which is based on information theory and is presented in Appendix B.

Before quoting the REM, we shall introduce a few relevant quantities.

d_n: The statistical weight of the n'th partition, given by the number of states with excitation energy less than the cutoff energy c(n) (eq.(120 and the preceding discussion).

$\Omega_{nm}(t)$: The mean matrix element for a transition from the n to the m partition,

$$\Omega^2_{nm} = \frac{1}{\sqrt{d_n d_m}} \sum_{\nu\mu} |\langle n\nu|V^{int}(R(t))|m\mu\rangle|^2 \tag{13}$$

$\Omega_{in}(t)$: Defined in a similar way to Ω_{nm}, but relates to the transition from a special state I to the "statistical" spectrum $|n\nu\rangle$.

$$\Omega^2_{In} = \frac{1}{d_n} \sum_{\nu} |\langle n\nu | V^{int}(R(t)) | I \rangle|^2 \tag{14}$$

Note that in this case V^{int} gets contributions from two kinds of interactions. The first, V(R(t)), corresponds to dynamical decay of the I states by inelastic particle transfer processes. Since the ground state of the original system is the I = 0 state, the mechanism proposed by Gross[14] to account for friction is included in the present formalism. The second contribution comes from the residual interaction W (eq. (17)) and it is responsible for the spreading widths of the excited I states over the "statistical" levels in the n = 0 partition. We shall make the plausible assumption that V and W are not correlated so that cross terms in eq.(14) will be neglected.

I(n, m): This quantity is the correlation function

$$I(n,\, m) = \frac{1}{\Omega^2_{nm}} \cdot \frac{1}{\sqrt{d_n\, d_m}} \sum_{\nu\mu} \left| V^{int}_{n\nu,m\mu}(R(t)\right|^2 \times e^{i(e_{m\mu}-e_{n\nu})t/\hbar} \tag{15}$$

which plays an important role in determining the behaviour of S in time. (See Appendix A for details). We expect it to decay exponentially with a time constant h/Γ_{nm} superimposed on an oscillatory pattern with a mean frequency ω_{nm}. The exact definitions of these qauntities is given in Appendix A, while their physical significance will be explained below.

The expressions for the quantities defined above for the application to the DIC will be given in the next section.

The reduced equations of motion read:

$$ih \frac{d}{dt} \rho_{IJ} = (e_I - e_J)\rho_{IJ} + \sum_L (V_{IL}\rho_{LJ} - \rho_{IL}V_{LJ}) - \delta_{IJ} 2i\mathrm{Im} \sum_n S_{In} \tag{16}$$

$$i\hbar \frac{d}{dt} S_{In} = - \Omega^2_{In}(t)(d_n \rho_{II} - P_n) + (V_{II} + \omega_{In} - i\Gamma_{IN})S_{In} + i\hbar\dot{R}f_{In} \tag{17}$$

$$\hbar \frac{d}{dt} P_n = 2\mathrm{Im} \left(\sum_I S_{In} + \sum_m S_{mn} \right) \tag{18}$$

$$i\hbar \frac{d}{dt} S_{nm} = - \Omega^2_{nm}(t) \left(\left(\frac{d_m}{d_n}\right)^{\frac{1}{2}} P_n - \left(\frac{d_n}{d_m}\right)^{\frac{1}{2}} P_m \right) + (\omega_{nm} - i\Gamma_{nm})S_{nm} + i\hbar \dot{R} f_{nm} \tag{19}$$

We supplement these equations with the relation between the forces f_{nm} and the coherences S_{nm},

$$f_{nm}(t) = S_{nm} \frac{1}{\Omega_{nm}} \frac{d}{dR} \Omega_{nm} \tag{20}$$

and a similar expression when m is replaced by I.

Equations (16)-(20) should be solved simultaneously with the trajectory equation of motion

$$m\ddot{R} = - \sum_{\alpha\beta} f_{\alpha\beta} \tag{21}$$

where α, β go over all the n and I indices.

The entire problem is subject to the initial conditions

$$\rho_{IJ} = \delta_{IO} \cdot \delta_{JO} \; ; \quad P_n = S_{nm} = 0 \quad \text{as } t \to -\infty \tag{22}$$

We shall now follow the development of the system in time and discuss in their turn the equations of motion relevant for each stage

The earlier stage of the collision is controlled by eq.(16). Apart from the last term on its r.h.s., this equation describes the coherent excitation of the special modes which build the states I by the time dependent (nuclear+Coulomb) interaction $V(R(t))$. The last term is responsible for the decay of these states due to the residual interaction W and the dynamical excitation and transfer to complicated states. In order to understand how this is brought about, we should turn to eq.(17). It is an inhomogenous equation for the coherences S_{In} driven by the first term on the r.h.s. In this term we can safely neglect P_n relative to $d_n\rho_{II}$ since $d_n \gg 1$, and during the early stages of the collision $\rho_{II} \gg P_n$. Eq.(17) therefore decouples from the equations which follow it and S_{In} can be solved as a functional of ρ_{II}. S_{In} is then substituted in equations (16) and (18). To illustrate the role of S_{In}, let us discuss a simple model in which eq.(17) can be solved analytically. Consider a state I as a GR which spreads over the dense "statistical" levels via the interaction W, and ignore its dynamical decay. As long as $e_I \gg \Gamma_{I0}$, we can neglect ω_{In} and V_{II}, and also take $\sqrt{d_0}\Omega_{I0} \sim \Gamma_{I0}/2$. When the solution of eq.(17) is substituted in eq.(16), the last term on r.h.s. becomes $-\delta_{IJ} i\Gamma_{I0}\rho_{II}$. That is, we get a "sink" term through which probablity is depleted from the I states and transferred to the complicated spectrum of the n=0 partition.

It is worthwhile mentioning that if we chose the states I to describe surface phonons, this simple model provides the link between our approach and the one proposed by the Copenhagen group[9]. The solution of the simplified eq.(16) and the corresponding equatiortions in ref.(9) coincide only to first order in V. The present model has the advantage that we do not have to impose any relation between the widths of states which differ by their "phonon numbers". Such correlations are inherent to the model of ref.(9) because of their strict adherence to the simple harmonic model. The main point of the present model (in its general form) is that it describes in a coherent way the decay of the I states due to dynamical processes as well as to the spreading due to W.

The "sink" term in eq.(16) appears with an opposite sign as a "source" in eq.(18). Thus, the total probability is conserved (unitarity). The fact that $\rho_{II}d_n \gg P$ also indicates that the probability flux is directed from the simple I space to the complicated $|n\nu\rangle$ space and that the return flux can be neglected.

As the two collision partners approach each other the interaction gets stronger and the populations P_n grow while the ρ_{II} are damped. Equations (16) and (17) become less important while eqs.(18) and (19) begin to take over. Eq.(18) is an exact relation which is due to our particular choice of dynamical variables. In order to investigate eq.(19), we should first discuss the physical meaning of the coherences S_{nm}. They can be looked upon as wave packets of off-diagonal matrix elements which oscillate with different frequencies. Therefore, once S_{nm} is prepared, it will die out after a time which is inversely proportaional to the width of the frequency band. The second term on the r.h.s. of eq.(19) takes care of the "free" dephasing of the S_{nm}; ω_{nm} is the mean transition frequency.

The first term on the r.h.s. of (19) is the driving term, and is proportional to the population flux in phase space. We can easily convince ourselves that the set

$$S_{nm} = 0$$
$$P_n = \text{const. and } P_n/P_m = d_n/d_m \tag{23}$$

is a solution of eq.(18)-(19) which describes the equilibrium state of the system. We see therefore, that our REM drive the system to equilibrium. When equilibrium is achieved the force f_{nm} (eq.20) vanishes, as would be expected.

The approach to equilibrium is characterized by the fact that the driving term in eq.(19) changes slowly relative to the dephasing rate $\hbar/\Gamma_{nm}$. During this stage, one can approximate the solution of (19) by

$$S_{nm} \approx \frac{1}{\omega_{nm} - i\Gamma_{nm}} \Omega^2_{nm} \left[\left(\frac{d_m}{d_n}\right)^{\frac{1}{2}} P_n - \left(\frac{d_n}{d_m}\right)^{\frac{1}{2}} P_m \right] \tag{24}$$

When eq.(24) is substituted in (18) one gets
When eq.(24) is substituted in (18) one gets

$$\hbar \frac{d}{dt} P_n = \sum_m \frac{2\Gamma_{nm}}{\omega^2_{nm} + \Gamma^2_{nm}} \Omega^2_{nm}$$
$$\times \left[\left(\frac{d_n}{d_m}\right)^{\frac{1}{2}} P_m - \left(\frac{d_m}{d_n}\right)^{\frac{1}{2}} P_n \right] + 2\mathrm{Im} \sum_I S_{In} \tag{25}$$

Thus, during the contact phase, the formalism naturally reduces to simple rate (master) equations. The rate by which equilibrium is achieved depends on Ω^2_{nm}, whose time dependence is governed by the rate of change of $V(R(t))$. Hence the equilibration process can be terminated once the relative motion forces the reactants to separate before equilibrium was reached.

VI. APPLICATION TO DIC

In this section we shall provide more information on the various quantities which appear as parameters in the equations of motion (16)-(20).

As far as the GR modes are concerned, we can use the surface phonons of ref.(9) to represent the salient features. The matrix elements $V_{IJ}(R)$ and the width of the dominant GR are given in detail in ref.(9) and we shall not discuss them here any further.

The statistical weights d_n are obtained by integrating the Fermi-gas level density formula[15]. After a few simplifications we get

$$d_n = e^{2\sqrt{ac(n)}}/(48ac(n))^{1/2}, \qquad a = (A_1 + A_2)/8 \tag{26}$$

In calculating the remaining parameters, we rely on the work of ref.(13), and use the basic relation

$$\overline{|\langle n\nu|V(R)|m\mu\rangle|^2} \cong \frac{1}{\sqrt{\rho(\varepsilon_{n\nu})\rho(\varepsilon_{m\mu})}} \cdot g(\varepsilon_{n\nu} - \varepsilon_{m\mu}) \cdot \phi(R)$$

where $\varepsilon_{n\nu}$ are excitation energies in the n partition,

$$g(x) = \frac{1}{\pi}\frac{\Gamma}{x^2 + \Gamma^2} \qquad \Gamma \sim 7 \text{ MeV} \tag{27}$$

and $\phi(R)$ is an overlap function which was determined numerically in ref.(13).

Notice that we use the Lorentzian rather than the Gaussian form for g(x) (see ref.(13)). With these forms, and some few semi-analytical steps we obtain the following estimates:

$$\Omega^2_{nm} \cong 2\left(\frac{c(n)c(m)}{a^2}\right)^{\frac{1}{4}} \cdot g(c(n) - c(m)) \cdot \phi(R) \tag{28}$$

$$\omega_{nm} \approx Q(n) - Q(m) \quad , \qquad \Gamma_{nm} = \Gamma \tag{29}$$

and

$$\Omega^2_{In} \simeq \frac{1}{d_n} \phi(R)/2 \quad , \qquad n \neq 0 \tag{30}$$

Numerical calculations of eq.(16)-(20) are now under way.

VII. SUMMARY

In the preceding sections we presented a theoretical framework for the description of the initial stages of the DI process. The theory should be now tested by comparing its prediction to measured QE cross-sections.

In the present state of the work, and for the purpose of this presentation, we developed the theory in terms of the minimal set of dynamical-variables. More details and accuracy can be achieved if the set is to be increased. We mentioned already that coarse graining in excitation energy for each partition may be called for. Our indices n will then stand also for the energy bins, and the formal structure of the theory will not be altered. The same is true also for the inclusion of angular momentum. By enlarging the space of dynamical variables we may soon encounter practical computational problems. It should be noted, however, that the present theory is built for the description of the initial phase of DI collision QE events, where the distributions did not yet develop to cover all the available phase-space.

The most serious limitation of the present theory is due to our use of the diabatic basis. Thus, deformations and neck formation are not taken into account and therefore the theory cannot describe the last stages of genuine DI events. A new version of the theory which is based in the adiabatic basis is now being prepared for publication[18].

APPENDIX A

In this Appendix we shall derive the REM for the expectation values of the operators a(n) and A(n,m) defined in eqs.(8) and (11). We shall use standard projection-operator techniques that are commonly used in statistical mechanics.

It is convenient to normalize the operators a(n) and A(n,m) in the following way:

$$\hat{a}(n) = \frac{1}{d_n^{1/2}} a(n) \tag{A-1}$$

$$\hat{A}(n, m) = \frac{1}{\gamma_{nm}} A(n, m) \tag{A-2}$$

where d_n is the number of $|n\nu\rangle$ states, as discussed in section 3. The γ_{nm} are related to Ω_{nm} of eq.(13), that is:

$$\gamma^2_{nm}(t) = \Omega^2_{nm}(t)d_n d_m \equiv Tr(A^+(n, m)A(n, m)) \quad \text{(A-3)}$$

An important ingredient in the present formalism is the statistical assumptions concerning matrix elements of V^{int} between various $|n\nu\rangle$ states. It allows us to neglect terms of the type $Tr[A(n,m)A(m,\ell)A(\ell,n)]$. (See eq.a-4). The normalized operators satisfy

$$Tr[\hat{A}^+(n, m)\hat{A}(k, \ell)] = \delta_{nk}\,\delta_{m\ell} \quad \text{(A-4)}$$

$$Tr[\hat{A}^+(n, m)\ a(\ell)] = 0 \quad \text{(A-5)}$$

Together with the trivial relation

$$Tr[\hat{a}(n)\ \hat{a}(m)] = \delta_{nm} \quad \text{(A-6)}$$

we see that the operators of interest form an orthonormal set under the trace operation. Unless otherwise indicated, we shall consider the indices n,m etc., as referring not only to partitions but also to the special states I,J etc. Care must be taken in order that (A-4)-(A-6) be satisfied also when the I states are involved. Indeed, we require that

$$\sum_\nu W_{I,n\nu}\, W_{n\nu,J} = \delta_{IJ} \cdot \text{const.} \quad \text{(A-7)}$$

$$\sum_\nu V_{I,n\nu}\, W_{n\nu,J} = 0 \quad \text{(A-8)}$$

The first relation is equivalent to the assumption of non-overlapping resonances, whereas the second relation expresses the assumption that V and W are not correlated.

We define a projection operator P on the space spanned by the operators with which we are concerned

$$P\ \cdot = \sum \hat{a}(n)\ Tr[\hat{a}(n)\ \cdot] + \sum_{nm} \hat{A}(n, m)\ Tr[\hat{A}^+(n, m)\cdot] \quad \text{(A-9)}$$

Thus

$$P_n(t) = Tr[\hat{a}(n)P\rho] \quad d_n^{1/2} \tag{A-10}$$

$$S_{nm}(t) = Tr[\hat{A}^+(n, m)P\rho]\gamma_{nm} \tag{A-11}$$

The starting point for the derivation of the REM is the Liouville equation (3) which governs the developemnt of the intrinsic system. Let $U(t,t_o)$ be the propagator such that

$$\rho(t) = U(t, t_o)\rho(t_o) \tag{A-12}$$

Since at $t=t_o$, $\rho(t_o) = |0><0|$, we can write

$$P\rho(t) = (PU(t, t_o)P)\rho(t_o) \qquad \text{A-13)}$$

and

$$\rho(t_o) = (PU(t, t_o)P)^{-1} \cdot P\rho(t) \tag{A-14}$$

On the other hand,

$$i\hbar\frac{d}{dt}(P\rho) = PLP\rho + PL(1 - P)\rho + i\hbar\,\frac{dP}{dt} \cdot \rho \tag{A-15}$$

where the last term appears because of the explicit time dependence of P. Since $\rho = U\rho(t_o)$, we make use of (A-14) and finally obtain the exact relation

$$\begin{aligned} i\hbar\,\frac{d}{dt}\,(P\rho) &= PLP(P\rho) + (PL(1 - P)UP)(PUP)^{-1}(P\rho) \\ &\quad + i\hbar\,\frac{dP}{dt}\,\rho \end{aligned} \tag{A-16}$$

If now we use the definition (A-9) of P, and apply (A-10) and (A-11) we can get the required REM. Hence, (A-16) is the basic equation in the present formalism and we would discuss it term by term.

The first term on the r.h.s. of (A-16) corresponds to the non-fluctuating driving of P . (Note that when P=1, only this term survives and eq.(A-16) becomes identical to eq.(3). This expresses the fact that fluctuations arise only when a genuine reduction of information ($P \neq 1$) is considered). The second term on the r.h.s. of eq.(A-16) depends on the exact propagator U and involves the operat (1-P). Hence it contains all the effects which are due to dynamica correlations between the P-space and the compelmentary (1-P) space. In other words, it is responsible for the delayed response which is present whenever one develops a scheme for reducing the amount of information. The calculation of U is synonymous to a complete solution of eq.(3) - which we try to avoid. Hence, this term will be further approximated in order to get a simpler but more tractabl form for the equations. The last term on the r.h.s. brings in a new source of trouble since dP/dt does not necessarily project on the P-space, and hence (A-16) is not a closed system. We shall introduce below the approximations which are necessary to simplify eq.(A-16) and bring it to a useful form.

The first step in this direction is to replace the exact propagator U by the free propagator $U_o(t, t_o)$,

$$U_o(t, t_o) = e^{-iL_o(t-t_o)/\hbar} \tag{A-17}$$

L_o is the Liouville operator corresponding to the unperturbed Hamiltonian h_o. This is a drastic approximation, but it affects only the 2nd term on the r.h.s. of (A-16). Therefore, our description of the response of the P space dynamics ot fluctuations in the 1-P space will be only approximately described. As will become clearer below, the main ingredients of this response are preserved.

Once (A-17) is substituted for U in (A-16) the REM are readily obtained. We shall demonstrate the process by deriving eq.(19). W shall make use of the following identities.

$$\mathrm{Tr}(\hat{a}(n)U_o\hat{a}(m)) = \delta_{nm} \tag{A-18}$$

$$\mathrm{Tr}(\hat{A}^+(n, m)U_o\hat{A}(k,\ell)) = \delta_{nk}\delta_{m\ell}I_{nm} \tag{A-19}$$

$$I_{nm} = \frac{1}{\gamma_{nm}^2} \sum_{\nu\mu} |V_{n\nu,m\mu}^{int}|^2 \exp(-i(e_{c\nu} - e_{m\mu})t/\hbar) \tag{A-20}$$

$$\mathrm{Tr}(\hat{A}^+(n, m)U_o\hat{a}(k)) = 0 \tag{A-21}$$

$$\mathrm{Tr}(\hat{a}(n)LU_o\hat{A}(m, \ell)) = \frac{1}{d_n^{1/2}} [\gamma_{nm} I_{mn} \delta_{n\ell} - \gamma_{\ell n} I_{n\ell} \delta_{nm}] \tag{A-22}$$

$$\mathrm{Tr}(\hat{A}^+(n, m)LU_o\hat{A}(k, \ell)) = i\hbar\delta_{nk} \delta_{\ell m} \dot{I}_{nm} \tag{23}$$

$$\overset{(\cdot)}{I}_{nm} = \frac{1}{\hbar\gamma_{nm}^2} \sum_{\nu\mu} (e_{n\nu} - e_{m\mu}) \, |V_{n\nu,m\mu}|^2 \qquad \text{(A-24)}$$

$$\times \exp(-i(e_{n\nu} - e_{m\mu})t/\hbar)$$

where in eq.(A-21) and (A-24) we used the statistical assumption that the mean value of odd powers of V vanish.

With the help of the above relations and equations (A-10) (A-11), we get

$$i\hbar \frac{d}{dt} S_{nm} = \Omega_{nm}^2 \left\{ P_m \left(\frac{d_n}{d_m}\right) - P_n \left(\frac{d_m}{d_n}\right) \right\} \qquad \text{(A-25)}$$

$$+ i\hbar \frac{\overset{(\cdot)}{I}_{nm}}{I_{nm}} S_{nm} + i\hbar \dot{R} f_{nm}$$

The last term is obtained from the (dP)/(dt) term in (A-16). Since the time scale for the relative motion is slow, we can consider the coefficient of S_{nm} as the logarithmic derivative of the correlation function I_{nm}. Equation (19) follows when we use the simplest possible ansatz for this logarithmic derivative ($\dot{I}_{nm}/I_{nm} \sim i[\omega_{nm} - \Gamma_{nm}/\hbar)$.

Equations (16)-(20) are time-local equations. In other words, we derived a Markovian approximation for the REM. The explicit appearance of the logarithmic derivative of I_{nm} indicates that the measure of the "coarse graining" of the time variable is finer than $\hbar/\Gamma_{nm}(\sim 10^{-22}$ sec).

The effect of the inclusion of higher order corrections to U would introduce memory effects on much smaller scale and therefore will be neglected. Higher order terms would also modify the parameters ω_{nm}, Ω_{nm} and Γ_{nm}. Still, the main feature that the coherences oscillate and decay on the time scale mentioned above is not expected to be altered. A more detailed discussion of these effects can be found in ref.(10).

APPENDIX B

In this appendix we shall derive the relation (eq.20) between the forces f_{nm} and the coherences S_{nm}. This will be achieved by writing an approximate expression for the density matrix, subject to the following requirements. (a) The expectation values of the operators a(n) and A(n,m) should take the values $P_n(t)$ and $S_{nm}(t)$ respectively, where P_n and S_{nm} are the solutions of eqs.(16)-(20) ate each time. (b) The construction of the approximate density matrix $\tilde{\rho}(t)$, does not require any other information beyond the constraints listed in (a). Information theory[16] provides us with the following method of construction. Consider

$$\tilde{\rho}(t) = \exp\left\{\Lambda - \sum_n \mu_n a(n) - \sum_{mn} \lambda_{nm} A(n, m)\right. \qquad \text{(B-1)}$$

where

$$\Lambda = -\ell g\left\{\mathrm{Tr}[\exp(-\sum_n \mu_n a(n) - \sum_{mn} \lambda_{nm} A(n, m))]\right\} \qquad \text{(B-2)}$$

The coefficients μ_n and λ_{nm} are Lagrange multipliers which are determined from the relations,

$$\frac{\partial\Lambda}{\partial\mu_n} = P_n(t) \qquad \text{(B-3)}$$

$$\frac{\partial\Lambda}{\partial\lambda_{mn}} = S_{nm}(t) \qquad \text{(B-4)}$$

When the solution of (B-3) (B-4) is substituted in (B-1), we get the desired approximation for the density matrix. Once this is done, the expectation value of any operator can be evaluated, and in particular, that of the force operators F(n,m). We solve eqs.(B-3,4) approximately by expanding (B-2) in power series in the λ_{nm}. Taking into account the orthogonality property of the a(n) and A(n,m) operators under the trace operation, we see that the terms linear in λ_{nm} vanish. Thus, to second order in the λ_{nm}

$$e^{-\Lambda} \simeq \sum_n d_n \exp(-\mu_n) - \sum_{n>m} \lambda_{nm}\lambda_{mn}\Omega_{nm}^2 T(n, m) \qquad \text{(B-5)}$$

where

$$T(n, m) = \left\{P_n \left(\frac{d_m}{d_n}\right)^{\frac{1}{2}} - P_m \left(\frac{d_n}{d_m}\right)^{\frac{1}{2}}\right\} \Big/ \left\{\ell g(P_m/d_n) - \ell g(P_m/d_n)\right\} \qquad \text{(B-6)}$$

Substituting (B-5) in (B-4) we get

$$S_{nm} = T(n, m)\Omega^2_{nm}\lambda_{mn} e^{\Lambda} \qquad \text{(B-7)}$$

which defines λ_{mn} in terms of the momentary values of S_{nm} and P_n as supplied by the REM. The expression for the μ_n multipliers is needed for the complete evaluation of Λ, but in the present application this will not be necessary. The approximate expression for f_{nm} is

$$f_{nm} = \mathrm{Tr}[\tilde{\rho}F(n, m)] \qquad \text{(B-8)}$$

To lowest order in the λ_{nm},

$$f_{nm} = T(n, m)\,\frac{\sum_{\nu\mu} V'_{n\nu,m\mu} V_{m\mu,n\nu}}{\sqrt{d_n d_m}}\,\lambda_{mn}\, e^{\Lambda}$$

comparing (B-7) and (B-9) we get the desired result

$$f_{nm} = \frac{1}{\Omega_{nm}} \cdot \frac{d\Omega_{nm}}{dR} \cdot S_{nm} \qquad \text{(B-10)}$$

It should be noted, that the relation (B-10) holds exactly when the interaction $V^{int}(R)$ can be factorized, that is, when

$$V^{int}_{n\nu,m\mu}(R) = U(R) \cdot \tilde{V}_{n\nu,m\mu} \qquad \text{(B-11)}$$

REFERENCES

1. T. Tanbe et al., GSI preprint 80-5 (1980).
2. M. Dakowski et al., GSI preprint 79-16 (1979).
3. W.H. Schroder, Proc. Int. Symp. on Continuum Spectra, in in H.I. Reactions, San Antonio, USA (1979).
4. H.A. Weidenmuller, Transport theories of DI reactions, in "Progress in Particle and Nuclear Physics", D. Wilsinson, editor, Pergamon Press.
5. D.H.E. Gross, Z. Physik A291 (1979) 145.
6. K. Mohring and U. Smilansky, Nucl. Phys. A338 (1980) 227.
7. K. Mohring and U. Smilansky, in preparation.
8. C.B. Willis and R.H. Picard, Phys. Rev. A9 (1974) 1343.
9. R.A. Boglia, C.H. Dasso and A. Winther, in Int. School of Physics LXXVII, Varenna 1979.
10. S. Mukamel, J. Chem. Phys. 70 (1979) 5834.

11. S. Mukamel, Adv. Chem. Phys. (1980), in press.
12. Y. Alhassid, R.D. Levine, J.S. Karp and S.G. Steadman, Phys. Rev. C20 (1979) 1789.
13. B.R. Barrett, S. Shlomo and H.A. Weidenmuller, Phys. Rev. C17 (1978) 544, and ibid, C20 (1979) 1.
14. D.H.E. Gross, Nucl. Phys. A240 (1975) 472.
15. A. Bohr and B.R. Mottelson, Nuclear Structure, Vol. I, Appendix 2B, W.A. Benjamin, Inc. (1968).
16. A. Katz, Principles of Statistical Mechanics, The Information Approach. W.H. Freeman (1967).
17. U. Smilansky, S. Mukamel, D.H.E. Gross and K. Mohring , Weizmann Inst. Report WIS-80/21/June-ph.
18. D.H.E. Gross, M. Sobel, U. Smilansky, S. Mukamel and K. Mohring K. Mohring, in preparation.

EXCITATION OF SHAPE-VIBRATIONAL MODES IN NUCLEI BY RELATIVISTIC HEAVY IONS*

J.O. Rasmussen

Nuclear Science Division
U.C. Lawrence Berkeley Laboratory
Berkeley, CA 94720

J.S. Blair

Physics Department
University of Washington
Seattle, WA 98195

X.J. Qiu

Nuclear Research Institute
Academia Sinica
Shanghai, People's Republic of China

I. ABSTRACT

Data on excitation of collective 2^+ and 3^- states in relativistic ($\geq$250 MeV/N) heavy-ion collisions are reviewed. Indirect evidence on excitation to the giant quadrupole region of ^{16}O at 2 GeV/N is put forward. Our theoretical approach uses the diffuse-edge diffraction model. We use a microscopic model to relate the elemental nucleon-nucleon total cross section to the nuclear transparency parameters for the diffraction model. The integrated inelastic cross sections are approximately inversely proportional to the thickness of the diffuse partial transmission region. Coulomb excitation of giant quadrupole states in the projectile appears to be negligible in comparison to the excitation produced by nuclear interactions.

*This work was supported in part by the U.S. Department of Energy under contract W-7405-ENG-48.

II. INTRODUCTION

With the availability in the early 70's of heavy ion beams in the 0.1-2 GeV/N range a new area of direct reaction studies opened. We want in this paper to concentrate on the simplest of these reactions, inelastic scattering to shape-vibrational states. Since the collision times will be short compared to the vibrational period, even for giant resonances, the sudden approximation can apply. Furthermore, for light nuclei the Coulomb field effects are negligible, so grazing trajectories can be taken with uniform velocity. Under these conditions the simple shadow scattering diffraction models should be quite applicable.[1,2]

The direct data are as yet rather sparse, since energy resolution of Bevalac beam and detectors have not been sufficient for direct measurement of inelastic scattering, and indirect measurements are required. Shibata, et al.[3] have measured gamma-ray yields from 0.25 and 0.40 GeV/N ^{12}C ions on a few targets. Their gamma transition yields give only an upper limit to the direct excitation of the state, since the fraction of yield due to cascade transitions from higher levels is typically quite uncertain, although cascade transitions themselves were generally too weak to be seen. The 3^-- to-ground state yield in ^{40}Ca is found to be less subject to uncertainties concerning the excitation and cascading of higher excited states and to thick target corrections than the 2^+ states of the nuclei measured. Thus, we examine just the 3^- data.

Before going into theoretical analysis of these results, we consider evidence for relativistic heavy ion excitation of shape vibrational states above the nuclear binding energy. Heckman and Lindstrom[4] have made a strong case for Coulomb excitation of ^{12}C and ^{16}O projectiles to the giant dipole resonance (GDR). They show excess yield for high-Z targets of fragmentation products derived from the projectile by loss of one nucleon.[5] The idea is that in addition to the normal one-nucleon knock-out yield there is extra yield from the 2-step mechanism via the isovector GDR.

III. GIANT QUADRUPOLE RESONANCE AS REACTION INTERMEDIATE

If the giant dipole resonance can be an important reaction intermediate, why not the isoscalar giant quadrupole strength under certain circumstances? In this connection we note interesting anomalies in the fragmentation yield and longitudinal momentum ($P_{||}$) dispersion of ^{12}C formed from ^{16}O projectiles at 2.1 GeV/N. The momentum dispersion, as measured by Greiner, et al.[6] is abnormally small compared to the systematic behavior of other neighboring products. Furthermore, the yield of ^{12}C is abnormally high and in excess of the quantitative abrasion-ablation calculations of Hufner, et al.[7] This high yield occurs with all targets from Z=1 to 82, so cannot be associated only with Coulomb excitation, as with the A_p-1 fragments. In contrast to the isovector giant dipole resonance, the isoscalar modes can undergo excitation in collision of T=0 partners via the nuclear force field or, equivalently, by shadow diffraction scattering.

The nucleus ^{16}O has been extensively studied with respect to isoscalar giant quadrupole strength and decay properties. Knöpfle, et al.[8] showed by inelastic scattering of 146-MeV alpha particles that there is a clear concentration of giant quadrupole resonance strength in ^{16}O excited states between 15.9 and 27.3 MeV, exhausting $\sim$65% of the isoscalar energy weighted sum rule. In a later paper Knöpfle, et al.[9] showed the dominant decay of the GQR states to be via alpha decay to the first excited state (4.44 MeV) of ^{12}C. Alpha decay to ground is also significant, with the proton decay mode very small.

With the above in mind Qiu and Rasmussen suggested[10] the anomalies in ^{12}C fragmentation products from ^{16}O could result from prominence of a two-step formation process via the GQR states. Very recently we have seen the ^{16}O fragmentation results of Sasagase, et al.[11] with 88-MeV ^{16}O on ^{27}Al. They measured both outgoing ^{12}C and α in coincidence and showed dominant contributions from discrete excited states at 10.0 (weak), 11.6, 13.2, 15.2, 16.2, and 21 MeV. It may be that the low (<5 MeV/N) bombarding energy skews the participation toward states at the low end of the GQR region. With relativistic energies it seems likely that all states in the GQR region would participate proportional to their strengths, as in the higher velocity (α,α') excitation[8,9].

Let us first test the hypothesis of significant GQR excitation in the 2.1 GeV/N ^{16}O beams by looking at consequences to momentum dispersion. Consider the states around 21 MeV in ^{16}O (centroid of the GQR). The energy release in alpha decay to ^{12}C (gd) is 13.8 MeV and that to ^{12}C (2^+) is 9.4 MeV. Three-fourths of this energy goes to the alpha particle, resulting in alpha particle momenta in the projectile frame of 278 MeV/c and 229 MeV/c, respectively, for decay to ground and excited state ^{12}C. For isotropic decay the dispersion in $P_{||}$ will be $\pi/4$ times the momenta above, respectively, 218 MeV/c and 180 MeV/c. This result at first appears unfavorable to the hypothesis of GQR participation, since the measured $P_{||}$ dispersion is only 120 $\pm$ 4 MeV/c with the expected systematics $\sim$150 MeV/c. However, in Ref.9 the anisotropy of the alpha decay of ^{16}O after (α,α') reactions was shown to be very large for both α_0 and α_1 groups groups, favoring decay parallel to the recoil direction. At 2.1 MeV/N the recoil direction of the $^{16}O^*$ will be essentially perpendicular to the beam. Hence, the GQR sequential process should show a small dispersion in $P_{||}$ and a large dispersion in $P_{\perp}$. Furthermore, the GQR mechanism should give a non-Gaussian shape with shoulders for the momentum distribution. The latter dispersion has yet to be measured, but it should provide a crucial test to the hypothesis of the GQR sequential process.

That there is excess yield of ^{12}C from ^{16}O can only be claimed by reference to specific theoretical calculations. We take for comparison Hüfner's version (C) of the abrasion-ablation model[7] with final state interaction, since it fits almost exactly the 29 mb production cross section of $^{9}Be\ (^{16}O,\ ^{13}C)\ X$ and is generally satisfactory. This theoretical calculation gives 33 mb for ^{12}C formation, in contrast to an experimental yield of 60.8 ± 4.9 mb. Thus we may surmise ∿30 mb cross section arising from other processes, perhaps in part the GQR excitation followed by alpha decay. We now explore this hypothesis.

IV. DIFFRACTION MODEL

Let us now apply the shadow diffraction scattering model of Refs.1 and 2. In this model inelastic scattering to isoscalar shape vibrational states increases with the strength function δ_L^2 (the square of the nuclear radius times the zero-point amplitude β_L of the 2^L pole vibration). The differential cross section is expressible in terms of Bessel functions, the period of the diffraction oscillations being inversely proportional to the product of wave number and nuclear radius. The fall-off of the envelope with angle is slow for a sharp absorption shadow surface and more rapid the more diffuse the shadow.

$$\frac{d\sigma_{0 \to L}}{d\Omega} = (kR)^2 \frac{\delta_L^2}{4\pi} \sum_{-L \leq M \leq L} [L:M]^2 J_M^2(kR\theta) F^2(y) \tag{1}$$

In this formula, δ_L ($\equiv \beta_L R$) is the familiar deformation length, k is the wave number for the relative motion of the projectile and the target, R is the strong absorption radius, and

$$[L:M] = \begin{cases} i^L \dfrac{[(L+M)!\ (L-M)!]^{1/2}}{(L+M)!!\ (L-M)!!}, & \text{for } (L+M) \text{ even} \\ 0, & \text{for } (L+M) \text{ odd} \end{cases} \tag{2a}$$

For a partial survival amplitude $\eta(b)$ of the Fermi form,

$$\eta_F(b) = \left[1 + \exp\left(\frac{R-b}{d}\right)\right]^{-1} \tag{2b}$$

so that d parameterizes the diffuseness of the shadow, the damping factor F(y) has the form

$$F(y) = \frac{\pi y}{\sinh \pi y} \qquad (2c)$$

with $y=kd\theta$.

This differential cross section can be integrated numerically over angle to obtain the total inelastic cross section, σ_L, for the transition in question. Alternatively, an approximate analytic exp- expression has been found[12] for σ_L when the Bessel functions are approximated by their (sinusoidal) asymptotic forms,

$$\sigma_L \approx \frac{\delta_L^2}{12} \frac{R}{d} \qquad (3)$$

Comparison to values obtained by numerical integration shows that this formula is amazingly accurate. The exact values can be written as the product of the cross section given by eq.(3) times a factor G_L which is a slowly varying function of the ratio d/R and L. For d/R=0.05, a value typical of low energy α scattering, G_L drops from unity to 0.9 as L varies from 0 to 5. For the more diffuse shadows, d/R=0.10 and 0.15, G_L varies from unity to 0.7 and 0.5, respectively, for the same range of L. In subsequent numerical work, the factor $G_L(d/R)$ is taken to account.

We now proceed to apply this model to the apparent excess cross section for producing ^{12}C in ^{16}O-^{9}Be collisions. The cross sections[8] for inelastically exciting quadrupole states in ^{16}O in the 11.5-27.3 MeV interval with 150 MeV α-particles and the branching ratios[9] for subsequent α-decay of these states imply that the square of the effective deformation length for this mode of excitation and decay is

$$\delta^2_{2,eff} \equiv \sum \left(\beta_2 R \frac{\Gamma_\alpha}{\Gamma}\right)^2 = 1.20 \text{ fm}^2$$

Attributing the apparent excess cross section for producing ^{12}C in ^{16}O-^{9}Be collisions, 30 mb, entirely to the process of quadrupole excitation and subsequent α-decay, we may use the above diffraction model expressions for angle-integrated cross sections to solve for the remaining unknown quantity, the ratio d/R in ^{16}O-^{9}Be collisions. The result is rather small value for the lower limit on the ratio, $(d/R) \gtrsim 0.033$. When one assigns to the radius parameter the value $1.2\ (16^{1/3}+9^{1/3})=5.52$ fm, one obtains $d \gtrsim 0.18$ fm.

Let us see how this limit for the diffuseness compares with that estimated previously[12] for ^{12}C-^{40}Ca collisions at 0.4 GeV/N. The inputs to this latter estimate have been the observed[3] cross section, 14.4 $\pm$ 4.4 mb, for producing the 3.73 MeV γ-rays which result from decay of the first 3^- state of ^{40}Ca, values of deformation lengths for the particle-stable states in ^{40}Ca obtained from analysis of low-energy scattering experiments, and the γ-ray branching ratios of the particle-stable states. When the diffraction model is used for the angle-integrated cross sections to the particl stable states, one finds the ratio, d/R=0.10 $\pm$ 0.02. With a guessed value for the radius parameter, R=6.0 fm, one has d$\sim$0.6 fm. Perhaps more meaningful is the quantity 3.69 d, the distance over which the transparency changes from 0.1 to 0.9, which is here estimated to be 2.2 fm. A similar analysis of the γ-ray cross section[3] for 0.4 GeV/N α-particles on ^{40}Ca yields a value for the ratio, d/R= 0.16 + 0.03.

There are also available differential cross section data[13] from the Saturne facility for 0.34 GeV/N α-particles scattered from ^{40}Ca. The elastic cross sections are adequately fitted[12] by the diffraction model with the parameters R=4.72 fm and d/R=0.135 so that d=0.64 fm. Similarly, the elastic differential cross sections to the lowest 3^- and 5^- levels are fitted rather well by eq.(1) with th the same values for R and d and with the deformation lengths, δ_3= 1.14 fm and δ_5=0.68 fm, respectively. Table I summarizes the determination of (d/R) and d.

The derived diffuseness parameters are generally consistent except for the 2.1 GeV/N entry. It is quite possible that the $\sim$30 m cross section inferred for excitation of ^{16}O to the giant quadrupole states is too large. This could come about if either the apparent excess cross section has been overestimated or if a major fraction of the excess cross section for production of ^{12}C is due to still other processes. When one assumes that the ratio of (d/R) for the ^{9}Be-^{16}O collision is the same as that surmised for ^{12}C-^{40}Ca collisions, 0.10, then the cross section for excitation of the GQR states of ^{16}O predicted by the diffraction model is only 10 mb.

Coulomb excitation after the GQR in ^{16}O at 2.1 MeV/N on ^{9}Be is quite negligible. Using standard formulas in the literature[14] and an impact parameter cut off of 4.68 fm, we estimate $\sigma_{coulex}\sim$0.11 mb. Certainly these results underscore the importance of new experiments to measure ^{12}C and α fragments in coincidence with sufficient momentum resolution to determine the excitation energy of the ^{16}O intermediate. The new HISS superconducting spectrometer system at the Bevalac will have that capability.

V. MICROSCOPIC SOFT-SPHERE MODEL

Can we estimate by model calculations how the sharpness of the

TABLE I

Shadow Diffuseness Parameters (d/R) And d Derived For The Diffuse-Edge Shadow Diffraction Model

BEAM ENERGY (GeV/N)	SYSTEM EXCITED	COLLISION PARTNER	EXCITED STATE E (MeV)	EXCITED STATE I_π	σ (exp) (mb)	d/R	DIFFUSENESS PARAMETER d (fm)	REFS.
2.1†	^{16}O	^{9}Be	11.5 - 27.3	2^+	≤ 30	≥ 0.033	≥ 0.18	5
0.40	^{40}Ca	^{12}C	3.736	3^-	14.4 ± 4.4	0.10 ± 0.2	0.6	3
0.40	^{40}Ca	α	3.736	3^-	7.2 ± 2.4	0.16 ± 0.03	0.76	3
0.34	^{40}Ca	α	3.736	3^-	$d\sigma/d\Omega$	0.135	0.64	13
0.34	^{40}Ca	α	4.4915	5^-		0.135	0.64	13

†In this case the system excited is the beam particle.

absorption shadow might change with bombarding energy? Certainly the elemental nucleon-nucleon total cross section changes over the energy range of Table I. The σ_{tot}(pp) reaches a minimum of 24 mb near 180 MeV and σ_{tot}(np) has a minimum of ∿33 mb near 390 MeV. Bot cross sections rapidly rise beyond this energy, reaching a ∿40 mb plateau around 1.0 GeV.

It is appealing now to make comparison with a microscopic model based on experimental pp and np scattering cross sections. Karol[15] has used Gaussian forms to approximate nuclear density distributions in the surface and has derived simple analytical expressions for the nuclear transparency T(b) as a function of impact parameter b. Karol's equation (16) has an apparent misprint in that the left hand side T(b) should actually be -ln T(b) and read as follows:

$$- \ln T(b) = \pi^2 \bar{\sigma}(E) \rho_T(0) \rho_P(0) a_T^3 a_P^3 (a_T^2 + a_P^2)^{-1} \cdot \exp[- b^2/(a_T^2 + a_P^2)] \quad (4)$$

where target (T) and projectile (P) nuclei have densities of

$$\rho_i(r) = \rho_i(0) \exp[- (r/a_i)^2] \quad (5)$$

The parameter $\bar{\sigma}(E)$ is a weighted average of pp and np cross sections (his Eq. (4)). Karol integrates to get the corresponding reaction cross section in terms of an exponential integral.

$$\sigma_{React} = 10\pi(a_T^2 + a_p^2)[Ei(\chi) + \ln(\chi) + \gamma] \quad (6)$$

where σ_{React} is given in mb and a_i in fm,

$$\chi = \pi^2 \bar{\sigma}(E) \rho_T(0) \rho_p(0) a_T^3 a_P^3/10(a_T^2 + a_P^2) \quad (7)$$

and γ = Euler's constant = 0.57721 ... The exponential integral is

$$Ei(\chi) = \int_{\chi}^{\infty} \frac{e^{-u}}{u} du$$

It is tabulated, but for the collisions we consider the χ values are so large that $Ei(\chi)$ is negligible.

One finds, on making numerical comparisons, that Eq.(4) is rather well approximated by the square of Fermi expression, Eq.(2b), in the middle of the transition region between complete and zero transparency. This suggests recasting Eq.(4) in terms of parameters appropriate to the Fermi form.

Using Eq.(7), and the abbreviation, $a^2 \equiv a_P^2 + a_T^2$, we note first the simplification

$$T(b) \equiv \eta\varepsilon^2(b) = \exp[-\chi \exp(-b^2/a^2)] \tag{4a}$$

Since the radius of the Fermi form, R_F, is defined by $\eta_F(R_F) = 0.5$, we define a radius parameter, R, for the soft-spheres model by requiring that

$$\eta(R) = \exp[-\frac{\chi}{2} \exp(-R^2/a^2)] = 0.5 \tag{4b}$$

so that

$$R = a[\ln(\chi/[2 \ln 2])]^{1/2} \tag{4c}$$

Similarly, since the diffuseness parameter of the Fermi form, d_F, has the property

$$\left.\frac{\partial\eta_F}{\partial b}\right|_{R_F} = \frac{1}{4d_F}$$

we define the diffuseness of the soft-sphere model through the condition,

$$\frac{1}{4d} \equiv \left.\frac{\partial\eta}{\partial b}\right|_R = \frac{1}{2}(-\ln 2)\left(-\frac{2R}{a^2}\right) \tag{4d}$$

which is rearranged to give

$$d = \frac{a^2}{R}\frac{1}{4 \ln 2} = 0.3607\frac{a^2}{R} \tag{4e}$$

In terms of the newly defined parameters. R and d, $\eta(b)$ can be written

$$\eta(b) = \exp\left\{-\ln 2 \exp[(R^2 - b^2)/(Rd\ 4 \ln 2)]\right\} \tag{4f}$$

To facilitate comparison to $\eta_F(b)$, it is convenient to introduce the dimensionless variable, $w \equiv (b-R)/d$, in terms of which Eq.(4f) becomes

$$\eta(b) = \exp\left\{-\ln 2 \exp[-w(1 + \frac{wd}{2R})/(2 \ln 2)]\right\} \quad (4g)$$

The numerical correspondence between $\eta_F(b)$ and $\eta(b)$, considered as functions of w, is quite close for $|w| \lesssim 1$; it is actually closer for values of $d/R \sim 0.15$ than it is for $d/R = 0$.

The parameters R, d, and d/R given by the "soft-sphere" model for the heavy ion systems being studied are listed in Table II. In obtaining these an unnormalized Gaussian density has been adopted for ^{40}Ca which gives a reasonable fit in the surface region to the matter density inferred from elastic electron scattering[16]. Normalized Gaussian densities are chosen for ^{4}He, ^{9}Be, ^{12}C, and ^{16}O which give the measured values[16] of the rms radius of the proton density.

Comparing the "experimental" d/R values of Table I to the microscopic "soft-sphere" model values of Table II, we see that the "soft" values for collisions with ^{40}Ca are of the order of the "experimental" values, though systematically smaller. The narrower transition regions of the "soft-sphere" model may be attributed to two oversimplifications of this model: (a) The unrenormalized Gaussian density adopted for ^{40}Ca is considerably more dense in the interior than that deduced from the scattering of electrons and protons. (b) The "soft-sphere" model may be viewed as a version of the "optical limit" to the Glauber multiple scattering theory[17]. As such it bears the deficiency of all optical limit treatments of nucleus-nucleus collisions, that no account is taken of shadowing corrections. Both oversimplifications result in greater attenuation as the impact parameter is reduced than is predicted by more sophisticated theories[18].

In contrast the lower limit on the "experimental" value of d/R for $^{16}O + {}^{9}Be$ is much less than that predicted by the "soft-spheres" model. This cannot be attributed to an incorrect value for the nucleon-nucleon effective cross section, since increasing that cross section above the free-space value to 60 mb decreases the value of d/R by less than 10 percent. Thus it appears that giant quadrupole excitation contributes to only a portion of the excess yield of ^{12}C in ^{16}O fragmentation. Rather than $\sim$30 mb, our calculations indicate that a value around 10 mb would be more reasonable for the angle-integrated cross section for giant quadrupole excitation.

TABLE II

Microscopic Model Values Of Diffraction Model Parameters

E_{lab} (GeV/N)	A_P	A_T	a_P (fm)	a_T (fm)	a^2	$\rho_P(0)$ (fm^{-3})	$\rho_T(0)$ (fm^{-3})	$\sigma(E)$ (mb)	χ	R (fm)	d (fm)	d/R
2.1	16	9	2.115	1.942	8.245	0.304	0.221	44	24.5	4.86	0.612	0.126
								60	33.4	5.12	0.581	0.113
0.40	12	40	1.884	2.520	9.900	0.322	0.658	30	67.8	6.21	0.575	0.093
0.34 - 0.40	4	40	1.187	2.520	7.759	0.430	0.658	30	28.9	5.45	0.576	0.119

Further experimental and theoretical work is called for to elucidate the role of giant resonances in relativistic heavy ion reactions.

REFERENCES

1. J.S. Blair, Lectures in Theoretical Physics, Vol.VIIIc, eds. P.D. Kunz, D.A. Lind, and W.E. Britten, (University of Colorado Press, Boulder, 1966), p.343.
2. N. Austern and J.S. Blair, Annals of Physics (N.Y.) 33, 15 (1965).
3. T. Shibata et al. Nucl. Phys. A308, 513 (1978).
4. H.H. Heckman and P.J. Lindstrom, Phys. Rev. Lett. 37, 56 (1976).
5. P.D. Lindstrom et al. "Isotope Production Cross Sections from the Fragmentation of ^{16}O and ^{12}C at Relativistic Energies." LBL Report No. 3650 (1974) (unpublished).
6. D.E. Greiner et al. Phys. Rev. Lett. 35, 152 (1975).
7. J. Hüfner, K. Schafer, and B. Schurmann, Phys. Rev. C12, 1888 (1975).
8. K.T. Knöpfle et al., Phys. Rev. Lett. 35, 779 (1975)
9. K.T. Knöpfle et al., Phys. Lett. 74B, 191 (1978).
10. X.J. Qiu , and J.O. Rasmussen, Contribution to Proceedings of the Int. Conf. Nucl. Phys., Berkeley (August 24-30, 1980), p.623
11. M. Sasagase et al. Ibid, p. 527.
12. J.S. Blair, Bull.-Am. Phys. Soc. 23, 932 (1978).
13. G.D. Alkhazov et al., Nucl. Phys. A280, 365 (1977).
14. L.C. Biedenharn and P.J. Brussaard, Coulomb Excitation (Clarendon Press, Oxford, 1965). See also Ch.2., K. Alder et al., Rev. Mod. Phys. 28, 432 (1956).
15. P.J. Karol, Phys. Rev. 11, 1203 (1975).
16. R.C. Barrett and D.F. Jackson, Nuclear Sizes and Structure, (Clarendon Press, Oxford, 1977).
17. W. Czyz, and L.C. Maximon, Ann. Phys. (N.Y.) 52, 59 (1969), See also V. Franco and G.K. Varma, Phys. Rev. C15, 1375 (1977).
18. G. Fäldt and I. Hulthage, Nucl. Phys. A316, 253 (1979).

NUCLEAR CHARGE AND MATTER DISTRIBUTIONS

P.E. Hodgson

Nuclear Physics Laboratory
Oxford, United Kingdom

I. INTRODUCTION

The radial distributions of the nuclear charge and nuclear matter, or equivalently of the neutrons and protons, are among the most basic of nuclear properties. Substantial advances in our knowledge of these distributions have been made in recent years, both experimental and theoretical. Experimentally, the accurate measurements of the differential cross-sections for electron elastic scattering and of the energies of muonic atom transitions, analysed by model-independent techniques, have given very precise charge distributions for many nuclei. At the same time, the elastic scattering of energetic protons by nuclei have been analysed by the Glauber theory and by the optical potential method of Kerman, McManus and Thaler to yield nuclear matter distributions of somewhat lower accuracy.

Parallel with this experimental work, several theories have been used to calculate the nuclear charge and matter distributions from our knowledge of nuclear structure. Among these, the mean field theories and Brueckner-Hartree-Fock theory enable nuclear densities to be calculated from various effective nucleon-nucleon potentials. These methods are of great generality and give an excellent understanding of overall trends but frequently encounter difficulties in accounting for the densities of particular nuclei. Another method, based on the shell model, uses empirical data for each nucleus to obtain precise fits to the distributions obtained from the analyses of experimental data. This method has been developed over the last few years in Oxford, and is the subject of the present review.

The earlier stages of this work have been reviewed at previous meetings[1] and so this review will concentrate on the more recent developments. Originally, the work grew out of a series of studies of the systematic behaviour of nuclear single-particle states[2-4]. These showed that the centroid energies of single-particle states determined from analyses of nucleon transfer and knock-out reactions, can be expressed as the eigenvalues of one-body potentials with depths depending rather simply on the atomic weight and on the nuclear asymmetry parameter. As a check of these potentials, the corresponding charge distributions were calculated by summing the squares of the single-particle wavefunctions, weighted by the appropriate occupation probabilities. The RMS radii of the resulting charge distributions agreed well with the experimental values, and the radial distributions themselves proved to be unexpectedly accurate, so that they were successfully used to calculate proton and heavy-ion potentials by a folding procedure[5,6].

The success of this work encouraged us to develop the method as a precise technique for calculating nuclear density distributions. As a severe test of the method, we first analysed the data for ^{208}Pb, because two model-independent analyses of electron elastic scattering were available, and also much information from nucleon transfer reactions to and from the single particle occupied and unoccupied states near the Fermi surface. The model-independent charge distribution showed considerable structure, and it was found possible to fit this structure with the single particle potential (SPP) calculations[1]. This established the usefulness of the method, but just as we were preparing an account of this work for publication some new experimental results became available that gave a significantly different model-independent charge distribution. The work on ^{208}Pb was therefore set aside until the experimental situation clarified.

One difficulty with the ^{208}Pb analysis was the lack of information on the more deeply-bound states. In the calculations it was assumed that the potential corresponding to these states is the same as that for the surface states for which data is available, but it is not certain that this is correct and indeed it is not consistent with the energy dependence of the potential that is found for the deeper states in lighter nuclei for which experimental binding energies have been determined.

With this in mind, the next nucleus was selected to satisfy the requirement that experimental binding energies are available for all the single-particle states. This, together with the need for the nucleus to be as heavy as possible to avoid complications due to centre-of-mass motion and deformation, indicated ^{58}Ni as a suitable choice. A detailed analysis was made of this nuclei, and the quantum-mechanical basis of the SPP method was described in the paper giving the results[8]. Subsequently the method was applied to ^{39}K, ^{40}Ca and ^{48}Ca, and also to the even zirconium isotopes . These analyses are briefly summarised in Section II.

All this work concerns nuclear charge distributions, but the same technique is also applicable to nuclear matter distributions, and these are particularly useful for calculating optical potentials by a folding procedure. The matter distributions may be calculated using the potentials found to give the best fit to the charge distributions, with depths adjusted to the neutron as well as to the proton separation energies. The calculations are somewhat less certain because there are no model-independent matter distributions for comparison.

It was expected that the matter distributions obtained in this way are particularly reliable in the far surface region, because the potentials are adjusted to fit the experimental separation energies, and this is the most critical region for optical potential tials, especially those of composite particles and heavy ions. To study this possibility, some nuclear matter distributions were calculated for several isotopic sequences and compared with the critical radii obtained from analyses of alpha-particle elastic scattering near the Coulomb barrier. It was found possible to reproduce the departures from the $A^{1/3}$ variation of the interaction radii through the various isotopic sequences, thus confirming the usefulness of the method . Subsequently the effective radii obtained in this way were used to improve the global heavy-ion optical potential of Christensen and Winter, giving improd fits to a wide range of heavy-ion elastic scattering data .

The matter distributions calculated by the SPP method may be compared with proton elastic scattering just as the charge distributions are compared with electron elastic scattering, although the accuracy of the comparison is less due to our less complete knowledge of the nuclear as compared with the electromagnetic interaction. The comparison with proton scattering is best made at high energies to minimise the effect of this lack of knowledge, for then the Glauber and Kerman-McManus-Thaler theories are applicable. Such comparisons were made with data on the elastic scattering of 800 MeV protons by several nuclei and gave additional information on the nuclear matter distributions . All this work on nuclear matter distributions is summarised in Section III.

These analyses treat the neutrons and protons as independent particles moving in rather similar potentials. In reality the neutrons and protons interact strongly with each other and this interaction is responsible not only for the similarity of the neutron and proton potentials but also for the isovector term in the nucleon potential. This term provides a way of treating the neutrons and protons together, by defining an isovector potential in terms of the neutron and proton densities. It is then possible to iterate the potentials and the densities until self-consistency is attained[11]. This method has been used to analyse the charge and matter distributions of some of the oxygen and calcium isotopes, and is described in more detail in Section IV.

II. ANALYSES OF NUCLEAR CHARGE DISTRIBUTIONS

As mentioned briefly in the introduction, the SPP method assumes that each nucleon is moving independently in a one body potential, and the proton and neutron density distributions are then obtained by summing their probability distributions, weighted by the occupation numbers of each orbit.

The essential feature of this method is that it is constrained at every point to fit selected experimental data for the nucleus whose charge and matter distributions are being calculated. Thus the potentials used to generate the single particle wavefunctions are adjusted to give the centroid separation energies and occupation numbers as determined by one-nucleon transfer and knock-out reactions,. and also to fit the charge distributions as determined by electron elastic scattettering and muonic atom analyses. The resulting distributions thus have a secure experimental basis, and because of the known systematic behaviour of single particle states they may be extended to a wide range of nuclei with some confidence.

The important feature of these calculations is that they unify the nuclear data obtained from nucleon transfer and knock-out reactions on the one hand, and from electron scattering and muonic atom data on the other. This increases the overall accuracy of our knowledge of nuclear charge distributions particularly in the region of the nuclear surface. Examination of the uncertainties in the model-independent analyses of electron scattering and muonic atom data shows that the resulting charge distributions are well-determined in the inner surface or knee region where the density begins to fall from its interior value but are relatively poorly determined in the far surface regions. The distributions calculated by the SPP method, however, are particularly accurate in the far surface regions because in this region they depend on the tails of the wave functions which are fitted to the experimental separation energies for each orbit.

The charge and matter distributions are given by expressions of the form

$$\rho(r) = \sum_i a_i |\psi_i(r)|^2 \qquad (1)$$

Where $\psi_i(r)$ is the wavefunction of the particle in the ith state, a_i the occupation numbers and the summation runs over all occupied states. The wavefunctions are bound solutions of the radial Schrodinger equation for a potential

$$V(r) = V_c(r) + Uf(r) + \left(\frac{\hbar}{m_\pi c}\right)^2 U_s \frac{df(r)}{r\,dr} L\cdot\sigma \qquad (2)$$

where $V_c(r)$, the electrostatic potential due to a uniformly-charged sphere, is included only for protons, and the form factor $f(r) = [1+\exp\{(r-R)/a\}]^{-1}$.

The proton charge distribution is obtained by folding the point distribution (Eq.1) with the charge distribution (r) of the proton itself[12].

$$\rho_{ch}(r) = \int \rho_p^c(\underline{r}') \, \rho_p(|\underline{r} - \underline{r}'|)dr'. \qquad (3)$$

The contribution to the nuclear charge distribution from the neutrons is evaluated similarly; this is a small but not negligible correction.

The occupation numbers a_i are taken to be (2J + 1) for the deeper states and for the surface states the experimental values determined from nucleon transfer reactions are used wherever possible.

Local potentials and wavefunctions are used in these calculations for convenience, although it is known that they are partly non-local in character. The effect of using effective local potentials is to enhance the wavefunction in the nuclear interior, subject of course to the overall normalization. An approximate relation between the local and non-local wavefunctions has been obtained by Perey in the form[13]

$$\psi_{NL}(r) = \psi_L(r) \exp\left\{\frac{m\beta^2 V(r)}{2h^2}\right\} \qquad (4)$$

where β is the non-locality parameter and m the reduced mass, and this is used to correct the calculated local wavefunctions.

Since the potential is allowed to vary from state to state to fit the measured binding energies, the resulting wavefunctions are not orthogonal. This inconsistency in the calculations is removed using the Gram-Schmidt orthogonalisation procedure. In practice this requires the extra assumption that the occupation numbers of all the states of a particular NLJ are unity except possibly that of the least bound one. Shell model considerations indicate that this approximation is a very reasonable one. Calculations with and without orthogonalisation show very little difference, as might be expected from the small differences between the potentials corresponding to the different states.

The early calculations of the charge distribution of ^{208}Pb gave a good fit to the model-independent charge distributions provided (1) the occupation number of the $3s_{1/2}$ orbit is reduced from 2 to 1.8; this is physically very reasonable and can be verified experimentally in principle, though it is difficult in practice, (2) the radius parameter is allowed to depend weakly on the principal quantum number N. The resulting fit is shown in Fig.1. As mentioned previously this work had to be abandoned because new data gave a significantly different model-independent charge distribution. Further work on ^{208}Pb has been carried out recently.

The first complete analysis was made for ^{58}Ni. A model-independent charge distribution had been determined by Sick et al[14], and the mean separation energies of most of the single-particle states have been determined experimentally. The centroid energies of the single-particle states are defined as the weighted mean of the energies of the hole and particle (T< and T>) fragments, according to the prescription of Baranger and Clement[15]. The experimental occupation numbers are used for the $2p_{3/2}$ and $1f_{7/2}$ states, and (2J+1) for the remainder. The fine adjustment of the potential to the charge distribution was made by adjusting the radius R of the potential, the non-locality parameter, the centroid of the 1d

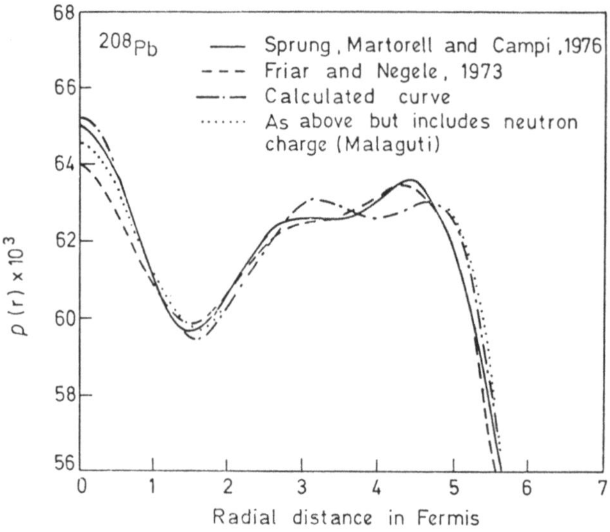

Fig.1 Model-independent charge distributions of ^{208}Pb compared with SPP model calculations with r_n=1.21, 1.23, and 1.244 for states with with N=1, 2 and 3.

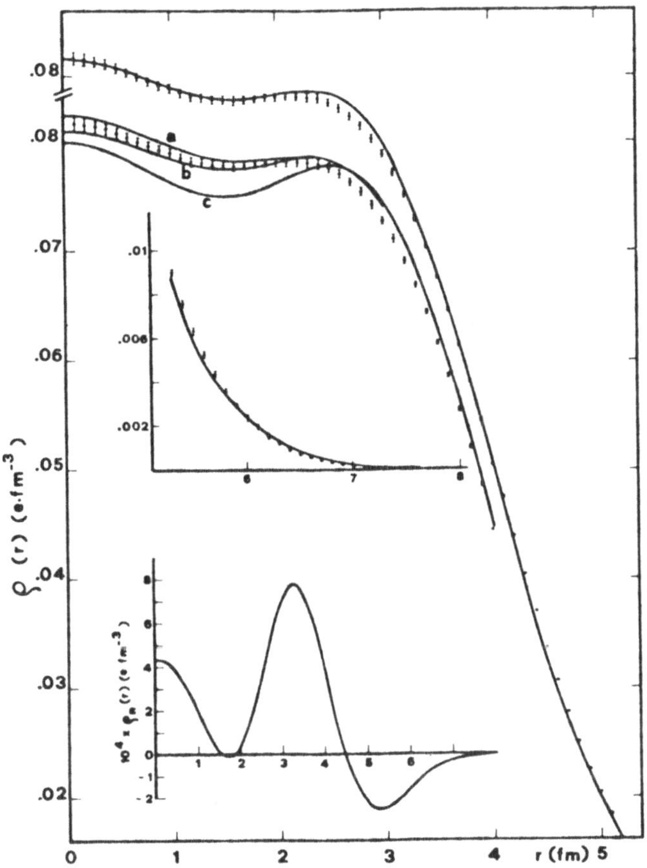

Fig.2 Comparison between SPP and model-independent charge distributions for ^{58}Ni. The tail is shown on an expanded scale in the upper insert, and the contribution of the neutron density in the lower. The curves a, b and c are sensitivity studies showing the effects of parameter variation[8].

states and the $2p_{3/2}$ occupation number (within the range permitted by the experimental data), subject to the RMS charge radius being constrained to 3.775fm. Subsequent work on the calcium isotopes showed that there is a correlation between the diffuseness parameter a_1 of the central potential and the occupation number of the p-state. Since a full parameter search varying both these parameters as well as those already listed proved ambiguous, a simplified calculation was made with the non-locality parameter fixed to 0.85 and the central potential of the 1d states to 60 MeV. The parameter search

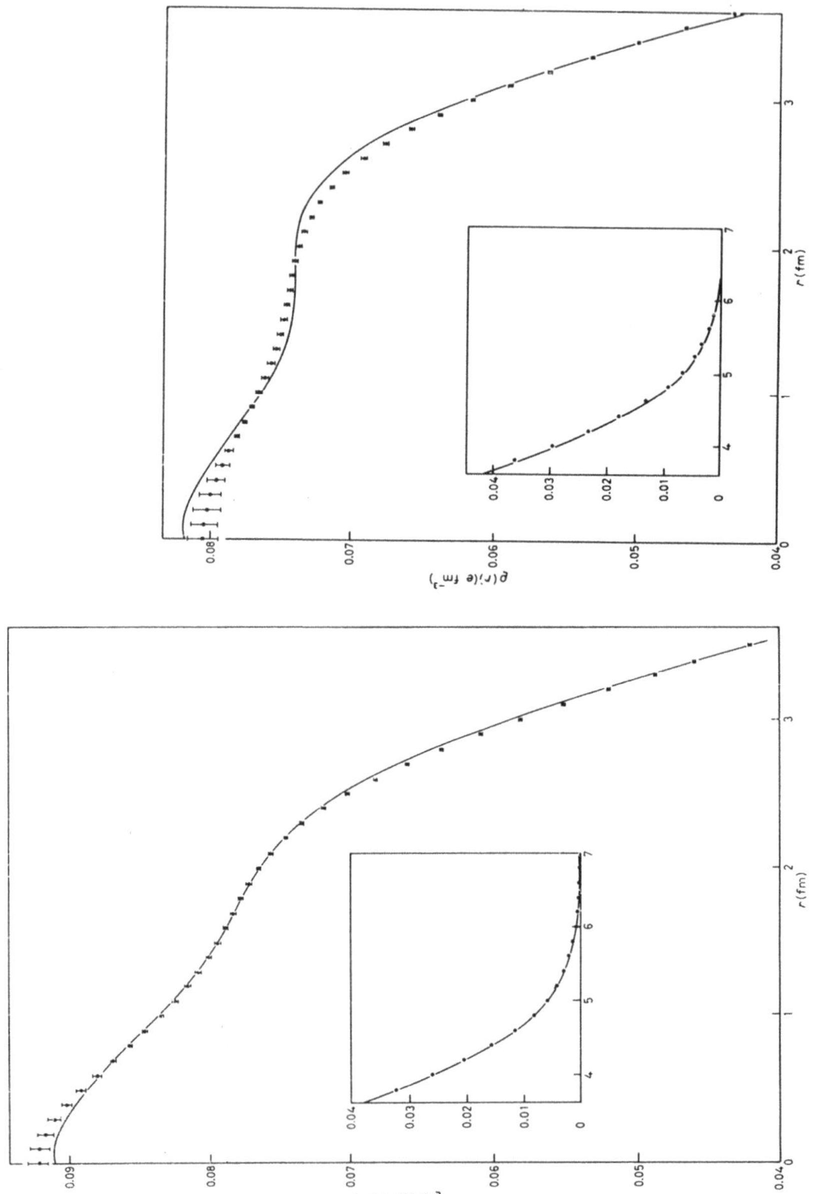

Fig.3 Comparison between SPP and model-independent charge distributions for ^{39}K and ^{40}Ca (reference 8).

in the subspace $\{R, a_1\}$ was then made for many fixed values of n_p. The best fit is obtained for $n_p \approx 1.5$, which is not acceptable experimentally. It must therefore be fixed to the optimum experimental value, and the resulting fit with the remaining parameters allowed to vary is shown in Fig.2.

The next series of analyses were made for ^{39}K, ^{40}Ca and ^{48}Ca, partly because adequate data are available on the single-particle states and partly because it is interesting to see if the method can account for charge distribution differences between isotopes, which are often better known than the charge distributions themselves. The analysis was made in essentially the same way as before, varying R, a_1, n_p ($2p_{3/2}$ occupation number) and n_s ($2s_{1/2}$ occupation number). The resulting charge distributions are compared with the experimental ones in Fig.3.

A further opportunity to compare the SPP method with the experimental charge distribution differences between isotopes was provided by the Mainz data on the Zirconium isotopes[16]. For these nuclei, the potentials for the deep states were calculated from the

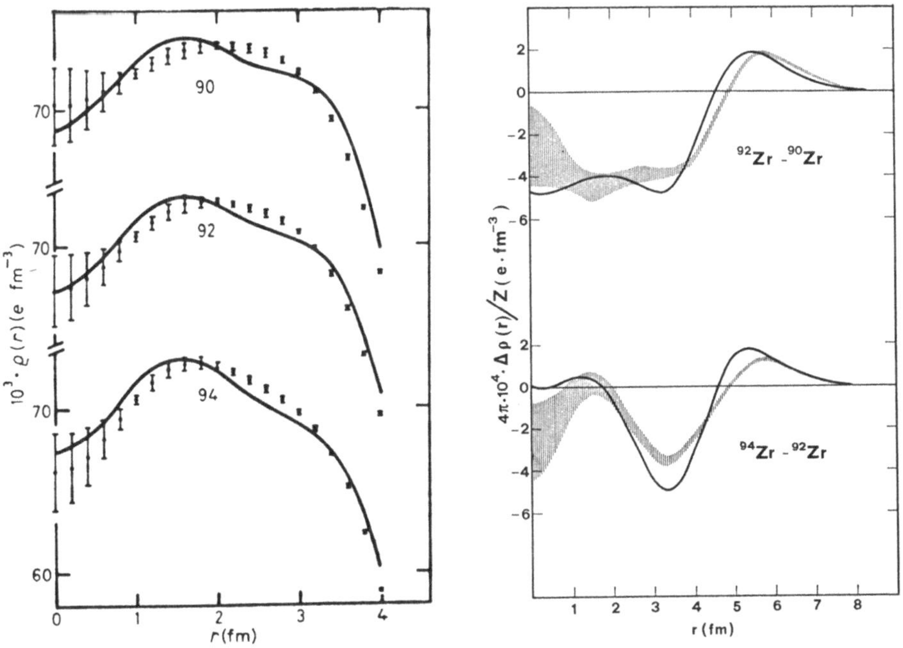

Fig.4 Comparison between SPP calculations and experimental data. (A) Charge distributions of zirconium isotopes (B) Charge distribution differences (a) ^{92}Zr-^{90}Zr, (b) ^{94}Zr-^{92}Zr (ref.8).

non-local potential fitted to the surface states, with the non-locality parameter $\rho = 0.85$fm. The variation of these potentials along the isotopic sequence was calculated using an isospin potential $U_1 = 36$ MeV. Several calculations were made to explore the sensitivity of the charge densities to variations of the binding energies, the occupation numbers of the surface states, the potential form factor and the non-locality parameter. These showed that the results are rather insensitive to changes in the binding energies within the statistical uncertainties for the surface states and within 3 to 4 MeV for the deeper states. Further calculations were therefore made varying only the form factor parameters R_1 and a_1, and the occupation number of the $2p_{1/2}$ proton state, the other occupation numbers being fixed to their experimental values. The resulting charge distributions and charge distribution differences are compared with the experimental data in Fig.4.

It is notable that the two curves are rather different, but both are well fitted by the SPP calculations. The first difference is what would be expected if the two extra neutrons produced just a uniform expansion of the core protons, and this is what might be expected from the strong neutron-proton attractive force. The second pair of extra neutrons act in a somewhat different way, however, and leave the central density practically unchanged while pulling out the rest of the protons.

III. NUCLEAR MATTER DISTRIBUTIONS

The methods described in the previous section for calculating nuclear charge distributions can also be used to calculate nuclear matter distributions. These matter distributions cannot be compared with experimental data so accurately as the charge distributions because the nuclear interaction is not so well understood as the electromagnetic interaction. The theories become more reliable at higher energies, so the most accurate comparisons are made using the Glauber and Kerman, McManus and Thaler theories to calculate the elastic scattering cross-sections of energetic protons. Earlier work[17,18] used functional forms for the matter distribution, and adjusted their parameters to optimise the fit to the proton elastic scattering data. Such functional forms are not, however, very satisfactory, especially in the interior and for surface regions, so the analysis has been repeated using matter distributions calculated by the SPP method[10].

These calculations were made for ^{40}Ca, ^{48}Ca, ^{58}Ni and ^{208}Pb for which detailed studies of the charge distributions are available, and also for ^{64}Ni, ^{116}Sn and ^{124}Sn. The resulting differential cross-sections for the elastic scattering of 800 MeV protons by these nuclei are compared with some of the experimental data in

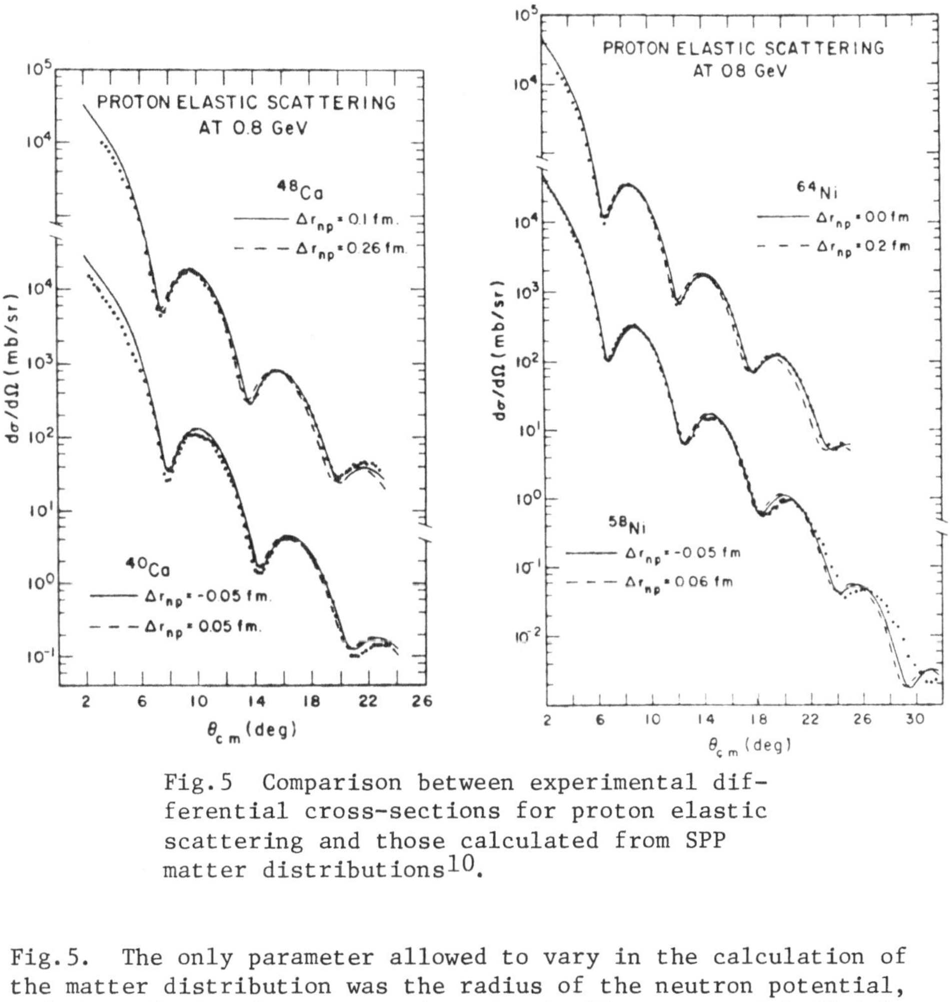

Fig.5 Comparison between experimental differential cross-sections for proton elastic scattering and those calculated from SPP matter distributions[10].

Fig.5. The only parameter allowed to vary in the calculation of the matter distribution was the radius of the neutron potential, and the calculated cross-sections in the figures are labelled by their value of $\Delta r_{np}^2 \equiv \langle r_n^2 \rangle^{\frac{1}{2}} - \langle r_p^2 \rangle^{\frac{1}{2}}$. On the whole the data is well fitted, and comparison between the experimental and theoretical cross-section enables Δr_{np} to be estimated to about ± 0.05fm.

These calculations provide neutron density distributions, and some of these are compared in Fig.6 with the corresponding neutron densities obtained by the density matrix expansion calculations of Negele[18] and Negele and Vautherin[20]. Particularly precise comparisons can be made for the differences between the neutron density distributions of isotopes, and one of these are compared in Fig.7 with the empirical neutron densities of Ray[18].

The nuclear matter distributions calculated by the SPP method are particularly reliable in the far surface region because the potentials of the surface states are adjusted to fit the measured

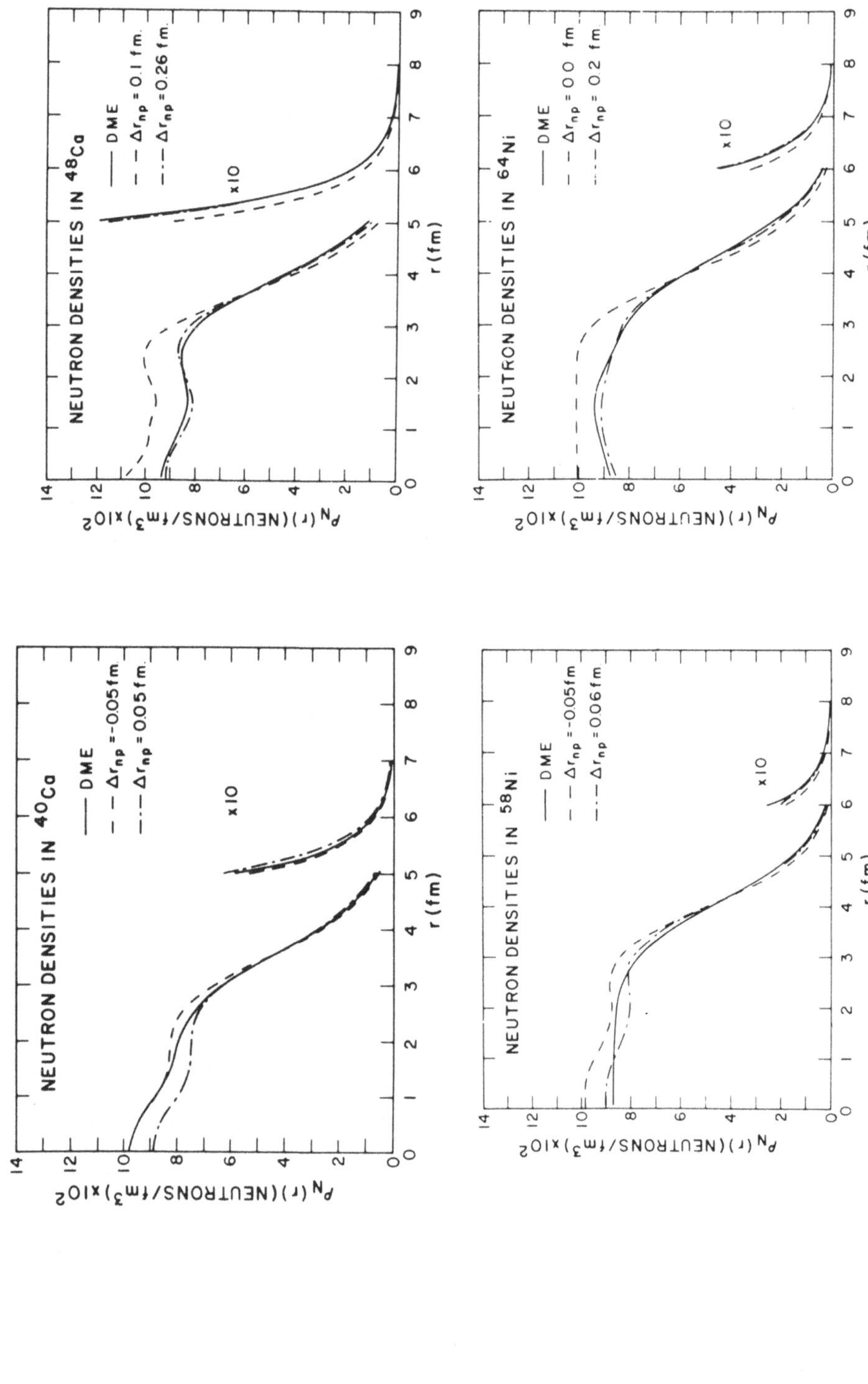
NEUTRON DENSITIES IN 40Ca
DME
Δr np =-0.05 fm
Δr np = 0.05 fm
x10
NEUTRON DENSITIES IN 48Ca
DME
Δr np = 0.1 fm
Δr np = 0.26 fm
x10
NEUTRON DENSITIES IN 58Ni
DME
Δr np = -0.05 fm
Δr np = 0.06 fm
x10
NEUTRON DENSITIES IN 64Ni
DME
Δr np = 0.0 fm
Δr np = 0.2 fm
x10
ρN (r) (NEUTRONS/fm3) x10²
r (fm)

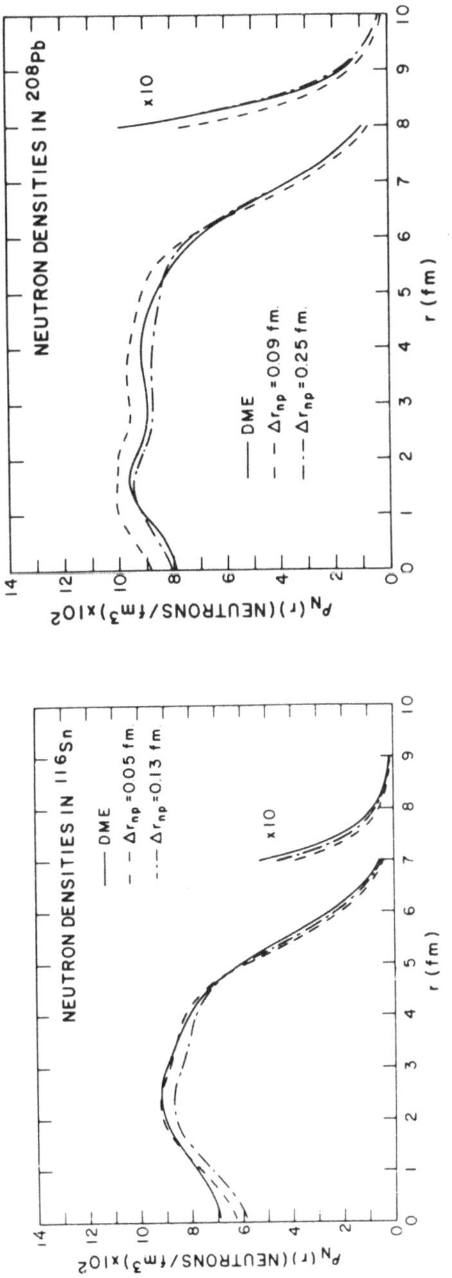

Fig.6 Comparison between neutron density distributions ealculated by the density matrix expansion (DME) and SPP methods with different values of Δr_{np} $\Delta\ r_{np}$(Ray and Hodgson[10]).

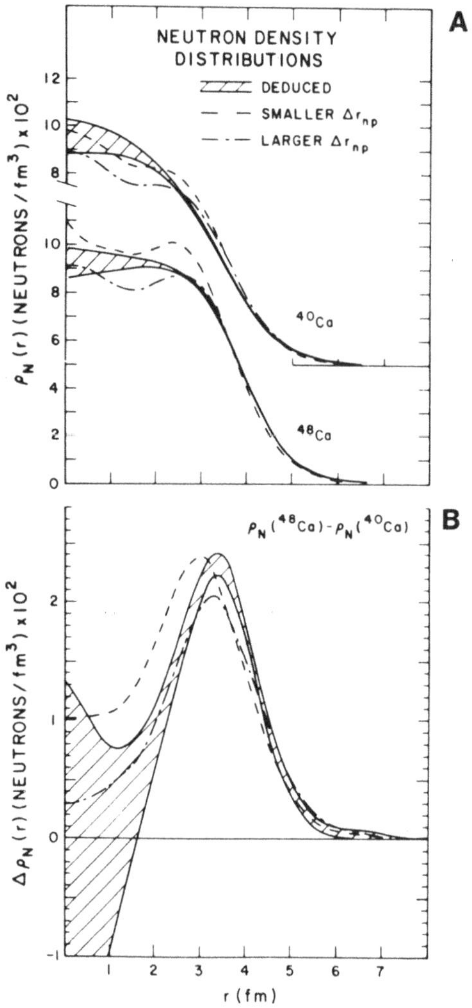

Fig. 7 Comparison between SPP and experimental neutron density distributions for ^{40}Ca and ^{48}Ca,

(a) Density distributions.
(b) Density distribution differences[10].

centroid separation energies. They are therefore very suitable for calculating optical potentials by the folding model, particularly for light and heavy ions whose elastic scattering is dominated by the far surface potential. This application of the SPP method was first made by Vary and Dover[6] and their work showed the usefulness of the method.

Since that time, SPP density distributions have been used in several calculations of alpha-particle and heavy-ion potentials. In the case of nucleons, it is possible to find global optical potentials that describe to good accuracy the scattering from many nuclei over a range of energies. This not so easy for light and heavy ions because their interactions are much more sensitive to the nuclear surface, so that the scattering is sensitive to details of nuclear structure. It is therefore interesting to see whether some information from SPP matter densities can be used to improve the optical potentials.

The first studies were made of the elastic scattering of alpha particles by several isotopic sequences at energies near the Coulomb barrier[21]. The cross-sections determine the radius R at which the nucleon density is 2×10^{-3} nucleons fm^{-3}. These radii are approximately proportional to $A^{1/3}$, but there are small systematic differences due to the structure of the nuclei concerned. In particular, the quantity $RA^{1/3}$ is found to increase slowly through some isotopic sequences and to decrease through others. This may be understood qualitatively in terms of the shell structure of the nuclei, since we expect the nuclear radius to increase more rapidly than $A^{1/3}$ when a major shell is just beginning to be filled than when it is nearly full.

To see if this effect could be reproduced by SPP densities, Lozano et al.[9] fitted them in the tail region by the expression

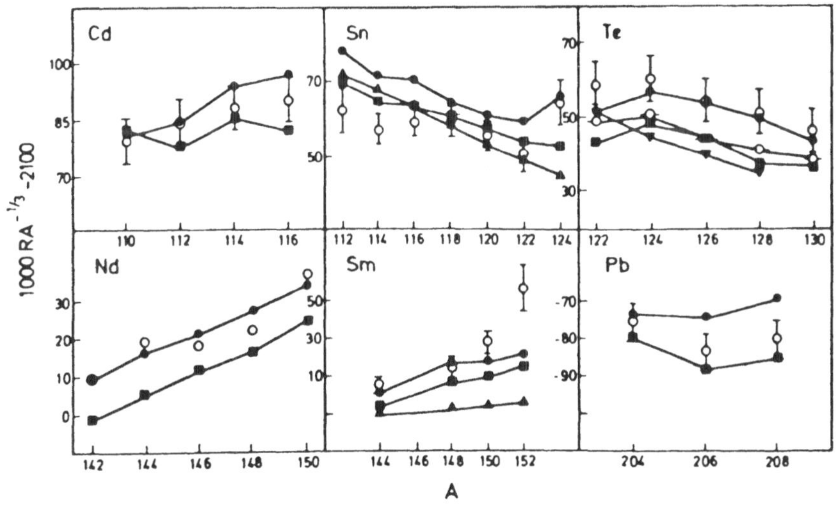

Fig.8 Nuclear radius parameter as a function of A for several isotopic sequences. The experimental data are that of Tabor et al.[22] for Nd and of Badawy et al.[21] for the other nuclei, and are indicated by open circles. The solid symbols show the results of SPP calculations and are connected by lines to show the trends of the variations with A (Lozano et al. ref.9).

$$\begin{aligned}\rho(r) &= \rho_o \left\{1 + \exp\,(r - R_\rho)/a_\rho\right\}^{-1} \\ &\approx \rho_o\left\{\exp\,(R_p - r)/a_p\right\}\end{aligned} \qquad (5)$$

Setting $\rho_o = 0.14$ gives

$$R = R_\rho + 4.25\, a_p \qquad (6)$$

Badawy et al.[21] showed that the real part of the alpha-particle potential can be obtained from the nuclear density distribution using the relation

$$R(0.2) = R_\rho + 5.54\, a_p + 2.37 \qquad (7)$$

where R(0.2) is the radius at which the real potential is -0.2 MeV. Thus

$$R(0.2) = R + 1.29\, a_p + 2.37 \qquad (8)$$

This enables the values of R obtained by Badawy et al.[21] to be compared with those calculated by the SPP method. As shown in Fig.8, this comparison shows the same overall trends for each isotopic sequence. It is therefore possible to use the SPP densities to calculate improved optical potentials.

This method was extended to heavy ions by Lozano et al.[9] They started from a global optical potential developed by Christensen and Winther[23] with parameters adjust to fit a range of heavy-ion elastic scattering data and altered it in two ways; firstly it was allowed to become complex, and secondly the radius parameter was related to the radius at which the nuclear density has a certain fixed value. The resulting potential has the form

$$V(r) = \frac{R_2+R_2}{R_1+R_2}\left\{50\,\exp\left(\frac{R_1+R_2-r}{0.63}\right) + iW\,\exp\left(\frac{R_1+R_2-r}{a_i}\right)\right\} \qquad (9)$$

The parameters W and a_i were determined by optimising the fit to twenty reactions, using Christensen and Winther's expression for the radius

$$R_i = 1.233\, A_2^{1/3} - 0.978\, A_i^{-1/3} \qquad (10)$$

This gave $W_o = 21.38$ MeV and $a_i = 0.686$fm.

The nuclear structure effects in the nuclear radii were then introduced by defining

$$R_i = R_i\,(\bar{\rho}) - A_i^{-1/3} \qquad (11)$$

where $R_i(\rho)$ is the radius at which the matter density has the value $\bar{\rho}$. The optimum values of the parameters $\bar{\rho}$ and $\bar{\eta}$ were determined by fitting several sets of data, giving $\bar{\rho} = 0.04$, $\eta = 0.903$. With these values, the overall fit to a large range of data is significantly improved. This is further evidence that the global heavy-ion optical potential can be improved by the incorporation of nuclear structure information.

A more fundamental calculation using the SPP densities has been made by Viñas et al.[24] They used a double-folding model and calculated several heavy-ion potentials with the density-dependent interaction.

$$\begin{aligned} V(r, r') = & - U_o(1 - S_{ij}\alpha)\left\{1 - c\rho(r)^{1/3}\rho(r')^{1/3}\right\} \\ & \cdot \exp\left[-\frac{(r - r')^2}{a}\right] \end{aligned} \qquad (12)$$

where r and r' are the positions of the two nucleons and $S_{ij} = +(-1)$ for the interaction between like (unlike) nucleons. The parameters of this interaction were chosen to give the best agreement with nuclear densities, binding energies and radii as described by Viñas and Madurga[25]. The resulting potentials were found to give excellent fits to several heavy-ion elastic scattering cross-sections.

IV. SELF-CONSISTENT CALCULATIONS

One of the main difficulties of the calculations described so far is that while the charge distributions can be compared with the experimental data to high accuracy, the neutron distributions are much less reliable, as they are calculated on the assumption that the neutron and proton potentials are the same, apart from an adjustment of the neutron potential radius to fit the proton elastic scattering data. In this respect the Hartree-Fock method is more satisfactory, since the neutron and proton wavefunctions are together iterated to self-consistency, and it is then very difficult to believe that the neutron distribution is not as reliable as the proton distribution, when the latter fits the experimental data.

In order to introduce a similar self-consistency into the SPP method, the nucleon potentials were written in a form that depends on the densities. The isospin term in the nucleon-nucleus potential was replaced by $V_1\rho_1(r)/\rho_o(r)$, where

$$\rho_o(r) = \rho_n(r) + \rho_p(r)$$

and (13)

$$\rho_1(r) = \rho_n(r) - \rho_p(r)$$

Thus the proton and neutron potentials become

$$V_p(r) = \left\{V_o + V_1 \frac{\rho_1(r)}{\rho_o(r)}\right\} f(r) + V_{so}(r) + V_c(r)$$

and (14)

$$V_n(r) = \left\{V_o - V_1 \frac{\rho_1(r)}{\rho_o(r)}\right\} f(r) + V_{so}(r)$$

In the first stage of the calculations these coupled equations are solved iteratively for a closed-shell nucleus using standard values of V_1, R_{s1}, a_s and a, and adjusting V and R to fit the single-particle centroid energies of orbits near the Fermi surface and the RMS charge radius.

To calculate the densities of other nuclei formed by adding one or more nucleons the potentials are written

$$V_p(r) = \left\{V_o\, \rho_o(r) + V_1\, \rho_1(r)\right\} F(r) + V_{so}(r) + V_c(r)$$

(15)

$$V_n(r) = \left\{V_o\, \rho_o(r) - V_1\, \rho_1(r)\right\} F(r) + V_{so}(r)$$

where $F(r) = f(r)/\rho_o^c(r)$ and $\rho_o^c(r)$ is the density of the closed shell or core particles. These equations are solved in the same way as before by iteration.[11]

This formalism is particularly appropriate for studying changes in the proton and neutron density distributions relative to core

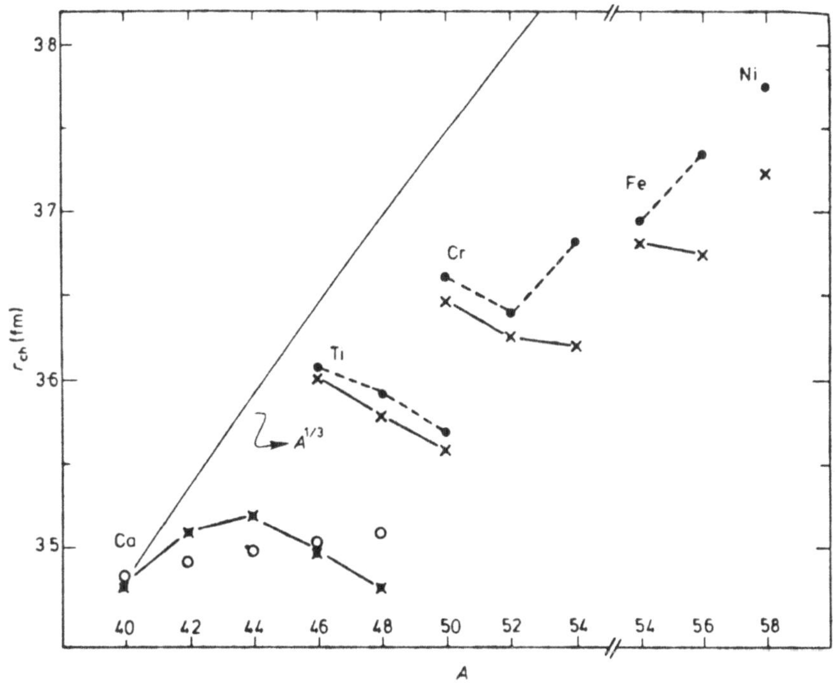

Fig.9 RMS charge radii for A = 40-58 nuclei. The experimental (•) and theoretical (oIIA; xIIB) values are shown for each isotopic sequence. Calculation IIA refers to a closed shell configuration and IIB allows for excitation out of the sd shell.[11]

nuclei such as ^{16}O and ^{40}Ca, and indeed in many experiments it is these distributions that are determined most accurately. There are no free parameters for the density differences as they are all fixed by the fit to the core nuclei. It is then possible to see if configuration mixing is able to account for the detailed structure of the charge density and also the RMS radii. The method is sensitive mainly to the changes in the configuration relative to the core nuclei. The model also includes the effects of polarisation of the core by the valence particles.

This method has been applied to the oxygen and calcium isotopes, and some of the results for the latter will be presented here. The first calculations were made using the occupation probabilities (90% $1f_{7/2}$ particles plus 10% $2p_{3/2}$ particles) of McGrory et al[26] and the results for RMS radii are shown by the open circles in Fig.9 normalised at ^{40}Ca by choosing R = 4.614fm. The isotopic dependence is small and smooth, in disagreement with experiments. The

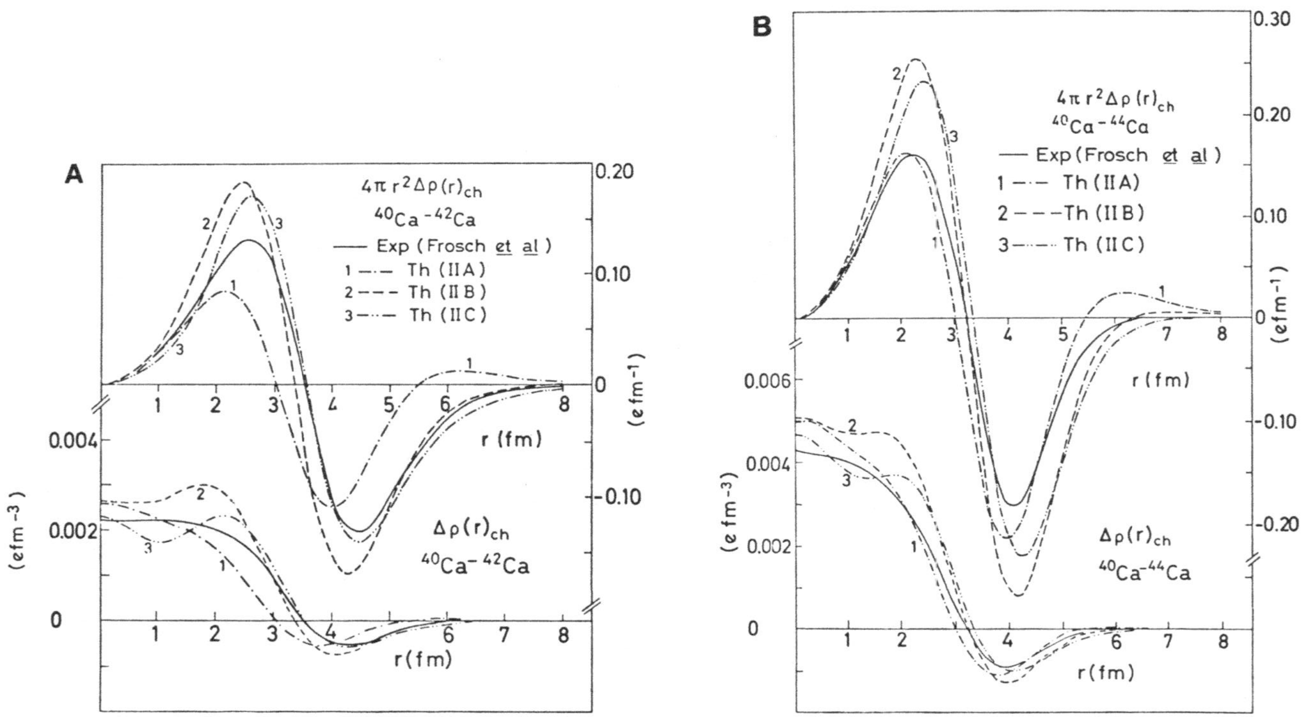

Fig.10 Comparison of experimental (full curves) and theoretical (broken curves) charge density differences between ^{42}Ca and ^{40}Ca and between ^{44}Ca and ^{40}Ca. The experimental curves are from the model-dependent fit to the electron scattering data by Frosch et al. The three theoretical curves correspond to a closed sd shell configuration (-·-·, IIA) to a non-closed shell configuration with 10% $2p_{3/2}$ (--- IIB) and to a non-closed shell configuration with 20% $2p_{3/2}$ (-··-, IIC). (a) $4\pi r^2\Delta\rho(r)_{ch}$; (b) $\Delta\rho(r)_{ch}$ (Brown et al.[11]).

corresponding density changes shown by the curves IIA in Fig.10 are however in fair agreement with experiment.

The calculations were then repeated with the following one-parameter wavefunctions for $n \geq 2$

$$|^{40+n}Ca> = \alpha|(\nu f_{7/2}\, p_{3/2})^n> + \beta|(\pi d_{3/2})^{-2}\, (\pi f_{7/2}\, p_{3/2})^2\, (\nu f_{7/2}\, p_{3/2})^n> \tag{16}$$

and for ^{40}Ca,

$$|^{40}Ca> = \alpha|o> + \beta|(\pi d_{3/2})^{-1}\, (\pi f_{7/2}\, p_{3/2})^{-1} \times (\nu d_{3/2})^{-1}\, (\nu f_{7/2}\, p_{3/2})^{1}> \tag{17}$$

TABLE I

Comparison Between The Numbers Of Proton Holes In The sd Shell Obtained From Electron Scattering And From Pickup And Stripping Reactions[11]

Nucleus	$\Delta + \delta$	(h, d)	(h, d)	(d, h)	(t, α)
^{40}Ca	0.7	0.4	0.27	0.73	0.3
^{42}Ca	1.7	0.9	1.12	1.03	0.4
^{44}Ca	1.8	1.9	1.98	1.01	0.6
^{46}Ca	1.05	0.4	-	-	0.2
^{48}Ca	0.3	0.15	0.44	0	0

where $|o\rangle$ is the closed-shell configuration. The same fp configurations were used as in the previous calculations. The numbers of proton holes in the $d_{3/2}$ orbit relative to ^{40}Ca is then

$$\Delta(n) = 2\beta^2 \, (^{40+n}Ca) - \beta^2 \, (^{40}Ca) \tag{18}$$

The $\Delta(r)$ were then chosen to reproduce the experimental RMS radii for the nuclei from ^{42}Ca to ^{48}Ca and the corresponding charge distribution differences are compared with the experimental data and with the previous calculations in Fig.10. These calculations including core excitation (labelled IIB) are in much better agreement with the data than the closed-shell calculations (IIA), especially in the important surface region. Also included in Fig.10 are the results of a calculation with 20% admixture $2p_{3/2}$ particles to show the sensitivity to this orbit.

The values of $\Delta(n)$ are compared with determinations of the number of proton holes in the sd shell obtained from analyses of various stripping and pickup reactions. Since $\Delta(n)$ is the number of proton holes relative to ^{40}Ca, the number of proton holes in $\Delta(n)$, $\delta \simeq .7$, is added to $\Delta(n)$ to give an estimate of the absolute number. It is apparent from Table I that there is not a satisfactory agreement among these quantities. This is being investigated further.

ACKNOWLEDGEMENT

This work has been carried out in collaboration with Franco Malaguti, Arnaldo Ugguzoni, Ettore Verondini, Manolo Lozano, Gonzalo Madurga, Alex Brown, Stelios Massen and Lanny Ray, and it is a pleasure to thank them for their stimulating discussions and essential contributions.

REFERENCES

1. P.E. Hodgson, Lecture Notes in Physics Vol.55 Ed. S. Boffi and G. Passatore (Springer-Verlag 1976) p88; Lecture at G.I.F.T. Seminar on Heavy-Ion Reactions, Seville 1979a; Atomki Kozlemnyck 21, 165 (1979b).
2. D.J. Millener and P.E. Hodgson, Nucl. Phys. A209, 59 (1973).
3. F. Malaguti and P.E. Hodgson, Nucl. Phys. A215, 243 (1973); A257, 37 (1976).
4. F. Malaguti, Nucl. Phys. A308, 125 (1978).
5. E. Kujawski and J.P. Vary, Phys. Rev. C12, 1271 (1975).
6. J.P. Vary and C.B. Dover, Phys. Rev. Lett. 31, 1511 (1973).
7. K. Bear and P.E. Hodgson, J. Phys. G4, L287 (1978).

8. F. Malaguti, A. Uguzzoni, E. Vernodini and P.E. Hodgson, Nucl. Phys. A297, 287 (1978); Nuovo Cim. 49A, 412 (1979a); Ibid 53A, 1 (1979b).
9. M. Lozano, G. Madurga and P.E. Hodgson, Phys. Lett. 82B, 170 (1979); (Preprint) 1980
10. L. Ray and P.E. Hodgson, Phys. Rev. C20, 2403 (1979).
11. B.A. Brown, S.E. Massen and P.E. Hodgson, J. Phys. G5, 1655 (1979).
12. H. Chandra and G. Sauer, Phys. Rev. C13, 245 (1976).
13. F. Perey, Proc. Conf. on Direct Interactions and Nuclear Reaction Mechanisms, Padua, (1962), p.125.
14. I. Sick, J.B. Bellicard, M. Bernheim, A. Bussiere de Nercy, B. Frois, M. Huet, Ph. Leconte, J. Mongey, Pham Xuan Ho, D. Royer and S. Turck, Phys. Rev. Lett. 35, 910 (1975).
15. C.F. Clement, Phys. Lett. 28B, 398 (1969).
16. H. Rothhaas, Dissertation, Mainz, 1978.
17. L. Ray, W.R. Coker and G.W. Hoffmann, Phys. Rev. C18, 2641 (1978).
18. L. Ray Phys. Rev. C19, 1855 (1979).
19. J.W. Negele, Phys. Rev. C1, 1260 (1970).
20. J.W. Negele and D. Vautherin, Phys. Rev. C5, 1472 (1972).
21. I. Badawy, B. Berthier, P. Charles, M. Dost, B. Fernandez, J. Gastebois and S.M. Lee, Phys. Rev. C17, 978 (1978).
22 S.L. Tabor, B.A. Watson and S.S. Hanna, Phys. Rev. C14, 514 (1976).
23. P.R. Christensen and A. Winther, Phys. Lett. 65B, 19 (1976).
24. F.J. Viñas, M. Lozano and G. Madurga, to be published (1980).
25. F.J. Viñas and G. Madurga, Anales de Fisica 73, 92 (1977).
26. J.B. McGrory, B.H. Wildenthal and E.C. Halbert, Phys. Rev. C2, 186 (1970).
27. D.W.L. Sprung, I. Martorell and X. Campi, Nucl. Phys. A268, 301 (1976).
28. J.L. Friar and J.W. Negele, Nucl. Phys. A212, 93 (1973).

A STUDY OF THE REACTION MECHANISM FOR ^{12}C PLUS ^{209}Bi AT E(C) = 61.1-73 MeV

Jin Gen-Ming, Xie Yuan-Xiang, Zhu Yong-Tai,
Shen Wen-Qing, Yu Jú-Sheng, Sun Xi-Jun,
Liu Guo-Xing, Sun Chi-Chang, Guo Jun-Sheng,
Wu Zhon-Gli, Xu Shu-Wei, Wu En-Chin and
Yang Chen-Zhong

Institute of Modern Physics
Lanzhou, The People's Republic of China

J.D. Garrett

The Niels Bohr Institute
Copenhagen Ø, Denmark

I. ABSTRACT

Elastic, evaporation-residue, fission, and light-particle cross section excitation functions have been measured for $^{12}C + ^{209}Bi$ at incident energies near and just above the Coulomb barrier. From these data a "direct ^{8}Be transfer" process has been established corresponding to about 10% of the total reaction cross section just above the Coulomb barrier and decreasing to about 5% of the total reaction cross section at about 10 MeV above the Coulomb barrier. It remains to be seen if the "massive transfer" mechanism can explain these transfer cross sections at such low angular momenta. An S-wave fission barrier of about 6.3 MeV is indicated for ^{221}Ac from a statistical-model analysis of the evaporation-residue excitation function. It is possible to account for the total reaction cross section obtained from the optical-model analysis of the elastic data by the measured reaction channels.

II. INTRODUCTION

In heavy-ion collisions the division of the reaction cross

TABLE I

Summary of Experimental Cross Sections for the $^{12}C + ^{209}Bi \rightarrow ^{221}Ac$ REaction.

E_{lab} (MeV)	$E^*(^{221}Ac)$ (MeV)	σ_{er} a) (mb)	σ_{fis} b) (mb)	σ_{cf} c) (mb)	σ_{α} d) (mb)
61.1	25.0	1.3±.2	-	-	2.4±0.4
62.5	26.3	12±2	-	-	-
64.0	27.7	34±5	93±4	127±9	14±2 e)
65.2	28.9	56±8	-	-	-
65.8	29.4	-	134±6	202±16 f)	-
66.4	30.0	79±12	-	-	-
67.5	31.0	109±16	199±8	308±24	29±4 e)
68.8	32.3	152±23	-	-	-
70.0	33.4	153±23	313±14	466±37	33±5 e)
73.0	36.3	178±27	520±23	698±50	34±5 g)

a) Evaporation residue cross section measured using He-jet technique and for 70 and 73 MeV using mica track detectors (see ref.6).

b) Weighted average of fission cross sections measured using mica track detectors and silicon surface barriers detectors (see ref.6)

c) $\sigma_{cf} = \sigma_{er} + \sigma_{fis}$.

d) Cross section corresponding to $(^{12}C, \alpha)$ direct ^{8}Be transfer.

e) Center of target beam energy for these measurements (64.3, 67.2, and 70.1 MeV) slightly different than the corresponding tabulated value.

f) σ_{er} obtained from extrapolation of values measured at $E(^{12}C) = 65.2$ and 66.4 MeV.

g) Also 1±1, 14±3, and 17±3 mb observed for $(^{12}C, Li)$, $(^{12}C, Be)$, and $(^{12}C, B)$ transfer, respectively.

section between the various competing reactions is a subject of considerable current interest. For example, it is possible, in principle, to obtain information on the angular momentum dependence of the fission barriers and the shape of the nucleus at the saddle-point deformation from the competition between particle decay and fission in the decay of an equilibrated system. Recently quite large cross sections have been observed for direct-like processes (the so-called "massive transfer" or "incomplete fission" reaction) in which apparently it is not possible for the complete mass of the projectile to fuse with the target nucleus even though the angular momentum is significantly less than that for which a fission barrier exists. Therefore, even at relatively low incident energies, it is important to make detailed measurements of the incident energy dependence of the cross sections corresponding to the various competing processes. At lower incident energies the peak in the angular distribution, which corresponds to direct transfer from a grazing trajectory, is located at a large scattering angle. Thus direct transfer to a heavy target nucleus can be distinguished unambiguously from both particle emission and transfer to a light target impurity by the angular distribution shape. It then is possible to search for such direct transfer cross section at lower incident energies and angular momenta where they are not yet well established.

We have measured the elastic scattering, fission, evaporation residue ($^{213,214,214}Fr$ and $^{209,210,211}At$), and transfer (σ_{α}, σ_{Li}, σ_{Be}, σ_{B}) cross sections for $^{12}C + ^{209}Bi$ at incident energies near the Coulomb barrier using $\Delta E - E$ telescopes, mica track detectors and He-jet techniques. These measurements are summarized in ref.6 which also contains a list of references to papers published (in Chinese) in Physica Energiae Fortis et Physica Nuclearis, describing the various experimental techniques in detail. The measured cross sections are summarized in Table I.

III. ELASTIC SCATTERING AND OPTICAL-MODEL ANALYSIS

The elastic scattering cross sections are compared with optical-model predictions in Fig.1. The optical-model parameters (shown in Table II) contain some of the usual features observed in the analysis of the "light, heavy-ion" elastic scattering : (i) an increase in the imaginary potential in the region of the nuclear surface as the incident energy increases from near to a few MeV above the Coulomb barrier; (ii) a preference for a smaller diffusivity in the imaginary potential than in the real nuclear potential; (iii) a incident energies near the Coulomb barrier a large discrepancy between both the value of the grazing angular momentum (ℓ_g) and the total cross section σ_R predicted by the optical model and in a "quarterpoint" analysis[8] (see Table II).

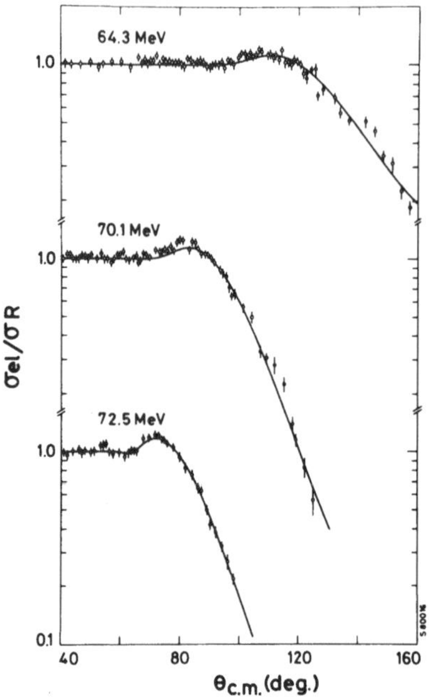

Fig.1 Comparison of measured and calculated elastic scattering cross sections for $^{12}C + ^{209}Bi$. The corresponding "best fit" opitcal-model parameters are given in Table II.

TABLE II

Optical-model parameters obtained from analysis of $^{12}C + ^{209}Bi$ elastic scattering
$V_N(r) = V[1+\exp(r-R_o/a)] + iW[1+\exp(r-R_i/a_i]$
where, $R_{o(i)} = r_{o(i)}[(12)^{1/3} + (209)^{1/3}]$
$= 8.224\ r_{o(i)}$

E_L (MeV)	V (MeV)	r (fm)	a (fm)	W (MeV)	r_i (fm)	a_i (fm)
64.3	100	1.146	0.60	10.0	0.900	0.45
70.1	100	1.169	0.55	10.0	1.278	0.45
72.5	100	1.262	0.45	22.5	1.345	0.30

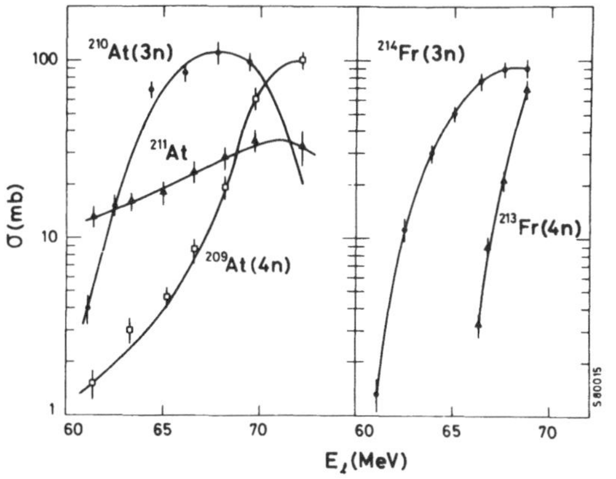

Fig.2 Excitation function of He-jet (right-hand side) and stacked-foil (left-hand side) activity cross sections for $^{12}C + ^{209}Bi$. The lines are only to guide the eye. The ^{214}Fr and ^{210}At and the ^{213}Fr and ^{209}At α-activities are associated with daughters of the 3n and 4n evaporation residues, respectively. The ^{211}At activity is thought to correspond to an ^{8}Be transfer to highly excited states in ^{217}Fr which emit two prompt neutrons before α-decaying to ^{211}At.

IV. STATISTICAL-MODEL ANALYSIS OF THE EVAPORATION-RESIDUE AND FISSION CROSS SECTIONS

The excitation function of the evaporation residues is shown for the individual channels in Fig.2, and for the total evaporation residue cross section in Fig.3. Except at the lowest incident energy the cross section, determined from the decay of ^{211}At is thought to be dominantly ^{8}Be transfer to the states in ^{217}Fr high enough in excitation to emit two neutrons before decaying to ^{211}At by α-emission (see Section V). The excitation function of the complete fission cross section, σ_{cf}, taken to be the sum of the fission and evaporation residue cross sections, also is given in Fig.3.

The code ALICE[9] has been used to calculate the competition between particle evaporation and fission through a rotating liquid-drop fission barrier[1]. The initial spin distribution of the compound system (see Fig.4) was calculated from the transmission coefficients using a parabolic approximation to a real nuclear (Wood-Saxon) plus Coulomb potential[10]. The parameters of the Wood-Saxon potential (V_0 = 67 MeV, r_0 = 1.193 f.m. and a = 0.57 f.m.) were adjusted to reproduce the energy dependence of the complete fusion

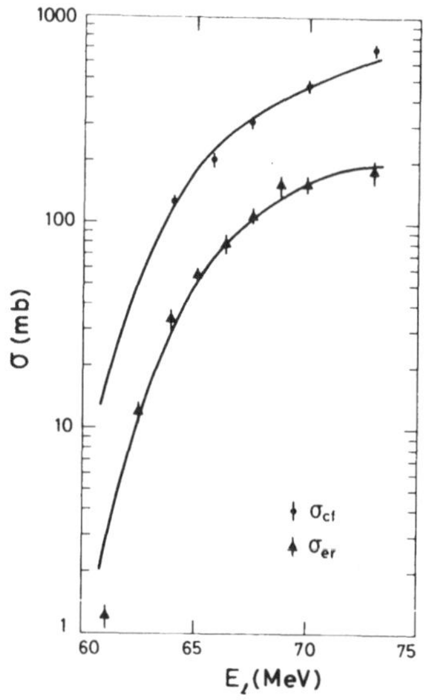

Fig.3 Excitation function for the complete fusion cross section ($\sigma_{cf} = \sigma_{er} + \sigma_{fis}$) and the evaporation residue cross section for $^{12}C+^{209}Bi$. The curve shown with the fusion cross section was calculated from transmission coefficients obtained from a parabolic approximation to a real Woods-Saxon plus Coulomb potential. The curve shown with the evaporation residue cross section was obtained from the calculated competition between particle emission and fission through a rotating liquid-drop fission barrier shown for ^{221}Ac in Fig.4. The parameters used in these calculations are discussed in Section III and in ref.6.

cross section (see Fig.3). The evaporation-residue cross section is concentrated at the lower values of angular momentum; therefore, the values of σ_{er} are particularly sensitive to the competition between fission and particle evaporation, which in the analysis is dependent on the fission barrier heights (which vary as a function of the angular momentum[1]- see Fig.4) and the level-density parameters at the fission saddle point, a_f, and the equilibrium deformation, a_n. The predicted energy dependence of the evaporation-residue cross section is compared in Fig.3 with the measured cross sections. The theoretical curve was calculated assuming a fission-barrier height kept equal to 0.905 times the rotating liquid-drop value , $B_f^{\ell d}$, for all angular momenta. Level-density parameters equal to A/8 were used at all angular momenta both for the particle emission, a_n, and for fission, a_f. Using such naive assumptions,

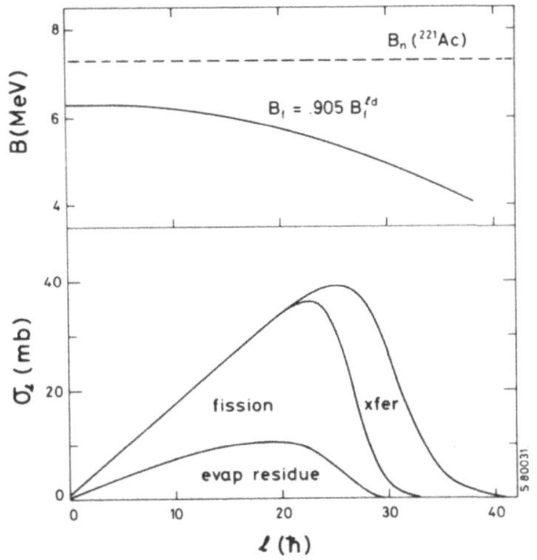

Fig.4 Actinium-221 fission barrier (0.905 times the rotating liquid-drop barrier) as a function of ℓ (top, which was used to reproduce the evaporation residue excitation function (see Fig.3)). The ^{221}Ac neutron binding energy also is shown. The resulting predicted evaporation residue cross section for 73 MeV is shown as a function of ℓ in the bottom portion of the figure. For comparison the complete fusion cross section calculated using the parabolic approximation as well as the total reaction cross section calculated using optical-model parameters obtained from the analysis of the 72.5 MeV elastic scattering data (Fig.1) also is shown for 73 MeV as a function of ℓ. The reaction processes associated with the various portions of the total reaction cross section are indicated.

it is possible to reproduce the energy dependence of the evaporation-residue cross sections.

Such an analysis yields a value of 6.3 MeV for the S-wave fission barrier of the ^{221}Ac compound system to be compared to 7.3 and 7.2 MeV obtained for ^{227,228}Ac respectively from an analysis of direct-reaction induced fission[11]. It is noted, however, that the values of the fission barrier heights and the level-density parameters are quite sensitive to the expression of the level density used in the calculations[6]. Furthermore, the barriers obtained from the analysis of heavy-ion induced evaporation-residue and fission cross sections are strongly dependent upon the angular-momentum dependence of the fission barrier and the level-density parameters, whereas direct-reaction and neutron-induced fission is sensitive to these parameters only at the lowest angular momenta. The present data, however, indicate a decrease in the fission barrier height for Ac nuclei with decreasing neutron number.

V. ESTABLISHMENT OF A DIRECT-LIKE ^{8}BE TRANSFER PROCESS

Probably the most interesting result of the present work is the establishment of a sizable (^{12}C, α) "direct ^{8}Be transfer" process at incident energies which extend down to the Coulomb barrier.

In the ^{209}Bi(^{12}C, α) spectra a broad group (≈5 MeV FWHM) is observed at an ejectile energy which corresponds to the optimum Q-value[12] for the (^{12}C, α) transfer reaction (see Fig.5). The angular distribution of this broad group is peaked at the grazing angle which moves from 165° to 80° as the incident energy is

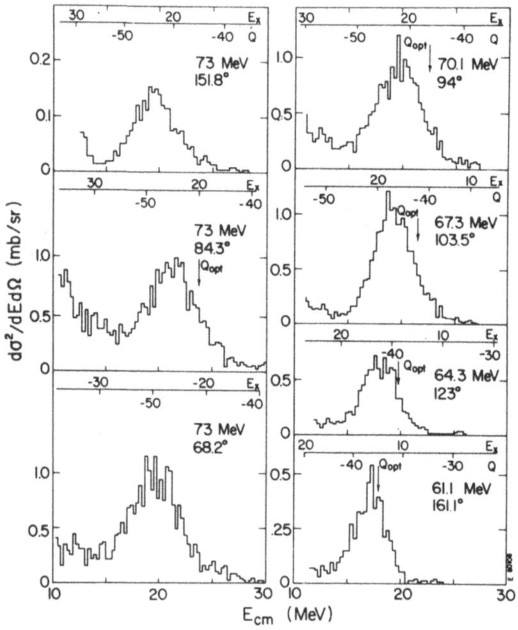

Fig.5 Bismuth-209(^{12}C, α) spectra for E(^{12}C) = 73 MeV at forward and backward of the grazing angle (left) and at the grazing angle as a function of the incident energy (right). An estimate of the optimum Q value[12] is shown by the arrows.

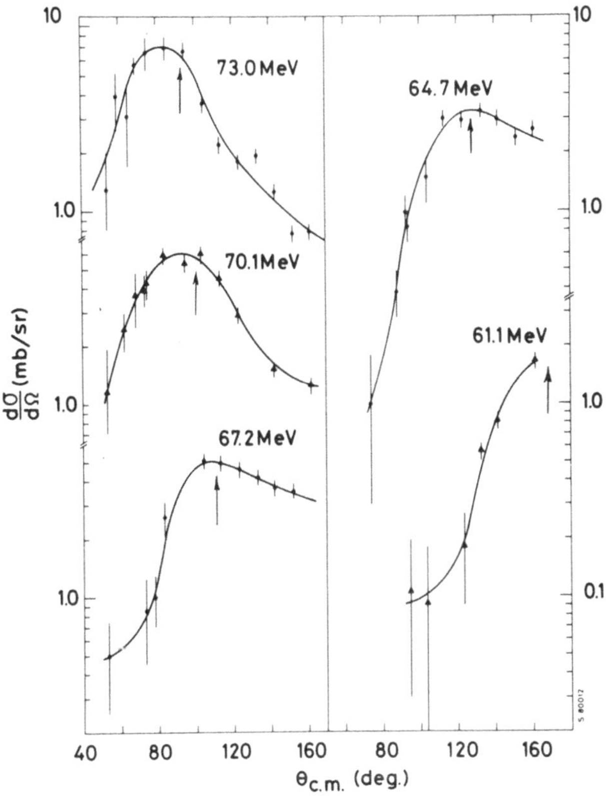

Fig.6 Angular distributions corresponding to the "direct" component (see Fig.5) in the ^{209}Bi(^{12}C, α) spectra as a function of incident energy. An estimage of the grazing angle which includes the effects of the nuclear potential, is shown for comparison by the arrows.

increased from 61 to 73 MeV (Fig.6). Such an angular and energy dependence of the ^{209}Bi(^{12}C, α) cross section is indicative of the direct transfer of ^{8}Be to the target. The maximum observed in the angular distributions corresponding to the (^{12}C, B) and (^{12}C, Be) reactions also is at the same angle as the (^{12}C, α) maximum (see Fig.7).

At these low incident energies where the direct-reaction α-particles are concentrated at large angles, there are no experimental problems associated with distinguishing the Be transfer cross section from statistical α-particles or α-particles originating from direct reactions on light impurities in the target. It

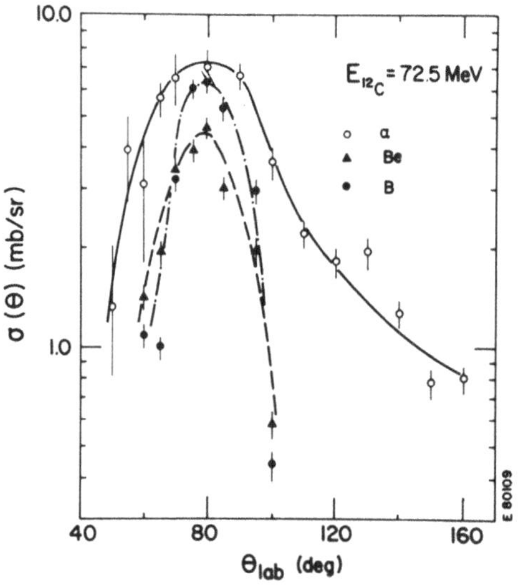

Fig.7 Comparisons of the angular distributions for the (^{12}C, B) and (^{12}C, Be) reactions on ^{209}Bi at E(^{12}C) = 73 MeV with the "direct" component of the ^{209}Bi(^{12}C, α) reaction at the same incident energy.

is necessary, however, to exclude the (^{12}C, ^{8}Be) reaction followed by the breakup of the ^{8}Be ejectile into two α-particles as the source of the direct α-particle production. The measured α-particle cross sections (Table I) agree in magnitude with the ^{211}At activity (Fig.2 and ref.13) except at the lowest incident energy where the 2n cross section should contribute to the ^{211}At activity. This agreement argues for direct ^{8}Be transfer as the source of the measured α-particles. The observed ^{211}At activity corresponds to the α-daughter of ^{215}Fr, which is formed by ^{8}Be transfer to states at sufficient excitation energies in ^{217}Fr (see Fig.5) so that two prompt neutrons are emitted before the α-decay to ^{211}At. If an α-particle had been transferred to the target followed by the detection of one of the α's from the breakup of ^{8}Be, then ^{213}At would have been formed with low excitation energy and the ^{211}At activity would not have been seen except at the lowest incident energy where it corresponds to the daughter of the 2n evaporation residue. Furthermore, at high incident energies, $\gtrsim$72 MeV, the α-activity of ^{212}At (τ=315 ms), the 2n daughter of ^{213}At, should have been observed in He-jet measurements if the (^{12}C, ^{8}Be) cross sections were large. This activity apparently was not observed in the Orsay He-jet measurements[14], which cover this range of incident energies.

Another experimental method of excluding the (^{12}C, $^{8}Be \rightarrow 2\alpha$) reaction is the detection of the two breakup α-particles in the same E-E telescope where they would identify as Li. Such measurements have been made for C + Bi at E(C) = 73 MeV. The resulting

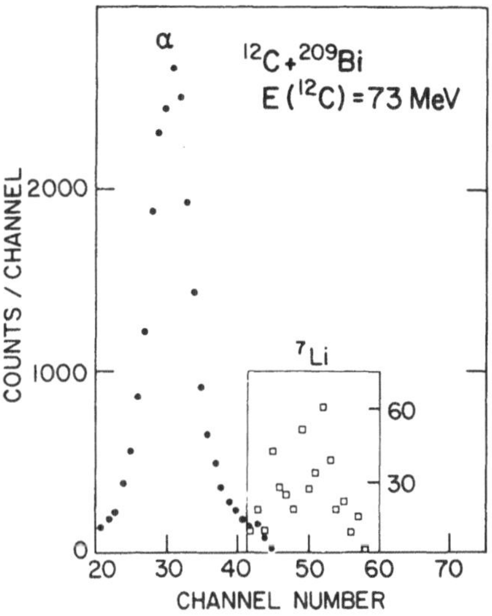

Fig. 8 Comparison of $^{209}Bi(^{12}C, \alpha)$ and $^{209}Bi(^{12}C, ^{7}Li)$ spectra for $E(^{12}C) = 73$ MeV at the angle of the maximum in the $^{209}Bi(^{12}C, \alpha)$ angular distribution (see Figs. 6 and 7). Since the detection of the two α-particles from the breakup of ^{8}Be identifies as ^{7}Li (see ref. 15), the ^{7}Li represents an upper limit for the $^{209}Bi(^{12}C, ^{8}Be)$ reaction. The ^{7}Li spectrum has been multiplied by a fractor of three to account for the efficiency for detecting both α-particles from ^{8}Be breakup in the ΔE-E telescope.

same ΔE-E telescope where they would identify[15] as ^{7}Li. Such measurements have been made for $^{12}C + ^{209}Bi$ at $E(^{12}C) = 73$ MeV. The resulting α-particle and ^{7}Li energy spectrum are shown in Fig.8. The ^{7}Li cross section is observed to be much less than the α cross section, thereby ruling out the $(^{12}C, ^{8}Be \rightarrow 2\alpha)$ reaction as the dominant source of the direct α-particles.

Similar measurements for the $^{197}Au(^{12}C, \alpha)$ reaction as a function of angle and incident energy, as well as for the $^{197}Au(^{12}C, ^{8}Be \rightarrow 2\alpha)$ process also establish a "direct-like ^{8}Be transfer" for $^{12}C + ^{197}Au$ at similar incident energies, which are just above the Coulomb barrier for $^{12}C + ^{198}Au$.

The present measurements definitively establish a "direct ^{8}Be-transfer" process for ^{12}C + heavy target at incident energies near the Coulomb barrier. Such a process has been established previously for the $^{12}C + ^{209}Bi$ system at higher incident ^{12}C ener gies from measured α-particle cross sections[16] and has been suggested at incident energies near the Coulomb barrier from radiochemical measurements[13]. The Be transfer cross section is as

TABLE III

Comparison of grazing angular momentum and total reaction cross sections obtained from the optical-model analysis and from the quarter point of the elastic data.

	Optical-model[a]		Quarter Point			
	[b]		[c]	[d]	[d]	
E_L (MeV)	ℓg (h)	σ_R (mb)	$\theta_{1/4}$ (deg)	ℓ_g ($\hbar$)	σ_R (mb)	$\sigma_R(om)/\sigma_R(qp)$
64.3	9.7	109	153.4	7.5	69	1.58
70.1	23.2	515	110.5	21.7	463	1.11
72.5	29.3	769	96.5	27.9	710	1.08

[a] Values from optical-model analysis using parameters given in Table II.

[b] Angular momentum for which the transmission coefficient, T_ℓ, equals 0.5.

[c] Center of mass angle, at which the experimental differential cross section falls to 1/4 of the Rutherford cross section.

[d] Defined as in ref.8.

large as $\approx$10% of the observed total reaction cross section at $E(^{12}C)$ = 64 MeV, the lowest incident energy for which complete cross section measurements are available (Table I), decreasing to about 5% of the total reaction cross section at 73 MeV. Large cross sections for "massive transfer" or "incomplete fusion" reactions have been observed (see e.g. refs.3-5) for "light heavy ion" induced α-particle producing reactions; however, the cross section corresponding to such reactions usually has been associated with rather large angular momentum. Applying the criteria suggested by Siwek-Wilczynska et al.[5] to the $^{12}C + ^{209}Bi$ system, such "massive-transfer" reactions would be expected to dominate the cross sections above $\ell \gtrsim 45$ for which it would be impossible to fuse ^{12}C with ^{209}Bi. The grazing angular momentum is well below this value for all the new data presented in this article (see Table III). It remains to be seen if the more sophicated analysis of the "massive transfer" reactions as multistep transfer processes[17] can explain the present measurements.

VI. DECOMPOSITION OF THE TOTAL REACTION CROSS SECTIONS INTO THE VARIOUS REACTION CHANNELS

From the measured cross sections and the present analysis, it is possible to decompose the ℓ-dependence of the total reaction cross section into the various composite reaction cross sections. This is shown for the 73 MeV case in Fig.4. The ℓ-distribution of the total reaction cross section is given by the optical-model analysis of the elastic scattering. The distribution in ℓ of the

TABLE IV

Comparison of measured $\sigma_{cf} + \sigma_{\alpha}$ with optical-model total reaction cross section.

E_L (MeV)	$\sigma_{cf} + \sigma_{\alpha}$ [a] (mb)	σ_R [b] (mb)
64.0	141±11	93
67.5	337±28	369 [c]
70.1	499±41	515
73.0	731±55 [d]	769

[a] Measured value - see Table I.

[b] Predicted total reaction cross section from optical model using parameter of Table II.

[c] Optical-model parameters obtained from 70.1 Mev elastic data used.

[d] An additional cross section of 32 mb has been measured for the Li, Be, and B exit channels.

complete fusion cross section was obtained from transmission through a parabolic approximation to the Coulomb plus real Woods-Saxon potential, which has been adjusted to reproduce the energy dependence of the $\sigma_{er} + \sigma_{fis}$ cross sections. The fusion cross section then i divided as a function of ℓ by the competition between particle evaporation and fission assuming the angular-momentum dependence of the fission barrier given by a rotating liquid drop[1]. In this mass region the S-wave fission barrier is approximately equal to the neutron threshold; therefore, already at ℓ = o there is competition between particle evaporation and fission. The fission competition increases with increasing ℓ, the result of the decreased fission barrier with ℓ, until at $\ell \gtrsim 20$ fission becomes the dominant decay process of the compound system. At 73 MeV incident energy the unitarity limit of the partial cross section i.e. $\sigma_{\ell} = \pi \lambda\!\!\!^{-2}(2\ell + 1)$, is accounted for by the compound processes up to $\ell \approx 22$ units; therefore, the "direct transfer" cross section is expected to correspond to the higher or "grazing" ℓ values. The localization at the grazing ℓ's of the remaining cross section is consistent with the Q and angular dependence of the transfer cross section, which are indicative of a direct mechanism for these processes. Except at the lowest incident energy (near the Coulomb barrier), where there may be problems resulting from a large variation in the measured cross sections or to the spread in incident energy, there is excellent agreement in the magnitude of the total reaction cross section predicted from the optical-model analysis of the elastic data and the sum of the measured cross section (see Table IV).

REFERENCES

1. S. Cohen, F. Plasil, and W.J. Swiatecki, Ann. Phys. (N.Y.) 82, 557 (1974).
2. S. Bjørnholm, Aa. Bohr, and B.R. Mottelson in Proceedings of the Third International Atomic Energy Symposium on Physics and Chemsitry of Fission, Rochester, 1973. (International Atomic Energy Agency, Vienna, 1974) Vol.I, p.367.
3. T. Inamura, M. Ishihara, T. Fukuda, T. Shimoda, and H. Hiruta, Phys. Lett. 68B, 51 (1977).
4. D.R. Zolnowski, H. Yamada, L.E. Cala, A.C. Kahler, and T.T. Sugihara, Phys. Rev. Lett. 41, 92 (1978).
5. K. Siwek-Wilczynska, E.H. du Merchie van Voerthuysen, J. van Popta, R.H. Siemssen, and J. Wilczynski, Phys. Rev. Lett. 42, 1599 (1979).
6. Jin Genming, Xie YuangXiang, Zhu Yongtai, Shen Wenqing, Sun Xijun, Guo Junsheng, Liu Guoxing, Yu Jusheng, Sun Chichang and J.D. Garrett, Nucl. Phys. (in press).
7. See, e.g., F. Videbæk, R.B. Goldstein, L. Grodzins, S.G. Steadman, T.A. Belote, and J.D. Garrett, Phys. Rev. C15, 954 (1977).
8. See, e.g., J.S. Blair, Phys. Rev. 95, 1218 (1954).

9. F. Plasil and M. Blann, Phys. Rev. C11, 508 (1975).
10. T.D. Thomas, Phys. Rev. 116, 703 (1959).
11. E. Konecny, H.J. Specht, and J. Weber, Phys. Lett. 45B, 329 (1973).
12. D.M. Brink, Phys. Lett. 40B 37 (1972).
13. R. Bimbot, D. Gardes, and M.F. Rivet, Nucl. Phys. A189, 193 (1972).
14. Y. Le Beyec. M. Lefort, and M. Sarda, Nucl. Phys. A192, 405 (1972).
15. G.J. Wozniak, H.L. Harney, K.H. Wilcox, and J. Cerny, Phys. Rev. Lett. 28, 19 (1972); and ibid. 29, 760E (1972).
16. H.C. Britt and A.R. Quinton, Phys. Rev. 124, 877 (1961).
17. See, e.g., T. Udagawa and T. Tamura, preprint 1980.

INELASTIC SCATTERING AND TRANSFER REACTIONS USING VERY HEAVY IONS

M.W. Guidry, R.E. Neese and T.L. Nichols

Department of Physics
University of Tennessee
Knoxville, Tennessee 37916

and

Oak Ridge National Laboratory
Oak Ridge, Tennessee 37830

I. ABSTRACT

The quasi-elastic population of discrete collective states with very heavy ions is a relatively unexplored area of nuclear physics. This is a consequence of the fact that standard experimental (charged-particle spectroscopy) and theoretical (DWBA) methods are not adequate for this problem. We may expect new features in such reactions which are not present in reactions with lighter ions or with non-collective nuclei. These unique features may be broadly separated into two categories: (1) the strong response of the collective modes allows the study of nuclear single-particle and correlation structure under the influence of multiple collective quanta; (2) strong collective excitation may lead to unique localizations in coordinates conjugate to quantum numbers which assume large values.

It now seems that this new area of heavy-ion physics can be studied experimentally using particle-γ coincidence spectroscopy, and theoretically using semiclassical coupled-channels and semiclassical coupled-channels Born approximations. In this paper we discuss the application of such methods to inelastic and transfer reactions involving very heavy ions and deformed target nuclei.

II. INTRODUCTION

Inelastic scattering and transfer reactions with light ions have long proved a useful source of information about nuclear structure[1]. In recent years there has been significant interest in extending such studies by using heavy ions ($A > 4$). Except for numerous Coulomb excitation experiments, these investigations have emphasized light heavy ions such as oxygen, and have concentrated on nuclei only a few nucleons removed from closed shells. For the most part, they have used the experimental and theoretical methods developed for light-ion reactions: high-resolution charged-particle spectroscopy and DWBA or quantum-mechanical coupled-channels calculations[2].

In this talk we wish to discuss the extension of such investigations to include much heavier ions and target nuclei appreciably removed from closed shells. Inelastic scattering and transfer reactions of this type represent a largely unexplored area of nuclear physics, due primarily to the inadequacy of the standard experimental and theoretical techniques in this context. However, it now appears that a combination of particle-γ coincidence spectroscopy and semiclassical coupled channels calculations allow access to this new area.

We may expect that in such experiments features will appear which play only a minimal role in previously investigated reactions. We separate these expected features into two broad categories:

1. Due to the strong coupling of the target collective modes to the Coulomb and nuclear fields of the heavy ion there will be appreciable collective excitation. This provides an opportunity to study those aspects of nuclear structure to which inelastic and transfer reactions are sensitive under the stress of the collective excitation.

2. Strong collective excitation may allow an uncertainty principle kind of localization in coordinates conjugate to quantum numbers which assume large values.

For our purposes we shall define a collective nucleus as one, usually heavy and appreciably removed from closed shells, for which collective and multiple collective excitations are dominant processes in a heavy-ion collision. By very heavy ion we mean an ion of sufficient charge and mass such that in a collision with a collective nucleus there is appreciable probability of multiple collective excitation.

As prototypes of such collective nuclei we shall consider strongly deformed rare-earth nuclei. However, strongly collective vibrators or even nuclei having more esoteric collective modes such as pairing rotations could also fall within this category. Our definition of "very heavy" clearly depends on the collectivity of the target nucleus, but in what follows we shall assume that A $\gtrsim$ 40 for the projectile.

III. INELASTIC SCATTERING

The use of heavy ions to place collective stress on nuclear systems has received significant attention. A well-known example is the area of 'yrast spectroscopy' where the manner in which the nucleus accomodates energy and angular momentum is studied[3]. Less well explored is the idea that excitation of strong collective degrees of freedom may lead to localizations in coordinates which are conjugate to quantum numbers acquiring large values. Inelastic scattering of very heavy ions from well-deformed nuclei provides a particularly clear example of such a localization effect[4,5]. If we consider experiments in which the heavy ions are detected at backward angles the classical limit description of the excitation is quite simple: the scattering corresponds approximately to collisions with zero impact parameter and the system is defined by a radial coordinate r, a target orientation angle χ, and the conjugate momenta P_r and $P_\chi = I/h$, where I is the rotor angular momentum in units of h (see Fig.1).

In the classical-limit S-matrix (CLSM) formalism the S-matrix element for the transition $0 \to I$ in the rotor in a zero impact parameter collision (J=0, where J is the total initial angular momentum) may be written[6-12]

$$S^{J=0}_{I\leftarrow 0} \approx \frac{\sqrt{2I+1}}{2} \int^{\pi} \sqrt{\sin\chi_o \, \sin\bar{\chi} \, \frac{d\bar{\chi}}{d\chi_o}} \; P_I \, (\cos\bar{\chi}) \qquad (1)$$

where the integral is over initial classical orientations χ_o, $\bar{\chi}$ is related to χ by a canonical transformation, and the phase $\Delta(\chi_o)$, defined in ref.7, is essentially the classical action over the

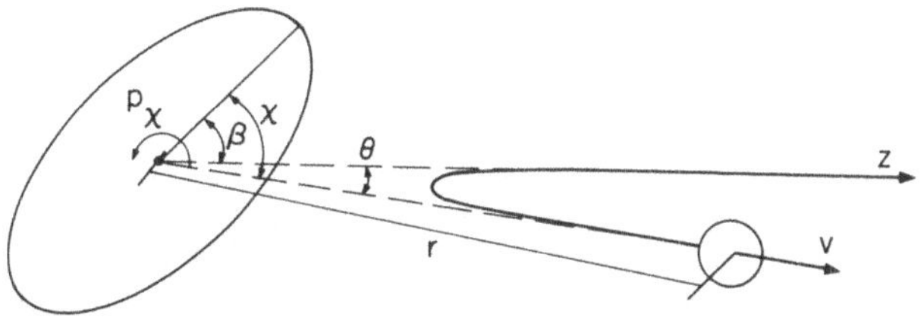

Fig.1 Coordinate system for inelastic backward scattering from a deformed nucleus.

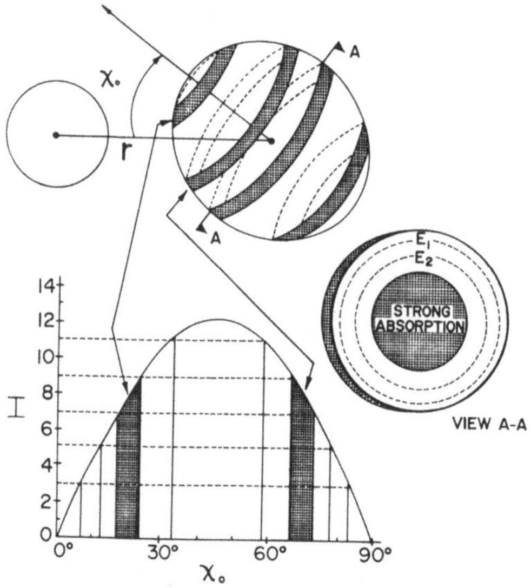

Fig.2 Schematic illustration of radial and angular localization in the collision of a very heavy ion with a deformed nucleus.

trajectory beginning with initial orientation angle χ_0. Although the quantum principle of superposition is implicit in the integral of Eq.(1), all quantities there are to be determined by solving the classical equations of motion for the system.

Eq.(1) may be integrated numerically. Alternatively, we may exploit the fact that in the classical limit $\Delta \gg \hbar$ and use stationary phase methods to evaluate the integral. The resulting expressions acquire a particularly simple and instructive form if the points of stationary phase are well separated. In that case the excitation probability for populating a state I may be written

$$P_I \sim |S^{J=0}_{I\leftarrow 0}|^2$$

$$\sim p_1 + p_2 + 2\sqrt{p_1 p_2}\, \sin(\Phi_2 - \Phi_1) \qquad (2a)$$

(classically allowed)

$$\sim p\, e^{-2\mathrm{Im}\Phi} \qquad (2b)$$

(classically forbidden)

where the p's are classical probabilities and Φ represents a classical action. The classically allowed and forbidden cases are conveniently distinguished on the basis of the quantum number function (final classical spin vs. initial orientation), as shown schematically in Fig.2. The classically allowed transitions are those

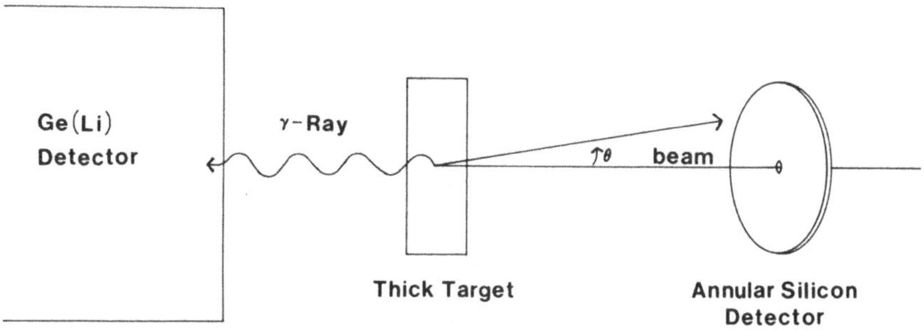

Fig.3 Schematic diagram of the particle-γ coincidence experiment used to study inelastic scattering

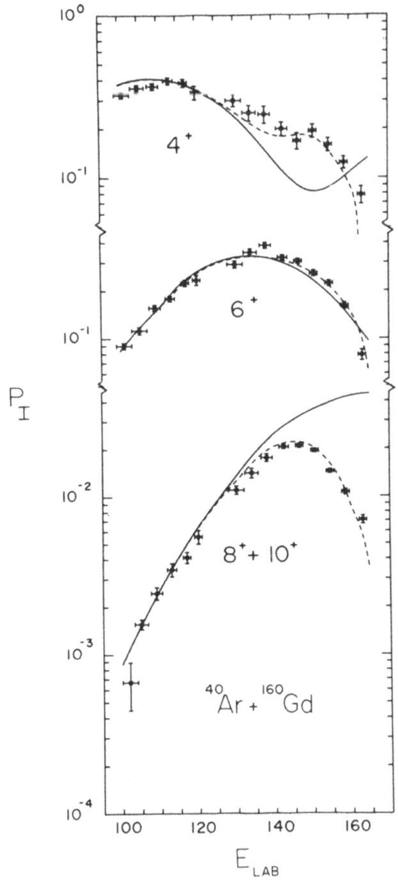

Fig.4 Some excitation functions in coincidence with backward scattered projectiles for the reaction $^{40}Ar + ^{160}Gd$ in the Coulomb-nuclear interference region.

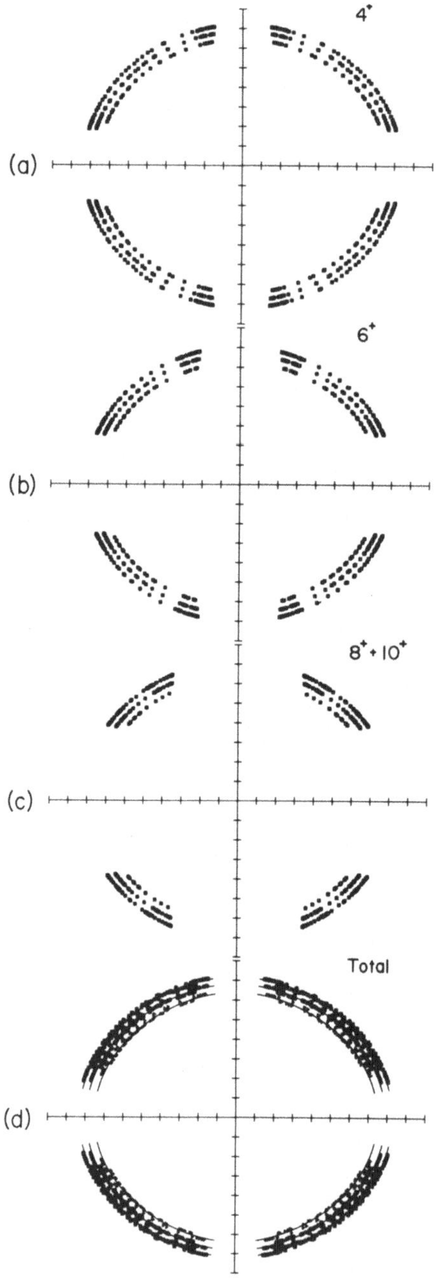

Fig.5 The deformed ion-ion potential contours for $^{40}Ar + ^{160}Gd$. Figs. (a)-(c) represent the contribution of each state and Fig. (d) is the composite nuclear potential energy surface for the ion-ion interaction. Each radial unit is 2fm. The solid lines are spherical harmonic representations of each contour (see Eq. 14 and Table II).

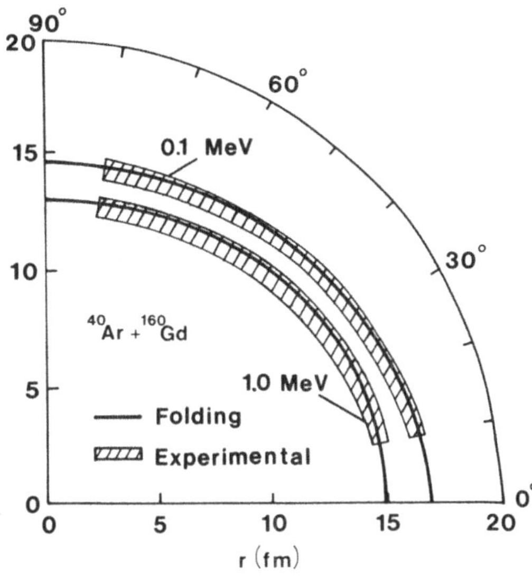

Fig.6 Comparison of experimental contours of the ion-ion potential with those calculated using a deformed folding potential, as described in the text. The calculations are shown using solid lines and the curved boxes denote the experimental uncertainties in the contours of Fig.5.

with I less than the maximum of the quantum number function. From them there are generally two initial orientations satisfying the semiclassical quantization conditions (I = even integer + 1/2) and the classical probabilities p_n and the phases Φ_n are evaluated for trajectories with these initial angles. For the classically forbidden transitions the quantum number function and phase are analytically continued in the complex χ_o plane, and only a single (complex) initial angle contributes. The oscillatory structure of the allowed transitions may be attributed to quantal interference between different orientations leading to the same final state. The exponentially damped structure of (2-b) represents the penetration of the system into classically forbidden regions by virtue of the uncertainty principle.

From Eq.2 and Fig.2, we see that in the classical limit (large average angular momentum transfer) particular spin states involve localized initial orientation angles. In typical heavy-ion collisions the target nucleus does not turn very much during the period of nuclear interaction, and this implies that a scattering event populating a given final spin state probes localized regions of the deformed nuclear surface[4-5].

We illustrate the use of these ideas to determine equipotential contours of the ion-ion potential for the inelastic scattering of $^{40}Ar + ^{160}Gd$ in Figs.(3)-(6). In Fig.3, we show schematically the particle-γ coincidence arrangement used to obtain the data. In Fig.4, excitation functions for exciting several members of the ground-state band of ^{160}Gd are shown. In accordance with the above arguments, it is assumed that each of the states probes particular localized regions of the deformed nuclear surface. Each state is separately fit with a potential consisting of monopole + quadrupole + hexadecapole Coulomb terms using moments taken from Coulomb excitation measurements, an adiabatic giant resonance polarization potential as described in ref.13, and a nuclear potential of the form U = V+iW with

$$V = \frac{V_o}{1 + e^{\frac{r - R(\chi)}{a}}}$$

and an analogous expression for W, and with $R(\chi) = R_0\{A_1^{1/3} + A_2^{1/3}[1+\beta_2 Y_{20}(\chi)+\beta_4 Y_{40}(\chi)]\}$. The resulting best-fit parameters are shown in Table I. By using the classical trajectories to determine which regions of orientation angle are being probed by a particular state, equipotential contours of the ion-ion potential may then be constructed segment by segment as shown in Figs.(5a)-(5c). A smooth

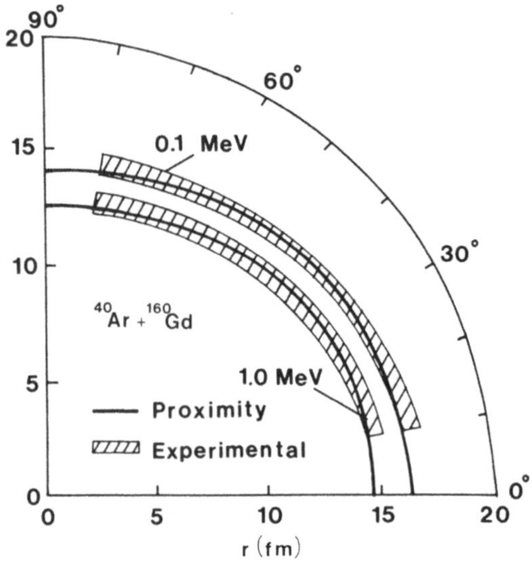

Fig.7 As in Fig.6, but for a deformed proximity potential

TABLE I

Best-Fit Parameters For Potentials

Parameter	4^+ State	6^+ State	$\Sigma 8^+$ - 10^+ States
V_o (MeV)	50.	50.	50.
R_o^R (Fm)	1.253	1.200	1.195
a_R (Fm)	0.649	0.671	0.703
W_o (MeV)	25.	25.	25.
R_o^I (Fm	1.321	1.269	0.981
a_I (Fm)	0.440	0.446	0.915
$\bar{\beta}_2$	0.34	0.34	0.34
$\bar{\beta}_4$	0.032	0.032	0.032

TABLE II

The Parameters $R_o(V)$, $\alpha_2(V)$, And $\alpha_4(V)$
For Deformed Ion-Ion Potentials.

V (MeV)	R_o (V) [Fm]	α_2 (V)	α_4 (V)
0.1	1.683	0.237	0.043
0.3	1.599	0.249	0.047
1.0	1.506	0.264	0.048

average of these contributions from various states yields the real ion-ion potential displayed in Fig.5d.

These contours may be compactly represented by defining each contour using an equation of the form

$$R(V) = R_o(V)(A_1^{1/3} + A_2^{1/3}(1 + \alpha_2(V)Y_{20}(\chi)$$
$$+ \alpha_4(V)Y_{40}(\chi)))$$

The lines shown in Fig.5d are curves fitted to the contours using this expression and the parameters in Table II.

The contours shown in Fig.5 do not represent a shape for the ^{160}Gd nucleus. The ion-ion potential must be unfolded in some manner to separate the projectile and target contributions. To do this we have applied deformed folding potential calculations and deformed proximity potential calculations. The folding calculations have used the Satchler formalism[14] with two-particle fermi densities and the ^{160}Gd density deformed by an amount $C' = C(1+\beta_2 Y_{20}(\chi))$ with β_2=0.33. The deformed proximity calculation follows the formalism of ref.15 with β_2=0.30. The experimental and theoretical comparison are shown in Fig.(6)-(7). The theoretical contours agree rather well with the experimental ones, with no parameter adjustment. Therefore, we conclude that folding and proximity potentials can give a correct description of the average field acting between heavy ions when one of the collision partners is strongly deformed. Since static and dynamical deformations undoubtedly play a significant role in processes such as deep inelastic ones, this provides a necessary (but not sufficient) test of the ability of these theoretical potentials to describe more general heavy-ion collisions.

IV. TRANSFER REACTIONS.

Elastic scattering between spherical ions is governed by a potential representing the average interaction of the projectile nucleons with the target nucleons. The potential governing inelastic and elastic scattering when one of the ions is deformed is a function of additional collective orientation coordinates. However, it too involves an average interaction of many projectile and target nucleons.

Few-nucleon transfer reactions constitute a more specific probe of nuclear structure. It is well known from light-ion reactions that 1-nucleon transfer reactions yield information about single-particle wavefunctions, while 2-nucleon transfer is sensitive to

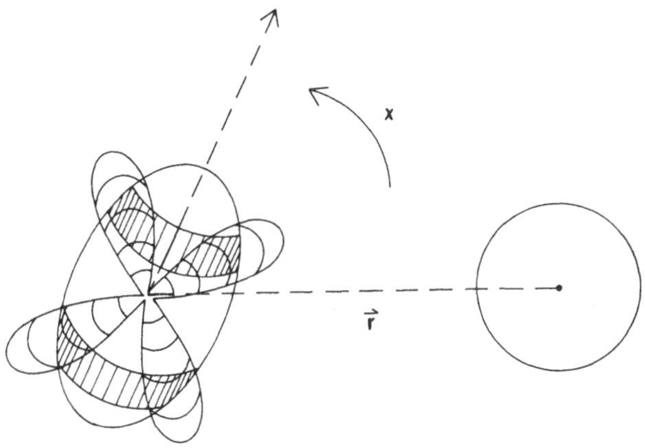

Fig.8 Schematic illustration of zonal localization for transfer reactions in deformed nuclei.

particle-particle correlations[1,2]. As was the case for inelastic scattering, we may expect qualitatively new features for heavy-ion transfer populating strongly collective states.

For a deformed nucleus the m-substate degeneracy is broken for single-particle quantization and there are preferred directions for motion in particular orbits. However, the intrinsic quantization axis executes collective rotation relative to a laboratory coordinate system. If there is no collective excitation in the transfer the anisotropy in the intrinsic frame is destroyed by the quantum fluctuation of this axis. If there is strong collective rotational excitation accompanying the transfer we may expect angular localization of the rotating frame, similar to that for inelastic scattering. In that case we may expect to see effects associated with anisotropy of the single-particle and pairing degrees of freedom in heavy deformed nuclei. This is illustrated in Fig.8. For inelastic scattering this effect led to a zonal partitioning of the average ion-ion potential. The analogous effect in 1-particle transfer implies zonal partitioning of the deformed orbit(s) involved in the transfer. For 2-particle transfer we may expect a zonal probe of those orbits participating in strong pairing correlations (i.e., localized in momentum space around the Fermi momentum).

In the collision of a heavy ion with a strongly deformed nucleus appreciable collective excitation may be expected by the time the ions reach minimum separation, as shown in Fig.9. Quantum mechanically all paths contribute but the transfer process will be dominated by those like the one illustrated schematically in Fig.10: strong inelastic excitation in the entrance channel, transfer, and strong

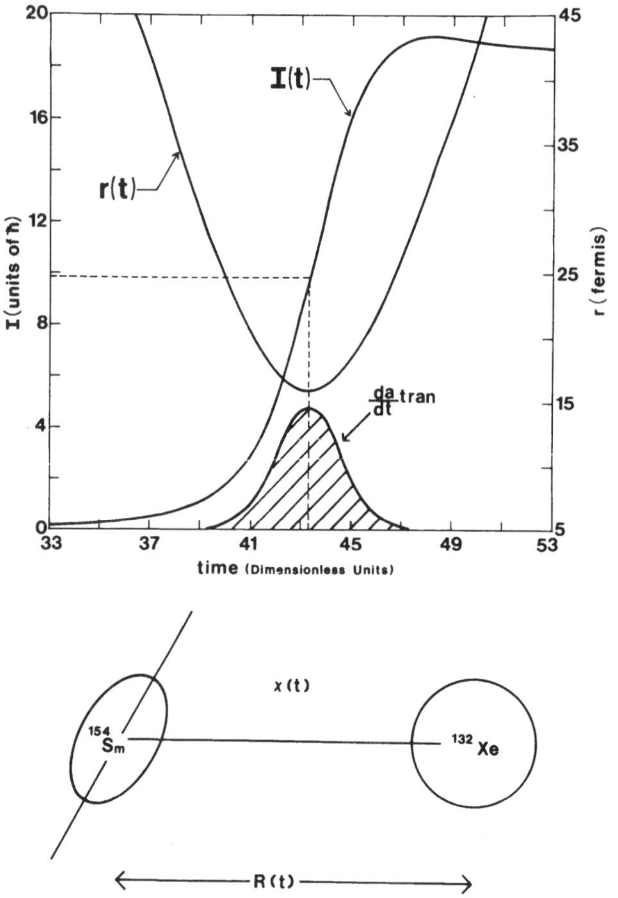

Fig.9 The time dependence of some important quantities in a heavy-ion collision. A Hamiltonian with parameters similar to those listed in Table I has been used. The collision was assumed to involve zero impact-parameter scattering with an initial orientation angle for the rotor of $\chi_0=30°$, The radial separation r and rotor angle momentum I, are shown as a function of time, which is given in dimensionless units of the time required at asymptotic velocity to cover half the distance of closest approach for a Rutherford trajectory. Also shown is a differential transfer amplitude calculated for transfer from a pure [642] 5/2+ 2- particle configuration using the semiclassical barrier penetration form-factor described in the text. Note that most transfer occurs within about 1 Fermi of the turning point, and that more than half of the final collective angular momentum is in the rotor when transfer is likely to take place.

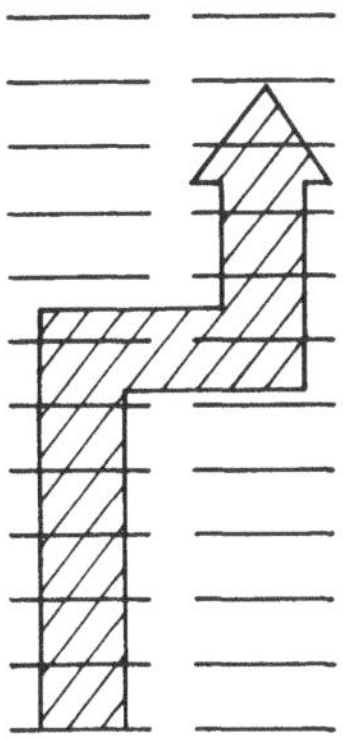

Fig.10 A schematic illustration of heavy-ion transfer reactions involving deformed nuclei, as suggested by Figs.(1) and (2). Although all possible paths contribute, in the classical limit we may expect paths such as that illustrated to dominate: strong inelastic excitation in the entrance channel, particle transfer, and strong inelastic excitation in the exit channel. Accordingly, we may expect such transfer reactions to probe nuclear structure under the influence of significant amounts of collective angular momentum.

inelastic excitation in the exit channel. Such reactions may be used to infer the transfer formfactors between collectively excited states, and hence to study the effect of collective motion on single-particle wave functions and pairing correlations.

There now exists preliminary experimental evidence suggesting that these transfer experiments can be done using sophisticated particle-γ coincidence methods. Theoretically, the problem appears tractable by using a semiclassical coupled-channels Born approximation. That is, the transfer itself is considered in perturbation theory but the attendant collective excitation is treated by methods exploiting the classical-limit nature of the problem.

We shall discuss such a formalism in future publications. In this paper we would like to confine ourselves to some simple calculations for 2-neutron pick-up reactions in even-even rotors. These calculations have been made using a modification of the formalism in section II and a simple model for the transfer formfactor.

In the backward scattering and no-recoil limits the transition amplitude between the ground state of the deformed nucleus in the entrance channel and the state I in the ground band of the transfer daughter nucleus is given by[16]

$$\tilde{S}^{J=0}_{I\leftarrow 0} \sim \frac{\sqrt{2I+1}}{2} \int_0^{\pi} \sqrt{\sin\chi_o \, \sin\bar{\chi} \, \frac{d\bar{\chi}}{d\chi_o}} \, P_I(\cos\bar{\chi}) \cdot F(\bar{\chi}(\chi_o)) \, e^{i/h\Delta(\chi_o)} \, d\chi_o \tag{3}$$

This equation differs from Eq.1 only in the presence of an additional transfer amplitude $F(\bar{\chi}(\chi_o))$, and in that the classical trajectories used are symmetrized between entrance and exit channels. By expanding the product of the transfer amplitude and the imaginary part of the phase factor in Legendre polynomials the S-matrix amplitude in this simple no-recoil limit can be approximated by a linear combination of amplitudes $S^{J=0}_{N\leftarrow 0}$ for inelastic excitation of the state N in a purely real potential (calculated from Eq.1 or stationary phase approximations to it).

$$F(\bar{\chi}(\chi_o)) \, e^{(i/h)I_m(\Delta)} = \sum_L a_L P_L(\cos\bar{\chi}) \tag{4}$$

$$\tilde{S}^{J=0}_{I\leftarrow 0} = \sum_{NL} \sqrt{\frac{2I+1}{2N+1}} \, |\langle L0I0|N0\rangle|^2 \, a_L \, S^{J=0}_{N\leftarrow 0} \tag{5}$$

The calculation of the inelastic amplitudes $S^{J=0}_{N\leftarrow 0}$ has been described a number of times[6-12]. The transfer amplitude $\bar{F}(\chi(\chi_o))$ will generally be a complicated function to calculate. It must embody recoil corrections (already neglected in this simple treatment), possible superfluidity of projectile and target, a dependence on the collective orientation coordinates, and a dependence of the single-particle wavefunctions and correlations on collective angular momentum. Although complicated, the required formfactor can be decomposed into linear combinations of the formfactors already employed in simpler transfer problems. We will discuss this elsewhere.

Here, we observe that while the exact calculation of the amplitude $(F(\bar{\chi}(\chi_o))\cdot e^{(i/\hbar)\mathrm{Im}(\Delta)}$ is complicated, its final form is likely to be simple and determined by the three factors illustrated in Fig.11. The penetration and damping effects conspire to define an optimum orientation for a grazing trajectory. The majority of the nuclear structure information will reside in the orbital geometry factor defining the orientation angle dependence of the wavefunctions tions participating in the transfer.

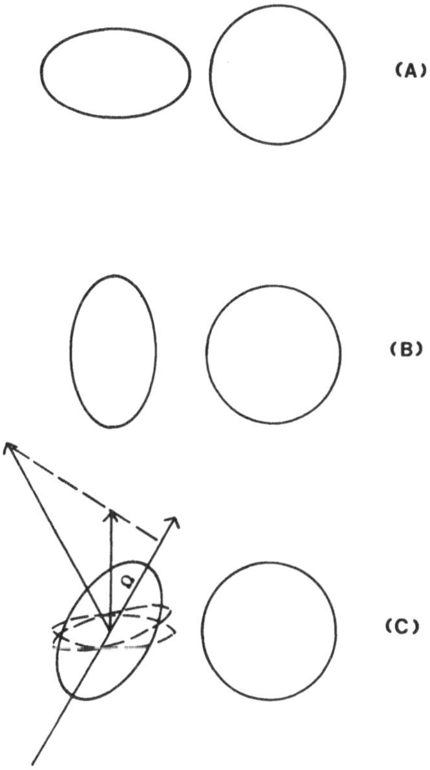

Fig.11 Illustration of the basic factors governing transfer reactions for heavy ions on deformed target nuclei. Polar collisions as in (A) favor penetration of nucleons through the effective barrier separating the ions relative to equatorial collisions as in (B). However, if the collision is too violent orientation (B) will experience less damping out of the transfer channel than (A). The penetration and damping effects compete to define an effective grazing angle. Fig.(C) represents schematically that orbits available for transfer in the deformed potential have preferential orientations with respect to intrinsic axes.

In Fig.12, a very simple model incorporating the above features is illustrated. A transfer formfactor $f(r, \chi)$ is defined by

$$f(r, \chi) = \sum_L f_L(r' = R, \theta' = \chi) \cdot f_r(r, \chi, L) \qquad (6)$$

where $f_L(r'=R, \theta'=\chi)$ represents an amplitude for finding a pair of

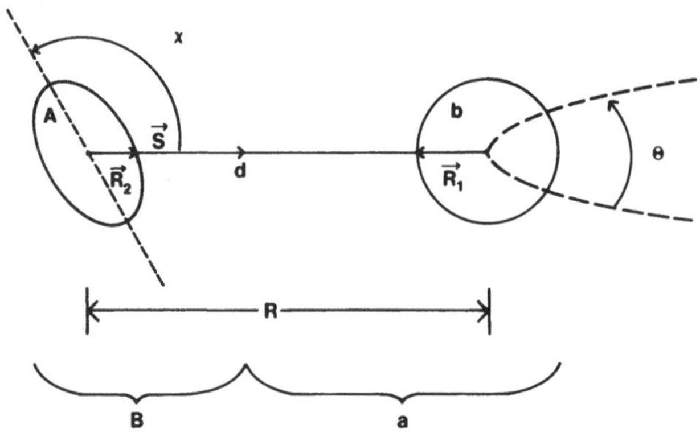

Fig.12 The simple model used to estimate transfer formfactors.

nucleons at the nuclear surface along the line of centers for the two ions, coupled to an angular momentum L and weighted by the (spectroscopic) amplitude for participation in the transfer. For example, if we consider 2-neutron pickup from an even-even deformed nucleus with Nilsson wavefunctions we have[16]

$$f_L(r', \theta') \simeq \sum_i U_i(A - 2)V_i(A)A_L^{(i)}(r)P_L(\cos\theta') \tag{7}$$

$$A_L^{(i)}(r) = \sum_{\ell\ell'} \sqrt{2\ell' + 1}\,\sqrt{2\ell + 1}\,(-1)^{\Lambda}\, R_{N\ell}(r')R_{N\ell'}(r')$$

$$\langle \ell 0 \ell' 0 | L0\rangle [\langle \ell - \Lambda \ell' \Lambda | L0\rangle\, C^{\Omega i}_{N\ell\Lambda}\, C^{\Omega i}_{N\ell'\Lambda} \tag{8}$$

$$- \langle \ell - (\Lambda + 1) \ell' \, (\Lambda + 1) | L0\rangle\, C^{\Omega i}_{N\ell\Lambda+1}\, C^{\Omega i}_{N\ell'\Lambda+1}]$$

Where U and V are BCS amplitudes, $R_{N\ell}(r')$ are spherical oscillator basis radial functions, the Nilsson coefficients $C^{\Omega}_{N\ell\Lambda}$ are labeled by the standard quantum numbers for an uncoupled basis, and the index i runs over Nilsson levels ($\pm\,\Omega$ degenerate)[†].

The quantity $f_r(r, \chi, L)$ defines the radial dependence of the formfactor. It may be approximated by parameterizing it in the form

[†]Due to the symmetries assumed, L must be an even integer and is not a conserved quantum number, and the 2-particle spin function must be singlet. The primed coordinates r', θ' denote intrinsic polar coordinates.

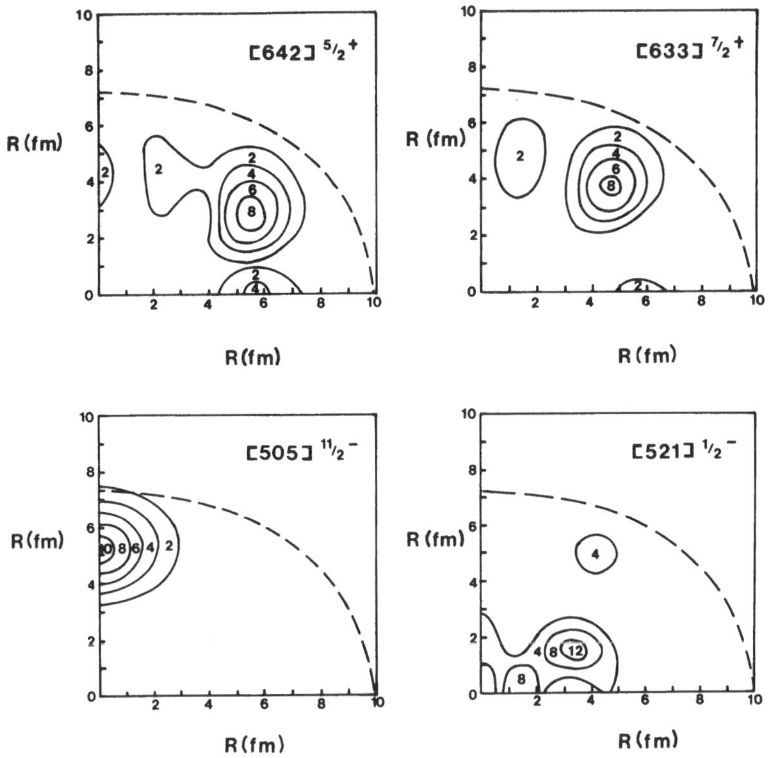

Fig.13 The amplitudes $f(r', \theta')$ defined by Eq.11 for some configurations involving only a single rare-earth neutron orbital (single term in sum over i). The plots are contour plots in the intrinsic r', θ' plane and the orbitals are labeled by the standard asymptotic quantum numbers. All contours are to be multiplied by 10^{-3}. The dashed line indicates the endpoints for the tunneling integral R for various orientations of the rotor in a $^{154}Sm(^{132}Xe, ^{134}Xe)\ ^{152}Sm$ collision. The angular localization of the probability for 2-nucleon transfer is clearly implied in these figures.

$$f_r(r, \chi, L) \sim \frac{1}{1 + e^{\frac{r - R(\chi)}{a_L}}} \qquad (9)$$

where $R(\chi) = R_0^L(A_1^{1/3}+A_2^{1/3}(1+\alpha_2^L Y_{20}(\chi))+\alpha_4^L Y_{40}(\chi))$ with the constants adjusted to reflect the deformation of the system and to give an asymptotic radial behavior consistent with a spherical system having the same binding energies for the nucleon(s) being transferred. Alternatively, f_r (r, χ, L) may be approximated by a WKB barrier penetration amplitude for tunneling between potential wells centered on the two ions. The two approaches give similar results.

The transfer amplitude $F(\bar{\chi}(\chi_o))$ is finally obtained by integrating the differential amplitude f(r, χ) over a classical trajectory beginning with an initial orientation angle χ

$$F(\bar{\chi}(\chi_o)) \sim \int_{traj(\chi_o)} f(r(t), \quad (t))\, e^{iQt}\, dt \tag{10}$$

where $Q = E_{reactants} - E_{products}$

The amplitudes f(r, χ) are shown in Fig.13 for several cases where only a single $\pm\Omega$ - degenerate Nilsson orbit contributes to the sum in Eq.7 (corresponding to weak pairing correlation). The concentration of the transfer amplitude in certain angular regions of the nuclear surface is obvious. For example, a transfer reaction involving the $[505]\frac{11}{2}^-$ orbital is strongly favored for a collision where the projectile grazes the equator of the rotor, while one proceeding via the $[521]\frac{1}{2}^-$ orbital is partial to polar collisior

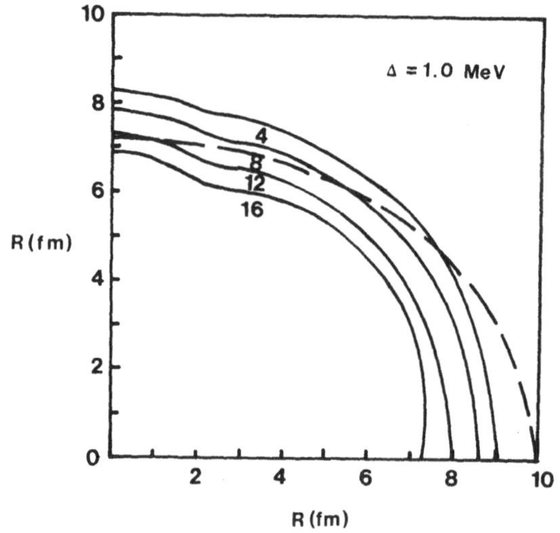

Fig.14 Same as Fig.13, but showing UV-weighted sum over orbitals for the Δ=1 MeV ^{154}Sm paired case. Note the absence of sharply localized angular structure in the surface region and the larger values of f(r', θ') due to coherence as compared to the single-orbit examples in Fig.13.

In Fig.14, the amplitude $f(r, \chi)$ corresponding to 2-neutron stripping from a superfluid deformed nucleus with a pairing gap Δ=1 MeV is shown. For this case many terms such as those plotted in Fig.13 contribute coherently to the sum (7) with weighting factors UV. Comparing Figs.(13) and (14) we see that $f(r, \chi)$ has been spread out in angle and pushed out radially for the latter case by the enhanced 2-particle correlations. The shape of the contours in Fig.14 may generally differ from the shape of the nucleus as inferred from inelastic scattering (section II). This difference is a measure of how much the average multipole moments of a nucleus differ from the multipole moments of those orbits lying near the Fermi surface (those with large UV factors).

In Fig.14, we show the amplitudes $F(\chi(\bar{\chi}_o))\ e^{(i/\hbar)\mathrm{Im}(\Delta)}$ and the quantum number function $I_f(\chi_o)$ for two of the cases shown in Fig.13 and for Δ=0.5, 1 MeV paired cases in the reaction ^{132}Xe(^{160}Gd, ^{158}Gd) ^{134}Xe. The preceeding formalism has been used except that the sudden collision approximation ($\bar{\chi}=\chi_o$) has been made with respect to rotational motion. This approximation is not essential and is introduced

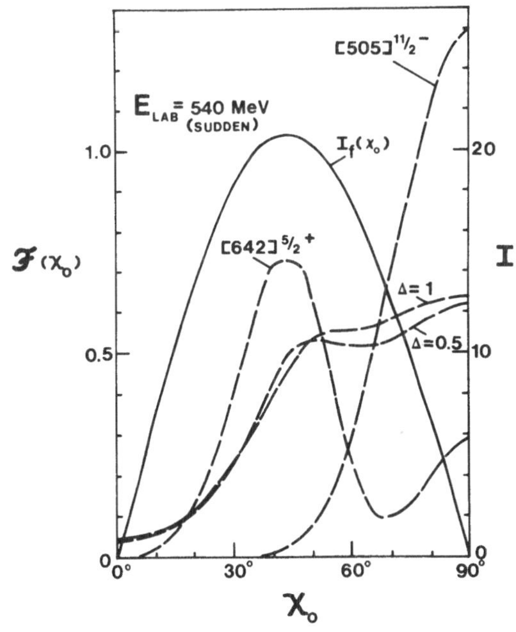

Fig.15 The quantum number function $I_f(\chi_o)$ and the amplitudes $F(\bar{\chi}\ (\chi_o))\ \exp(\mathrm{Im}(\Delta))$ (normalized) for the examples discussed in the text involving the reaction ^{154}Sm(^{132}Xe, ^{134}Xe)^{152}Sm in the sudden limit.

here only for conceptual and computational simplicity. The corresponding rotational signatures are shown for backward scattering at a fixed beam energy in Fig.16. The structure of these signatures and their relation to Fig.15 is easily understood.

Consider the inelastic signature. It oscillates in the classically allowed region because of two different initial orientations contributing coherently to a given final spin [cf. Fig.2 and Eq.2]. For example, from Fig.15, initial orientation angles $\chi_o \sim 20°$ and $70°$ contribute to the inelastic excitation of the rotational state $I_f=10$

For the $[642]\frac{5}{2}^+$ case $F(\chi_o)\ e^{(i/\hbar)\mathrm{Im}(\Delta)}$ is peaked near the maximum of the quantum number function (Fig.15). Therefore the $[642]\frac{5}{2}^+$ transfer strongly favors maximum collective rotational excitation, and the corresponding rotational signature in Fig.16 is heavily weighted toward high spin states.

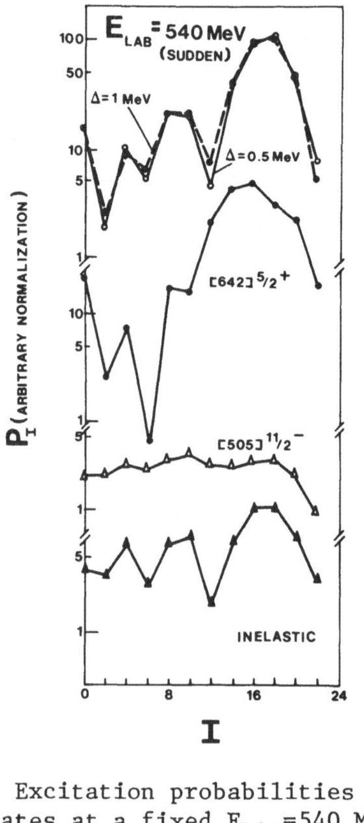

Fig.16 Excitation probabilities for ground band states at a fixed E_{LAB}=540 MeV for the examples described in Figs.(13)-(14). As discussed in the text, the diverse patterns have simple explanations in terms of the semiclassical formfactor $F(\chi_o)$.

For the $[505]\frac{11}{2}^-$ case the transfer amplitude $F(\chi_o)$ is peaked around $\chi_o \sim 90°$, largely due to the concentration of $f(r, \chi)$ near the equator of the deformed nucleus (cf. Fig.13). Accordingly, the corresponding rotational signature does not favor high spins. Furthermore, it exhibits little oscillatory structure because $F(\chi_o)$ selects only the large orientation angle solution from the two which would contribute to inelastic scattering (e.g., $\chi_o \sim 70°$ for the 10^+ state). This strongly suppresses the quantum interference between orientation angles which is responsible for the marked oscillations in the inelastic scattering signature. The reader can verify that the other cases shown, which correspond to the more realistic situation of a finite pairing gap, may be understood by similar arguments.

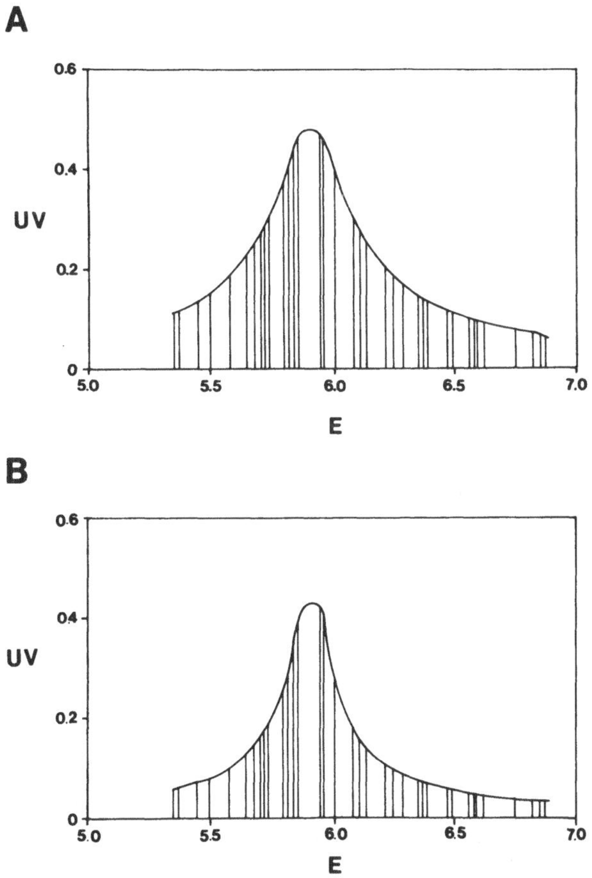

Fig.17 The amplitudes UV for various orbits near the neutron Fermi surface and for values $\Delta = 0.5$ and 1 MeV for the BCS gap parameter.

It should be noted that the probability signatures in Fig.16 have arbitrary normalization. The Δ=1 MeV probabilities are actuall enhanced because of pairing correlations by a factor of $\sim$50 relative to the single orbit cases. Although the signatures have similar shape for the Δ=0.5 and 1 MeV examples, the latter probabilities are about a factor of 4 larger than the former due to increased coherence for the larger gap.

In the preceeding examples we have assumed that the transfer formfactor $f(r, \chi)$ is independent of collective angular momentum. A more realistic example is now discussed where the formfactor changes with the collective rotational frequency ω. We assume that the pairing gap Δ varies quadratically with the collective angular momentum I ($I=\mathcal{J}\omega$ where $\mathcal{J}$ is the collective moment of inertia),

$$\Delta \sim \Delta_o(1 - \lambda_2 I^2) \tag{11}$$

A variation of Δ changes the contributions of the deformed orbits in in the sum (7) because of its influence on the UV factors. This is illustrated in Fig.17. In Fig.18, rotational signatures for several possible variations of Δ with I (see inset) are shown for the reaction ^{132}Xe(^{160}Gd, ^{158}Gd')^{134}Xe. In this example the differences among the cases result primarily from a single effect: for classical trajectories leading to large final collective angular momentum there is significant angular momentum in the rotor near the point of closest approach. For larger values of λ_2 the gap is strongly reduced and with it the coherence in the 2-particle transfer. Therefore, those trajectories leading to large collective excitation have smaller transfer amplitudes, and the population of high spin states is suppressed in the transfer reaction. In the general case the variation of the gap may also effect the angular shape of the function $F(\chi_o)$. In the particular example discussed here, this effect is minimal (cf. Fig.15).

This simple discussion neglects some important and difficult questions such as those concerning the competition between successive and simultaneous pair transfer. Nevertheless, it suggests that a sophisticated experiment and theoretical analysis could allow one to determine the variation of pairing with rotational frequency in heavy deformed nuclei.

We may also consider the possibility of transfer into the two quasi-particle (2QP) superbands responsible for backbending in the rare earth region[17]. The 2QP wavefunction will be of the form

$$\Psi(\omega)_{2QP} \sim \sum_{\substack{k>0 \\ k'<0}} C_k^{(\omega)} C_{k'}^{(\omega)} \alpha_k^+ \alpha_{k'}^+ |0\rangle \tag{12}$$

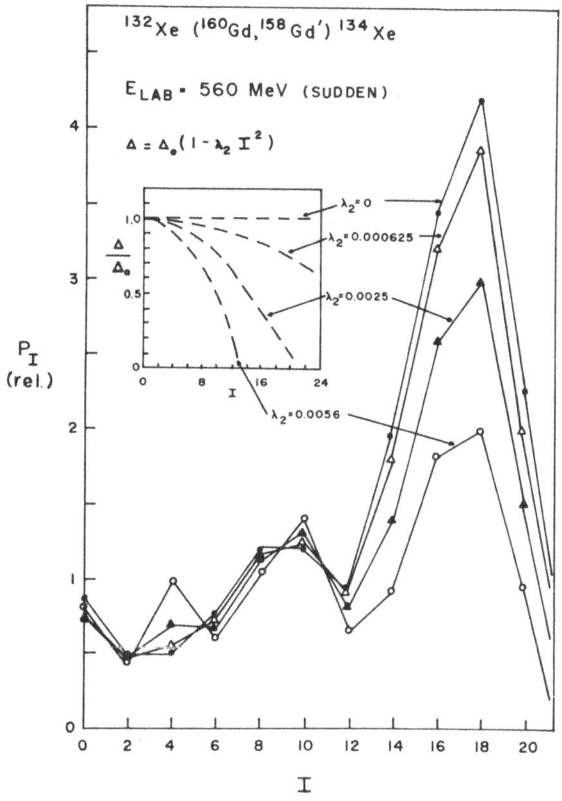

Fig.18 Similar calculation as that of Fig.16, but for a gap parameter Δ varying quadratically in the collective angular momentum. The behavior of Δ is shown in the inset.

where the quasiparticle creation operators α_k^+ operate on the quasiparticle vacuum $|0\rangle$. The coefficients $C_k(w)$ depend on the collective rotational frequencey. They may be determined from a cranked shell model diagonalization, but for the calculations presented here we have used a simpler approximation.

In Fig.19, a single j-shell model of quasiparticle motion originally introduced by Mottelson[18,19] is illustrated. The quasiparticles responsible for backbending are assumed to be $i_{13/2}$ neutrons, and j is considered a conserved quantity since there are no near-lying shell model orbits of positive parity with which it can mix. Therefore, the quasiparticle motion is restricted to the surface of a sphere of radius j in the angular-momentum space. Fig.19 shows the projection of this motion onto the j_x-j_z plane, where x is the cranking axis and z is the nuclear symmetry axis, with $j_z=\Omega=K$.

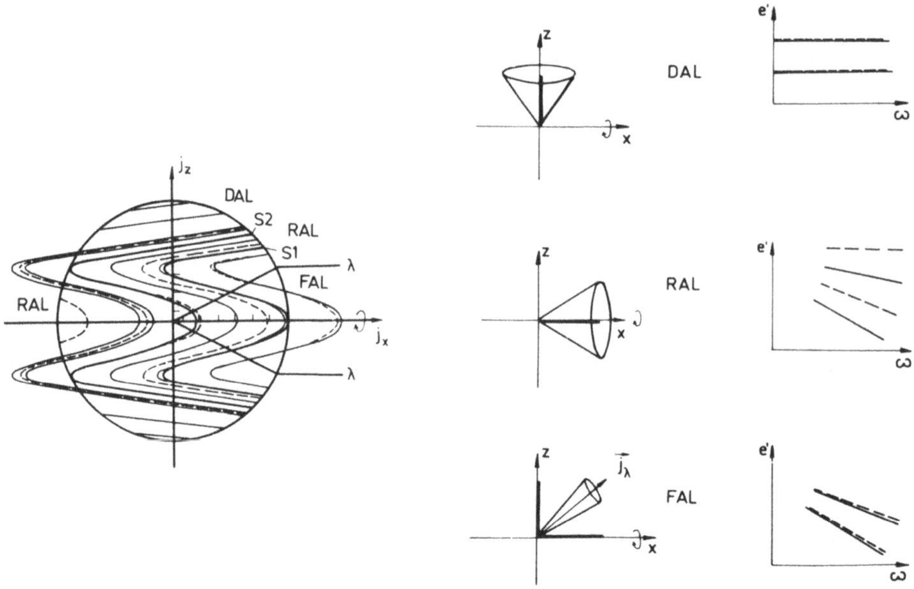

Fig.19 Quasiparticle motion in the j_x-j_z plane of the angular momentum space (where $j_z \equiv \Omega$). The quasiparticles are assumed to be $i_{13/2}$ neutrons with j conserved. Hence the quasiparticle motion is restricted to the surface of a sphere in the j-space. The diagram represents the projection of this motion onto the j_x-j_z plane. Three distinct topologies emerge, separated by the heavy lines (separatrices). The three kinds of quasiparticle motion correspond to the schematic couplings depicted in the figure. The cranking axis is the x-axis with cranking frequency ω.

The diagram is typical of intermediate cranking frequencies in the middle of the rare-earth regions where pairing is significant. Three distinct topologies appear, corresponding to three basic kinds of quasiparticle motion: (1) Deformation aligned (DAL), exhibiting precession around the z-axis with j_z=K approximately conserved. (2) Rotation aligned (RAL), characterized by precession about the x-axis, with j_x conserved but not K. (3) Fermi aligned (FAL), in which neither j_x nor j_z is conserved and precession is about the "Fermi axis". We consider the nucleus ^{164}Er where Fermi alignment is thought to be the dominant mechanism in the first backbend of the ground band.

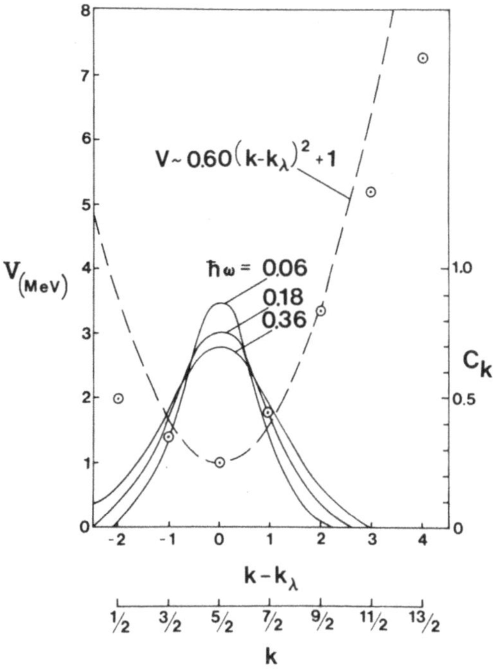

Fig.20 Plot of the term $V \equiv \sqrt{(\varepsilon_2 - \varepsilon_\lambda)^2 + \Delta^2}$ vs. k (the points) for the cranked quasiparticle Hamiltonian[13]. As discussed in the text, this term is interpreted as a potential energy and approximated by a parabola (dashed line). The Hamiltonian may then be cast in the form of an harmonic oscillator in the angular momentum space. The lowest oscillator solution is shown for several values of the cranking frequency. From this solution (solid lines labeled by $\hbar\omega$) the coefficients $c_K(\omega)$ in Eq.12, may be inferred.

The cranked quasiparticle Hamiltonian is[19]

$$h' \sim \sqrt{(\varepsilon_k - \varepsilon_\lambda)^2 + \Delta^2} - h\omega\hat{j}_x \tag{13}$$

Where Δ is the pairing gap, $\varepsilon_k - \varepsilon_\lambda$ is the single-particle energy relative to the Fermi surface, and ω is the cranking frequency about the x-axis. In Fig.20, a plot of the term $\sqrt{(\varepsilon_k - \varepsilon_\lambda)^2 + \Delta^2}$ is shown vs. k for ^{164}Er $i_{13/2}$ neutrons, assuming the $\Omega=5/2^+$ orbit to lie at the Fermi surface. We approximate it by a parabola (dashed line in Fig.20) and interpret the ωj_x term as a kinetic energy operator. Then (13) can be cast in the form of a generalized

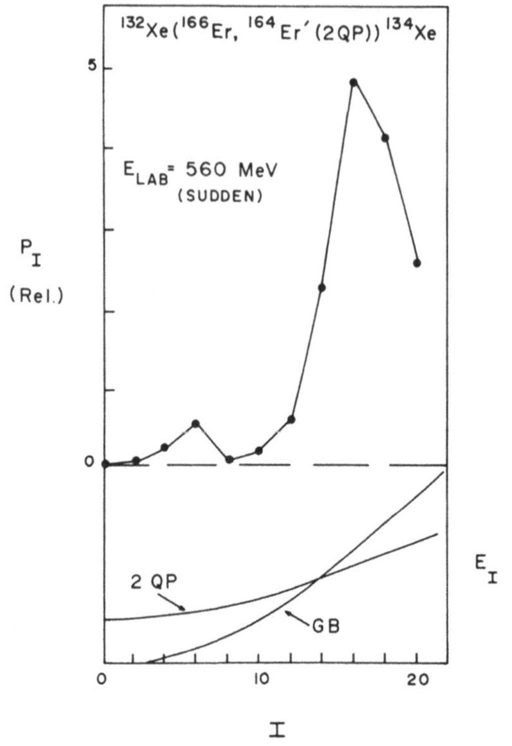

Fig.21 Rotational signature for populating the superband in ^{164}Er for the reaction $^{132}Xe(^{166}Er, ^{164}Er)^{134}Xe$. The calculation employs the same assumptions as for Fig.16. In addition, no mixing of the ground and superband has been included, and only K=0 components have been retained, where K $\equiv$ k+k'.

harmonic oscillator with lowest energy solution

$$\psi_o = \left(\frac{\sqrt{\frac{2\eta}{\hbar\omega B}}}{\pi}\right)^{1/2} \exp\left\{-\frac{1}{2}\sqrt{\frac{2\eta}{\hbar\omega B}}\,(k - k_\lambda)^2\right\} \tag{14}$$

where η is the curvature of the parabola at the origin in Fig.20, and $B=j_x^2-k_\lambda^2$. This solution is shown in Fig.20 for several values of the cranking frequency. From these curves we may deduce the coefficients $C_k(\omega)$ appearing in Eq.12, and proceed as in the approximate treatment of the ground band transfer discussed earlier. In Fig.21, we show a rotational signature calculated for population of the ($\sim$2QP) superband in the reaction $^{132}Xe(^{166}Er, ^{164}Er')^{134}Xe$.

As in the other calculations discussed here, the sudden rotational limit has been introduced as a non-essential simplification. Since we are concerned primarily with qualitative features we have also neglected mixing between the ground and superbands and have retained only K=0 (where K = k+k') components in the wavefunction. The first assumption only effects the one or two states nearest the band crossing due to the small Coriolis matrix elements between the bands. The second one still accounts for the majority of the wavefunction since those components with K≠0 (i.e., k ≠ -k') constitute only ∿20-30% of $|\psi_{2QP}|^2$, as may be inferred from Fig.20.

With the assumptions introduced here, the reaction favors population of the 2QP band above the backbend. In a more realistic calculation without the sudden-limit approximation and including the K≠0 components, the rotational signature would favor somewhat lower spins.

In the preceding examples we have considered 2-particle transfer. Very similar arguments may be applied to 1-particle reactions. In fact, the theory and experiment for those reactions is likely to be simper than that for 2-nucleon transfer. We may expect 1-nucleon transfer with strong rotational excitation to produce zonal localization in individual Nilsson orbits, and to provide information on the effect of collective angular momentum on deformed single-particle wavefunctions.

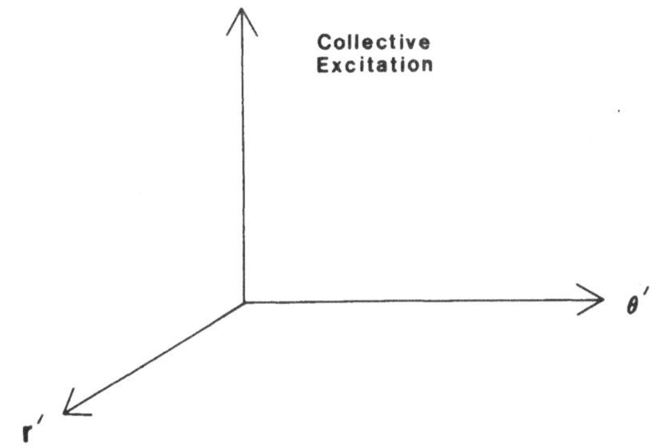

Fig.22 Schematic diagram of the multidimensional space in which experiments of the kind advocated here could sample single-particle and pairing degrees of freedom. For the particular examples discussed the collective axis would represent collective angular momentum of the intrinsic frame. For the general case it could represent other collective degrees of freedom, of some combination of collective modes.

Schematically, the situation is as depicted in Fig.22. The reactions discussed here provide the possibility of studying single-particle motion and pairing forces in a multi-dimensional space consisting of two (or possibly even 3 if triaxial degrees of freedom are admitted) spatial dimensions and dimensions corresponding to collective degrees of freedom. Many Fermion systems have been studied with respect to a single radial degree of freedom (e.g., atoms). Anisotropy of single particle motion is well known for molecules, atoms in external fields, deformed nuclei, etc. Considerably less is known about the anisotropy of particle-particle correlations. The anisotropic pairing of superfluid He III provides a notable exception, and 2-particle transfer reactions to collective states in deformed nuclei should provide another.

The transfer reactions discussed here would appear to provide a rather unique laboratory for studying the response of anisotropic single-particle and pairing modes to collective degrees of freedom. Some information may be obtained from other sources, such as the response of a superconductor to lattice vibrations. However, in those cases, one studies an effect on a macroscopic number of particles. The nuclear case is relatively unique in that only a few Fermions in well-defined quantal orbits respond to the collective mode. This, coupled with the localization effects already discussed, implies a precise probe of the response to collective modes in heavy-ion collisions.

Finally, we note that consideration need not be restricted to rotational collective modes. For example, the response of single-particle and pairing degrees of freedom to collective vibrations can be studied in 1 and 2-nucleon transfer reactions populating multiple-phonon vibrational states, using many of the arguments employed here. A discussion of such ideas would take us too far afield, however, and we must deal with those things in future papers.

V. CONCLUSIONS

The calculations discussed here have been intended to provide a a simple introduction to a new area of heavy-ion physics. Because the discussion has been qualitative we have ignored a few substantial problems, some experimental and some theoretical in nature. However, these problems seem to be tractable and will be dealt with in future work.

We believe that heavy-ion transfer and inelastic scattering reactions populating strongly collective states are likely to exhibit features not easily studied in any other field of physics. A concerted experimental and theoretical investigation should yield significant information concerning nuclear structure, as well as basic insights into the general nature of classical-limit mechanics and the detailed behavior of quantal many-body systems under the influence of strong collective excitation.

ACKNOWLEDGEMENTS

R.A. Broglia, R. Donangelo, E. Maglione, L. Oliveira, J.O. Rasmussen, and A. Winther have collaborated in parts of the theoretical work discussed here. The experimental work has been done in collaboration with C.R. Bingham, I.Y. Lee, N.R. Johnson, and L.L. Riedinger. Discussions with D.H. Feng, Stefan Frauendorf, and C.H. Dasso have proved to be particularly useful. The hospitality extended to MWG and TLN by the Niels Bohr Institute during some of this work is appreciated, as is the financial support of the Danish Research Council.

Research at the University of Tennessee is supported by the U.S. Department of Energy (DOE) under contract No. DE-AS05-76ER04936. The Oak Ridge National Laboratory is operated by Union Carbide Corporation for the U.S. DOE under contract W-7405-eng-26.

RERERENCES

1. R. Elbek and P. Tjøm, Adv. in Nucl. Phys. 3, 259 (1969); R. Broglia, O. Hansen and C. Riedel, Adv. in Nucl. Phys. 6, 287 (1973), and refs. therein.
2. W.R. Phillips, Rep. Prog. Phys. 40, 345 (1977); S. Kahana and A.J. Baltz, Adv. in Nucl. Phys. 9, 1 (1977) and refs. therein.
3. See, e.g., A. Bohr and B. Mottelson, Proc. Int. Conf. on Nucl. Struct., Tokyo, (1977), J. Phys. Soc. Japan 44, (Suppl. p. 157 (1978).
4. R.E. Neese, M.W. Guidry, R. Donangelo and J.O. Rasmussen, Phys. Lett. 85B, 201 (1979); M.W. Guidry, P.A. Butler, R. Donangelo, E. Grosse, Y. El Masri, I.Y. Lee, F.S. Stephens, R.M. Diamond, L.L. Riedinger, C.R. Bingham, A.C. Kahler, J.A. Vrba, E.L. Robinson and N.R. Johnson, Phys. Rev. Lett. 40, 1016 (1978).
5. M.W. Guidry, R.E. Neese, C.R. Bingham, L.L. Riedinger, J.A. Vbra I.Y. Lee, N.R. Johnson, P.A. Butler, R. Donangelo and J.O. Rasmussen, to be submitted to Nucl. Phys. A.
6. S. Levit, U. Smilansky and D. Pelte, Phys. Lett. 53B, 39 (1974).
7. H. Massmann and J.O. Rasmussen, Nucl. Phys. A243, 155 (1975); H. Massmann, Ph.D. Thesis, University of California at Berkeley (1975), unpublished.
8. M.W. Guidry, H. Massmann, R. Donangelo and J.O. Rasmussen, Nucl. Phys. A274, 183 (1976).

9. R. Donangelo, M.W. Guidry, J.P. Boisson and J.O. Rasmussen Phys. Lett. 64B, 377 (1976); R. Donangelo, Ph.D. Thesis, University of California at Berkeley (1977), unpublished.
10. M.W. Guidry, R. Donangelo, J.O. Rasmussen and J.P. Boisson, Nucl. Phys. A295, 482 (1978).
11. R. Donangelo, L.F. Oliveira, J.O. Rasmussen and M.W. Guidry, Nucl. Phys. A308, 136 (1978).
12. P. Frobrich, Q.K.K. Liu and K. Möhring, Nucl. Phys. A290, 218 (1977).
13. J.O. Rasmussen, P. Møller, M.W. Guidry, and R.E. Neese, Nucl. Phys. A341, 149 (1980).
14. G.R. Satchler and W.G. Love, Phys. Rep. 55C, 184 (1979); G.R. Satchler, Nucl. Phys. A329, 233 (1979).
15. J. Randrup and J.S. Vaagen, Phys. Lett. 77B, 170 (1978).
16. M.W. Guidry, T.L. Nichols, R.E. Neese, J.O. Rasmussen, L.F. Olivera, and R. Donangelo, Nucl. Phys. A., A361, 275 (1981)
17. See, e.g., R.M. Lieder and H. Ryde, Adv. in Nucl. Phys. 10, 1 (1978) and refs. therein.
18. B. Mottelson, Proc. Symp. on High-Spin Phenomena in Nuclei, Argonne, (1979), Report ANL/PHY - 79-4.
19. See the contribution by S. Frauendorf in this volume.

PROBING TRANSITIONAL REGIONS WITH NUCLEAR TRANSFER REACTIONS

Jan S. Vaagen

Department of Physics
University of Bergen
Bergen, Norway

I. ABSTRACT

Experimental probes which may provide ways to assess differences between presently competing theories for transitional nuclei are of great current interest. In this paper one-neutron transfer data for $_{52}$Te nuclei and one-proton transfer data for a long chain of $_{61}$Pm nuclei is discussed, with special emphasis on what may be learned from cross sections for low-lying weakly excited high-spin states with the parity of the intruding $h_{11/2}$ orbitals in these regions. The data for the Pm nuclei covers the full range from the closed N = 82 shell to the good rotors (N = 92) and exhibits how the proton spectrum responds to increasing the neutron number of the system. The population of the states is discussed within the framework of the coupled-channels-Born-approximation (CCBA), including a critical evaluation of current recipes for calculation of transfer and scattering form factors.

II. INTRODUCTION

Quasielastic transfer reactions have provided a main bridge to our present understanding of the nuclides. Although 20 years have elapsed since the first attempts to carry out DWBA calculations, the field is by no means exhausted, either experimentally or theoretically. The fact that DWBA (or extensions like CCBA) seem to work so well (contrary to the case in atomic physics), has not ceased to puzzle us. Improved computer facilities have vastly strengthened our computation power, and new generations of accelerators and sources have opened up new energy regions and supplied our arsenal of projectiles so as to comprise a sizeable fraction of the stable elements.

TABLE I

S_p (b, a)	(keV)	S_n (b, a)	(keV)
(d, n)	2224.6	(d, p)	2224.6
(τ, d)	5493.6	(t, d)	6257.3
(α, t)	19814.0	(α, τ)	20577.8

Still, however, light-ion projectiles have kept their leading role in spectroscopy, being the most precise way to induce one-nucleon transfer. Having received the information on a general (democratic) cut-back on the time allowed for each speaker, I have decided to focus on recent results with such reactions in transitional regions. Being at a workshop it seems less offending, perhaps even in the correct spirit, to speak mainly about work done in my nearest environment. I will already at the out-set thank my colleagues in Bergen and Copenhagen for generously letting me bring along a selection of research results, largely independent of the degree of my own participation. A major part of my talk will consist of showing you data, due to my colleagues, which in my opinion represents a potential basis for assessments of presently competing nuclear structure theories. My own part as a theorist has mainly been on elucidation of peculiarities in the observations, such as anomalous cross section magnitudes and angular distribution shapes.

One-nucleon transfer reactions have traditionally been used primarily to identify and elucidate strongly populated states dominated by one- (quasi-) particle motion in a spherical or a deformed average field. As a reminder, Table I lists the most popular light-ion induced reactions A(a, b)B and the corresponding projectile separation energies. (We have used the notation $\tau = {}^3$He).

To illustrate the remarks to be made below, I give in Figs.1 and 2, observed spectra for the neutron pick-up reactions[1] ${}^{126}_{52}\mathrm{Te}\ {}^{(d,t)}_{(\tau,\alpha)}\ {}^{125}_{52}\mathrm{Te}$ and the proton pick-up reaction[2] ${}^{154}_{62}\mathrm{Sm}\ (\vec{\tau},\alpha)\,{}^{153}_{61}\mathrm{Pm}$, the former populating a transitional nucleus in the $Z \gtrsim 50$, $N < 82$ region, the latter a rotational nucleus, bordering the lower rare-earth transitional region ($N \sim 88$). The kinematics relative to the Coulomb barriers $V^{CB} = Z_1Z_2e^2/R$ in the various reaction channels is given in Table II, all numbers in MeV. The jump in Coulomb barrier makes the (t, α) reaction very different from (τ, α) although the Q-values are comparable, hence the need for a higher τ-energy.

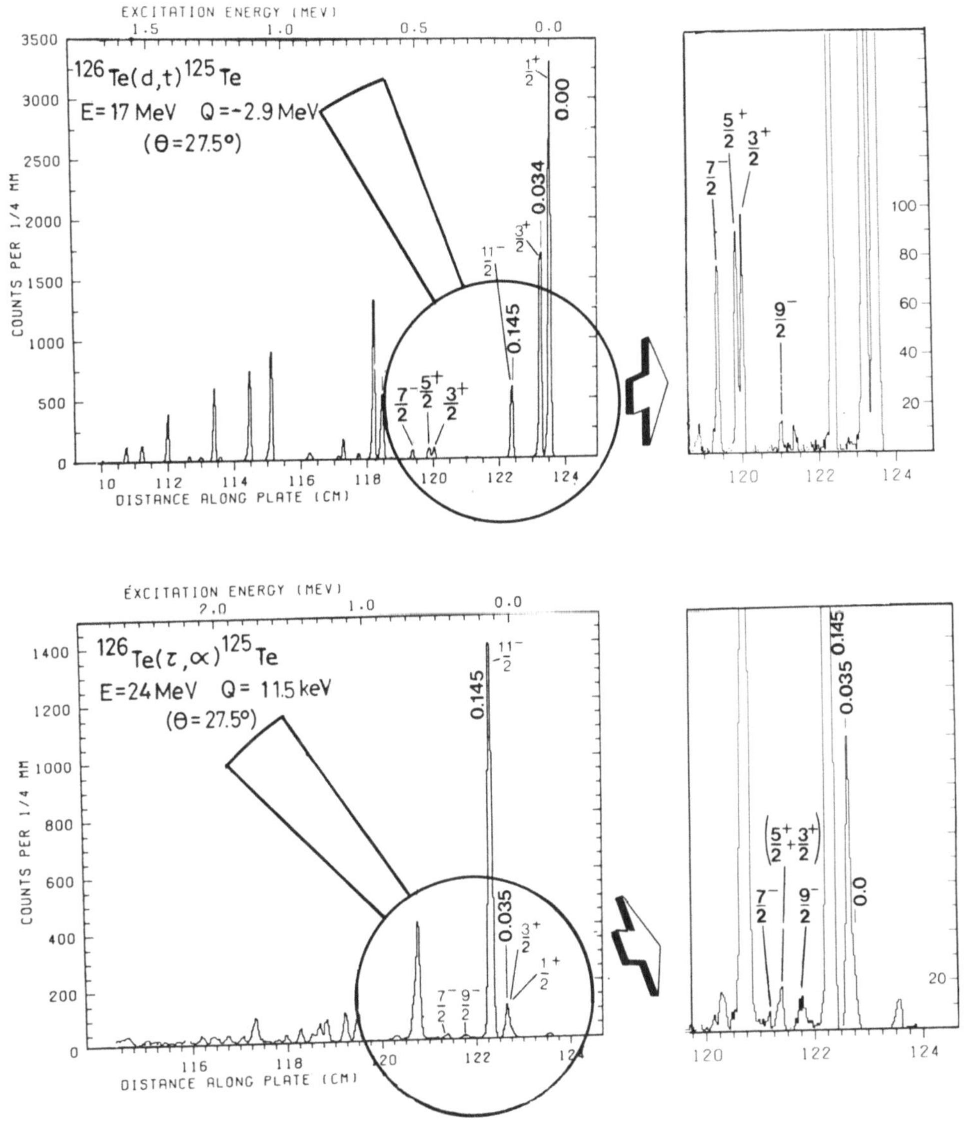

Fig.1 Levels below $E_x \sim 2$ MeV in $^{125}_{52}Te_{73}$ observed by means of (d,t) and (τ,α) pick-up reactions at beam energies E_d = 17 MeV and E_τ = 24 MeV respectively. The presence of weak low-lying high-spin negative-parity states (discussed in tha main text) is emphasized by showing a blow-up of the lowest spectral region. A weak positive parity doublet in this region is also indicated.

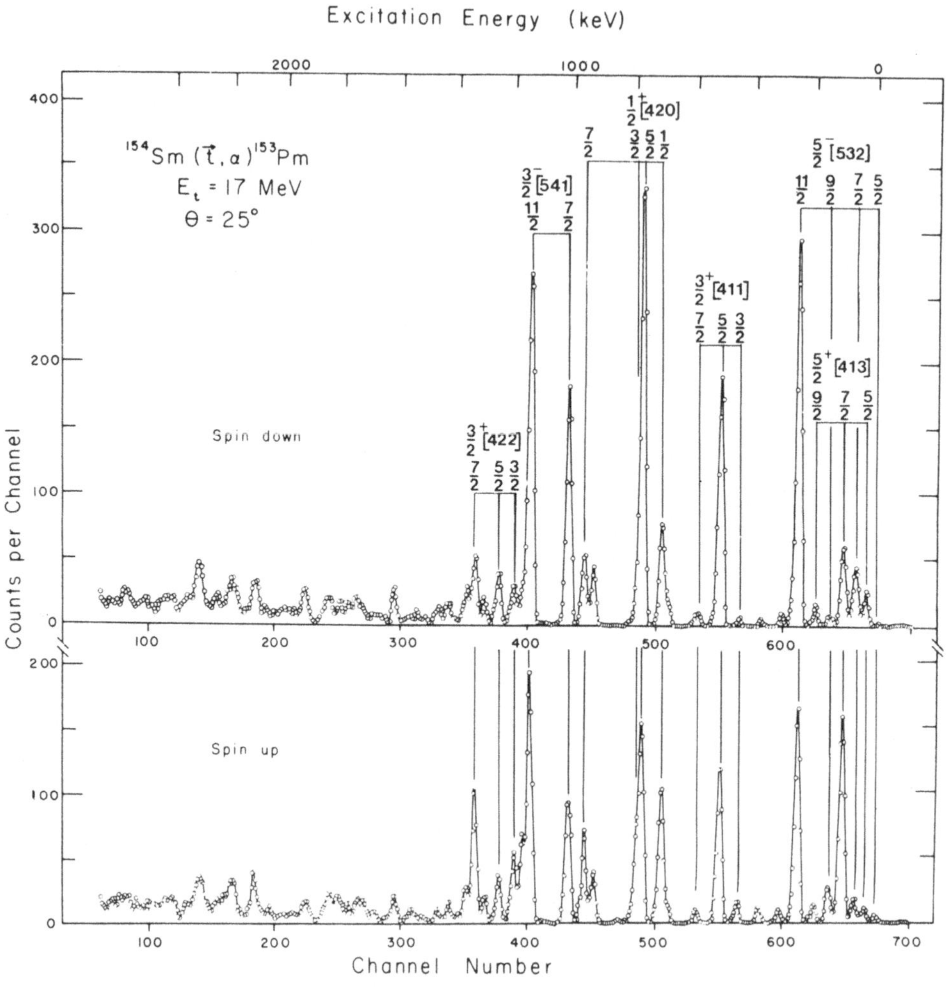

Fig.2 "Spin up" and "spin down" spectra from the ^{154}Sm($\vec{t}$,α)^{153}Pm reaction at E_t = 17 MeV. Nilsson assignments for the various bands are indicated. The analyzing power effects are clearly seen as peaks for the j = ℓ + 1/2 states usually dominate the "spin down" spectrum while those for j = ℓ - 1/2 states are also large in the "spin up" case. The maximum transfer cross sections observed in the (d,t), (τ,α) and (t, α) reactions of Figs.1, 2 were approximately 10000 μb, 1000 μb and 300 μb respectively.

A rough estimate shows the (d, t), (τ, α) and (t, α) reactions favour an angular momentum transfer of Δℓ = 2, 7 and 3 respectively. Fig.1 also displays the ℓ-transfer selectivity of the various

TABLE II

		E_{in}	V_{in}^{CB}	$E_{in}-V_{in}^{CB}$	Q	V_{out}^{CB}	E_{out}	$E_{out}-V_{out}^{CB}$
(-n)	${}^{126}_{52}Te(d,t){}^{125}_{52}Te$	17	11.5	5.5	-2.9	11.5	13.1	1.6
(-n)	${}^{126}Te(\tau,\alpha){}^{125}_{52}Te$	24	23	1	11.5	23	35.5	12.5
(-p)	${}^{154}_{62}Sm(t,\alpha){}^{153}_{61}Pm$	17	12.5	4.5	10.8	25	35.8	10.8
(+p)	${}^{150}_{60}Nd(\tau,d){}^{151}_{61}Pm$	24	25	-1	1.5	12.5	25.5	13.0

reactions, so strikingly reflected in the relative populations of the three lowest-lying levels, $1/2^+$ (ℓ = 0), $3/2^+$ (ℓ = 2) and $11/2^-$ (ℓ = 5) in ^{125}Te.

Now, returning to the observed spectra of Fig.1, we notice the presence of a large number of weakly populated levels in addition to the traditional strong ones. A magnification with a more fair count-number scale for the smaller and small peaks (some hardly visible in the main figures) shows a clear peak separation form a low background. Careful data gathering with long exposures has thus provided full angular distributions for a number of these weakly populated states. That "dwarfs" of this type can be quite interesting with a lot of individuality is reflected in their often anomalous angular distributions. We will have the opportunity to come back to examples in detail. For the rotational promethium nucleus (Fig.2) even the grand structure depends on the identification of the dwarfs. Here the "fingerprint patterns", reflecting the distribution of multipole strength†.

$$||\mathcal{R}_{\ell j}^{\nu\pi k}||^2 = \int_0^\infty dr\ r^2 |\mathcal{R}_{\ell j}^{\nu\pi k}(r)|^2 \qquad (1)$$

for each participating deformed orbital (body-frame)

$$\phi_{\nu\pi k}(x) = \sum_{j\geq k} \mathcal{R}_{\ell j}^{\nu\pi k} [Y_\ell(\hat{r}) \otimes \chi_{\frac{1}{2}}(\sigma)]_{jk} \qquad (2)$$

(x = ($\tilde{r}$, σ)) were of limited help (in particular for negative parity). This is because the relevant orbitals are dominated by <u>one</u> multipole component, that of their spherical limit ($1h_{11/2}$ for

†The DWBA cross section magnitude for an (ℓj)-transfer on a 0^+ target is essentially proportional to the multipole strength.

TABLE III*

	$(1h_{11/2})$ $1/2^-(550)$	$(1h_{11/2})$ $3/2^-(541)$	$(1h_{11/2})$ $5/2^-(532)$
$p_{1/2}$	-0.0413		
$p_{3/2}$	0.1440	0.0701	
$f_{5/2}$	0.0528	0.0650	0.0402
$f_{7/2}$	-0.4177	-0.3536	-0.2583
$h_{9/2}$	-0.0567	-0.1165	-0.1396
$h_{11/2}$	0.8928	0.9232	0.9551

	$(1g_{7/2})$ $1/2^+(431)$	$(1g_{7/2})$ $3/2^+(422)$	$(1g_{7/2})$ $5/2^+(413)$	$(2d_{5/2})$ $1/2^+(420)$	$(2d_{5/2})$ $3/2^+(411)$
$s_{1/2}$	-0.2615			-0.4276	
$d_{3/2}$	-0.5253	-0.3071		-0.0541	-0.1844
$d_{5/2}$	0.1835	0.2071	0.1659	0.6387	0.8723
$g_{7/2}$	0.7240	0.8900	0.9647	-0.5167	-0.3519
$g_{9/2}$	0.3127	0.2660	0.2043	0.3731	0.2852

*Nilsson coefficients for proton orbitals typical for ^{153}Pm and a deformation parameter $\delta = 0.3$. The spherical origins of the orbitals are indicated.

negative parity), a point illustrated by Table III, which contains the results of the Nilsson-type one-oscillator-number approximation $||\mathcal{R}_{\ell j}^{\nu\pi k}||^2 \approx |C_{N\ell j}^{\nu\pi k}(\text{Nilsson})|^2$. Here, the identification of the weak band members became vital for the establishment of the band structures, a feature which is shared by many odd-Z nuclei and in the lower rare-earth region. This becomes even more critical when one enters the transitional regions and the traditional structures dissolve and intermediate structures emerge. Our discussion of a long chain of Pm isotopes will testify to this.

We would also like to emphasize that transfer reactions in some cases provide one of the few ways to reach low-lying states associated with an intruder orbital, like in the telluriums and the promethiums $^{141-151}$Pm where the ground-state parity is positive, while that of the intruder orbital is negative, ($\nu h_{11/2}$ in Te and $\pi h_{11/2}$ in Pm). Hence, the interesting negative parity states cannot be reached by quadrupole excitation from the ground state. (In addition, the promethiums are unstable, i.e. cannot be used as targets). In tellurium, furthermore, most decay processes enter at rather low excitation energy.

Weakly populated states may in general be populated by transfer in an intertwined complexity of ways. Sometimes, however, the pattern is reasonably transparent, usually connected with the presence of an opposite-parity high-spin intruder state, like the $h_{11/2}$. Since this radial function is only moderately affected by deformations it will provide the dominant direct transfer strength in the multistep processes and we may often describe dominant reaction features by a CCBA diagram of the type shown in Fig.3 involving only one transfer multipole ℓj (= $h_{11/2}$) and the quadrupole (λ = 2) excitation mode. The (weak) one-step transfer couplings will enter on this two-step back-ground, the relative importance of the direct routes depending on ℓ-matching properties, as examplified by the tellurium case.

The need to go beyond the one-step DWBA scheme is no definite disadvantage since it provides an opportunity to prove the strength of inelastic transitions (for example $11/2^- \rightarrow I$) in the final system which may be hard (or impossible) to get at by other means, as alluded to above. It also gives a chance to assess the transfer strength from excited target states, mainly the 2_1^+ state. Thus CCBA involves both the single-particle and collective nature of the participating states. Furthermore, the relative role between the various transfer routes may be varied by changing beam energies and/or projectiles, in the study of tellurium the (d,t) and (τ,α) reactions turned out to be useful complementary tools.

One should obviously not expect the light-ion induced one-nucleon transfer reactions on even-even targets to provide much information on very high-spin states in odd-A nuclei. The most we may hope for is to elucidate states up to spin $I = j + 2$, where j is the spin of the high-spin orbital. To go beyond this would require a one-step transfer of unreasonably high (ℓj) for the shell model, or multistep processes via repeated quadrupole excitation or excitation with multipolarity $L > 2$; detection of such cross section would be a true challenge to precision even for strongly deformed nuclei.

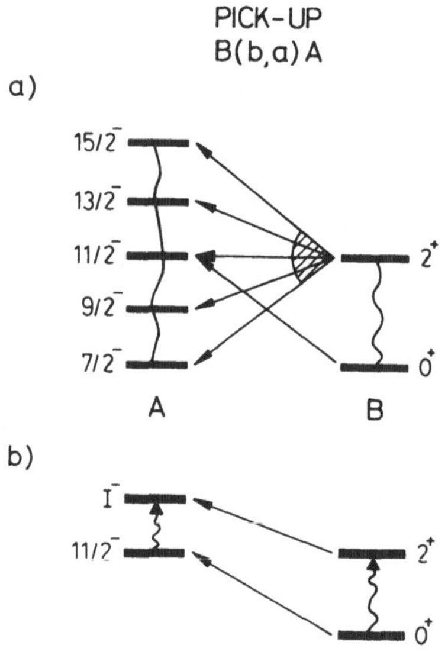

$(\wr\ \lambda = 2,\ \leftarrow (lj) = h_{11/2})$

Fig.3 Schematic CCBA (Coupled-channels-Born-approximation) transfer (pick-up) diagrams for a situation totally dominated by one transfer multipole and quadrupole excitation (de-excitation). Diagrams (a) and (b) correspond to large and moderate deformations respectively.

The (d, t) cross section for $15/2^-$ in Te is already as low as $\sim$10 μb which is also roughly the cross section for (t, α) populated 15/2 states in strongly deformed rare-earth nuclei.

The discussion I am going to present to you has the character of being only semi-quantitative and rather preliminary in what concerns transitional promethium nuclei, the CCBA analysis being in its earliest phase. Some solid quantitative understanding has however been obtained for the rotational limit ^{153}Pm and for the tellerium nuclei, within the framework of the CCBA which for collective nuclei represents the most obvious step beyond DWBA. A number of elements in the practical calculation scheme are open to question, we will only have time to scartch the surface of a few.

The CCBA equations are expressed in terms of transfer and inelastic couplings. A fundamental building block in the transfer coupling between two states $|AM_A\rangle$ and $|BM_B\rangle$ of nuclei A and B = [A+1] is the nuclear overlap function $\Theta_{AM_A}^{BM_B}(x) = \langle BM\ |a^\dagger(x)|AM\ \rangle^*$,

$x = (\underset{\sim}{r}, \sigma)$, with radial multipole components

$$\Theta_A^B(\ell j, r) = \langle B|[a^\dagger_{\ell j}(r) \otimes |A\rangle]^*_{IB} \tag{3}$$

the multipole expansion being derived from that of the field operator, $a^\dagger(x) = \sum_{\ell jm} a^\dagger_{\ell jm}(r) \mathcal{Y}^\dagger_{\ell jm}(\hat{x})$. The spin-angle functions are the same as in eq.2, while the radial operator $a^\dagger_{\ell jm}(r)$ creates a nucleon at a radial distance r with angular momentum quantum numbers (ℓjm) and parity $(-1)^\ell$.

Having a norm $||\ \Theta_A^B(\ell j)\ ||$ (defined by eq.1), which in general deviates from unity, we find it convenient to write the radial overlap functions as

$$\Theta_A^B(\ell j, r) = \beta_A^B(\ell j)\ \hat{\Theta}_A^B(\ell j, r) \tag{4}$$

where we choose the normalized functions $\hat{\Theta}$ to be positive for large radii. Thus, the generalized spectroscopic amplitude $\beta_A^B(\ell j)$ equals the norm to within a sign.

From general arguments we know that the asymptotic form of eq.3 in the case of a neutron (proton) can be written as an amplitude $\beta_A^B(\ell j)\ \hat{\Theta}_A^B(\ell j)$ times a Hankel (Whittaker) function with a decay constant $\kappa_{AB} = (\sqrt{2\mu\ S_{AB}})/\hbar$ determined by the separation energy for taking the nucleon out of state $|B\rangle$ leaving the residual system in state $|A\rangle$. This property which is the theoretical basis for the so-called well-depth-methods (both for a spherical and a deformed field[3]), is most easily seen from the Berggren-Pinkston-Satchler equations[4] which constitute the equations of motion for the overlap functions. These equations may be generated by taking matrix elements of the commutator $[H, a(x)]$ (or $[H, a^\dagger(x)]$) between states of the neighbouring systems, H being the Hamiltonian. An alternative formulation is that of Kerman and Klein[4] involving both even-even neighbours, hence both commutators $[H, a^\dagger(x)]$, $[H, a(x)]$ are involved. This procedure is most natural for treatment of the pair-field. The solution of these equations is by no means trivial for transitional nuclei and is only now being attempted. The analysis presented below employs the well-depth methods and collective model scattering form factors. We will comment on its limitations.

III. NEUTRON TRANSFER IN THE TRANSITIONAL REGION $Z \gtrsim 50$, $N < 82$; LOW-LYING NEGATIVE-PARITY STATES IN ^{125}Te, STUDIED WITH THE (d, t) AND (τ, α) REACTIONS.

The sequence of tellurium nuclei $^{A}_{52}$Te corresponds in the shell model to filling of the $N_{osc} = 4$ neutron shell with the $\nu 1h_{11/2}$ unique-parity orbital ($N_{osc} = 5$) embedded. Fig.4 gives single-particle energies and occupation numbers extracted by Rødland et al[1] from their transfer data. The large number of stable isotopes (A_{even} = 120-130, A_{odd} = 123, 125) and the presence of a number of relatively easily prepared long-lived β-decay parents, have made accumulation of a wide variety of experimental data possible. Recently also substantial amounts of data on high-spin states in odd-N nuclei in this region has become available via (h, xn), h = p, d, τ, α, HI, reactions, for tellurium nuclei mainly due to the Rossendorf group[5]. The known negative parity states below the 2-phonon energy are shown in Fig.5. The states for ^{125,129}Te are those seen

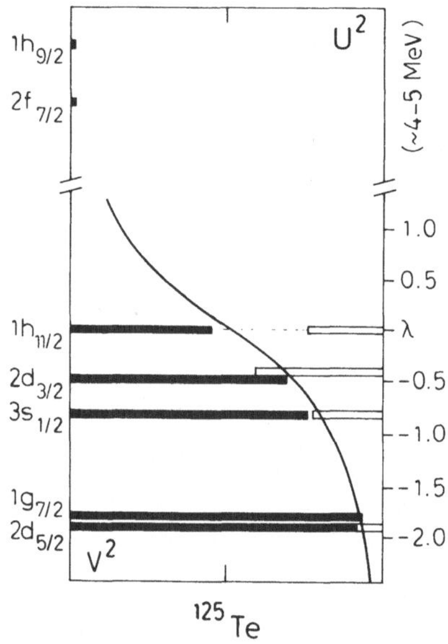

Fig.4 Single-neutron energies for $^{125}_{52}Te_{73}$ (relative to the Fermi surface) and occupation numbers deduced from (τ, α) and (d, p) transfer data by solving the inverse pairing gap equations. (The values given contain self-energies). Similar pick-up information was extracted from the (d, t) study (see main text), which however disclosed somewhat more strength to the p-f-h states in the next shell.

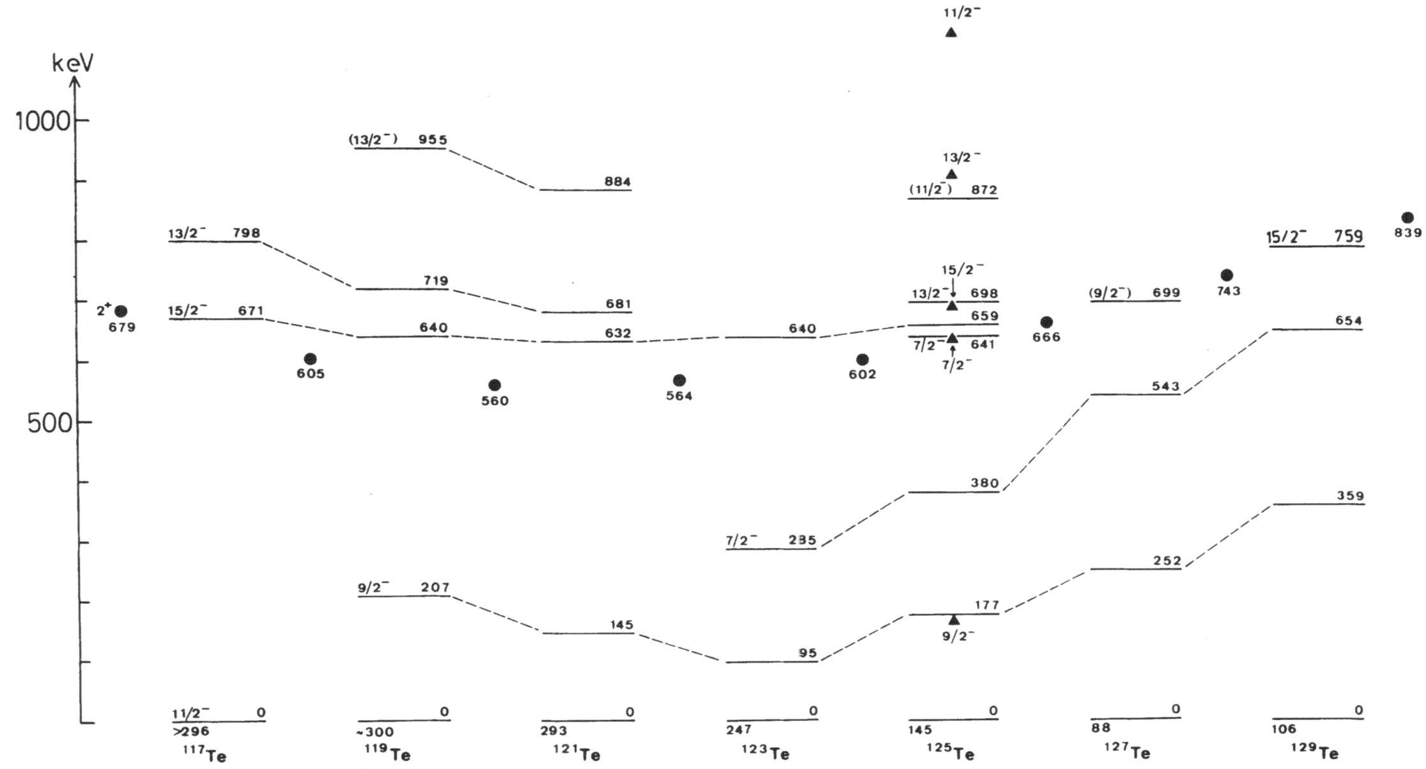

Fig.5 Energy systematics relative to the $11/2^-$(1QP) state for negative parity states in odd-A tellurium isotopes observed at energies less than the double phonon energy. These energy differences as well as the $11/2^-$(1QP) energies are given and compared with the 2_1^+ core energies. The levels and assignments for ^{125,129}Te are those of the transfer studies. The levels for ^{125}Te are compared with the predictions of the dressed 3QP theory (marked by triangles).

in the transfer studies[1,6], among these only the three lowest-lying ones are well known from decay work. The additional transfer information on low-lying states with the parity (some also the spin) of the next major shell evidences the usefulness of transfer as a complementary tool to map out the full level patterns. For higher excitation energies ref.5 contains (h, xn) information on states with spin $17/2^-$, $19/2^-$, $21/2^-$ and $23/2^-$. As alluded to above such states have a rather low population probability in light-ion induced transfer, and none of these states could be separated from the background in the transfer studies of ref.1. For completeness we add that these experiments were carried out using the tandem accelerator facility and the multiple-gap magnetic spectrometer at the Niels Bohr Institute, Risø.

From Fig.5 we conclude that the multiplet states are spread over a wide energy range 322-1017 keV, indicating strong deviations from a simple perturbation picture with a quasiparticle coupled to a core phonon, $\{11/2^- \otimes 2^+\}_{7/2^-,\ 9/2^-,\ 11/2^-,\ 13/2^-,\ 15/2^-}$. In fact, the splitting is as large as the phonon energy. In addition the transfer studies revealed one $7/2^-$ state in excess to form a quintuplet. (It is, however, of interest to note that the centroid energy $\{\Sigma(2I + 1)E_I/\Sigma(2I + 1)\} - E(1QP)$ comes out to within a few percent of the excitation energy of the 2_1^+ (phonon) state of the ^{124}Te core nucleus). Significant deviations from the simple core-excitation picture are not surprise, both due to its neglect of the Pauli principle and as indicated by the large quadrupole moments of the 2_1^+ states in the neighbouring even-even nuclei. Notice (Fig.5) the variation with N of the 2_1^+ energy in the tellurium isotopes (contrary to the rather constant value through the $_{50}$Sn isotopes) emphasizing the role of the neutron-proton interaction as a deformation-driving agent.

A characteristic feature of the spectra in Fig.5 is the strong lowering of the $9/2^-$ member of the multiplet. Such extraordinary low-lying states with spin $I = (j - 1)$, where j is the unique parity orbit, have been observed in a number of nuclei and are usually referred to as anomalous coupling (AC) states. In the microscopic theory of Kuriyama et al.[7], the AC states are regarded as manifestations of dressed three-quasiparticle (3QP) modes, which under the special shell structure conditions for their appearance (ε_j close to λ) are relatively pure eigenmodes with little or no coupling to 1QP modes. In the same manner as the 2^+ phonon mode (dressed 2QP mode) is regarded as an elementary excitation in spherical even-even nuclei, the AC states are phenomena associated with the elementary modes in "spherical" odd-mass nuclei. In tellurium the $9/2^-$ mode is imagined to be brought about through the 3QP correlations between quasiparticles in the $h_{11/2}$ neutron orbital, breaking the quasiparticle-phonon-coupling picture.

The observation of a second low-lying $7/2^-$ level in the transfer studies of ^{125}Te suggested[1] a need for extending the picture so as to include a (j - 2) = $7/2^-$ component of dressed 5QP nature, also known to be lowered by the interactions. Such a calculation has so far not been available. The data for the two $7/2^-$ states indicates a substantial sharing of the dressed 3QP strength between the two states. The one-quasiparticle content of these low-lying $7/2^-$ states, as supplied by the transfer data, provides interesting information on the role of the next major shell.

An alternative understanding of the level structures in tellurium nuclei has been attempted within the quasiparticle-core coupling model, a quasiparticle due to the position of the Fermi level (see Fig.4) close to the intruder orbital. The observed energies of the $15/2^- \to 11/2^-$, $19/2^- \to 15/2^-$ and $23/2^- \to 19/2^-$ transitions (in particular in ^{117}Te) are similar to the yrast $2^+ \to 0^+$, $4^+ \to 2^+$ and $6^+ \to 4^+$ transition energies in the core nuclei, suggesting that the yrast sequence $11/2^-$, $15/2^-$, $19/2^-$ and $23/2^-$ forms a "decoupled" band. The aligned coupling of particles in the unique-parity orbital $h_{11/2}$ seems to be one of the most fundamental excitation modes found in many odd-Z and odd-N nuclei around Z = 50. Some hints for deviations from complete alignment can be found in the odd-mass Te nuclei with A > 118, where in particular the ratio $r(I, I-2) = \{E(I) - E(I-2)\}/\{E(R) - E(R-2)\}$ for I = $23/2^-$ starts to substantially exceed unity. This behaviour may be understood[5] in terms of a blocking effect due to the Pauli principle in the $h_{11/2}$ neutron orbital (see Fig.4), present when the states in the even-Te yrast sequence contain a substantial contribution of the $\nu(h_{11/2})^2$ configuration. In this case an $I_{max} - 1$ coupling should be energetically preferred as is also observed for the $21/2^-$ states in Te isotopes with A > 118.

Calculations[5] with various schematic core models have indicated that a γ-unstable potential model for the core nuclei is the most favourable one in Te nuclei. A dynamic deformation theory (DDT)[8] calculation for ^{124}Te which correctly reproduced the substantial negative 2_1^+ quadrupole moment gives a potential energy surface $V(\beta, \gamma)$ which is rather γ-soft, but slightly favouring prolate shape. A calculation with a γ-soft DDT-core and the coupling approach* of Dönau and Frauendorf[9] which has the advantage that the particle-hole composition accords with the transitional shape of the underlying (two neighbouring) core nuclei accounted for the energy data as satisfactory as the dressed quasiparticle model, in fact better for the $7/2^-$ states (giving two low-lying states) and the excited $11/2^-$ state. This latter state was found to have a large parentage with the core-excited 0^+ state. We would also like to point out that inclusion of the of the next shell $2f_{7/2}$ orbital induced significant spectral changes, the $1h_{9/2}$ orbital being of less importance.

*This represents a computational procedure for solving an orbital representation of the Kerman-Klein equation, hence it gives spectroscopic amplitudes, but not radial distributions.

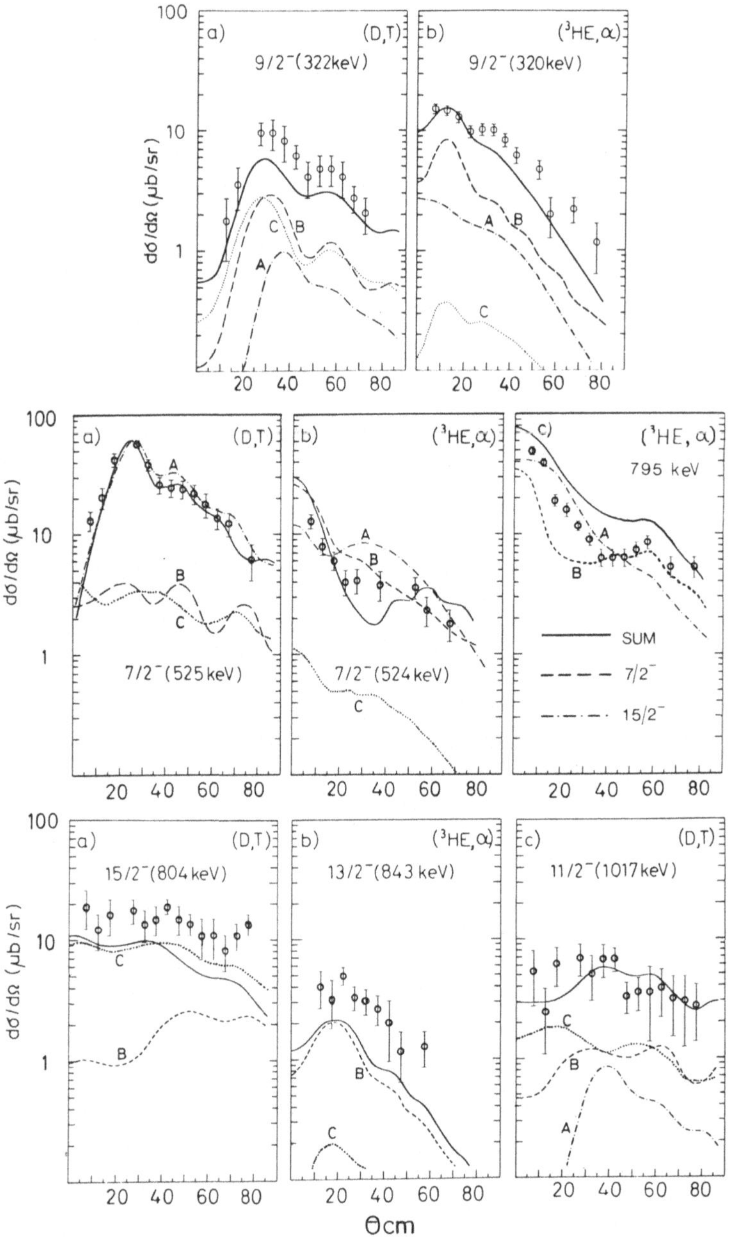

Fig.6 I, II, III Angular distributions for low-lying negative-parity states in ^{125}Te. The solid curves correspond to full CCBA calculations. The partial calculations shown are as follows (a) The one-step contribution (dash-dot curve), (b) the two-step contribution via final nucleus excitation (dashed curve) and (c) the two-step contribution via target excitation (dotted curve).

Fig.6 contains a selection of the transfer angular distribution data for ^{125}Te, as well as the results of CCBA calculations. These calculations allowed for the following transfer routes: (A) one-step transfer $0^+ \rightarrow I$; (B) two-step transfer via quadrupole excitation of the final nucleus $0^+ \rightarrow 11/2^-(1QP) \Rightarrow I$; and (C) two-step transfer via quadrupole excitation of the target nucleus $0^+ \Rightarrow 2^+ \rightarrow I$. (See labels on individual curves). As is evidenced by the calculations displayed in the figures the process (B) is by far the dominant two-step route in the (τ, α) reaction, partly due to the increased exit-channel energy. The well-depth-method for a spherical field was employed to generate the radial shapes of the transfer form factors, i.e. one employs the recipe of simulating $\hat{\Theta}_B^A(\ell j, r)$ by a normalized radial Woods-Saxon solution $\hat{\mathcal{R}}_{n\ell j}(r, S_{AB})$ determined so as to have the correct separation energy S_{AB} and the "physical" number of nodes. The spectroscopic amplitudes were calculated from the expression

$$\beta_B^A(\ell j) \approx \beta_A^B(n\ell j) = \langle B|[a_{n\ell j}^\dagger \otimes |A\rangle]_{IB}^* \qquad (5)$$

the operator $a_{n\ell j}^\dagger$ creating a particle orbital belonging to the set of orbitals used to generate the nuclear states involved in eq.5. Obviously, this procedure amounts to an amalgamation of separate approximations for the two parts of eq.4. This recipe seems to successfully account for strong transitions in nuclei near closed shells (or moderately collective nuclei); it also provided as satisfactory description of such transitions in the tellurium data of ref.1. In the standard BCS-pairing-theory the spectroscopic amplitude for a pick-up $(I_A \leftarrow 0_B^+)$ transition to a fairly pure 1QP state, is approximately $\beta_I^{0^+}(\ell j = I) = \sqrt{2j+1}\, V_j$.

For weak transitions this procedure might be more questionable, one might for example anticipate a need for a state-dependent adjustment of the Woods-Saxon potential geometry so as to simulate individual variations in the overlap functions $\hat{\Theta}_A^B(\ell j, r)$. No simple rule of thumb exists however, emphasizing the need for further work on the equations of motion. Refering again to Fig.6 we conclude that a good overall description of the multiplet states seems to result from CCBA. Notice in particular the complementary role of (d, t) and (τ, α): For the $9/2^-$-AC-state we have experimental information on the dominant $0_B^+ \rightarrow 11/2_A^-(1QP) \Rightarrow 9/2_A^-$ two-step-route in (τ, α), thus we can determine the one-step amplitude $\beta_{9/2^-}^{0^+}(1h_{9/2})$ so as to fit the experimental data. The route via the 2_B^+ state may next be assessed by fitting the (d, t) data where also this route is essential. For the $7/2^-$ states the one-step transfer dominates the (d, t) cross sections (ℓ = 3 being well matched), thus the direct transfer strength can be determined from the experiment. The (τ, α) data is then used to assess the strength of the $0^+_B \rightarrow 11/2^-_A$(1QP) $\Rightarrow 7/2^-_A$ routes. For a detailed discussion I refer to ref.1. From the analysis of all $9/2^-$, $7/2^-$, $11/2^-$ multiplet states one finds that the direct transfer amplitude needed is about 1.5-2 times larger than the theoretical[7] estimate based on a 0QP vacuum description of the 0^+_B ground state. This finding may (i) reflect shortcomings of the theoretical estimates, (ii) suggest an important role of 4QP and 6QP vacuum components in the 0^+_B state, (iii) indicate short-comings of the well-depth-method, or a combination of all three points. The excitation process also needs to be addressed more carefully, especially in the odd-A system. Some progress has already been made, within the framework of DDT. Regrettably, time does not allow me to enter into any details.

IV. RESPONSE IN THE PROTON SPECTRUM TO AN INCREASING NUMBER OF NEUTRONS; SPECTROSCOPY IN THE $_{61}$Pm ISOTOPES WITH (τ, d), (α, t) AND ($\vec{\tau}$, α) REACTIONS

Nuclei in the A $\sim$ 150 region have been devoted considerable attention through the post-war era of nuclear physics, as they form

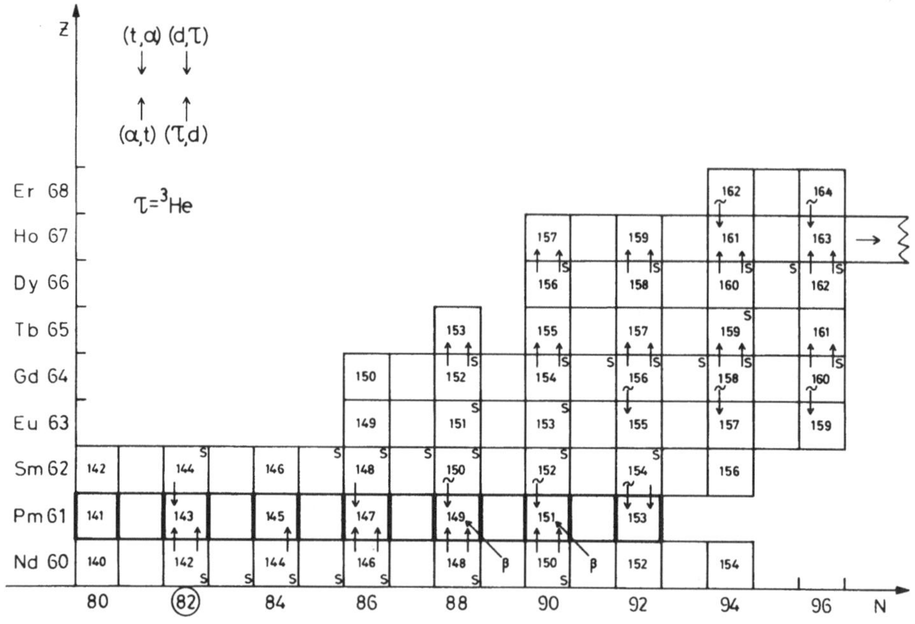

Fig.7 Local geography in Chart of the Nuclides, emphasizing the Pm isotopes. Experimental transfer work on even-even targets to these nuclei and to their Tb and Eu neighbours has been indicated by arrow code.

the doorway to the best established region of deformed nuclei. An inspection of the Table of Isotopes[10] for the even ${}^{A}_{60}Nd_N$ and ${}^{A}_{62}Sm_N$ nuclei reveals a rapidly changing pattern starting at the closed neutron-shell (N = 82) nuclei and ending up in the good rotor (N=92) nuclei. As we shall discuss below, the changes in the patterns for odd-A nuclei are even more dramatic. An explanation of this almost pathological entrance to the astonishing simple rotational rare-earth region has been (and is still) one of the great challenges of nuclear structure theory.

In this part of my talk the focus is on odd-Z transitional nuclei. Nature has provided an unbroken pathway from the spherical region well into the deformed region, the chain of promethium isotopes ${}^{A}_{61}Pm_N$, A(N) = 141 (80), 143 (82), 145 (84), 147 (86), 151 (90) and 153 (92). Paradoxically, none of these are stable enough to be targets, the whole chain is, however, sandwiched between ${}_{60}Nd$ and ${}_{62}Sm$ nuclei (see Fig.7), most of which are stable. Hence, the promethiums may be reached from their neighbours, allowing for a study of how the proton spectrum responds to an increase in the neutron number beyond the closed N = 82 shell. This increase is accompanied by a polarization which eventually affectuates the full shape transition.

1. Experimental Situation

The available experimental information on the Pm isotopes almost fully dates from the second part of the 1970's and stems from decay work and high quality transfer studies[11,2] with the ${}_{60}Nd(\tau, d)Pm$ and ${}_{62}Sm(\vec{t}, \alpha)Pm$ reactions, the $(\vec{t}, \alpha)$ study being limited to ${}^{149,151,153}Pm$ at the present time while all isotopes except ${}^{153}Pm$ have been populated by (τ, d). Fig.8 shows the single particle level diagram and occupation numbers extracted from the (τ, d) data for ${}^{143}Pm$.* The τ-induced transfer studies have been carried out at the Tandem Laboratory of the McMaster University and those with $\vec{t}$-beam at the Los Alamos Scientific Laboratory, as a collaboration between these laboratories and the University of Bergen.

For the rotational limit ${}^{153}Pm$ no published information existed prior to the $(\vec{t}, \alpha)$ study, a situation found for most nuclei on the neutron-rich side of the line of β stablity, reasons for this fact are given inf ref.2. The above mentioned collaboration has also provided a substantial amount of new information on neutron-rich ${}_{63}E$, ${}_{65}Tb$ and ${}_{67}Ho$ isotopes.

For the isotopes ${}^{143-151}Pm$ some previous information existed,

*With a common overall normalization factor the observed stripping strength fulfilled the sum rule rather well for the ${}^{143-149}Pm$ isotopes, contrary to what has been found for odd-neutron nuclei in this region.

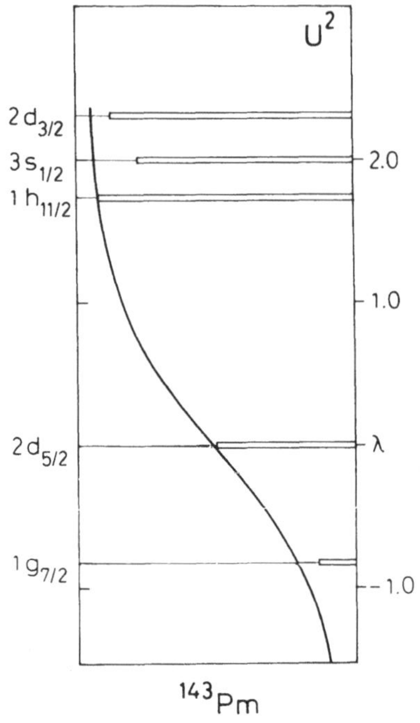

Fig.8 Single-proton energies for $^{123}_{61}\mathrm{Pm}_{82}$ (relative to the Fermi surface) and occupation numbers deduced from the (τ, d) data. (The energy value contain self-energies). Notice the gap of almost 2MeV between the two groups of levels supporting existing evidence for a shell-gap at Z = 64 making $^{146}_{64}\mathrm{Gd}_{82}$ a doubly-magic nucleus. The solid line represents the U^2 prediction of the BCS pairing theory.

restricted to the lower part of the spectrum. The present information, exhibited in Fig.9 for the negative parity levels, almost all stems from the transfer studies[2,11] and (for $^{141-149}$Pm) in-beam γ-ray and electron spectroscopy by the Jyväskylä group[12], using (p, xn) and (τ, xn) reactions. The nucleus ^{151}Pm has also been studied[13] by β-decay of ^{151}Nd.

Again, our discussion will have to be limited to the negative parity states (Fig.9). Regrettably only scanty (mostly indicative) information exists on high-spin states of both parities, in fact none on negative parity states with spin higher than 19/2. One could hope for an improvement in this situation using heavy-ion beams on, for example, Ce targets. (One may think of reactions like $^{140}\mathrm{Ce}(\binom{6}{7}\mathrm{Li}, \binom{3}{4}\mathrm{n})^{143}\mathrm{Pm}$, $^{142}\mathrm{Ce}(\binom{6}{7}\mathrm{Li}, \binom{3}{4}\mathrm{n})^{145}\mathrm{Pm}$, $^{142}\mathrm{Ce}(\binom{6}{7}\mathrm{Li}, \binom{1}{2}\mathrm{n})^{147}\mathrm{Pm}$, (the latter one being perhaps prohibitively difficult),

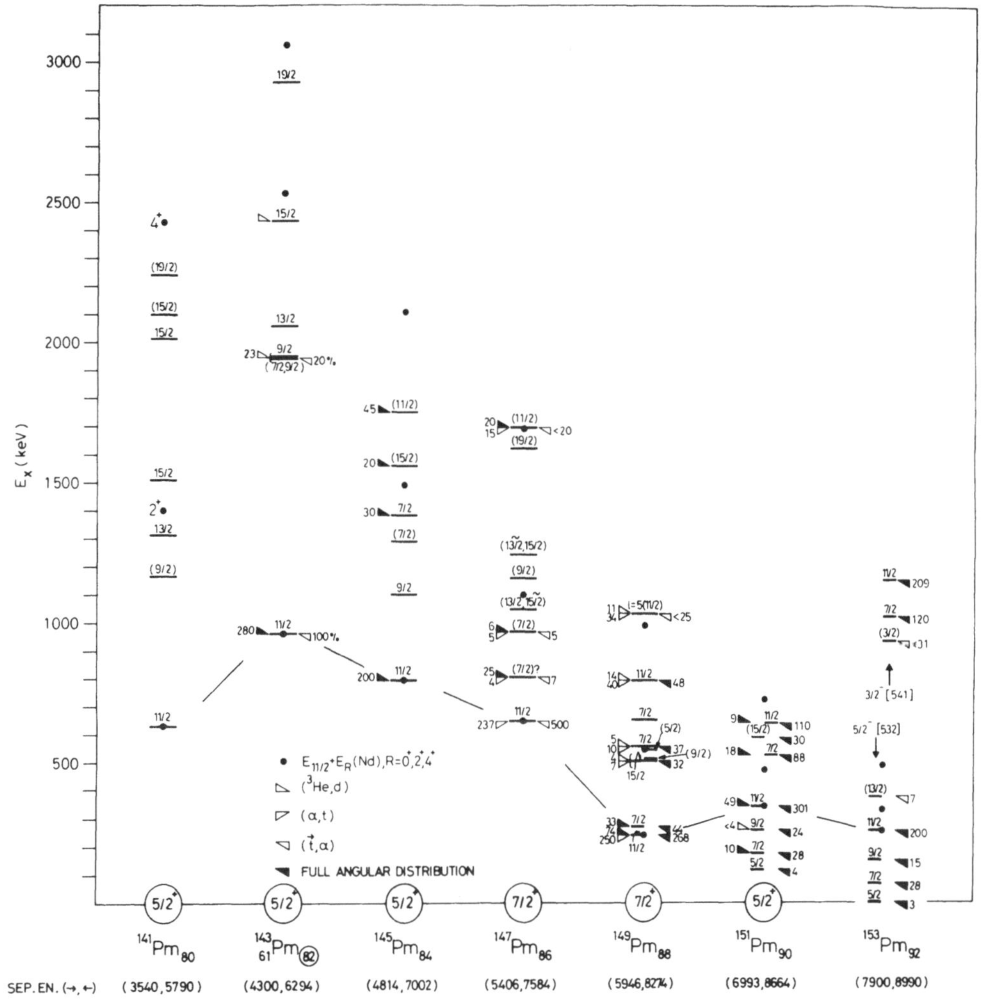

Fig.9 Observed negative parity states in the isotopes $^{141-153}$Pm. Levels populated in transfer reactions are indicated by a triangular arrow code, the neighbouring numbers giving the observed cross section at 45°, 45° and 25° for (τ, d), (α, t) and (t, α) respectively. Separation energies for stripping and pick-up are given at the bottom of the figure, as well as the ground state spin-parity of each isotope. The circular dots indicate the positions of the yrast 0^+, 2^+, 4^+ states in the lower Nd neighbour, measured from the lowest $11/2^-$ state in each promethium isotope.

and also $^{150}Nd(p, n\gamma)^{151}Pm$.). For the moment one has to lean on such information on $_{63}Tb$ and $_{65}Eu$ isotones, where regrettably the transfer spectra have a poorer quality due to more closely lying positive and negative parity states.

2. Map Of The Negative Parity States

Most of the states given in Fig.9 have been seen in transfer with full angular distributions in spite of often rather low cross sections. For some of the assignments the present author takes full responsibility, having made his "personal choice" based on the existing information. Additional data will be taken for $^{145,147}Pm$ this year with the powerful $(\vec{t}, \alpha)$ reaction. The richness of states

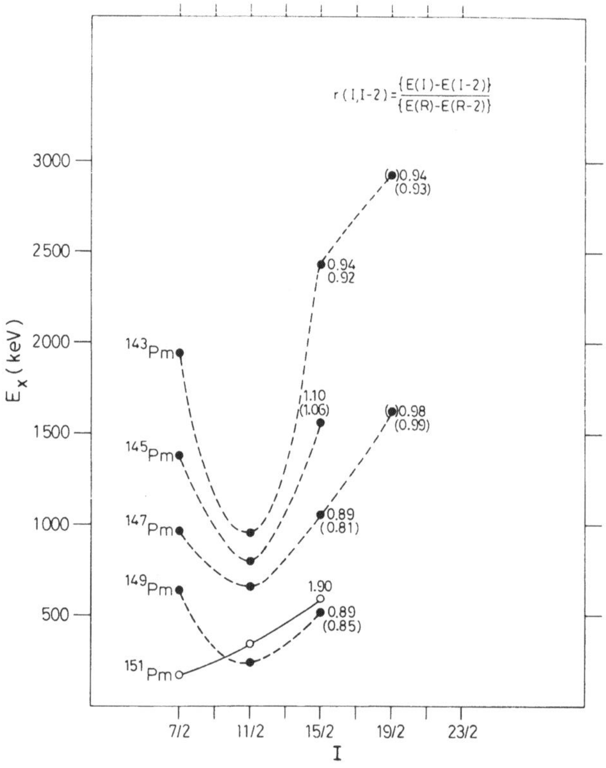

Fig.10 Energy versus angular momentum for states in the promethium isotopes, assigned negative parity, and tentatively associated with the $h_{11/2}$ shell model intruder level. The numbers (in parantheses) are with respect to Nd (averaged Nd-Sm) energies.

in Fig.9 again emphasizes the usefulness of the transfer reactions.

A first look at the level diagram shows a strong coupling-type scheme for $^{153,151}Pm$ which seems to dissolve at ^{149}Pm. The structures of $^{145,147}Pm$ more resemble that of tellurium with a low-lying $9/2^-$ state. The spectrum of ^{143}Pm is more sparse and obviously reflects the closed N = 82 shell property.

Inspired by successes[14] in classification of spectra for $_{63}Eu$ and $_{65}Tb$ isotones, we looked for structures in the Pm spectra possibly reflecting decoupled bands. Fig.10 shows the result of such an attempt. Our efforts are obviously limited by the sparse data on very-high-spin states. Values for the ratio r(I, I-2) have been given for the yrast parallel-spin-coupling states and also for the $7/2^-$ anti-parallel coupling, choosing among the low-lying $7/2^-$ states the one with smallest transfer cross section thought to be the best collective candidate. The numbers found for the r(15/2, 11/2) ratios in $^{143-149}Pm$ are quite similar to those found in the isotones of europium and terbium and close to the decoupled limit of 1.0. The ratio rises rapidly at ^{151}Pm reflecting the (deformation induced) change in the coupling structure. The values for our r(19/2, 15/2) ratios also compare with those of the isotones, and are again close to unity. As stated above our $19/2^-$ assignments are rather speculative and should be considered with due skepticism. In fact, more information is needed for a proper assessment of possible favoured as well as unfavoured states.

3. The Rotational Limit

A substantial amount of time has been spent studying the supposedly good rotational nucleus ^{153}Pm within the conventional CCBA, employing rotor model form facotrs. Such studies have also previously been carried out with success[15] for neutron orbitals. The results of ref.16 for the $5/2^-(532)$ g.s.b. in ^{153}Pm are given in Fig.11. As is seen from Table III, this orbital is especially simple, being dominated by its $1h_{11/2}$ component. A description of similar quality has also recently been obtained for the I = 5/2, 7/2, 9/2 members of the low-lying $5/2^+(413)$ band. These studies in the rotor limit are of great importance for establishing optical potentials for the CCBA. In the analysis of the (simpler) positive parity band all quadrupole couplings were included. Limitations in the local version of the computer program CHUCK prevented inclusion of the upward quadrupole couplings to the $13/2^-$, $15/2^-$ states in the analysis of the negative parity band, we hope to include these in the near future. It is of importance to find out (i) if the same deformed field can be used to generate both positive and negative parity orbitals, (ii) if a common set of optical potentials can be used and (iii) if a common overall normalization factor can be employed in the CCBA analysis. This is to set a standard against which expected surprises in the future analysis of the transitional nuclei can be evaluated.

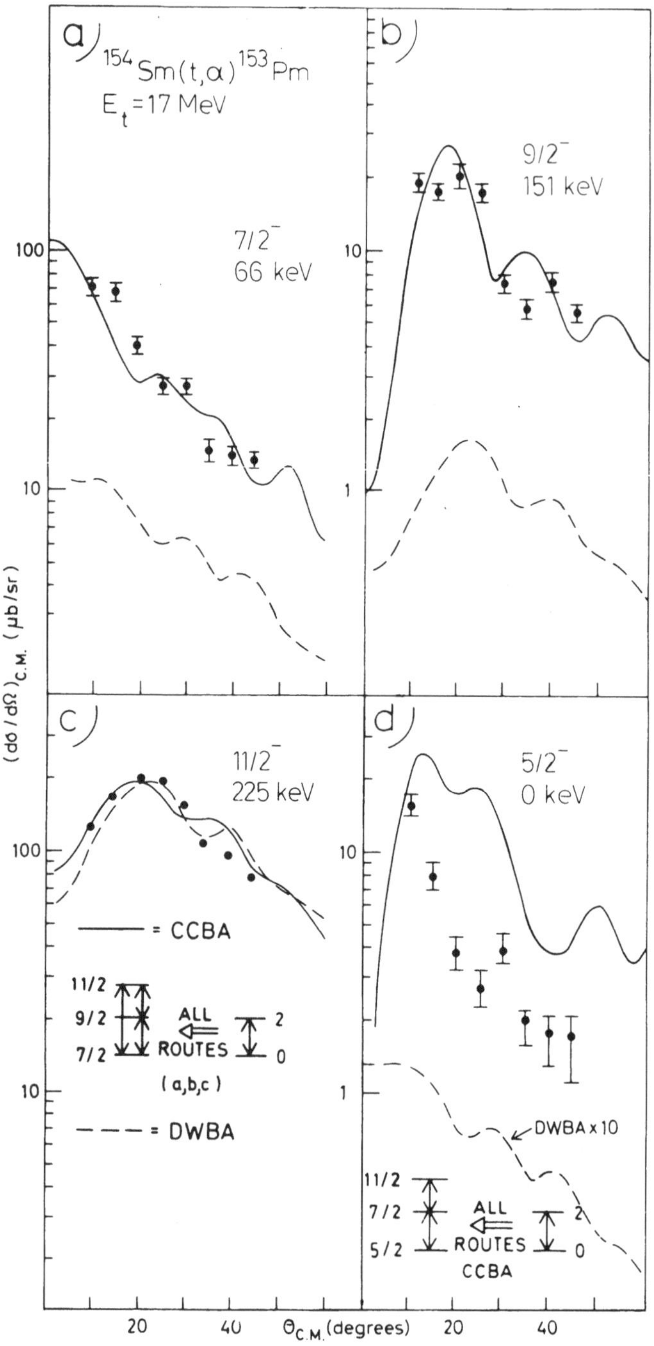

Fig.11 Data and theoretical DWBA and CCBA angular distributions for members of the 5/2$^-$(532) band in 153Pm. The 11/2$^-$ angular distributions are normalized to the data.

Next, a few details from the analysis of ^{153}Pm. In the adiabatic rotor model the overlap functions reduce to a geometry factor $\langle jk\ I_B 0 | I_A k\rangle$ times an intrinsic function $\sqrt{2\pi}\ V_{\nu\pi k} \mathcal{R}^{\nu\pi k}_{\ell j}(r)$, where the expressions correspond to pick-up from an even-even target (B), V being the pairing-occupation-amplitude. To produce the correct asymptotic decay, $\mathcal{R}$ is supposed taken to correspond to a deformed orbital generated for example by the Sturmian well-depth-method[3] so as to have the experimental separation energy S_{AB}. In the past most practitioners have employed the much cruder approach of replacing $||\mathcal{R}||$ by a single Nilsson coefficient and $\hat{\mathcal{R}}$ by a solution produced by the spherical well-depth-method, $\hat{\mathcal{R}}^{\nu\pi k}_{\ell j}(r) \approx \hat{\mathcal{R}}_{n\ell j}(r)$. An assessment of this recipe for a number of examples is given in ref.3. To describe the bands in ^{153}Pm non-adiabatic effects from the Coriolis coupling had to be included in the strong coupling scheme. The spectroscopic amplitudes are approximated by

$$\beta^B_A(\ell j) \approx \sum_{\nu k>0} \text{phase} \cdot \sqrt{2}\langle jkI_B 0 | I_A k\rangle a_{\nu\pi k}(A) \times V_{\nu\pi k} ||\mathcal{R}^{\nu\pi k}_{\ell j}|| \tag{6}$$

where $a_{\nu\pi k}$ (A) are the Coriolis-mixing amplitudes. The radial function $\hat{\Theta}^B_A(\ell j;\ r)$ is again approximated by $\mathcal{R}^{\nu\pi k}_{\ell j}(r)$ determined by the well-depth-method, where $\nu\pi k$ is the orbital with the largest $a_{\nu\pi k}$ (A) component. This procedure is obviously in general not a reliable one. For the $5/2^-$(532) band it seems reasonable, since this band gains strength mainly at the expence of the lowest-lying (k = 1/2 (hole)) member of the orbital-fan originating in $h_{11/2}$. The k = 1/2 orbital corresponds to a larger binding, hence a radially more rapidly falling multipole function.

The identification of the $7/2^-$ states in the rotational nucleus ^{153}Pm was made based on energy positions and analyzing powers and later on supported by the CCBA analysis (see Fig.11), which (contrary to DWBA) accounts rather well for relative cross sections as well as observed shapes of the angular distributions. In ^{149,151}Pm the spin-parity of the lowest $7/2^-$ level was known from decay work which also gave tentative assignments for the next $7/2^-$ states. The characteristic shape of the observed $(\vec{t}, \alpha)$ angular distributions (and the analyzing power) seem to leave no doubt about the $7/2^-$ assignments, the shape seen in Fig.11 being rather persistent. Details do, however, require a careful calculation; notice most importantly that the cross section ratio $\sigma(7/2^-_1)/\sigma(7/2^-_2)$ increases with decreasing neutron number, possibly reflecting the increasing non-adiabaticity.

I would also like to draw attention to the third 7/2$^-$ (655.2 keV) state in ^{149}Pm, which has been identified in decay work, but which is not seen either in one-proton stripping nor pick-up. It is, however, seen with significant transfer strength in three-nucleon transfer[18] in the reaction ^{152}Sm(p, α)^{149}Pm at E_p = 17 MeV. Thus, this state seems to have purely collective structure, probably mainly [$\pi h_{11/2} \times 2^+$] (supporting our labeling (Fig.10) of this state as decoupled).

As pointed out, non-adiabatic effects are of importance already for ^{153}Pm, influencing the transfer form factors. The question of the state dependence of form factor shapes truly enters focus in the transitional region, a clarification of this question is a challenge to future research.

4. Position Of 11/2$^-$ States Versus 0$^+$ Core States

The positions of observed 0$^+$ states in Nd and Sm isotopes are given in Fig.12 as well as the relative positions of levels in the Pm isotopes with spin-parity assignments (some tentative) 11/2$^-$. It is well known that the (p, t) and (t, p) transfer strength to excited 0$^+$-states in ^{150}Sm is exceptionally large, being comparable to that for the ground-to-ground transfer[17]. This observation,

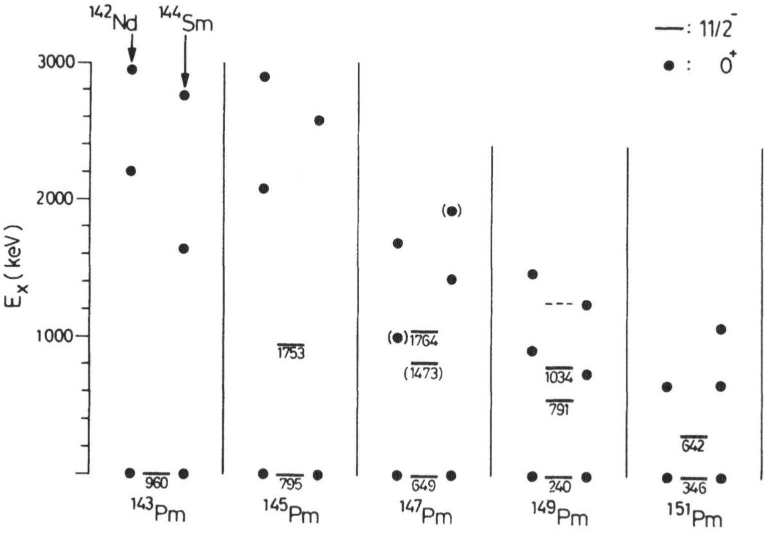

Fig.12 Comparison of states assigned 11/2$^-$ in the Pm isotopes with the relative positions of 0$^+$ states in the neighbouring Nd and Sm nuclei. For a discussion of the possible fourth 11/2$^-$ state in ^{148}Pm (dashed line) see main text.

usually referred to as a shape coexistence, provides by itself an interesting possibility to probe differences between existing nuclear structure theories, like the IBM or the DDT. Thinking about ^{149}Pm as being partly a proton-hole in ^{150}Sm, one might consider adding $11/2^-$ holes to the three known 0^+ states, i.e. building quasiparticle motion on each of the 0^+ states. The energy data for ^{149}Pm is not in disagreement with such a weak-coupling idea (involving ^{148}Nd, ^{150}Sm), further analysis to verify the $11/2^-$ assignment at $\sim$1450 keV is however needed. The relatively small one-nuclear transfer cross sections to the excited $11/2^-$ states also accord with this idea, being zero if the parentage was complete. Furthermore, the idea has some support from the (p, α) three-nucleon transfer data referred to above[18]. Thinking about such a reaction as a sequential process ^{152}Sm(p, t)^{150}Sm(t, α)^{149}Pm one mightlook for indications in $11/2^-$ cross sections of the second step starting out from an excited intermediate 0^+ state (here in ^{150}Sm). The (p, α) data does indeed give relative cross sections to peaks at the positions of the excited $11/2^-$ states. A more detailed analysis of the data is in progress.

V. SUMMARY AND OUTLOOK

From our survey of transfer data and analyses for transitional regions we conclude that transfer reactions do indeed represent a useful way to provide additional information on level patterns and properties. To analyze population strength distributions, however, one often needs to go beyond the simple DWBA; as we have pointed out this does not always represent a drawback. The question of appropriate transfer form factors is a crucial one, were a substantial amount of work remains to be done. As a theorist I would also hope for more future experiments were the experimentalists have the patience to let the accelerator run long enough to let weakly excited states (my pretty flowers) grow above the grass of background.

A current challenge for theorists is that of understanding the connections between the Dönau-Frauendorf approach and the IBM + fermion approach to odd-A nuclei described by O. Scholten at this workshop. The latter theory has most beautifully reproduced spectra for Eu isotopes, it is of great interest to find out how well it reproduces the trends in the transfer data for Eu and Pm isotopes.

REFERENCES

1. T. Rödland, J.S. Vaagen and J.R. Lein, Nucl. Phys. A338, 13 (1980); T. Rödland, (Thesis) University of Bergen (1979).
2. D.G. Burke, G. Løvhøiden, E.R. Flynn, and J.W. Sunier, Phys. Rev. C18, 693 (1978).
3. J. Bang and J.S. Vaagen, Z. Physik (in print) and references therein.

4. T. Berggren, Nucl. Phys. 72, 337 (1965); W.T. Pinkston and G.R. Satchler, Nucl. Phys. 72, 641 (1965); A. Kerman and A. Klein, Phys. Rev. 132, 1326 (1963); Phys. Rev. 138, 1323 (1965).
5. U. Hagemann, H.-J. Keller, Ch. Protochristow and F. Stary, Z. Phys. A290, 399 (1979); Nucl. Phys. A329, 157 (1979).
6. K. Ören, (Thesis) University of Bergen (1980).
7. A. Kuriyama, T. Marumori and K. Matsuyanagi, Suppl. Prog. Theor. Phys. No. 58, 53 (1975).
8. K. Kumar, B. Remaud, P. Aguer, J.S. Vaagen, A.C. Rester, R. Foucher and J.H. Hamilton, Phys. Rev. C16, (1977) 1235; and Kumar's contribution to this meeting.
9. F. Dönau and S. Frauendorf, Phys. Lett. 71B, 253 (1977).
10. C.M. Lederer and V.S. Shirley, Table of Isotopes, (John Wiley & Sons, N.Y. 1978).
11. O. Straume, G. Løvhøiden and D.G. Burke, Z. Physik A295, 259 (1980); Z. Physik A290, 67 (1979); Nucl. Phys. A266, 390 (1976); O. Straume G. Løvhøiden, D.G. Burke, E.R. Flynn and J.W. Sunier, Z. Physik A293, 75 (1979); Nucl. Phys. A322, 13 (1979).
12. M. Piiparinen, M. Kortelahti, A. Pakkanen, T. Komppa and R. Komu, Nucl. Phys. A342, 53 (1980); M. Kortelahti, M. Piiparinen, A. Pakkanen, T. Komppa, R. Komu, S. Brant, L.J. Udovicic and V. Paar, Nucl Phys. A342, 421 (1980); M. Kortelahti, A. Pakkanen, M. Piiparinen, T. Komppa and R. Komu, Nucl. Phys. A288, 365 (1977); M. Kortelahti, Doctoral Dissertation, Dept. of Phys., University of Jyväskyla 1979, Research Report No. 10/1979.
13. T. Seo, Nucl Phys. A282, 302 (1977).
14. J.G. Fleissner, E.G. Funk, F.P. Venezia and J.W. Mihelich, Phys. Rev. C16, 227 (1977); G. Winter, J. Doring, L. Funke, P. Kemnitz, F.Will, S. Elfstrom, S.A. Hjorth, A. Johnson, and Th. Lindbald, Nucl. Phys. A299, 285 (1978).
15. R.J. Ascuitto, C.H. King, L.J. McVay and B. Sørensen, Nucl. Phys. A226, 454 (1974).
16. T.F. Thorsteinsen, J.S. Vaagen, G. Løvhøiden, D.G. Burke, and E.R. Flynn, Phys. Lett. 93B, 223 (1980).
17. P. Debenham and N.H. Hintz, Nucl. Phys. A195, 385 (1972).
18. M.A.M. Shahabuddin, J.C. Waddington, D.G. Burke, O. Straume, G, Løvhøiden, Nucl. Phys. A307, 239 (1978).

NUCLEAR REACTIONS NEAR THE COULOMB BARRIER

J.S. Lilley

Science Research Council, Daresbury Laboratory

Daresbury, Warrington WA4 4AD, UK

I. INTRODUCTION

The most important feature of a low-energy collision is that it is gentle. Reaction processes are fewer, weaker, and simpler than at high energies, and lend themselves to rather precise theoretical analysis. The Coulomb barrier exerts a strong influence on charged projectiles, and, if it is not exceeded to any great extent by the bombarding energy, it restricts collisions to those which leave all but the outer regions of the colliding nuclei undisturbed. Indeed in some of the processes, which will be described, the nuclei barely touch each other. This restriction of the interaction to large radii is an important simplification which makes it possible to obtain detailed and unambiguous information about the nuclear surface and about the reactions which take place there.

This paper describes recent work which has contributed to both these areas. Illustrative examples will be taken mainly from experiments on nuclei in the lead region, some of which have not yet been completed. These fall into three categories:

(a) Heavy-ion sub-Coulomb nuclear transfer.
(b) The α-nucleus interaction.
(c) The two-neutron transfer reaction.

II. SUB-COULOMB NUCLEON TRANSFER

Nuclear shapes and sizes have been studied for many years using many different techniques. A common parameter which is quoted is the root-mean-square (rms) radius of the distribution. The electromagnetic force has enabled this to be determined accurately and reliably for charge distributions. Neutron rms radii, inferred from elastic scattering of hadronic probes, are less accuretely known.

Many reactions, particularly those involving heavy ions, are sensitive to details of nuclear structure in the outer regions of the nuclear surface, where the nucleon density is a small fraction of its central value, and contributes relatively little to the rms radius. An extreme example, described in detail below, is neutron or proton transfer on ^{208}Pb induced by a low-energy oxygen or carbon ion. The maximum transfer probability occurs when the interacting ions are about 15 fm apart[1].

Densities at these large radii are not reliably determined by elastic scattering of nucleons or even electrons, and, while extrapolation from measurements at smaller radii may be adequate in some cases, independent measurements are of great value, particularly those which give information about how the individual nucleons contribute to the total.

Such information has been forthcoming from low-energy reaction studies. Specific details about valence neutron orbitals have been reported from analyses of single-neutron transfer at sub-Coulomb energies[2-5]. Similar information about proton orbitals has begun to emerge from light-ion induced reactions[6-7]. A preliminary analysis of heavy-ion induced proton transfer on ^{208}Pb is presented below. As a result of this work, it is possible to build up most of the total surface matter distribution of ^{208}Pb from such data.

Apart from elastic and inelastic scattering, single-nucleon transfer is one of the simplest reactions to interpret particularly if it is carried out at sub-Coulomb energies on closed or near closed shell nuclei. Transfers take place at long range and so cross sections are small. This creates experimental difficulties, but theoretically it is a great advantage. DWBA should be an excellent approximation, and this, and the removal of the optical model ambiguity mean that this low-energy reaction can be used with some confidence to deduce nuclear structure information.

The DWBA transition amplitude for the reaction A(a,b)B is given schematically by

$$T_{fi} \sim \int \chi_f^{+} \langle \phi_2 | V | \phi_1 \rangle \chi_i \, d\tau_i \, d\tau_f \qquad (1)$$

where χ_i and χ_f are distorted waves, and ϕ_1 and ϕ_2 are single-particle wave functions for the transferred nucleon a and B (for stripping) or A and b for pick-up. In the sub-Coulomb regime, the cores do not interact and T_{fi} depends essentially only on the tails of the two wave fucntions ϕ_1 and ϕ_2. The cross section can be written

$$\frac{d\sigma(\theta)}{d\Omega} = F_1 F_2 \left(\frac{d\sigma}{d\Omega}\right)_{DWBA} \quad (2)$$

where $(d\sigma/d\Omega)_{DWBA}$ is calculated using an exact finite range code and F_1 and F_2 are multiplicative factors representing the unknown spectroscopic factors relating levels in the nuclei (a,b) and (A,B).

Reactions can be found which enable sub-Coulomb transfer reactions to be calibrated, and individual normalization or F-factors determined[4]. This has been done for neutron transfers[5], it has yet to be done for proton transfers.

II.a *NEUTRON TRANSFER*

The best studied example is ^{208}Pb, and transfers involving many different orbitals have been observed[5]. Spectroscopic factors are well known and competing multistep reactions have been shown to be small and tend to cancel each other. With this example the validity of the simple picture described above has been confirmed. Many different neutron transfer reactions have been calibrated and tail sizes and rms radii have been determined for neutron states in ^{208}Pb and neutron holes states in ^{207}Pb.

TABLE I

Measured Single Neutron Transfer Reactions

Reaction	Q(MeV)	E_{inc} (MeV)
$^{208}Pb(^{12}C,^{13}C)^{207}Pb$	-2.428	54.07
$(^{13}C,^{14}C)$	+0.801	52.07, 54.07
$(^{16}O,^{17}O)$ $(^{16}O,^{17}O^{*})$	-3.232 -4.103	69.05, 70.95, 72.95
$(^{17}O,^{18}O)$	+0.671	66.95
$(^{18}O,^{19}O)$	-3.416	66.95
$^{208}Pb(^{13}C,^{12}C)^{209}Pb$	-0.998	52.07, 54.07
$(^{17}O,^{16}O)$	-0.194	66.95
$(^{18}O,^{17}O)$ $(^{18}O,^{17}O^{*})$	-4.097 -4.968	66.95
$^{12}C(^{17}O,^{16}O)^{13}C$	-0.804	13.09, 12.55

TABLE II

Normalization factors for the ^{208}Pb neutron particle states and the ^{207}Pb neutron hole states for a Woods-Saxon bound-state well, with $r_o = 1.25$ fm, a = 0.65 fm and $V_{s.o.} = 7$ MeV.

State	$(^{13}C,^{12}C)$	$(^{14}C,^{13}C)$	$(^{17}O,^{16}O)$	$(^{18}O,^{17}O)$	$(^{19}O,^{18}O)$	Final Values
^{209}Pb						
$g_{7/2}, d_{3/2}$	1.02±0.09		1.01±0.08			1.01±0.06
$s_{1/2}$	0.89±0.07		0.88±0.08			0.89±0.05
$j_{15/2}, d_{5/2}$	0.84±0.06		0.85±0.07	1.05±0.26		0.85±0.04
$i_{11/2}$	1.1 ±0.3		0.95±0.11			0.97±0.10
$g_{9/2}$	0.81±0.06		0.85±0.07	0.85±0.13		0.83±0.04
^{207}Pb						
$p_{1/2}$	2.3 ±0.2	2.4±0.2	2.4 ±0.2	2.4 ±0.4	2.3±0.4	2.4 ±0.1
$f_{5/2}$	7.0 ±0.1	6.5±0.7	5.8 ±0.5	6.5 ±1.2	6.5±1.3	6.2 ±0.3
$p_{3/2}$		4.0±0.4		4.4 ±1.2	5.3±1.6	4.1 ±0.4
$i_{13/2}$	8.9 ±1.5	7.7±1.0				8.1 ±0.8
$f_{7/2}$	4.7 ±0.8	4.9±0.5				4.8 ±0.4

The reactions studied are listed in Table 1. The data were taken at the University of Minnesota and details are given in ref.[5]. The last reaction in the table enables all the individual light ion and ^{208}Pb and ^{207}Pb normalization factors to be determined. Those for the lead states are listed in Table 2. They correspond to bound-state well parameters $r_o = 1.25$ fm, a = 0.65 fm for the neutrons bound in the lead core. F-factors for the different particles or holes leading to final states in ^{209}Pb or ^{207}Pb respectively are listed according to the different reactions in the top row. The final column gives the average values. Within errors, F-values are independent of the reaction used to derive them, which justifies the simple approach used to describe the reactions.

The product of normalization factor and the square of the appropriate single-particle wave function calculated in the standard well gives the probability for finding a neutron in that orbital in ^{208}Pb, if it is assumed that ^{208}Pb is doubly magic and that the final states have pure particle or hole configurations. With this assumption, contributions to the neutron excess in ^{208}Pb were calculated and are plotted in Fig.1. The curve for the 1h9/2 orbital is an estimate using parameters extrapolated from the trend given by the other orbitals. Its contribution to the total is small. If spectroscopic factors are less than (2j + 1) on the average, (as they must be) then there will be additional contributions to this distribution.

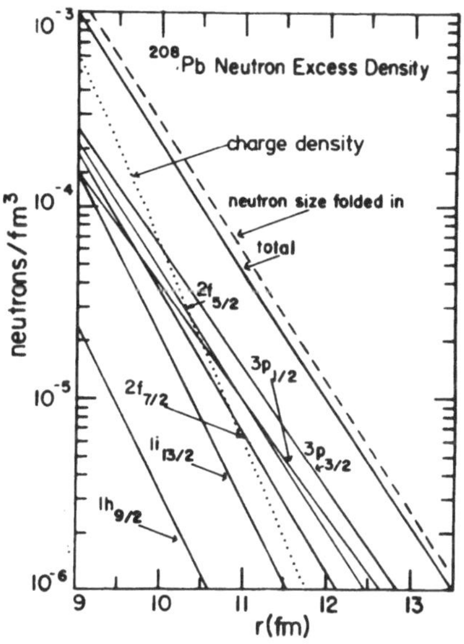

Fig.1. Distribution of centres (solid curves of the ^{208}Pb neutron excess and of the individual orbitals. The dashed line includes the effect of the finite size of the neutron and the dotted line is the ^{208}Pb charge distribution extrapolated from the data of ref.26.

This work agrees well with a similar study carried out earlier by Korner and Schiffer[2] using the sub-coulomb (d,t) reaction, and confirms their value used for the (d,t) normalization factor. Hartree-Fock calculation[8] and a simple shell model calculation using a Saxon-Woods potential[9] are able to reproduce the distribution and indicate that the neutron excess constitutes most of the total neutron density at these radii.

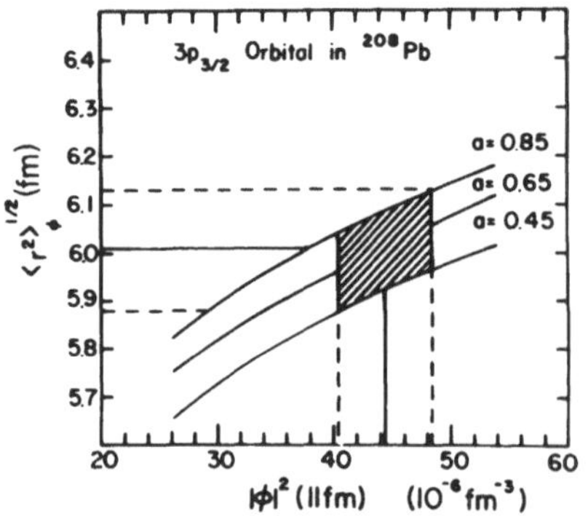

Fig.2. Rootmean square radius of the 3 p1/2 orbital in ^{208}Pb as a function of $|\phi|^2$, the square of the wave function at a radius of 11 fm. Calculations were done for different Woods-Saxon shapes of potential well and varying the diffuseness parameter a from 0.45 to 0.85 fm. The cross-hatched area indicates the uncertainty in the rms radius due to the uncertainty in $|\phi|^2$ for the chosen range of well shapes.

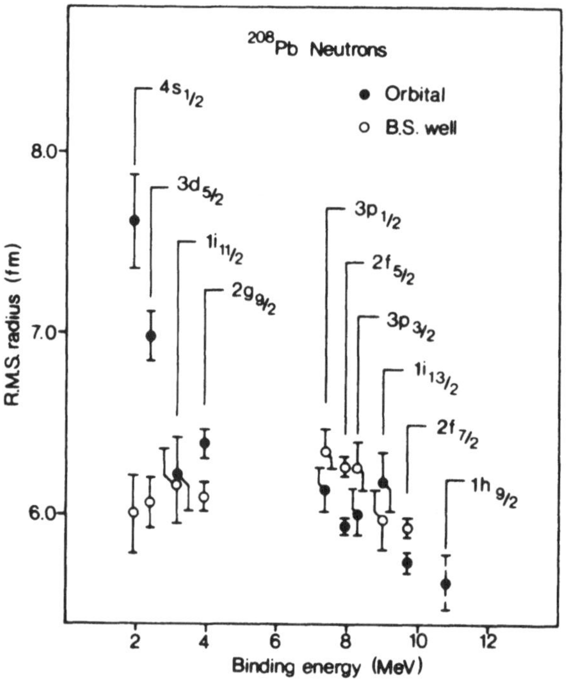

Fig.3. Root mean square radii of neutron orbitals in ^{208}Pb and ^{209}Pb and their potential wells versus neutron binding energy assuming pure shell-model spectroscopic factors. The value for the $h_{9/2}$ orbit (B.E. = 10 MeV) was estimated as described in ref.5.

If the tail of a particular orbital is known, as well as its overall normalization, then its rms radius can be obtained within narrow limits, subject to small corrections due to, for example, non-locality, and uncertainty in the shape of the bound-state well. This is illustrated in fig.2 for the $3p_{1/2}$ orbital in ^{208}Pb. The cross-hatched area indicates the range of uncertainty arising from the uncertainty in $|\phi^2(11\ \text{fm})|$ and from taking values for the b.s. well diffuseness parameter ranging from 0.45 to 0.85 fm.

Fig.3 shows the rms radii of the different single-particle orbits in ^{208}Pb and ^{209}Pb and the local b.s. wells used to calculate them as a function of the binding energy of the neutrons in those orbitals. The most weakly bound orbits are considerably more spread out, as expected. There is a noticeable effect on the $\ell = 6$ orbitals of the spin-orbit potential, which increases the size of the total potential for the $(\ell + 1/2)$ orbitals and decreases it for the $(\ell - 1/2)$ orbitals. The effect is more pronounced for higher ℓ-values. Root mean square radii of the bound-state wells are approximately the same. Variations could arise from variations in spectroscopic factors and from differential effects of non-locality. Taking ^{208}Pb to be a good closed-shell nucleus, the rms radius of the neutron excess is $r_{ex} = 5.93 \pm 0.13$ fm. fm. This increases by about 0.04 fm if the average spectroscopic factor is reduced to 90% of the maximum value.

II.b *COMPARISON WITH ELECTRON SCATTERING*

Root-mean-square radii of certain high-spin valence neutrons have been measured using backward-angle electron scattering [10,11]. So far, the nuclei studied are different to those measured using sub-Coulomb transfer and, since the two methods contain different interpretational assumptions, it is important to compare results taken by the two methods on the same nucleus.

This has been done in a recent experiment[12] for the $1g_{9/2}$ orbital in ^{87}Sr. The ^{86}Sr(^{18}O,^{17}O)^{87}Sr (g.s.) reaction was carried out at the near Coulomb energies of 45 and 46 MeV. The rms radius of the $1g_{9/2}$ wave function in ^{87}Sr of 4.60 ± 0.11 fm measured in this way is in good agreement with the value obtained from electron scattering of 4.66 ± 0.04 fm[10].

This result suggests that the two techniques are giving reliable rms radii in spite of the different assumptions made in the analysis. It also indicates that, since the electrons and heavy ions probe different regions of the nucleus, the radial shape of the wave function is well represented by a local bound-state well with a Woods-Saxon geometry. Effects of non-locality, which are small in this particular example, could modify this conclusion in general.

It is worth noting that the largest source of error in the

sub-Coulomb transfer result is due to the spectroscopic factor. Recent measurements of proton rms radii in the Sn isotopes[6,13] reveal an effect which also suggests a deficiency in our knowledge of spectroscopic factors. This is shown in fig.4, which plots ratios of rms radii and spectroscopic factors relative to the ^{124}Sn values as a function of mass number. The variation in rms radii is not accounted for by Hartree-Fock calculations, but suggestively follows a trend which would result if the variation of the spectroscopic factors is only apparent.

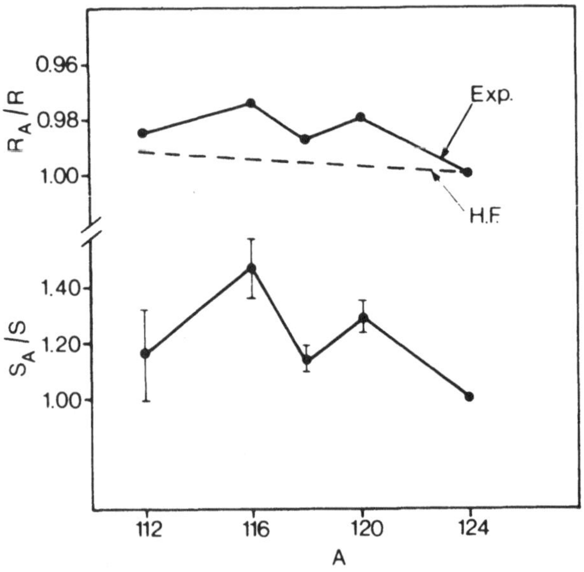

Fig.4. Systematics of $2p_{1/2}$ proton r.m.s. radii and spectroscopic factors in isotopes of tin.

II.c *PROTON TRANSFER*

To extend the study of the surface of ^{208}Pb, data are needed on the proton orbitals. The following results, which are incomplete, represent the first attempt to provide this information.

Sub-Coulomb proton transfer data are sparse for several reasons. Cross sections are even smaller than for neutron transfer, and yet must be measured at large angles with thin targets. It is difficult to find a reaction which is properly Q-matched, and none has been calibrated. The interaction peaks at somewhat smaller radii than does neutron transfer and more complex reaction processes may have to be taken into account in the analysis.

TABLE III

Measured Single-Proton Transfer Reactions on ^{208}Pb

Reaction	Q-value (MeV)	E_{inc} (MeV)
$(^{12}C,^{11}B)$	-12.16	55.95 - 65.95
$(^{16}O,^{15}N)$	-8.329	69.10 - 85.95
$(^{11}B,^{12}C)$	+7.826	44.95

The reactions used to populate levels in ^{209}Bi and $^{207}T\ell$ are listed in table 3. The $^{16}O(^{11}B,^{12}C)^{15}N$ reactions which would be needed to calibrate these reactions has not been completed, and the present results are based on a normalization for the (α,t) reaction and the results of a study of $^{209}Bi(t,\alpha)^{208}Pb$ which is soon to be published[7].

TABLE IV

Normalization Factors for ^{209}Bi Proton Particle States for a Bound-State Well with r_o = 1.25 fm a = 0.65 fm and V_{so} = 7 MeV

^{209}Bi State	$(^{12}C,^{11}B)$	$(^{16}O,^{15}N)$
$1h_{9/2}$	1.0^a	1.15 ± 0.20
$2f_{7/2}$	0.89 ± 0.07	0.80 ± 0.10
$1i_{13/2}$	0.58 ± 0.12	0.56 ± 0.12
$2f_{5/2}$	1.1 ± 0.22	-
$2p_{3/2}$	0.71 ± 0.13	0.76 ± 0.14

a) Overall error of 15% from the $^{209}Bi(t,\alpha)$ experiment of ref.[7]

Normalization factors deduced from the two proton stripping reactions are shown in table 4. They agree with each other, within the errors, suggesting that the simple picture used so successfully for describing neutron transfers is working satisfactorily here. A value for the (^{12}C,^{11}B) normalization factor was obtained and used to deduce normalization factors, tail sizes and rms for proton hole states in $^{207}T\ell$.

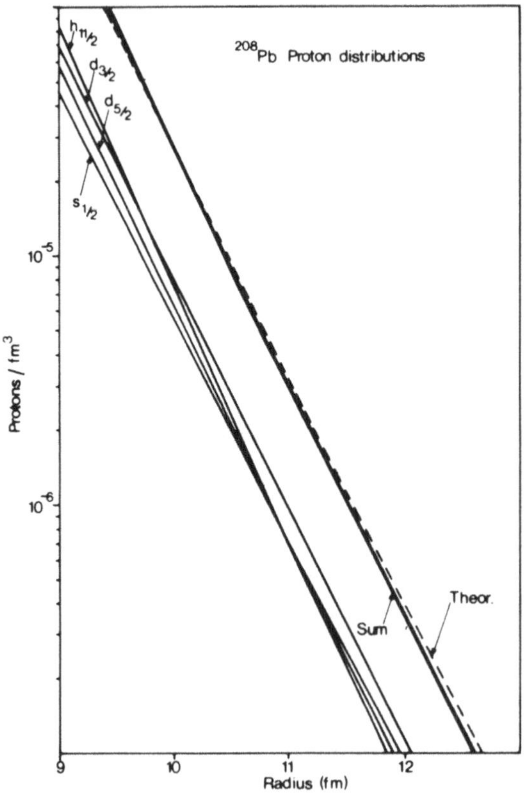

Fig. 5. Point proton distributions for measured orbitals in ^{208}Pb. The heavy solid curve is the sum, and the dashed curve is the sum of ortitals calculated in the Saxon-Woods potential well of ref. 9.

Fig.5 shows the tails of the four proton orbitals which were observed in the $^{208}Pb(^{11}B,^{12}C)^{207}T\ell$ reaction. Spectroscopic factors of (2j + 1) were assumed, as was done for the neutrons. The heavy solid curve is the sum and the dashed curve is a sum of orbitals calculated using the Woods-Saxon potential well of Nolen and Schiffer[9]. Wave functions calculated using this well give a reasonable fit to the neutron excess distribution (fig.1) and also reproduce the total rms proton radius of ^{208}Pb. In view of the overall error of about 15% from the (α,t) normalization, in addition to other uncertainties, the agreement is better than could be expected.

Root mean square radii of all the measured proton orbitals in ^{208}Pb and ^{209}Bi and their b.s. wells are shown plotted against binding energy in fig.6. The orbitals generally are smaller than their neutron counterparts reflecting the confining effect of the Coulomb barrier. The effect of the spin-orbit potential on the $i_{13/2}$ and $h_{11/2}$ orbital sizes is pronounced. The rms radii of the wells are all comparable and about 0.25 fm larger than the neutron wells suggesting that neutrons are indeed slightly bigger than protons in ^{208}Pb.

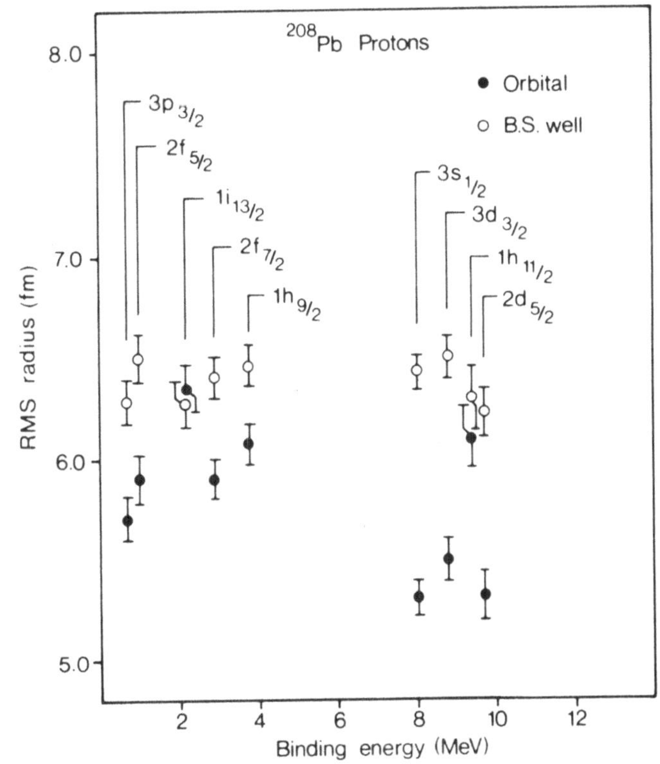

Fig.6. Root mean square radii of measured proton orbitals in ^{208}Pb and ^{209}Bi and their potential wells versus proton binding energy assuming pure shell model spectroscopic factors.

II.d *SURFACE DENSITY OF ^{208}Pb*

The preceding data for both protons and neutrons are combined in fig.7 to build up the total nucleon distribution in the surface of ^{208}Pb. The unobserved nucleons were calculated using the potential well of ref. , which as already described, gives a reasonable representation of the observed orbitals. The error arising from this procedure, therefore, should not be too great, particularly at larger radii, where the unobserved nucleons contribute a minor and decreasing fraction to the total density.

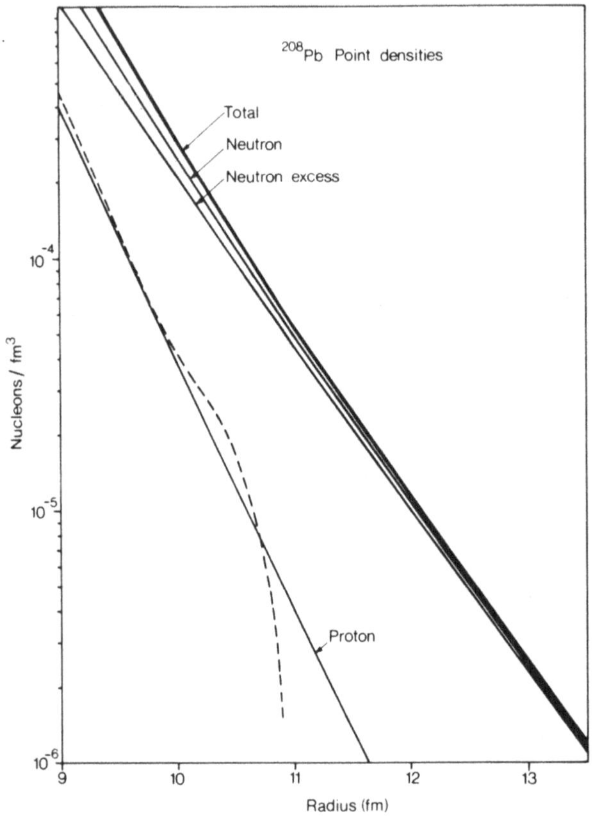

Fig.7. Point distributions for the total proton, neutron excess and total neutron densities in ^{208}Pb. The heavy solid curve is the total nucleon distribution and the dashed curve is the total proton distribution deduced from the measured charge distribution of ref.18. The contributions determined from sub-Coulomb transfer measurements assume spectroscopic factors of 90% of the pure shell model values.

The experimental data were modified to take into account the centre of mass correction[14], which is small. They also were increased by about 10% to correspond to an average spectroscopic factor of 90% of the full value for the observed transitions. This is likely to be more realistic than the maximum, which is assumed in figs.1 and 5. Spectroscopic factors must be at least several percent below the maximum due to effects of the hard-core potential[15]. Also some fragmentation of the hole states, particularly the deeper-lying ones, is known to occur[16]. The resulting total proton distribution is in good agreement with that (shown as the dotted curve) deduced[17] from the latest empirical charge distribution[18]. Errors in the latter account for the differences beyond about 10 fm. This agreement is, to some extent independent justification for the present procedure. In principle, a correction also should be made for (2p,2h) configurations in the ground state of ^{208}Pb. These are known to be small, however, and would have little effect on the distributions presented here.

TABLE V

Root Mean Square Radii in ^{208}Pb (fermis)

Total protons	$r_p = 5.47$
n-p difference	$r_n - r_p = 0.14 \pm 0.04^a$
neutron excess	$r_n(ex) = 5.97 \pm 0.13^b$
core neutrons	$r_n(core) = 5.41 \pm 0.13$

a) Ref.[17]. b) Present work.

There is clearly a neutron halo outside the main ^{208}Pb distribution which is due almost entirely to the neutron excess. The contribution of the core (N = 0 - 82) neutrons is more comparable with that of the protons. Furthermore, as may be seen in table 5, the rms radii of the protons and core neutrons also are approximately the same. It appears that by accepting the neutron excess, the nucleus does reasonably well in compensating its core for the electrical repulsion of the protons. The result is that the effect of the extra binding energy on the core neutrons is similar to that of the Coulomb barrier on the protons.

The distribution presented here constitutes a set of data which can be used to test nuclear structure calculations and also to unravel details of nuclear reaction mechanisms. I will give two examples of the latter category: α-particle scattering and the two-neutron transfer process.

III. THE α-PARTICLE-NUCLEUS INTERACTION

Elastic and inelastic scattering are even simpler reactions than nucleon transfer, but, above the barrier, the nuclear interaction is important. The object of analyzing low-energy scattering data is to obtain the surface interaction form factors and relate them to the density and transition densities via the folding model

In its simplest form the folding model gives an expresssion for the real nuclear potential:

$$V(\vec{r}) = \int \rho_m(\vec{r}\,') \, V_{eff}(\vec{r}\,' - \vec{r}) \, d\vec{r}\,' \qquad (3)$$

where ρ_m is the distribution of nucleons in the target and V_{eff} represents the effective interaction between the projectile and the target nucleons. The imaginary potential W either is taken to have the same form as V, or, more commonly, is independently parametrized. If the density distribution ρ_m is known, then, in principle, one can learn something about the effective interaction, and vice versa.

Ambiguities constitute the main source of uncertainty in this work, and data must be accurate and extensive. Even so, fits to elastic differential corss sections [20] (fig.8a) determine the real potential only in the region of 11 fm, which is close to the peak of the Coulomb barrier. Both the slope of V, characterized by the real diffuseness parameter a_R, and the surface absorptive potential W are not well determined individually. However, there is a correlation between a_R and W(r = 11 fm), as is shown in fig.9. This appears to suggest that as the real potential barrier to α-particles is thickened (large a_R), there must be an increase in surface absorption to offset the reduction in flux penetrating to the interior.

By including measurements of the reaction cross section σ_R, over a wide range of energy in the data analysed, (see fig.8(b)), the a_R - W(11) ambiguity was reduced to the extent that a_R could be constrained to lie between 0.6 and 0.75 fm[21]. This improves the determination of the real potential tail and places limits on the imaginary potential (see fig.9). Furthermore, in order to fit $\sigma_R(E)$ properly, W(r = 11 fm) must vary with energy as shown in fig. fig.10.

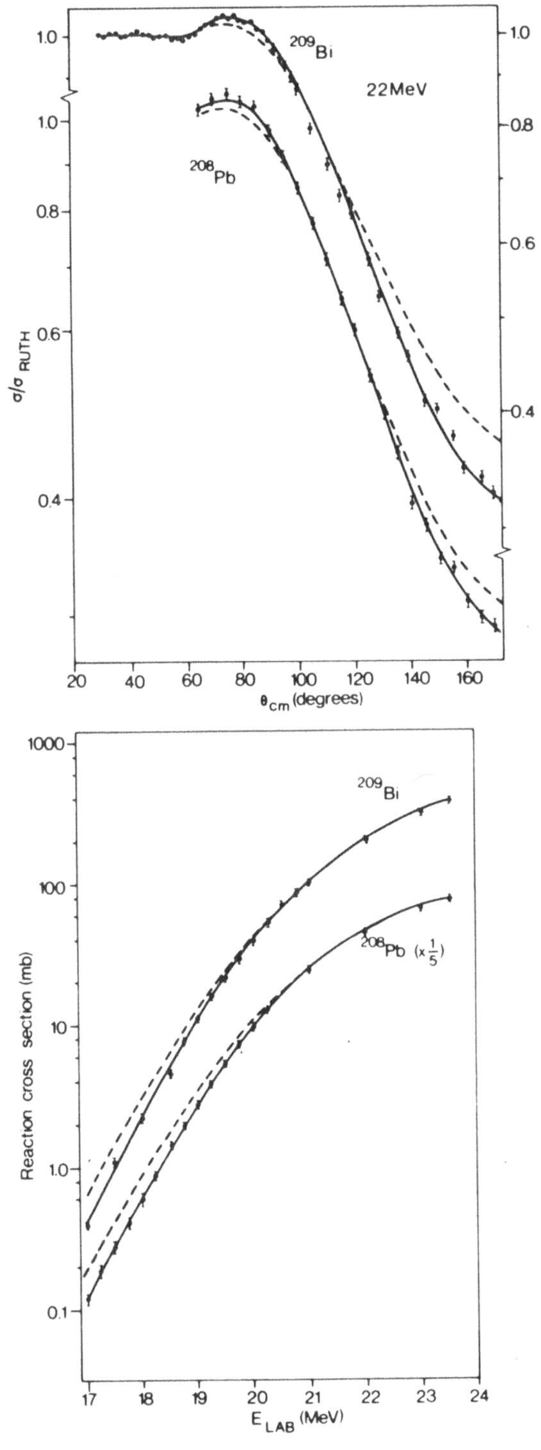

Fig.8. 22-MeV α-elastic scattering data on ^{208}Pb and ^{209}Bi and reaction cross-section excitation functions. The solid curves are given by an optical potential with a_R = 0.65. The broken curves to elastic data show the best fits that could be obtained with a_R = 0.555 if the reaction data are well fitted. Similarly, the broken curves to the reaction data are the best fits that could be obtained (with a_R = 0.555) if the elastic data are well fitted.

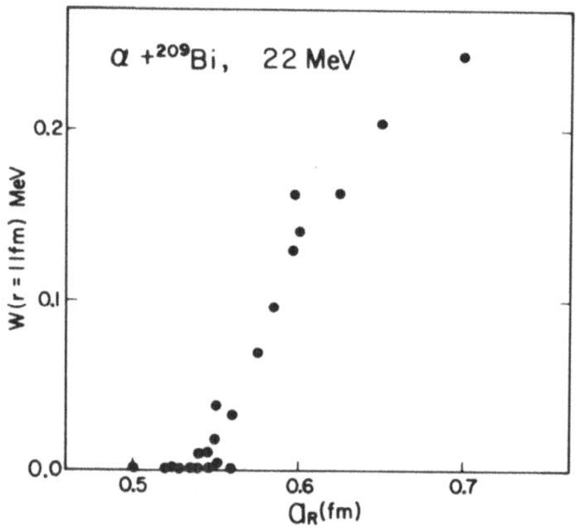

Fig.9. The dependence of the imaginary potential evaluated at 11 fm as a function of the real diffuseness parameter a_R, for equivalent potentials obtained in the analysis of the 22-MeV α-^{209}Bi data.

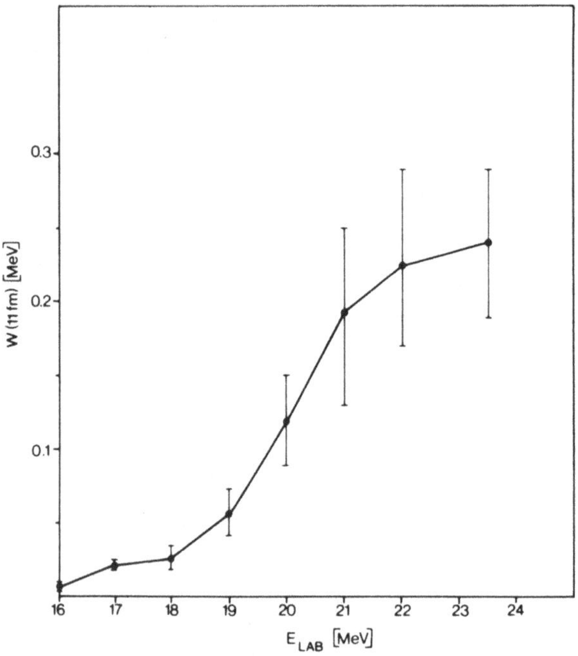

Fig.10. Energy dependence of the surface absorption potential required to fit the elastic scattering and total reaction cross section data of Fig.8.

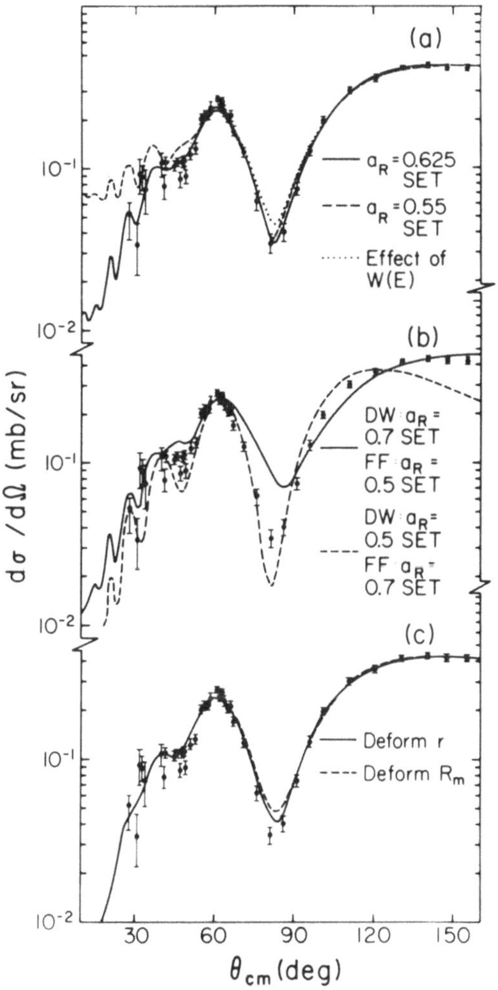

Fig.11. Differential cross sections of 23.5 MeV -particles inelastically scattered by ^{208}Pb to the 3^- state at 2.62 MeV. The curves are DWBA predictions as described in ref.22 using (a) the deformed potential model, (b) inconsistently derived distorting potential and form factor and (c) folding model.

This last observation is supported by an analysis[22] of inelastic scattering of 23.5 Mev α-particles exciting the 2.6 MeV (3^-) state of ^{208}Pb (see fig.11). A good fit could be achieved only if the surface imaginary term used in the exit channel optical potential was smaller than that used in the entrance channel.

At 23.5 MeV, the Coulomb and nuclear inelastic scattering amplitues are comparable, and a strong interference minimum is seen near 80°. Coulomb excitation dominates the peak near 60°, and nuclear excitation is responsible for the yield at large angles. Thus it is possible to obtain reliable Coulomb and nuclear information simultaneously from fits to such data. The analysis gave information about V and W similar to that obtained from analysing the data of fig.8(a). In addition, the tails of the Coulomb and nuclear transition potentials to the 3^- state were determined.

The nuclear densities and transition densities given by the folding model analysis are shown in figs.12 and 13. The interaction v_{eff} was taken to be either a Woods-Saxon potential, obtained from fits to alpha-nucleon scattering data betwen 1 and 10 MeV[23], or its gaussian equivalent[24].

The solid curve in fig.12 is the experimentally deduced nuclear matter distribution. It depends slightly on the average energy of the alpha-nucleon interactions since v_{eff} is weakly energy dependent. For the distribution shown here, which agrees with that of fig.7 to better than 5%, the interaction energy was taken to be close to zero. The dashed curve is a theoretical calculation[25] which also matches the empirical distribution very well. Other curves and data in fig.12 refer to the ^{208}Pb charge distribution, which is not at all critical for fitting α-scattering. In fact uniformly-charged sphere would be just as effective.

Fig.13 shows the nuclear ρ_m^{tr} and charge ρ_m^{tr} transition densities obtained from fitting the inelastic scattering data. They are derivatives of the matter and charge densities scaled by the appropriate deformation parameters needed to fit the data. The tail of ρ_m^{tr} is well determined and again agrees remarkably well with the theoretical calculation of ref.25. The charge transition density is by no means as well determined in the surface as is the matter distribution. However, the B(E3) value, which is proportional to the r^3-moment of ρ_c^{tr}, is determined simultaneously with the nuclear information and agrees well with other reliable values for this quantity[22]. Furthermore, additional information can be obtained about the form factors by carefully fitting the interference minimum in cases where the nuclear and Coulomb excitations are comparable.

These results paint a consistent and reassuring picture that the low-energy α-nucleus interaction can be well represented by a

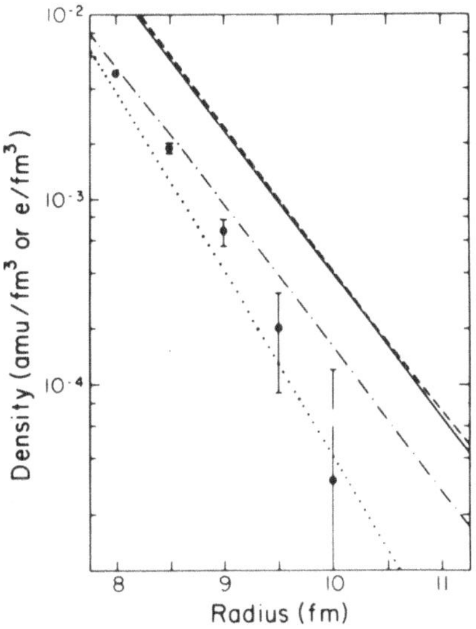

Fig.12. ^{208}Pb matter ρ_m and charge ρ_C ground state density distributions. Solid curve: ρ_m of ref.20 and ref.22; dashed curve: theoretical ρ_m of ref.25. Dot-dashed curve: ρ_C = solid curve x (Z/A); dotted curve: theoretical ρ_C of ref.25; experimental points: ρ_C of ref.26.

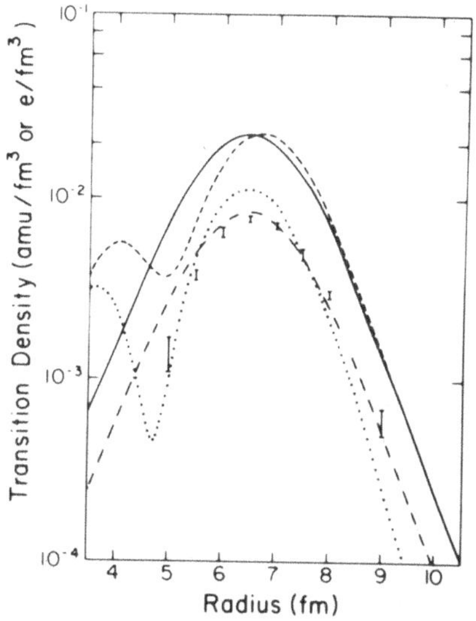

Fig.13. Matter (ρ_m^{tr}) and charge (ρ_C^{tr}) transition densities for the 2.62 MeV(3) state of ^{208}Pb. Solid curve: ρ_m^{tr} obtained in ref.22; dashed curve: theoretical ρ_m^{tr} of ref.25; dot-dashed curve: ρ_C^{tr} obtained in ref.22; dotted curve: theoretical ρ_C^{tr} of ref.25; experimental points ρ_C^{tr} of ref.27.

simple folding model. In the surface region, the nucleons interact with the α-particle as if they were free, which is to be expected in the limit of very low density.

Using the empirical ^{208}Pb matter distribution, the surface form of the folded potential matches that of a Woods-Saxon potential with a_R between 0.6 and 0.65 fm. This is within the preferred range for a_R given by the analysis of the combined elastic scattering and reaction cross section data[21]; and also establishes the form of the energy dependence of the surface imaginary potential.

IV. THE TWO-NEUTRON-TRANSFER PROCESS

The final example is the reaction $^{208}Pb(^{16}O,^{18}O)^{206}Pb$ to the (0^+) ground state, measured at sub-Coulomb energies[28].

Two nucleon transfer has been studied widely with many different target-projectile combinations, but a detailed understanding of the way the reaction proceeds is far from having been achieved. It is much more complicated than the other reactions which have been discussed. Cross sections are low and highly sensitive to a proper description of the wave functions, two-step processes and optical model ambiguities. The present example was designed to investigate the process in a situation free from optical potential ambiguities, where the wave functions are among the best known, and for which, as described earlier, much information is known about the relevant single-neutron transfers. This latter information enables the amplitudes of important second-order paths corresponding to successive neutron transfers to be calculated with confidence.

The results are shwon in fig.14. Data were taken at 155° (lab) at three incident ^{16}O energies. In the calculations all the direct single-step and two-step paths shown in fig.15 were considered. Other two-step paths do not contribute because of the structure of the states involved.

The one-step contribution (path 1) calculated using an exact finite range, full recoil, DWBA code with microscopic form factor, is about an order of magnitude below the observed values. Improved agreement is obtained only when the two-step processes are taken into account. This was done using both semi-classical[29] and DWBA approaches. Amplitudes are listed in table 6. Contributions from the six sequential transfers are all approximately in phase and add constructively. Also included is the important non-orthogonality (NO) correction which must be made with both one-step and two-step processes are considered.

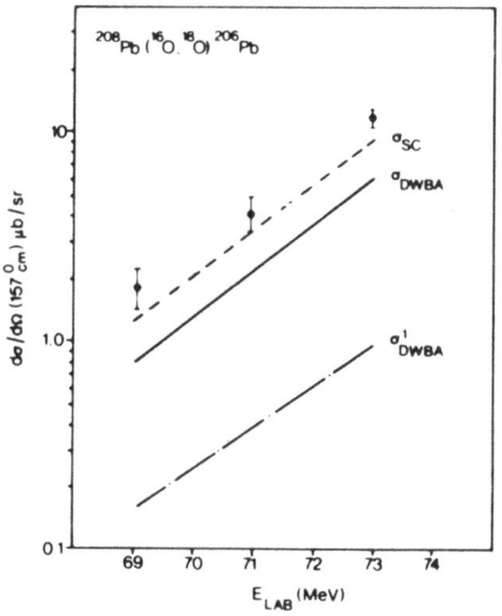

Fig.14. The differential cross section for the reaction $^{208}Pb(^{16}O,^{18}O)^{206}Pb$ to the ground state in the exit channel observed at 157° (c.m.) and at three ^{16}O bombarding energies. The point at 73 MeV includes the data from the reaction $^{206}Pb(^{18}O,^{16}O)^{208}Pb$ taken at 71.62 MeV ^{18}O energy. The lines with the various labels correspond to the different theoretical predictions described in the text and in ref.28.

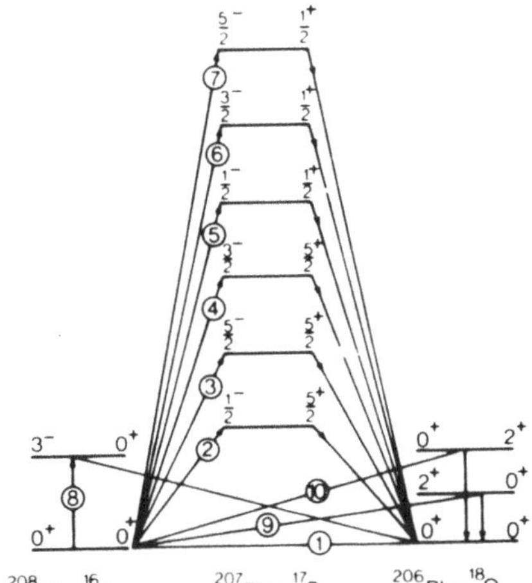

Fig.15. The direct and two-step paths considered in the predictions of the cross section for the reaction $^{208}Pb(^{16}O,^{18}O)^{206}Pb$. The levels in ^{207}Pb and ^{17}O concerned in the sequential transfer paths are the $p_{1/2}$, $f_{5/2}$ and $p_{3/2}$ single-hole states and the $d_{5/2}$ and $s_{1/2}$ single-particle states, respectively.

The essential result of all these calculations is that a one-step DWBA calculation will usually be inadequate for describing two-nucleon transfer in heavy-ion reactions, and that it will be necessary to take sequential transfer into account. The important point is that this result could be obtained quantitatively and unambiguously at low energies.

TABLE 6

Semiclassical (s.c.) and DWBA reaction amplitudes for the 2NT paths shown in fig.15. Calculations are for 73 MeV and $\theta_{c.m.} = 157°$

Path	Magnitude $(\mu b/sr)^{1/2}$		Phase (rad)	
	S.C.	DWBA	S.C.	DWBA
1	1.48	0.96	0.00	0.16
NO	0.97	0.63[a]	π	3.30[a]
2	0.76	0.63	0.70	0.70
3	0.35	0.28	0.57	0.54
4	0.37	0.33	0.43	0.46
5	0.51	0.40	0.42	0.47
6	0.12	0.09	0.32	0.36
7	0.40	0.34	0.24	0.30
8		0.22		
9		0.32		
10		< 0.04		
$2 \rightarrow 7$	2.5	2.1	0.49	0.52
$1 \rightarrow 7$ + NO	3.0	2.4	0.41	0.47

[a]Assuming the same relative NO amplitude as given by the semiclassical calculation.

V. FINAL REMARKS

There are many other reactions which could be discussed or areas into which the present ones could be extended. For example, the calibrated reactions can be exploited for extracting information about nucleon distributions in many other nuclei. The resulting data are critical raw material for testing theories of nuclear structure and, possibly, for investigating non-locality effects, effective charges and details of other reactions. Indeed even sub-Coulomb nucleon transfer on nuclei far from closed shells may require a more complex reaction description than the simple one-step DWBA.

The folding model also can be tested in detail to learn about the interaction potential at different energies and for different projectiles. Certainly the free alpha-nucleon potential is not the correct one to describe the α-particle interaction with denser nuclear matter at the smaller radii reached at higher energies. Also the simple approach, which appears to work for low-energy alphas may not do so for other, less tightly bound, projectiles, even at low energy.

Even the studies of the present examples are far from complete. Sub-Coulomb proton transfers should be calibrated, and the assumption that multistep processes are unimportant needs to be verified. Some data on the deeper neutron-hole states in ^{208}Pb are available[16], and more would be desirable but difficult to obtain. Our knowledge of spectroscopic factors needs to be improved, which may require old work to be repeated and improved.

The great advantage of a low-energy interaction is the gentleness of the collision and the simplifications that arise therefrom. One can expect to understand the reactions in great detail and obtain microscopic information about the extreme surface regions to which the direct processes by and large are restricted. The picture that emerges from the study of ^{208}Pb is a most consistent one, and the body of data built up already on this nucleus can be used to great advantage in furthering our understanding of nucleur structure and reactions in other systems and at higher energies.

VI. ACKNOWLEDGEMENTS

It is a pleasure to acknowledge the collaboration of many colleagues who participated in different aspects of this work: Michael Franey and Ben Bayman (University of Minnesota), Bill Phillips, John Durell and Ross Barnett (University of Manchester), Da Hsuan Feng (Drexel University), Brian Fulton and Neil Sanderson (Daresbury Laboratory), Trevor Ophel, Peter Clark and Charles Attwood (A.N.U.Canberra) and Brian England (University of Birmingham).

I am also grateful to the members of the Nuclear Physics Laboratories at the University of Minnesota, the University of Rochester, New York, the Niels Bohr Institute, Denmark, and the Australian National University, Canberra, who in recent times made their facilities available to us and also offered help, hospitality and many stimulating discussions.

REFERENCES

1. P.J.A. Buttle and L.J.B. Goldfarb, Nucl. Phys. A176, (1971) 299.
2. H.J. Körner and J.P. Schiffer, Phys. Rev. Lett. 27, (1971) 1457; J.P. Schiffer and H.J. Körner, Phys. Rev. C8, (1973) 841.
3. G.D. Jones, J.L. Durell, J.S. Lilley and W.R. Phillips, Nucl. Phys. A230, (1974) 173; J.L. Durell, C.A. Harter, J.N. Mo and W.R. Phillips, Nucl. Phys. A334, (1980) 144; R. Chapman, J.N. Mo, J.L. Durell and N.H. Merrill, J. Phys. G: 2 (1976) 951; R. Chapman et al. Nucl. Phys. A316, (1979) 40.
4. J.L. Durell, et al. Nucl. Phys. A269, (1976) 443.
5. M.A. Franey, J.S. Lilley and W.R. Phillips, Nucl. Phys. A324, (1979) 193.
6. A. Warwick et al. Phys. Lett. 87B, (1979) 335.
7. A. Warwick, R. Chapman, J.L. Durell and J.N. Mo, to be to be published.
8. J.W. Negele, Phys. Rev. C9, (1974) 1054.
9. J.A. Nolen and J.P. Schiffer, Ann. Rev. Nucl. Sci. 19, (1969) 471.
10. I. Sick et al. Phys. Rev. Lett. 38, (1977) 1259.
11. De Witts Hubert et al. Phys. Lett. 71B, (1977) 317.
12. J.L. Durell, W.R. Phillips, B.R. Fulton, J.S. Lilley and N.E. Sanderson, Phys. Lett. 94B, (1980) 335; and Proc. Int. Conf. on Nuclear Physics, Berkeley, (1980).
13. J.L. Durell, I.O.P. Conference on 'Trends in Nuclear Physics', University of Manchester (1980) and private communication.
14. C.F. Clement, Nucl. Phys. A213, (1973) 493.
15. M.C. Birse and C.F. Clement, I.O.P. Conference on 'Trends in Nuclear Structure Physics', University of Manchester (1980), unpublished.
16. S. Gales, G.M. Crawley, D. Weber and B. Zwieglinski, Phys. Phys. Rev. C18, (1978) 2475.
17. L. Ray, Phys. Rev. C19, (1979) 1855; G.W. Hoffman et al. phys. Rev. C21, (1980) 1488.
18. B. Frois et al. Phys. Rev. Lett. 38, (1977) 152.
19. A.M. Bernstein, Advances in Nuclear Physics ed. M. Baranger and E. Vogt (Plenum, New York, 1969) Vol.3.
20. A.R. Barnett and J.S. Lilley, Phys. Rev. C9, (1974) 2010.
21. R.M. De Vries, J.S. Lilley and M.A. Franey, Phys. Rev. Lett. 37, (1976) 481.

22. J.S. Lilley, M.A. Franey and D.H. Feng, Nucl. Phys. A342 (1980) 165.
23. P. Mailandt, J.S. Lilley and G.W. Greenlees, Phys. Rev. C8, (1973) 2189.
24. C.J. Batty, E. Freidman and D.F. Jackson, Nucl. Phys. A175, (1971) 1.
25. I. Hamamoto, Phys. Lett. 66B, (1977) 410; and private communication.
26. C.W. de Jager, H. de Vries and C. de Vries, Atomic and Nuclear Data Tables 14, (1974) 501.
27. H. Rothhaas, J. Friedrich, K. Merle and B. Dreher, Phys. Lett. 51B, (1974) 23;
J.H. Heisenberg and I. Sick, Phys. Lett. 32B, (1970) 249.
28. M.A. Franey, B.F. Bayman, J.S. Lilley and W.R. Phillips, Phys. Rev. Lett. 41, (1978) 837.
29. R.A. Broglia and A. Winther, Phys. Rep. 4C, (1972) 153.

HEAVY ION REACTION MECHANISMS

C.F. Maguire and J.H. Hamilton

Physics Department
Vanderbilt University
Nashville, Tennessee 37235

I. INTRODUCTION

Here we wish to report on some recent work in Heavy-ion scattering involving Vanderbilt University. The studies concern quasi-elastic experiments and their reaction mechanism interpretations. Specifically, we present a striking new example of backangle resonant behavior for the odd mass system $^{19}F + ^{12}C$. At the opposite extreme of backward rising cross sections we discuss exponentially falling rainbow scattering data and the effects of inelastic coupling on the extraction of optical model parameters.

II. QUASI-ELASTIC REACTION MECHANISMS IN HEAVY-ION COLLISIONS

Quasi-elastic processes in heavy-ion collisions involve low energy and few (if any) particle transfer between the projectile and the target nucleus. Such processes are intrinsically peripheral in nature and, since they excite few degrees of freedom, might have been expected to be among the simplest experiments to interpret theoretically. In this section we describe a number of systematic studies, to which we have contributed, demonstrating that our understanding of these supposedly simple processes is not as good as it should be.

Elastic scattering is the most trivial of all quasi-elastic collisions but nonetheless the most important since information about all flux removed to other channels is contained implicitly in the elastic differential cross-section. In Fig.1 is shown a systematic study[1] of the $^{16}O + ^{28}Si$ elastic scattering from 32 to 215.2 MeV incident energy. A single, global optical model type potential

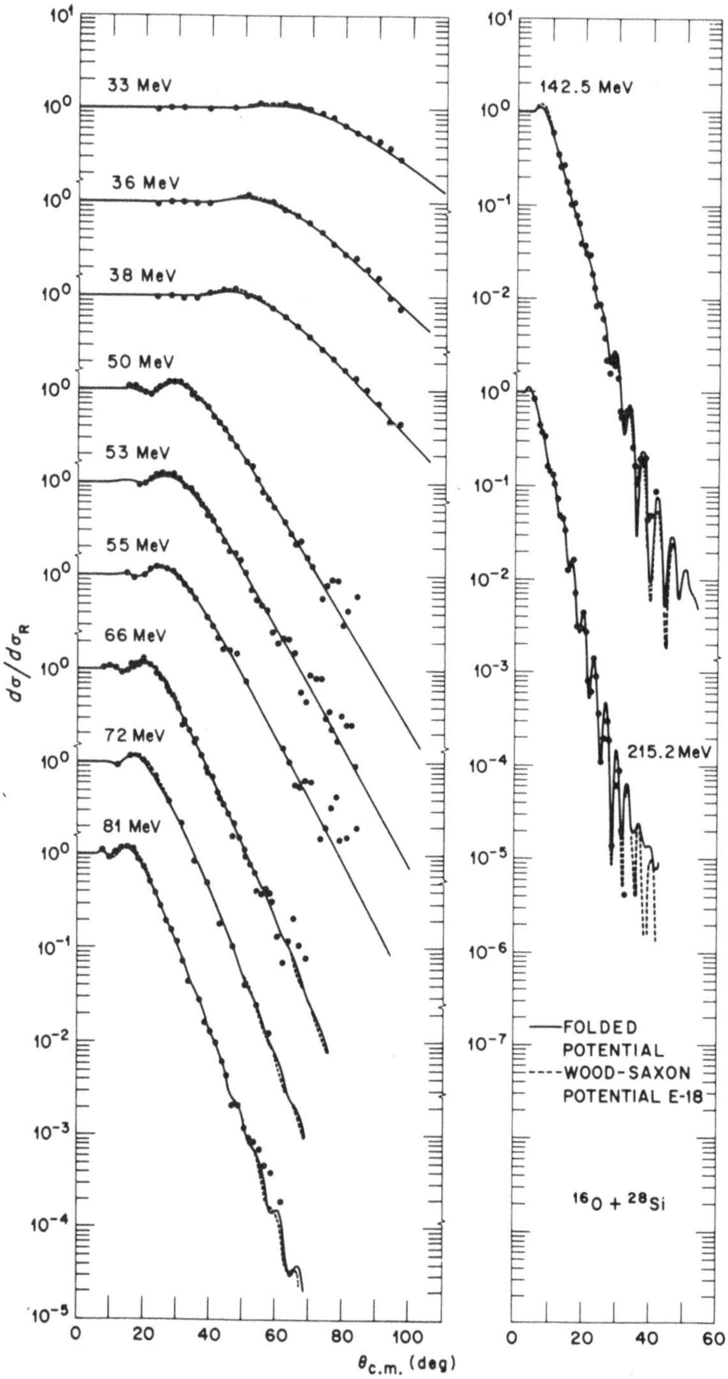

Fig.1 Global survey of the elastic scattering of $^{16}O + ^{28}Si$.

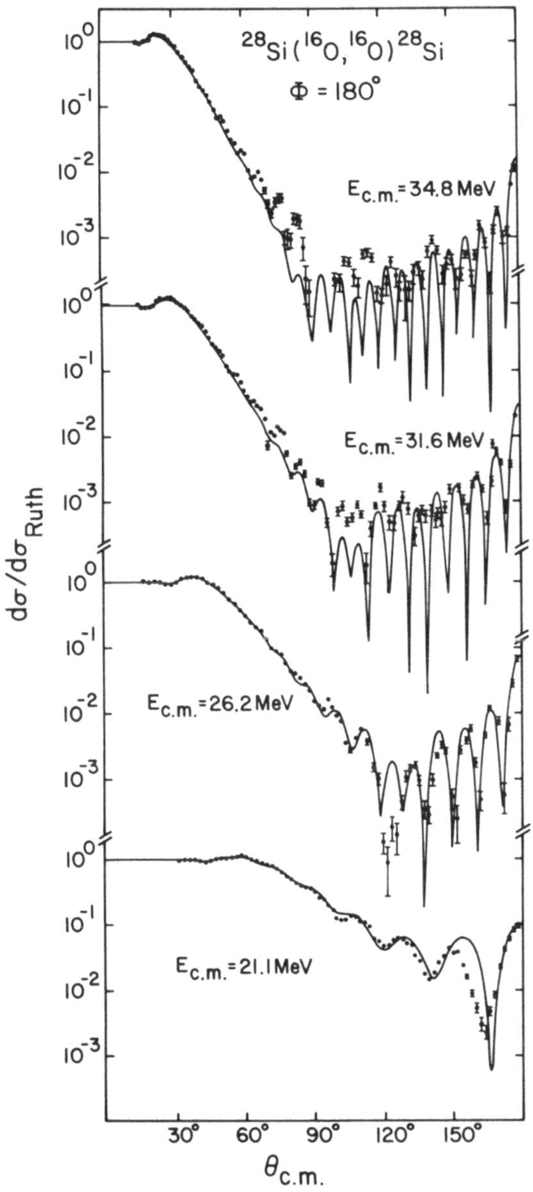

Fig.2 Angular distributions of $^{16}O + ^{28}Si$ elastic scattering. These energies correspond to E_{LAB} = 55, 50, and 33 MeV for which the forward angle data and conventional optical potential fits are shown in Fig.1, ref.2.

("E18") is found to fit well the general features of the data. Closer inspection of the mid-energy data between 50 and 66 MeV energy, however, hints of a serious discrepancy between the theory and the data. Subsequent experiments reveal a surprising picture when the full angular range of the data was measured at just these energies. As seen in Fig.2, the data here do not fall off to near vanishingly small values at large angles as potential set E18 would predict.

Instead, the more backward angle yield rises to as much as several per cent of the Rutherford value. In fact, the more backward angle data can be fit by a simple $|P_L(\cos\theta)|^2$ shape indicating the dominance of a single partial wave. More generally one can make a Regge pole parameterization of the nuclear S-matrix:[2]

$$S_\ell = S_\ell^o \left\{1 + iD_\ell e^{2i\phi}/(\ell - \ell_o - i\Gamma/2)\right\} \tag{1}$$

The background S-matrix term for each partial wave, S_ℓ^o, is obtained from the global fit potential E18 as are the values for Γ and D_ℓ, the total and the elastic widths. The real part of the Reggee pole ℓ_σ and the mixing angle ϕ come from fitting the back and mid-angle ranges of Fig.2 data.

Further information is gained by doing 180° angle excitation function of this system which is no easy experimental task[3]. The ^{28}Si is the beam projectile and the ^{16}O is the target which recoils forward at 0° into a magnetic spectrometer. Since there is a charge state of ^{28}Si which degenerate with the primary charge state of the ^{16}O, a clever system of absorbers had to be designed to preserve the magnet's focal plane counter. The data then obtained exhibit a remarkable series of resonant-like (see Fig.3) structures heretofore found only in light systems such as $^{16}O + {}^{16}O$, $^{12}C + {}^{16}O$, and

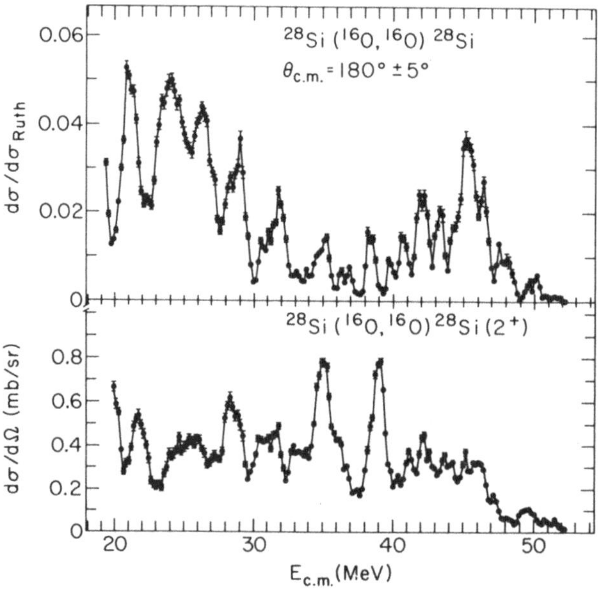

Fig.3 Back angle excitation function of $^{16}O + {}^{28}Si$ elastic and inelastic (2^+) scattering[3].

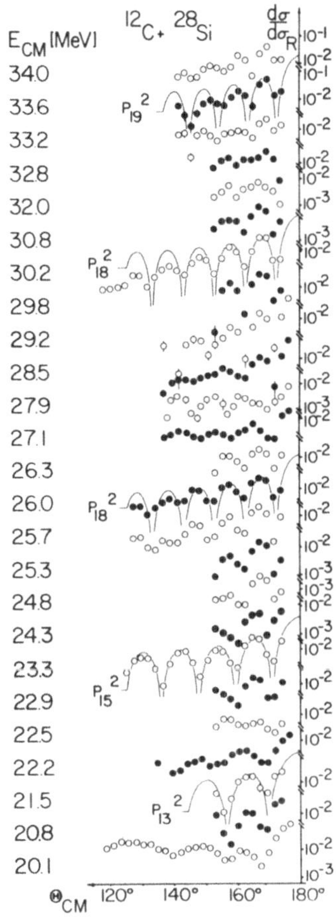

Fig.4 Back angle angular distributions for $^{12}C + ^{28}Si$. "On resonance" the shape is fit-fitted well by a single Legendre polynomial squared whereas "off resonance" the shape is featureless and the magnitude depressed[5].

$^{12}C + ^{12}C$ (ref.4). From the angular distribution data, as in Fig.2, one can determine the particular ℓ value which is resonating and for $^{28}Si + ^{16}O$ it appears that this ℓ value tracks with energy as the grazing partial wave.

This effect has been observed in other systems; the most extensively studied of which is $^{28}Si + ^{12}C$ for which the data[3,5] are shown in Figs.4 and 5. A significant result about the $^{28}Si + ^{12}C$ study is that for this system the resonating partial waves do not track regularly with the grazing partial wave values. This has also been seen in the systems[6] $^{24}Mg + ^{12}C$ and $^{20}Ne + ^{12}C$ whose data are

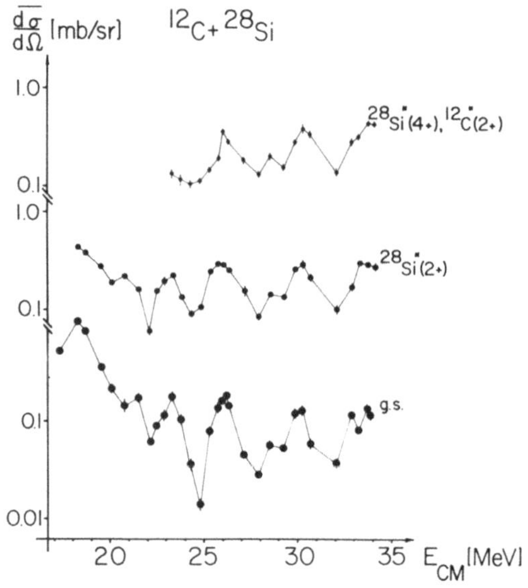

Fig.5 Integrated back angle yields of $^{12}C + ^{28}S$ scattering showing gross structure effects[5].

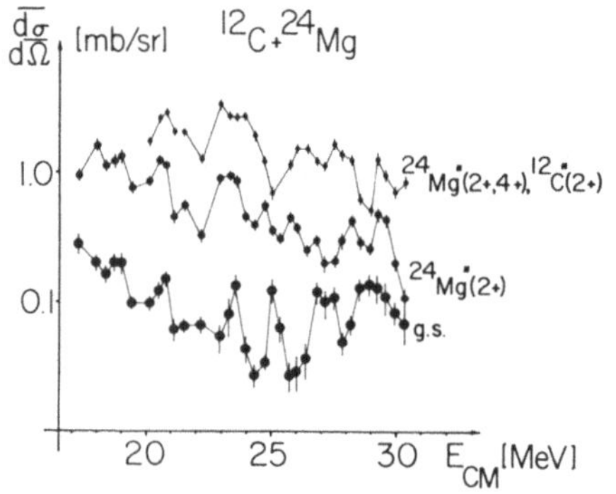

Fig.6 As in Fig.5 for $^{12}C + ^{24}Mg$ excitation function[6].

shown in Fig.6 and 7. For $^{20}Ne + ^{12}C$ the backangle yield at 74 MeV (Lab) is best fitted by $|P_L(\cos\theta)|^2$ with $L_o = 14$, whereas the grazing partial wave is $L_{gr} = 20$ or 21 ħ. Similarly, the resonance bumps at 72.6 and 75.2 MeV are best fitted by $L_o = 15$ and 19, respectively. In $^{24}Mg + ^{12}C$ the resonant bumps have L_o values of 15, 16, 16, 16, 18 and 19 corresponding to energies of 20.8, 23.6. 25, 27 and 29 MeV (c.m.). It is clearly going to be very difficult to associate these resonant-like structures as being generated directly by grazing, surface partial waves.

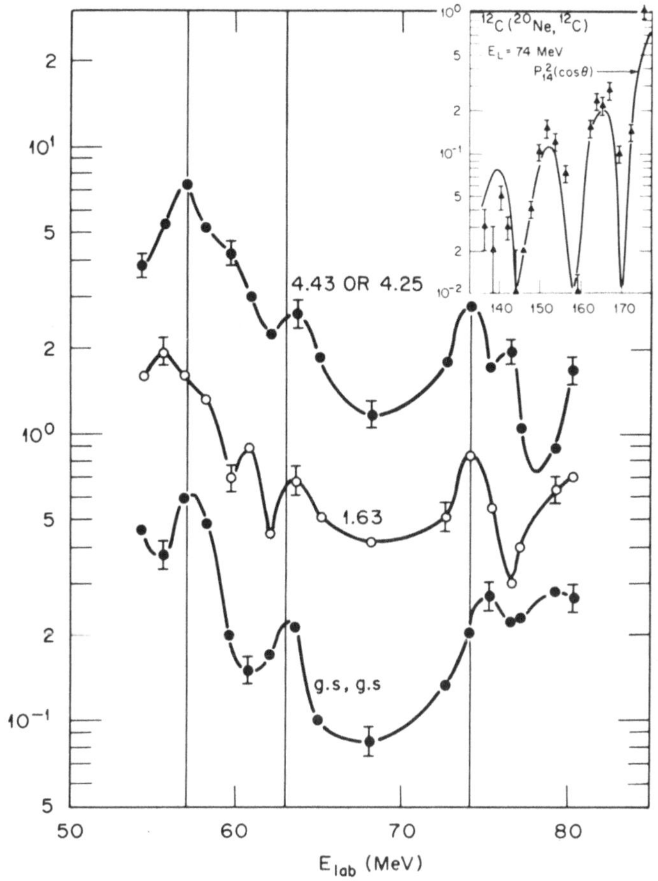

Fig.7 Excitation function of ^{12}C + ^{20}Ne (ref.6) showing correlated structures in the elastic and inelastic channels. A squared Legendre polynomial fit to the back angular distribution is shown in the inset.

Up until this point, one trend which could be counted on was the substantial weakening or virtual disappearance of the resonances when an odd mass or a non-alpha cluster even mass nucleus was introduced. That is, for the systems ^{9}Be + ^{28}Si, ^{13}C + ^{28}Si, ^{18}O + ^{28}Si, ^{12}C + ^{27}Al, ^{16}O + ^{29}Si, and ^{16}O + ^{30}Si (refs.5, 7, 8, 9) the resonant structures were reduced significantly both in size and peak-to-valley ratio by factors of 5 to 100 compared to the neighboring α cluster system (see Figs.8, 9 and 10).

We report in this section a dramatic exception to this erstwhile trend for the system ^{19}F + ^{12}C. These data were measured[10]

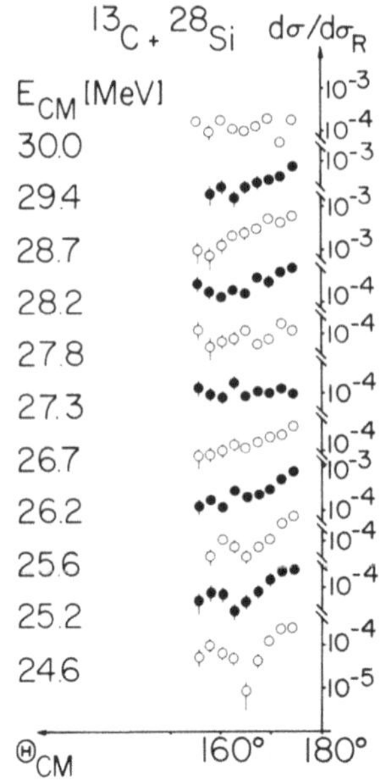

Fig.8 Angular distributions at back angles for $^{13}C + ^{28}Si$ (ref.7). Note the absence of oscillations as seen in Fig.5 for $^{12}C + ^{28}Si$.

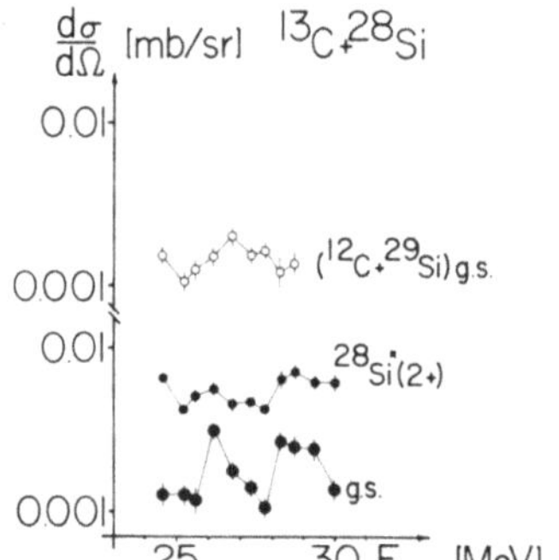

Fig.9 Integrated back angle yields for $^{13}C + ^{28}Si$ elastic and inelastic scattering[7]. Note the structures present are at least an order of magnitude smaller smaller than in $^{12}C + ^{28}Si$ (see Fig.5).

at the Brookhaven National Laboratory QDDD facility using a thin ($\sim$ 10 μg/cm^2) natural carbon target on which was evaporated a small ($\sim$ 1 μg/cm^2) amount of gold for excitation function normalization. The target had to be this thin in order to resolve the low-lying excited states of ^{19}F (110 keV, 1/2$^-$ and 197 keV, 5/2$^+$) from the

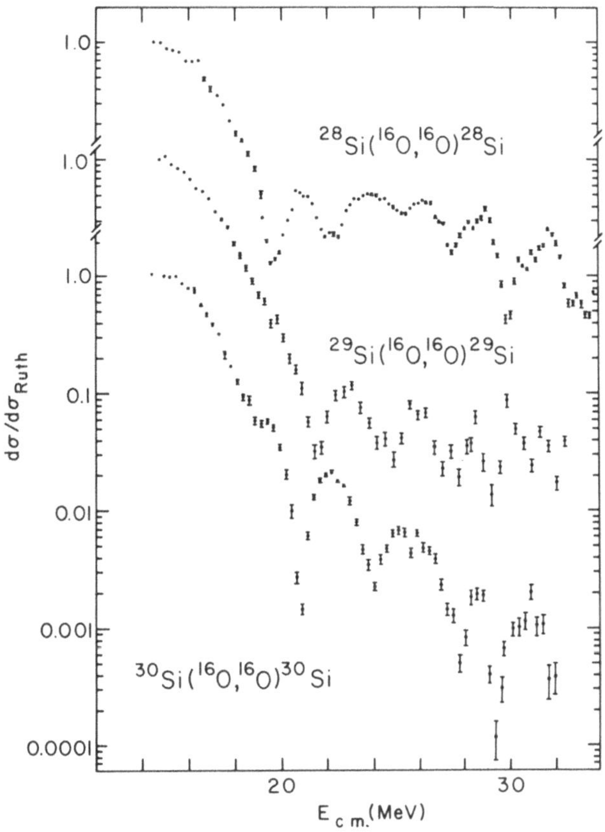

Fig.10 Back angle (180° ± 5°) excitation function for $^{16}O + ^{28,29,30}Si$ (ref.9). Note the systematic weakening of the gross structure as the target neutron number is increased.

ground state ($1/2^+$) elastic scattering. As seen in Figs.11 and 12, the data for this system show resonances as prominent in magnitude and peak-to-valley ratio as any heretofore observed in alpha cluster nuclei. Both the ground $1/2^+$ and the second excited $5/2^+$ state at 197 keV resonate more or less in phase while the first excited state does not participate. Angular distribution data taken earlier at 47 MeV (lab) confirm a highly oscillatory, rising backangle distribution, but more systematic measurements remain to be conducted. Detailed comparisons with the neighboring system $^{20}Ne + ^{12}C$ may shed further light on this resonance mechanism. We show in Fig.13 a preliminary example of such a comparison in which the excitation function data of $^{20}Ne + ^{12}C$ and of $^{19}F + ^{12}C$ are plotted together. The $^{20}Ne + ^{12}C$ abscissa data values have been adjusted slightly (by 1.02) so that along the center-of-mass energy scale one is

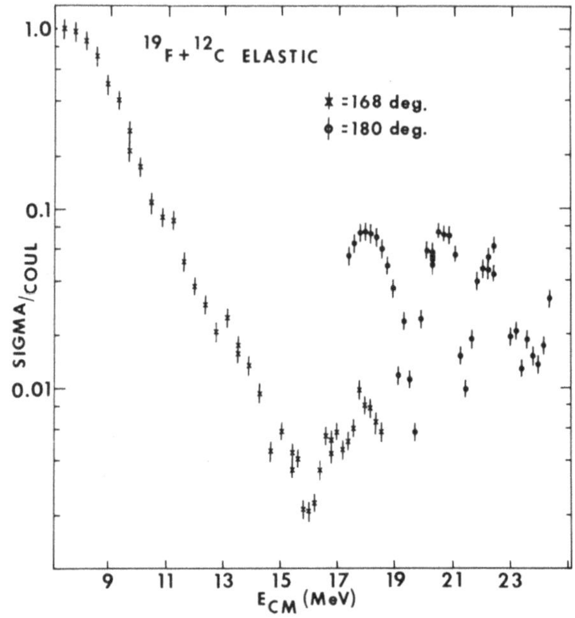

Fig.11 Back angle integrated (±5°) excitation function for ^{19}F + ^{12}C elastic scattering[10] (ref.10). Note the rise in the cross section between 168° and 180° around E_{cm} = 17 MeV.

looking at equivalent entrance channel angular momenta for these two data sets. It is seen that there is a striking correlation in both magnitude and phase for these two excitation function in the energy region where the data overlap. Extension of the ^{19}F + ^{12}C measurements to higher energies is clearly warranted. For now, it appears that this correlation supports the belief[3] that these resonances are dynamically induced by the interaction potential rather than being of structural origin in the compound nuclear system.

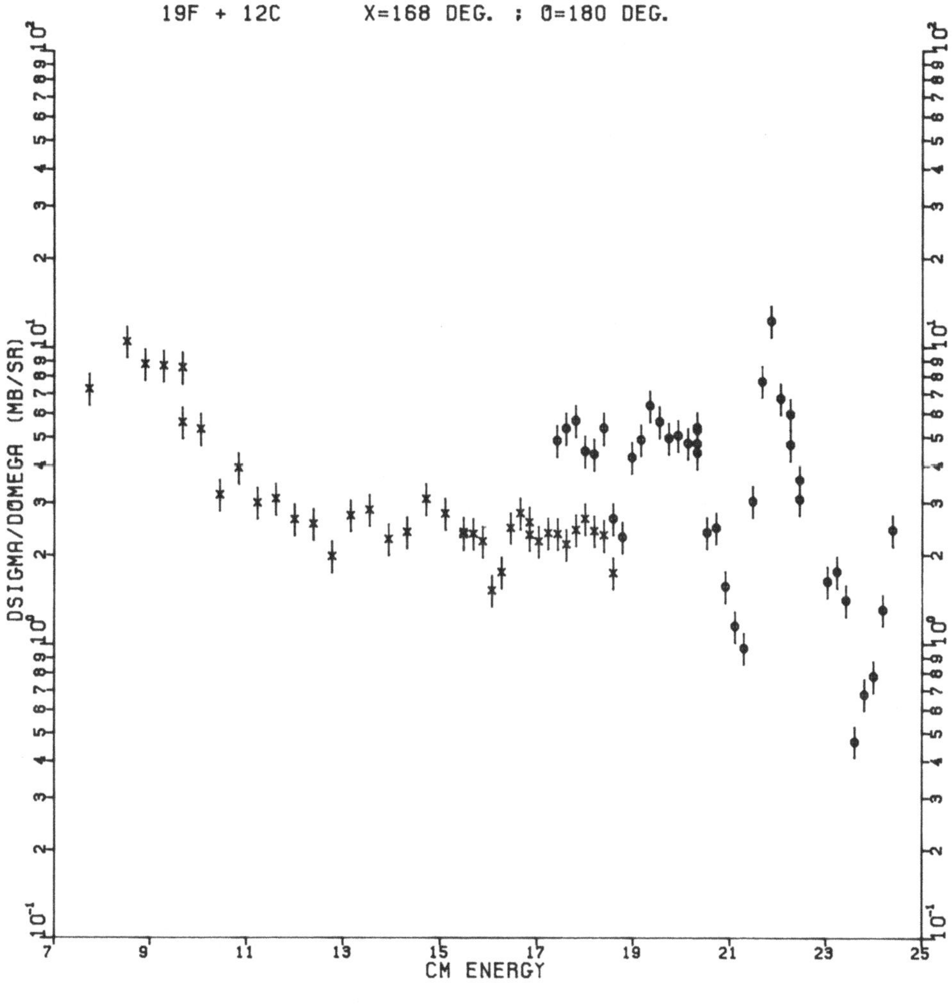

Fig.12 As in Fig.11 for the $^{19}F + ^{12}C$ scattering to the 197 keV, $5/2^+$ state in ^{19}F.

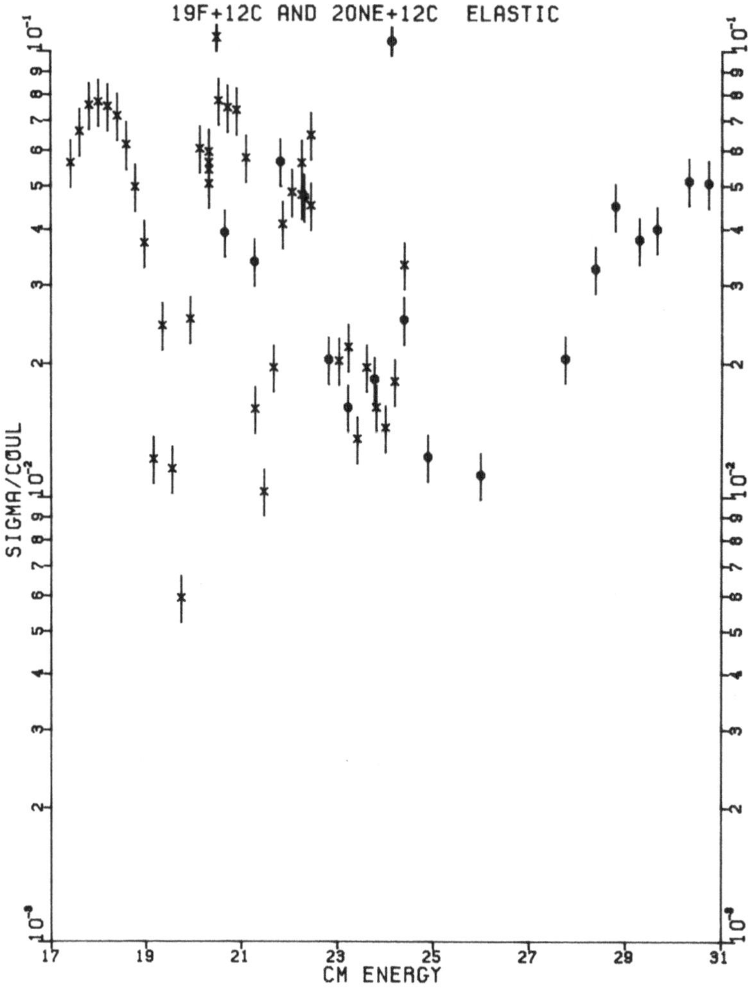

Fig.13 Comparison of ^{12}C + ^{20}Ne and ^{12}C + ^{19}F back angle excitation functions. The ^{20}Ne points points are for 168° ± 5° and the ^{19}F points are for 180° ± 5°.

At the opposite extreme from rising backward angle distributions are exponentially falling backward angle shapes. In fact the high energy 215.2 MeV ^{16}O + ^{28}Si measurement shown in Fig.1 was originally intended to determine if this system was sensitive to the rainbow angle effect first seen in α + ^{58}Ni scattering[11] . In Fig.14 is shown an expanded view of these data along with theoretical predictions of two different types of optical potentials. The refractive potential curve (which does not fit the data) has at smaller

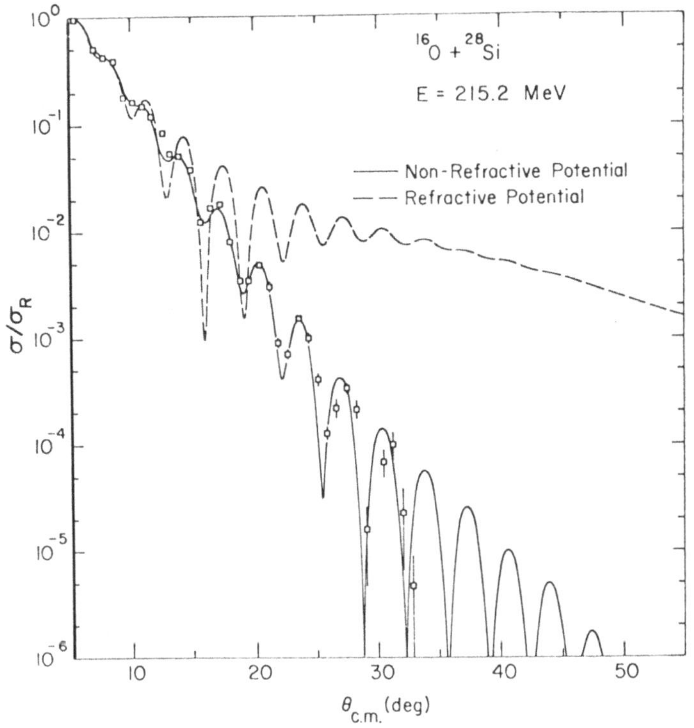

Fig.14 Elastic scattering of ^{16}O + ^{28}Si at E_{LAB} = 215 MeV. The solid curve is an optical model fit prediction using a strongly absorptive potential while the dashed curve is a prediction from a shallow, absorptive rainbow angle producing potential.

angles a typical oscillatory behavior while at large angles the oscillations are completely damped away and the shape is exponentially falling. Such predicted behavior can be easily explained as a rainbow angle phenomenon, that is, if the real nuclear potential is deep enough there exists a certain maximum angle of scattering which classically cannot be exceeded. As the incident particle energy is increased, this nuclear rainbow angle decreases since the particle's trajectory becomes more difficult to bend. Of course, there exists an analogous Coulomb rainbow angle having the same effect but not as physically interesting. Unfortunately, for the system ^{16}O + ^{28}Si, it appears that the imaginary absorption is too strong for the rainbow angle effect to be useful as seen in the solid curve of Fig.14. For the original light ion system α + ^{59}Ni (ref.11) shown in Fig.15, this is not the case. A rainbow angle effect is clearly evident in the data, and moreover, χ^2 fits to the differential cross section manifest a definite preference for an

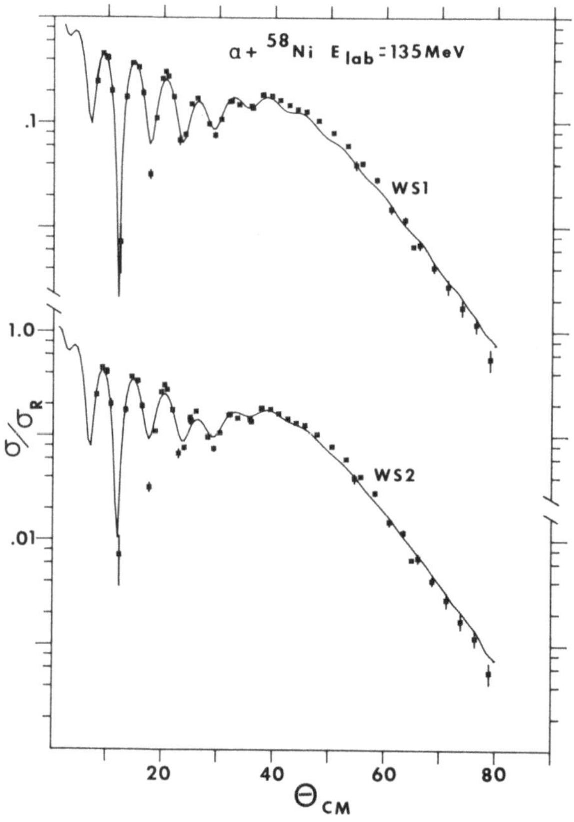

Fig.15 · Elastic scattering of α + ^{58}Ni at E_{LAB} = 139 MeV. The fits are predictions based on a first degree, WS1, and a second degree WS2, Woods Saxom potential[11,20].

optical potential real well depth of V $\sim$ 118 MeV as opposed to V $\sim$ 180 MeV (see Table I). Similar results are found for other targets ^{12}C, ^{40}Ca, and ^{90}Zr when an alpha particle is incident[12]. An interesting questions then is whence the transition from "light" ion scattering with high energy rainbow-like behavior to heavy ion systems such as ^{16}O + ^{28}Si where the imaginary absorption severally limits our knowledge of the real potential. Experiments have thus been carried out on the ^{28}Si with beams of ^{6}Li and ^{9}Be at the highest energies now available[13,14,15]. It would seem that the transition region takes place at ^{9}Be as projectile since ^{6}Li + ^{28}Si shows clear refractive behavior at the two highest energies so far measured (see Fig.17 and Table I) and ^{9}Be does not (see Fig.17).

TABLE I

One Channel Optical Potentials.[a,b]

Set	V_o	r_R	a_R	W_o	r_I	a_I	χ^2/F
			$\alpha + {}^{58}Ni$ 139 MeV				
WS1[a]	118.2	1.240	0.796	20.47	1.595	0.571	6.0
WS2[b]	143.5	1.391	1.218	21.40	1.540	0.686	1.6
			${}^{6}Li + {}^{28}Si$ 135 MeV				
CA1[a]	178.2	1.161	0.848	39.454	1.583	0.921	9.6
			${}^{6}Li + {}^{28}Si$ 154 MeV				
IN1[a]	183.4	1.031	0.959	33.40	1.569	1.025	16

a) $$V(r) = -V_o \left\{ 1 + \exp\left(\frac{r - R_R}{a_R} \right) \right\}^{-1} \quad ; \; R_R = r_R \, A_{Target}^{1/3}$$

$$W(r) = -W_o \left\{ 1 + \exp\left(\frac{r - R_I}{a_I} \right) \right\}^{-1} \quad ; \; R_I = r_I \, A_{Target}^{1/3}$$

b) $$V(r) = -V_o \left\{ 1 + \exp\left(\frac{r - R_R}{a_r} \right) \right\}^{-2}$$

W(r), as above

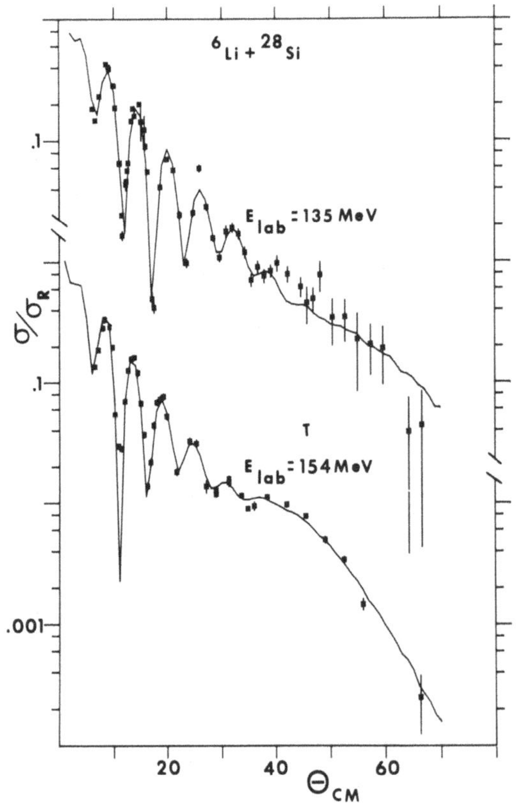

Fig.16 Elastic scattering of ^{6}Li + ^{28}Si at E_{LAB} = 135 and 154 MeV[13,14]. Fits are from one channel optical model calculations

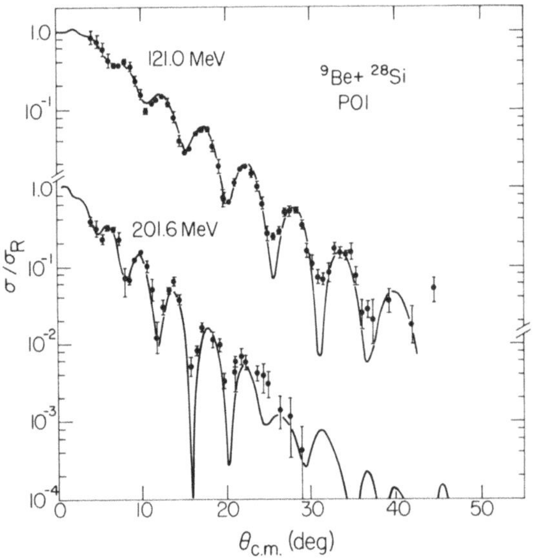

Fig.17 Elastic data of ^{9}Be + ^{28}Si scattering at E_{LAB} = 202 MeV[15]. Fit is from a strongly absorptive non-refractive potential. No evidence of a rainbow angle pattern is seen.

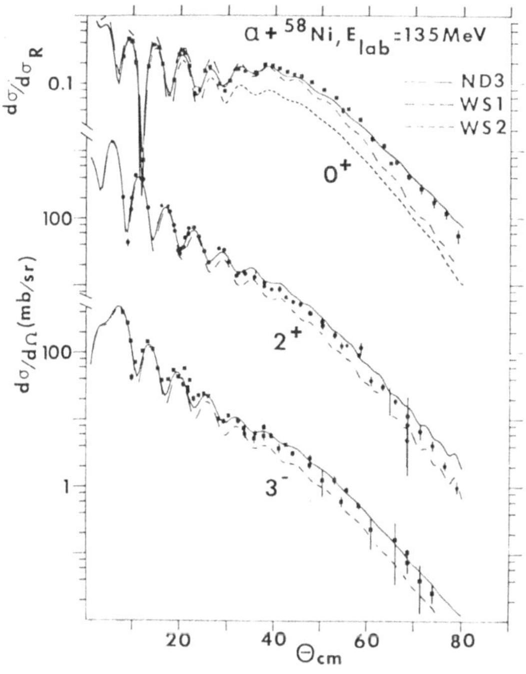

Fig.18 Elastic and inelastic scattering data for α + ^{58}Ni at 139 MeV[11]. The curves are coupled channels predictions using sets WS1 and WS2, which were one channel derived potentials, and set MD118 which was obtained in a coupled channels search.

One of the most useful results to come from these high quality rainbow angle experiments is the apparent depth of the real well potential at about 30 MeV/A for α and ^{6}Li. Such a value is about 25% below recent theoretical estimates of folding model interactions[16]. Arguments are also now being presented in favor of a Wood-Saxon squared shape as being a better parameterization of the data. It should be pointed out now that all of these mentioned analyses have been made with a direct reaction optical model, that is flux is allowed to leave the elastic channel (via the imaginary term) but no back coupling to the ground state is taken into account. As it happens, equally high quality inelastic data exist (unpublished) for these systems, and we have decided to investigate if any modifications in the above conclusions must be made if back coupling from the inelastic channel is permitted. Our method was to require a simultaneous fit to both the inelastic and the elastic angular distributions with the same optical potential using standard CCBA theory[17]. The numerical fitting program was the sequential interaction code EC1S by Raynal[18]. Our preliminary results[19] portend far reaching consequences for the derivation optical model

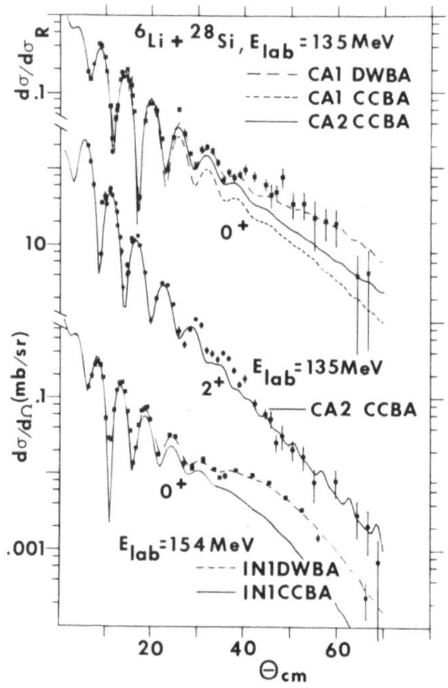

Fig.19 Elastic and inelastic scattering data for $^{6}Li + ^{28}Si$ at 135 MeV[13] and elastic scattering at 154 MeV[14]. The short dashed curves are one channel ("DWBA") predictions as first shown in Fig.16. Also shown are coupled channels predictions for the elastic data using the same optical potentials and for the 135 MeV data CCBA fits using a deeper potential CA184.

potential parameters (see Figs.18, 19 and Table II). In the sensitive, transrainbow angle data region, the inclusion of coupling reduces the predicted yield by between a factor of 3 to 10. One should be aware, also, that this is a self-consistent calculation. That is, the amount of inelastic flux perturbing the elastic yield is the correct amount to fit the measured inelastic cross section. In order to recover a semblance of a good fit to these data the CCBA optical potentials must be significantly changed from their direct model values.

We consider first the $\alpha + ^{58}Ni$ scattering data at 139 MeV with a one channel optical model fit shown in Fig.15, using the optical potential parameters of set WS1 given in Table I. An even better fit to these data may be obtained by using a squared Woods-Saxon shape[20] for the real part of the potential as in parameter set WS2.

Although the real well parameters of WS1 and WS2 are naturally different, the volume integrals per target-projectile nucleon pair are equivalent for the two sets. Consideration of which parameter set is better becomes irrelevant, however, when inelastic coupling to the 2^+ and the 3^- states is allowed to affect the elastic angular distribution. The chi square per datum point drastically worsens to more than 100 for both sets WS1 and WS2 (see Table II and Fig. Fig.18). As it turns out for $\alpha + {}^{58}Ni$ the one channel real well depth of 118 MeV can be preserved in a CCBA fit merely by adjusting the imaginary depth downward by about 16% or the imaginary volume integral downward by about 12% (see Table II at MD3). If one forces a fit with a deeper potential, say V = 134 MeV, the resulting χ^2 is significantly worse (see Table II, set MD4). It appears then that the self-consistent CCBA calculation is required to fit these data and this calculation reinforces the conclusion that the real well depth of the $\alpha + {}^{58}Ni$ system is shallower than at first predicted.

The other system which we have considered is ${}^{6}Li + {}^{28}Si$ at 135 MeV for which the one channel fit is shown in Fig.16 using the parameters of set CA1 given in Table I. Not unexpectedly, the inclusion of inelastic coupling to the ${}^{28}Si$ 2^+ state has a dramatic effect on the elastic prediction as illustrated in Fig.8. Unlike the vibrational ${}^{58}Ni$ case, the change for the more collective rotational ${}^{28}Si$ scattering prediction is one of magnitude and slope. This would indicate that simply adjusting the imaginary potential (which effects only the magnitude of the data in the rainbow region region[11]) will not be sufficient in recovering the fit to the ${}^{6}Li + {}^{28}Si$ data. Coupled channels search calculations yield parameter sets CA2 and CA3 given in Table II which reproduce the data with substantially the same χ^2. Both the real and imaginary parts of the potential are appreciably altered, and it does not appear that a particular real well depth (184 or 309 MeV) can be singled out as being a significantly better representation of the data. A convergence to a more shallow real well depth (∿160 MeV) has been reported for a one channel fit to ${}^{6}Li + {}^{28}Si$ data at 154 MeV incident energy, but our own calculations on those data show inelastic effects to be just as important in the rainbow region as at 135 MeV incident energy. Hence, the proposed convergence to a more shallow real well depth cannot be supported in a CCBA calculation.

ACKNOWLEDGEMENT

We would like to thank our collaborators at Vanderbilt and Brookhaven National Laboratory for permission to use the results of various experiments in this paper. This work was supported in part by a grant from the U.S. Department of Energy, Contract No. DE-AS05-76ER05034.

TABLE II

CCBA Optical Potentials

Set	V_o	r_R	a_R	W_o	r_I	a_I	χ^2/F	J^π
$\alpha + ^{58}Ni(\beta_2 = 0.159,\ \beta_3 = 0.128)$139 MeV								
WS1	118.2	1.240	0.796	20.47	1.595	0.571	103	0^+
							28	2^+
							88	3^-
WS2	143.5	1.391	1.218	21.40	1.540	0.686	121	0^+
							16	2^+
							74	3^-
MD3	118.2	1.241	0.822	17.14	1.624	0.563	8.9	0^+
							14.8	2^+
							16.4	3^-
MD4	134.0	1.145	0.921	16.72	1.597	0.630	21	0^+
							26	2^+
							25	3^-
$^6Li + ^{28}Si(\beta_2 = -0.316,\ \beta_4 = 0.15)$135 MeV								
CA1	178.2	1.161	0.848	39.45	1.583	0.921	30	0^+
							25	2^+
CA2	184.6	1.219	0.796	48.84	1.489	0.924	11	0^+
							12	2^+
CA3	309.9	1.045	0.821	63.6	1.361	0.945	13	0^+
							13	2^+
$^6Li + ^{28}Si(\beta_2 = -0.316,\ \beta_4 = 0.15)$ 154 MeV								
IN1	183.4	1.031	0.959	33.4	1.569	1.025	51	0^+
							(no 2^+ data)	

REFERENCES

1. J.G. Cramer, R.M. DeVries, D.A. Goldberg, M.S. Zisman and C.F. Maguire, Phys. Rev. C14, 2158 (1976).
2. P. Braun-Munzinger, G.M. Berkowitz, T.M. Cormier, C.M. Jachcinsk C.M. Jachcinski, J.W. Harris, J. Barrette, and M.J. Levine, Phys. Rev. Lett. 38, 944 (1977).
3. J. Barrett, M.J. Levine, P. Braun-Munzinger, G.M. Berkowitz, M. Gai, J.V. Harris, and C.M. Jachcinski, Phys. Rev. Lett. 40, 445 (1978).
4. See D.A. Bromley in Proc. of the Int. Conference on Reactions Between Complex Nuclei, eds., R.L. Robinson, F.K. McGowan, J. Ball and J.H. Hamilton, (North Holland Publ. Co., 1974), p. 603.
5. M.R. Clover, R.M. DeVries, R. Ost, N.J.A. Rust, R.N. Cherry, Jr R.N. Cherry, Jr., and H.E. Gove, Phys. Rev. Lett. 40, 1008 1008 (1978).
6. J.L.C. Ford, Jr., J. Gomez del Campo, D. Shapira, M.R. Clover, R.M. DeVries, B.R. Fulton, R. Ost and C.F. Maguire, Phys. Lett. 89B, 48 (1979).
7. R. Ost, M.R. Clover, R.M. DeVries, B.R. Fulton, H.E. Gove and N. J. Rust, Phys. Rev. C19, 740 (1979).
8. P. Braun-Munzinger, G.M. Berkowitz, M. Gai, C.M. Jachcinski, T. Renner, C.D. Uhlohrn, J. Barrette and M.J. Levine, Bull. Am. Phys. Soc. 24, 571 (1979).
9. C. Roy, A.D. Frawley and K.W. Kemper, Phys. Rev. C20, 2143 (19 1243 (1979).
10. C.F. Maguire, G;L. Bomar, R.B. Piercey, J.H. Hamilton and P.P. Bond, unpublished data.
11. D.A. Goldberg and S.M. Smith, Phys. Rev. Lett. 29, 500 (1972).
12. D.A. Goldberg, S.M. Smith and G.F. Burdzik, Phys. Rev. C10, 1362 (1974).
13. R.M. DeVries, D.A. Goldberg, J.W. Watson, M.S. Zisman, and J.G. Crammer, Phys. Rev. Lett. 39, 450 (1977).
14. P. Schwandt, S. Kailas, W.W. Jacobs, M.D. Kaitchuck, W. Ploughe, and P.P. Singh, Phys. Rev. C21, 1656 (1980).
15. M.S. Zisman, J.G. Crammer, D.A. Goldberg, J.W. Watson, and R.M. DeVries, Phys. Rev. C21, 2398 (1980).
16. D.F. Jackson and R.C. Johnson, Phys. Lett 49B 249 (1974).
17. T. Tamura, Reviews of Modern Physics, Lett. 49B, 249 (1964).
18. J. Raynal, unpublished.
19. C.F. Maguire, Bull. Am. Phys. Soc. 25, 742 (1980); C.F. Maguire, to be published.
20. D.A. Goldberg, Phys. Lett. 55B, 59 (1975).

HEAVY-ION INDUCED TRANSFER REACTION TO HIGH-J ORBITAL STATES*

J. Barrette

Brookhaven National Laboratory
Upton, New York 11973

I. INTRODUCTION

The importance of high-J orbitals predicted by the shell-model[1] in the level structure of nuclei has been recognized for a long time. The most dramatic confirmation of these high-J orbitals is certainly the observation of islands of isomers in nuclei with proton or neutron number immediately below the major shell closure[2].

More recently a renewed interest in the influence of high-J orbitals has been generated by the observation of high-spin isomers in even-even rare earth nuclei. These isomers have been attributed to a sudden alignment to the maximum available spin of two nucleons in one of the orbits with high angular momentum[3]. This alignment is accompanied by a reduction of the excitation energy creating the high-spin isomeric state.

Up to now, however, the understanding of the importance of these orbitals in the structure of these levels mainly originates from the study of the decay properties of the levels as observed in in standard gamma spectroscopy. More direct information on the high-J orbital components of some high spin states has been obtained by the measurement of g factor[4] and recently, the use of the reaction (^{3}He, α) has also been successfully used to pick up the $i_{13/2}$ neutron strength from ^{161}Dy[5].

However, standard stripping reactions such as (d, p) have been up to now unsuccessful[6] in locating strength based on high angular

*Work supported by Division of Basic Energy Sciences, U.S. Department of Energy under Contract No. DE-AC02-76CH00016.

momentum level. In this short presentation I want to demonstrate that properly selected heavy ion induced stripping reactions are a very suitable tool to locate this strength.

II. REACTION MECHANISM

Besides the nuclear structure of the states involved, the cross section for heavy ion induced transfer reaction is governed by the energy and angular momentum matching conditions between the entrance and exit channels. This is due to the strong surface lo localization of such reactions.

The condition of angular momentum matching states that the favored transfer angular momentum will be equal to the difference between the grazing angular momentum in the entrance and exit channel. As an example, Table I compares the grazing angular momentum in the entrance and exit channel for three one neutron stripping reaction to the ground state of ^{149}Sm target. The grazing angular momenta were calculated using standard radius parameters. One advantage of heavy ion induced transfer reactions is immediately visible. By changing the projectile it is possible to change considerably the matching condition or the favored angular momentum transfer. This is due to the large changes in Q value between the various reactions. As shown in the table, heavy ion induced transfer reaction with large negative Q value favor large transfer of angular momentum whereas well match reactions such as (^{13}C, ^{12}C) will favor transfer of lower angular momentum.

TABLE I

REACTION	Q (MeV)	E_{lab} (MeV)	l^{gr}_{in}	l^{gr}_{out}	
^{148}Sm(^{16}O,^{15}O)^{149}Sm	- 9.8	120	71	61	-10
^{148}Sm(^{12}C,^{11}C)^{149}Sm	-12.8	90	52	40	-12
^{148}Sm(^{13}C,^{12}C)^{149}Sm	0.9	90	55	53	- 2

Grazing angular momentum in the entrance and exit exit channels for three one-neutron stripping reactions on ^{148} Sm.

III. EXPERIMENTAL RESULTS

Odd Residual Nuclei, ^{149}Sm

An example of the selectivity of heavy ion induced reaction is presented in Fig.1. In this figure is displayed the ^{149}Sm spectra observed following the three one neutron stripping reactions (^{16}O, ^{15}O), (^{12}C, ^{11}C) and (^{13}C, ^{12}C) at $E_{lab} \approx 7.5$ MeV/amu. The spectra were obtained using beams from the BNL MP tandem facility and the projectiles were detected in the QDDD spectrometer. The energy resolution of 80-120 keV is mainly due to the non-uniformity of the targets. In targets with large Z such as Sm the Coulomb interaction dominates giving rise to structureless bell-shaped angular distributions. Contrary to forward peak angular distributions, the present angular distributions are not sensitive to the transferred angular momentum. However, one advantage of the bell-shaped angular distribution is that the measurement of a single spectrum on the peak of the angular distribution at the grazing angle such as in Fig.1 provides a precise value of the relative integrated cross section over the whole excitation energy range.

The spectra in Fig.1 corresponding to the (^{16}O, ^{15}O) reaction has only two strong peaks and two weaker peaks. The strongest peak corresponds to the $13/2^+$ state at 0.88 MeV (normal $\Delta L = 7$, ΔL = tran transferred angular momentum) whereas the other large peak is produced by transfer to the $7/2^-$ ground state (normal $\Delta L = 4$). The center of mass cross section to these two states is 1.0 and 1.5 mb/sr, respectively.

Based on our systematic study of the (^{16}O, ^{15}O) reaction on many Nd isotopes[6] which shows that the $i_{13/2}$ strength is always divided between two $13/2^+$ states, we can conclude that the state at 1.82 MeV observed in the (^{16}O, ^{15}O) reaction is a previously unobserved $13/2^+$ state. The small peak at 0.29 MeV corresponds to a $9/2^-(h_{9/2})$ state. This spectrum is pretty spectacular considering that ^{149}Sm has 25 known states below 1.0 MeV. The (^{16}O, ^{15}O) reaction selects only the few states states based on the $f_{7/2}$, $h_{9/2}$, and $i_{13/2}$ single particle neutron configuration.

The spectrum from the (^{12}C, ^{11}C) reaction demonstrates another interesting and useful aspect of heavy ion induced reaction. The three lowest peaks observed in the (^{16}O, ^{15}O) reaction are still visible but the relative cross section to the various states has changed considerably. The cross section for the $13/2^+(i_{13/2})$ and $7/2^-(f_{7/2})$ states is reduced by a factor of 10 or more whereas the cross section for the $9/2^-(h_{9/2})$ state at 0.29 MeV has hardly changed. The relative enhancement of the $h_{9/2}$ state in the (^{12}C, ^{11}C) reaction can be easily understood in terms of simple classical arguments[7,8]. In the present case, where large transferred angular momenta are favored, these arguments lead to the

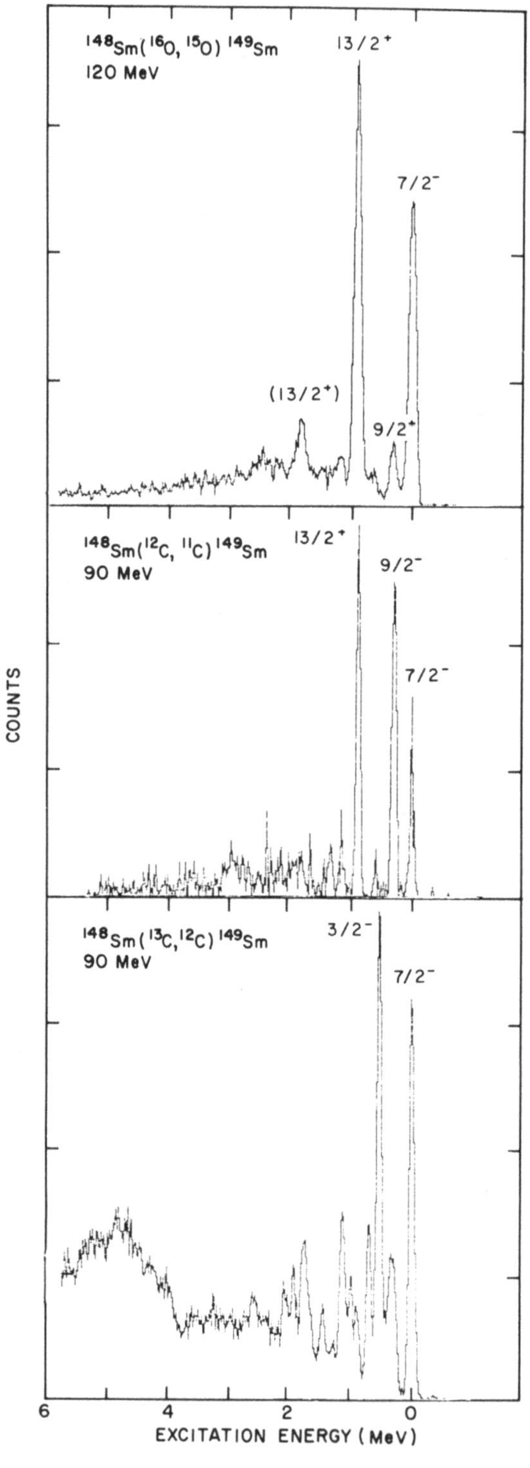

Fig.1 Energy spectra for the reactions $^{148}Sm(^{16}O,^{15}O)^{149}Sm$, $^{148}Sm(^{12}C,\ ^{11}C)^{149}Sm$, and $^{148}Sm(^{13}C,\ ^{12}C)^{149}Sm$. Note the enhancement of transfer to high spin state in the first two reactions and the increase yield of the $9/2^-$ state in the $(^{12}C,\ ^{11}C)$ reaction.

conclusion that the intrinsic spin s of the transferred nucleon or cluster does not flip whereas the orbital angular momentum changes sign. So a particle with initial total spin $J = \ell + s$ or $J_>$ will transfer preferentially to a state with $J = \ell - s$ or $J_<$ and vice-versa. Since in the (^{16}O, ^{15}O) reaction a $p_{1/2}$ particle is transferred the J states such as $i_{13/2}$ or $f_{7/2}$ are favored, whereas in the (^{12}C, ^{11}C) reaction the transfer of $p_{3/2}$ particle favored $J_<$ states such as $h_{9/2}$. The cross section for a given state in a given reaction will not depend only on the reaction dynamic but also on the spectroscopic properties of that state. However, a simple comparison of the relative cross section observed in various reactions can give us valuable information on the possible spin of the observed states. As stated previously, this information cannot be obtained from the angular distributions.

The third spectrum in Fig.1 corresponding to the reaction (^{13}C, ^{12}C) is very similar to the spectrum observed following the (d, p) reaction[9]. The strongest peak is due to a $p_{3/2}$ state at 0.54 MeV. Many other low spin states are populated, including the $7/2^-$ ground state, but the $13/2^+$ and $9/2^-$ states are completely buried below the large number of low spin states.

Even Residual Nuclei ^{148}Sm

Spectra for the single neutron transfer reaction (^{16}O, ^{15}O) and (^{12}C, ^{11}C) to levels in ^{148}Sm are presented in Fig.2. Similarly to ^{149}Sm, the nucleus ^{147}Sm has its last unpaired nucleon in the $f_{7/2}$ subshell. Extrapolating from what we observed in the transfer to the odd nucleus ^{149}Sm, the (^{16}O, ^{15}O) reaction should populate the configuration $(f_{7/2})^2$ and $f_{7/2}\cdot i_{13/2}$ whereas the (^{12}C, ^{11}C) reaction should enhance transfer to the $f_{7/2}\cdot h_{9/2}$ configuration.

In the (^{16}O, ^{15}O) reaction the ground state band is observed up to a spin 6^+ which is the maximum spin that can be reached by a $(f_{7/2})^2$ configuration. The fact that the relative intensity of the state in the ground state band does not follow 2J + 1 pattern indicates that these states are not pure $(f_{7/2})^2$. A more quantitative conclusion necessitates a knowledge of the absolute spectroscopic factors which have not yet been extracted from the data. We can already conclude from the very large cross section for the 6^+ state at 1.91 MeV that it is predominantly $(f_{7/2})^2$ in contrast to the nearby 6^+ state at 2.10 MeV which is only weakly populated.

In the (^{16}C, ^{11}C) reaction the strongest peak is the 8^+ state from the ground state band indicating that this state has a large $f_{7/2}\cdot h_{9/2}$ particle configuration.

The (^{16}O, ^{15}O) reaction should also lead to the levels with large $f_{7/2}\cdot i_{13/2}$ components. Due to residual interaction, the states of odd spins 3^-, 5^-, 7^- and 9^- are expected to form a band

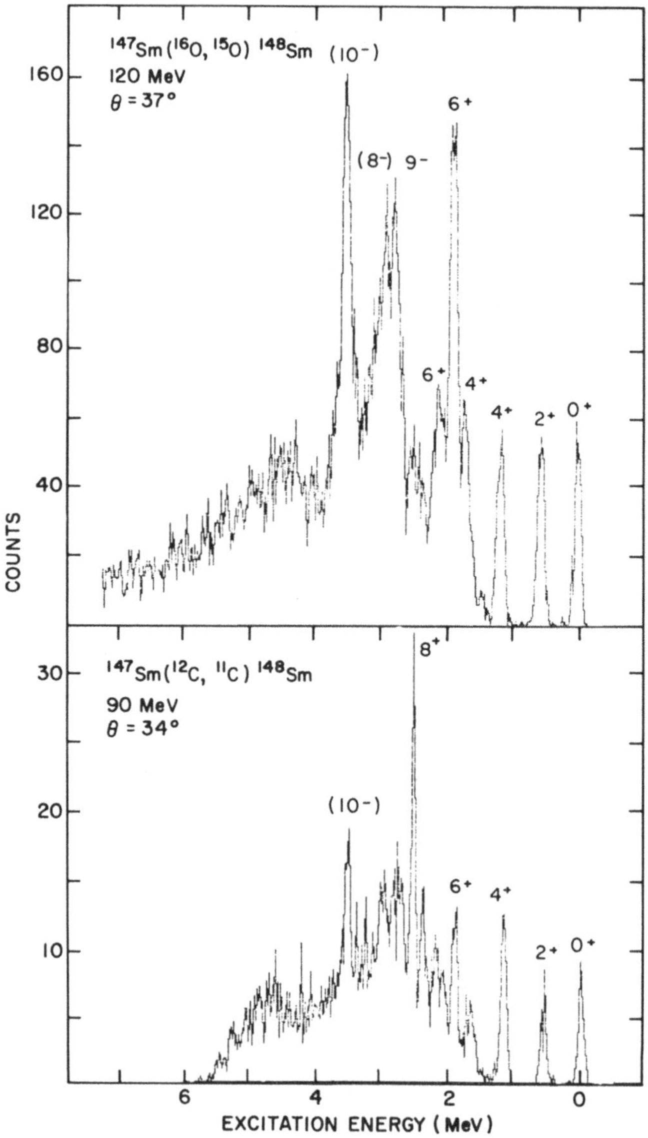

Fig.2 Energy spectra for the reactions $^{147}Sm(^{16}O, ^{15}O)^{148}Sm$ at E_{lab} = 120 MeV and $^{147}Sm(^{16}O, ^{15}O)^{148}Sm$ at E_{lab}=90 MeV.

of ascending energy whereas the state of even spin 4^-, 6^-, 8^-, and 10^- should be moved at higher energy. The reaction mechanism should also favor the transfer to state of high spin. It is impossible to evaluate the cross section to the 3^- state from the particle spectrum since it is degenerate within the energy resolution with the first 4 state. The first 5^- and 7^- states are only weakly excited and appear in the spectrum as shoulder on the second 4^+ and second 6^+ peaks. However, three strong peaks are observed at 2.80, 2.96 and 3.55 MeV. The first peak is the known 9^- level at 2.807 MeV and based on the previous arguments it is reasonable to assume that the two other levels are respectively the 8^- and 10^- states based on the $f_{7/2} \cdot i_{13/2}$ configuration. Only one 8^- state has been proposed previously in the level scheme of ^{149}Sm[10] and no 10^- level has up to now been observed. One of the main reasons that these states have not been observed previously is that (HI, xn) reactions which are used to study the high spin structure of rare earth nuclei have a strong preference for populating Yrast levels. The even spin states which are pushed up by the residual interaction are then suppressed. Because of their simple configuration, these states are, however, as important as the Yrast states in the study of the nuclear structure of rare-earth nuclei.

As explained above, it is probably impossible to obtain more definitive information on the spin of these states or on the transferred angular momentum from the particle angular distribution. Some insight on the spin of the observed states can be obtained, however, by measuring their decay properties. This was first attempted for the reaction $^{147}Sm(^{16}O, ^{15}O)^{148}Sm$ at E_{Lab} = 120 MeV. The outgoing ^{15}O detected in the Q3D spectrometer were observed in coincidence with the gamma rays emitted in the decay of the ^{148}Sm levels. Such experiment is particularly difficult due to the very small cross section of the specific reaction studied. To obtain results in a reasonable time we had to take full advantage of the large solid sngle of the Q3D spectrometer (12mrs) and a special scattering chamber with thin wall and small radius was built so that the Ge(li) detector could be approached at 10 cm from the target.

The measured coincidence gamma-ray spectrum is presented in Fig.3. This spectrum is for coincidences with ^{15}O ions corresponding to all excitation energies in ^{148}Sm. All the known transitions associated with the ground state band up to the 6^+ state and with the first odd spin negative parity states up to the 9^- states are clearly visible. This is a confirmation of our interpretation of the particle spectra which was based on our understanding of the reaction mechanism alone. The most interesting aspect of this spectrum is that we observed a weak transition corresponding to the decay of the postulated 10^- state at 3.55 MeV to the 9^- level. This transition which is weak in Fig.3 becomes evident when we gate the particle spectrum at the appropriate excitation energy. This transition adds support to the assumption that the state at 3.55 MeV belongs to the negative parity multiplet. For reasons that are not clear at present we did not observe any transitions corresponding to the postulated 8^- state at 2.96 MeV.

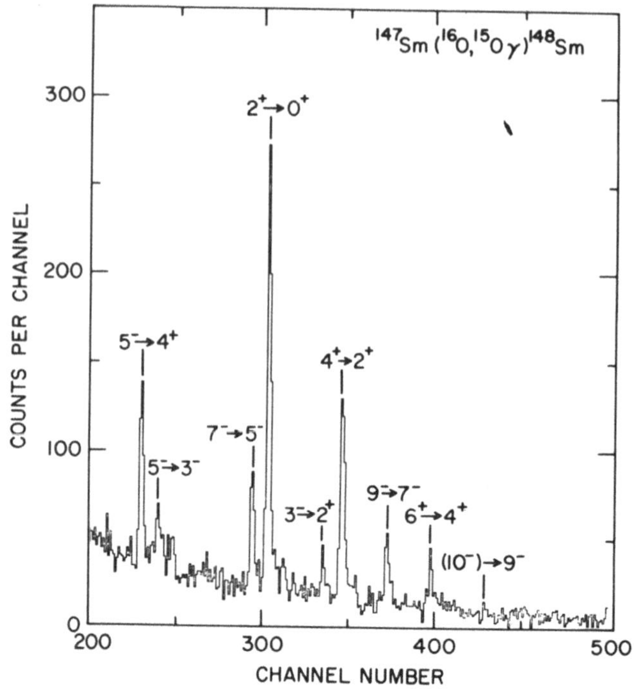

Fig.3 Coincidence gamma ray spectrum for the reaction $^{147}Sm(^{16}O,^{15}O)^{148}Sm$. The spin and parity of the levels connected by the observed transitions are indicated above the peaks.

This preliminary result clearly shows that the particle-gamma coincidence can be very useful to ascertain the spectroscopic properties of the states observed in heavy ion direct reaction on heavy target. It is also evident that the same technique can be used to evaluate the distribution of strength in peak which contains more than one level due to the limited particle resolution. An example of that is the first 4^+ and 3^- state on ^{148}Sm which are not separated in Fig.2. By gating the particle spectrum near 1.2 MeV we could determine that this peak is almost entirely due to the 4^+ state.

Deformed Nuclei, ^{167}Er.

The previous examples were concentrated on spherical samarium nuclie. However, the same approach can be applied to the study of deformed nuclei. The level structure of deformed nuclei can often be described as a series of rotational bands built on Nilsson configurations. It is well known that the Nilsson states based on particles with high angular momentum play a dominant role in the backbending phenomenon and in the presence of high spin isomers. The

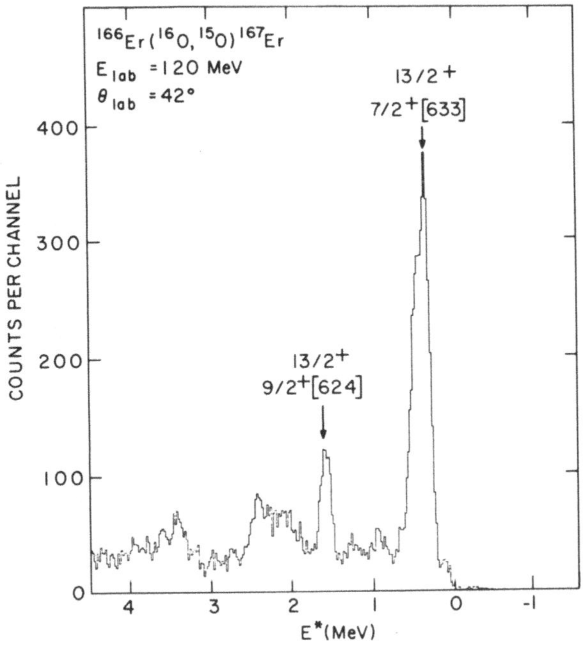

Fig.4 Energy spectra for the reaction $^{166}Er(^{16}O, ^{15}O)^{167}Er$.

present information on the location of these Nilsson states is often indirect and limited to relatively low energy above or below the ferm fermi energy. Heavy ion induced transfer reactions similar to those described previouṡly, offer the possibility to extend our knowledge to much higher excitation energy and to provide absolute spectroscopic factors.

Preliminary results for the reaction $^{166}Er(^{16}O, ^{15}O)^{167}Er$ are presented in Fig.4. The largest peak at 0.294 MeV is the $13/2^+$ state of the band based on the $7/2^+(633)$ Nilsson orbit originating from the deformation of the $i_{13/2}$ single particle level. The most interesting aspect of these data is, however, the strong peak at 1.56 MeV which can be associated with the previously unobserved $13/2^+$ state of the rotational band based on the $\Omega = 9/2^+$ level also originating from the deformation of the $i_{13/2}$ orbit. The $9/2^+$ and $11/2^+$ member of that rotational band had been proposed previously from a study of the (d, d') reaction[11]. The present measurement is an unambiguous confirmation of these previous results and of the nature of this rotational band. It is possible that one of the small peaks at 2.4 and 3.4 MeV are due to the $\Omega = 11/2^-$ level orbit. We do not have at present sufficient evidence to confirm this.

From particle-gamma coincidence similar to that described above it was possible to establish that the shoulder on the $13/2^+$ state at 0.293 MeV is due to the two $7/2^-$ states at 0.413 and 0.430 MeV clearly indicating that these levels contain a large amplitude from Nilsson levels based on the deformation of the $f_{7/2}$ orbit. More quantitative analysis of these data is in progress but the present data already show that heavy-ion transfer reactions are well suited to the study of high spin levels in deformed nuclei.

IV. CONCLUSION

The few experimental results presented here are a clear demonstration that heavy-ion transfer reactions can be a very useful tool in the study of states based on high-J orbitals which are particu larly important in the understanding of nuclear structure at high spin. The large number of available projectiles allows to change considerably the kinematic conditions of reactions as well as the quantum number of the nucleon to be transferred leading to an unmatched selectivity and often to the determination of the spin of the observed levels. Heavy-ion induced reactions on heavy targets suffer from a limited resolution and from a lack of sensitivity of the angular distribution on the transferred angular momentum. However, some techniques like particle-gamma coincidences described above can be used effectively to overcome some of these limitations.

The present data demonstrate the potential of heavy-ion induced reactions for obtaining original spectroscopic information on the structure of nuclei and much experimental and theoretical work remains to be done to exploit this potential.

ACKNOWLEDGEMENTS

This work was initiated by P.D. Bond and done in collaboration with C. Baktash, C.E. Thorn and A. Kreiner. They should receive a large part of the credit.

REFERENCES

1. M.G. Mayer and T.H.D. Jensen, "Elementary Theory of Nuclear Shell Structure", (London, Wiley, 1955).
2. M. Goldhaber and J. Weneser, Ann. Rev. Nucl. Sci. 5, (1955) 1.
3. S. Frauendorf, See invited talk to this conference.
4. S.A. Hjorth, Contribution to this conference and S.A. Hjorth, I.Y. Lee, J.R. Beene, C. Roulet, D.R. Haenni, N.R. Johnson, F.E. Obenshain and G.R. Young, Phys. Rev. Lett. 45 (1980) 878.
5. Jin Gen-Ming, J.D. Garrett, G. Lovhoiden, T.F. Thorsteinsen, J.C. Waddington and J. Rekstad, preprint 1980.

6. P.D. Bond, J. Barrette, C. Baktash, A. Kreiner and C.E. Thorn, to be published.
7. D.M. Brink, Phys. Lett. 40B (1972) 37.
8. D.K. Scott in "Classical and Quantum Mechanical Aspects of Heavy Ions Collisions", p.65 Lecture Notes in Physics, 33 (Springer-Verlag, New York 1975).
9. E. Veje, Nucl. Phys. A103 (1967) 188.
10. B.A. Brown, T.L. Khoo, and C.H. King, Proceedings of the International Conference of Nuclear Structure, Contributed papers, p.376, Tokyo 1977.
11. F. Sterba, J. Sterbova and M. Rozkoz, Czech.I. Phys. B23 (1973) 601.

PERTURBATION THEORY FOR A SYSTEM OF FERMIONS IN A DEFORMED BASIS

Daniel R. Bes

Department of Physics
Comision Nacional de Energia Atomica
Buenos Aires, Argentina

I. MOTIVATION

The total Hamiltonian $H(b_m^+, b_m)$ of a system of fermions may be written in two parts $H=H_0+xH_{res}$ $(0\leq x\leq 1)$. We consider the case in which the basic set of states (eigenfunctions of H_0) is deformed, i.e., $[H,L]=0$ and $[H_0,L]\neq 0$, where L can be components of the angular momentum; also, the number operator may follow the same rules.

1. Since the inclusion of H_{res} $(x=1)$ restores the original symmetry of the problem, a perturbative treatment of H_{res} would imply a perturbative transformation from the deformed to the spherical basis. This cannot exist, since the opposite transformation (from the normal to the deformed basis) is not possible to perform perturbatively.

2. In simple system (particles moving in a degenerate shell and coupled by pairing interaction) we have verified that the radius of convergence of the corresponding perturbative series in x is always smaller than one (and tends to one as the degeneracy of the shell increases.

3. In a deformed basis, there exist zero frequency modes (Goldstone bosons) which couple with the fermion degrees of freedom with a coupling strength that diverges as $w^{-1/2}$.

The origin of these difficulties can be traced back to the abscence of restoring force in the angular direction for a freely rotating system, giving rise to infrared catastrophes.

II. THE CONSTRAINED FERMION-ROTOR HAMILTONIAN

In ref.1 a procedure to overcome these difficulties is derived.

Up to a constant, the Hamiltonian H is equivalent to

$$H' = \lim_{D\to 0} H(a_m^+,a_m) + \frac{1}{2D}\sum_v (I_v - L_v(a_m^+,a_m))^2 + \frac{Q^2}{6A}\sum_v \Theta_v(a_m^+,a_m)^2 + \frac{w}{2}\sum_n (1 - \det(i[L_v,\Theta_u])^n/n \tag{1}$$

where $a_m^+(a_m)$ are fermion creation (anihillation) operators in the rotating system, I_v are collective angular momentum operators in the body fixed frame; $L_v(a_m^+,a_m)$ are their microscopic expressions; $\Theta_v(a_m^+,a_m)$ are, in principle, functions of the fermion variables that do not commute with the L_v's. In practice it is convenient to choose Θ_v as the microscopic expression for the angular conjugate to L_v, at least to leading order in a small parameter Q^{-1}. The following points may be stressed

1. The equivalence between H and H' can be demonstrated by the fact that the amplitude for ground to ground state transitions in a problem involving only fermion fields b_m^+,b_m in the laboratory system

$$Z = \prod_m \int D(b_m^+)D(b_m)\, e^{-\int (\sum_m b_m^+ b_m^\circ + iH(b_m^+,b_m))dt} \tag{2}$$

may also be expressed as a path-integral including constraints and involving both fermion a_m^+,a_m and collective I_v,ϕ_v degrees of freedom

$$Z = \prod_{m,v} \int D(a_m^+)D(a_m)D(I_v)D(\phi_v) \times e^{-(\sum_m a_m^+ a_m^\circ - i\sum_v \Pi_v\phi_v + iH(a_m^+,a_m))dt} \cdot\delta(I_v - L_v(a_m,a_m))\cdot\delta(\Theta_v(a_m,a_m) - \bar{\Theta}_v)\cdot\det(U_{vu}(\phi_v)) \cdot\det([L_v(a_m^+,a_m),\ \Theta_u(a_m^+,a_m)]_{PB}) \tag{3}$$

One obtains the Hamiltonian H' through exponentiation of the constraints, replacement of the fields by operators and Poisson brackets $[\,]_{PB}$ by quantum mechanical commutation brackets.

2. The Hamiltonian H' acts on wave functions belonging to the product space

$$\Psi_{I,n}(\phi,a_m^+,a_m) = \frac{1}{\sqrt{2\Pi}} e^{iI\phi} f_{I,n}(a_m^+,a_m)|0\rangle$$
$$\Psi_{I,M,K,n}(\phi_v,a_m^+,a_m) = \sqrt{\frac{2I+1}{8\Pi^2}} D^I_{M,K}(\phi_v) f_{I,K,n}(a_m^+,a_m)|0\rangle \tag{4}$$

for two and three dimensional rotations, respectively. One has in mind an independent-particle basis for the intrinsic states. Thus the basic set of states is identical with the one used in the unified model.

3. The expansion parameter Q^{-1} and the angular variable θ_v. The quadrupole operators in configuration space are written

$$Q_v = 2.3^{1/2} \sum_i (x^2_{v+1} - x^2_{v+2})_i \qquad (v = x,y,z = 1,2,3)$$
$$S_v = -2.3^{1/2} \sum_i (x_{v+1}\, x_{v+2})_i \tag{5}$$

The following commutation relations are satisfied

$$[L_v,S_u] = -iQ_v\,\delta_{u,v} - iS_{v+2}\,\delta_{u,v+1} + iS_{v+1}\,\delta_{u,v+2}$$
$$[L_v,Q_u] = -2iS_v(1-3\delta_{v,u}) \tag{6}$$

The usual choice of intrinsic frame implies that the expectation values $\bar{S}_v$ vanish. On the contrary, the expectation values $\bar{Q}_v$ are assumed to be large, at least in the (determinantal) ground state. Therefore, the operators Q_v are split into a large constant term $\bar{Q}_v$ and a term Q_v' that has matrix elements of the order of single-particle values and therefore are of O(1) (much smaller than the $\bar{Q}_v$s). The operators S_v have particle-hole matrix elements and can be considered to be of order $\bar{Q}_v^{1/2}$, since the expectation values of S_v^2 are of order $\bar{Q}_v$. Today, the same is assumed for the angular momentum components L_v. Therefore, the expansion parameter Q^{-1} is taken to be the inverse of any of the three static moments $\bar{Q}_v$.

Using these general estimates in connection with eq.(6), the angular function Θ_v can be approximated by

$$\Theta_v = S_v/\bar{Q}_v \tag{7}$$

Due to the absence of a restoring force in the angular direction, there is no contribution from H to the frequency of the angular oscillation (Goldstone boson). On the contrary, the appearance in the constrained Hamiltonian of the three harmonic oscillators

$$h_v = (1/2D)L_v^2 + (Q^2/6A)\Theta_v^2 \tag{8}$$

yields oscillations along the three axis of rotation with the Oco (common) frequency

$$w = Q/(3AD)^{1/2} \tag{9}$$

Consequently, the spurious states are eliminated from the spectrum as D vanishes. However, they must always be used as intermediate states. The limit $D\to 0$ can only be taken in the final physical results.

Using the expression (7) for the angular functions, the determinant appearing in the constrained Hamiltonian is written

$$1 + \sum_v Q'_v/\bar{Q}_v + \sum_v S_v^2/\bar{Q}_{v+1}\,\bar{Q}_{v+2} + O(Q^{-3/2}) \tag{10}$$

4. Treatment of the constrained Hamiltonian. The original Hamiltonian H is written in the rotating frame of reference, thus defining a single-particle (deformed) basis and a residual Hamiltonian acting among the Nilsson particles. Neither the independent-particle terms, nor the residual Hamiltonian, separately have the spherical symmetry of the total H. This step is common to most treatments of deformed systems. Next, one writes the other contributions to the effective Hamiltonian in the representation thus defined. The resultant Hamiltonian includes arbitrarily large residual interactions ($D\to 0$). These may be conveniently treated within the NFT formalism, which also allows to cast the results into a power series in Q^{-1}.

III. APPLICATION TO THE ELLIOT'S MODEL

IIIa. Exact solution

Let us consider a single harmonic oscillator shell with particles interacting through a residual quadrupole interaction such that, the matrix elements of the quadrupole operator between states belonging to different h.o. shells vanish[2]. In representation carrying the SU3 quantum numbers $(\bar{Q}_\beta, \bar{Q}_\gamma)$ this interaction

$$H_q = -\frac{V}{2}\sum_\mu Q_\mu Q_\mu^+ \tag{11}$$

is diagonal with eigenvalues

$$\langle\bar{Q}_\beta,\bar{Q}_\gamma,I,M,K\,|H_q|\,\bar{Q}_\beta,Q_\gamma,I,M,K\rangle = -\frac{V}{2}C_2(\bar{Q}_\beta,\bar{Q}_\gamma) + \frac{3V}{2}I(I+1) \tag{12}$$

where C_2 is the eigengalue of the Casimir operator, namely

$$C_2(\bar{Q}_\beta,\bar{Q}_\gamma) = \bar{Q}_\beta^2 + \bar{Q}_\gamma^2 + 6\bar{Q}_\beta + 2\sqrt{3}\,\bar{Q}_\gamma \tag{13}$$

IIIb. Solution in a deformed basis

It is convenient to add a term $\frac{3V}{2}\sum_v(I_v^2 - L_v^2)$ to the quadrupole Hamiltonian (II). We are allowed to do it, since this term becomes negligible in the constrained Hamiltonian, We write

$$H = H_{rot} + H_{coup} + H_{intr}$$

$$H_{rot} = \left(\frac{1}{2D} + \frac{3V}{2}\right)\sum_v I_v^2 \;;\quad H_{coup} = -\frac{1}{D}\sum_v I_v L_v \tag{14}$$

$$H_{intr} = -\frac{V}{2}C_2 + H_o + H'_{res}$$

The term H_{intr} depends only on the fermion degrees of freedom. The constant C_2 is given in (3). The independent-particle term H_o determines the basic set of intrinsic states. In this case, the Hartree or Nilsson solution is given by the single-particle harmonic oscillator wave functions in the cartesian basis. The relevant particle-hole excitations are obtained by acting on the ground state with the operators S_o. For each v, these states are degenerate, with excitation energy

$$E_v = 3^{1/2}V|\bar{Q}_v| \tag{15}$$

The RPA treatment of the residual two-body Hamiltonian H'_{res} yields the same collective frequencies (9) for each of the three

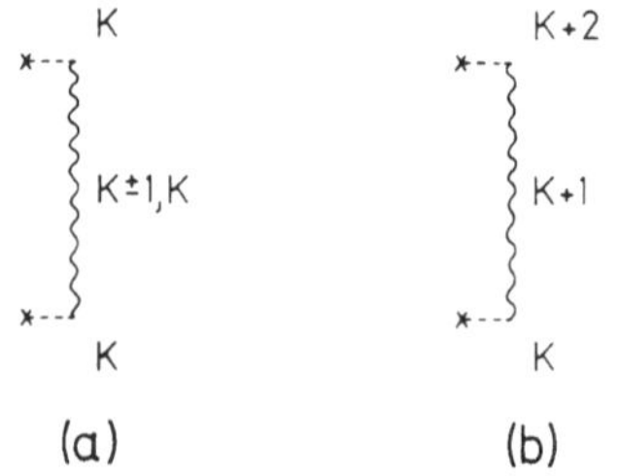

Fig.1 Diagrammatic contributions to the rotational energies and matrix elements.

degrees of freedom. Particle-phonon and phonon-phonon vertices arising from[14] are obtained from the standard NFT procedure. In particular, the NFT version of H_{coup} includes a linear term in boson creation and anihilation operators

$$H_{coup} \rightarrow \frac{1}{2}\left(\frac{W}{2D}\right)^{1/2}\Big(I_+\,(\Gamma_x^+ + \Gamma_x - \Gamma_y^+ + \Gamma_y) + I_-\,(\Gamma_x^+ + \Gamma_x + \Gamma_y^+ - \Gamma_y) - 2I_z(\Gamma_z^+ + \Gamma_z)\Big) \tag{16}$$

This term gives rise to a correction to the rotational energy through diagrams of Fig.1. The contribution of graph 1(a) is $-I(I+1)/2D$ which, if added to H_{rot}, cancels the divergent term and thus leaves only the exact rotational energy[12]. If the system is axially symmetric (no Γ_z^+, Γ_z phonons), there appears an additional contribution $-I_z^2/D$ which effectively eliminates from the ground state rotational band states with $I_z \neq 0$. The graphs 1(b) vanish, which is necessary in order to maintain the degeneracy between states with the same value of I in the Elliot's model.

The NFT transcription of the quadrupole operators in the laboratory frame yields

$$\begin{aligned} Q_m^{\ell} &= D^2_{m,o}Q_\beta + 2^{-1/2}\,D^2_{m,\pm 2}\,(Q \pm iS_z) + 2^{-1/2}\,D_{m\pm 1}(iS_x \pm S_y) \\ &\rightarrow D^2_{m,o}Q_\beta + 2^{-1/2}D^2_{m,\pm 2}\Big(Q_\gamma \pm \frac{1}{V}\left(\frac{1}{6WD}\right)^{1/2} E_z(\Gamma_z^+ - \Gamma_z)\Big) \\ &\quad \pm \frac{2^{-1/2}}{V}\left(\frac{1}{3WD}\right)^{1/2} D^2_{m,\pm 1}\,\big(\pm E_x(\Gamma_x^+ - \Gamma_x) + E_y(\Gamma_y^+ + \Gamma_y)\big) \end{aligned}$$

The leading order terms in the transition matrix elements $\bar{Q}_\beta$,

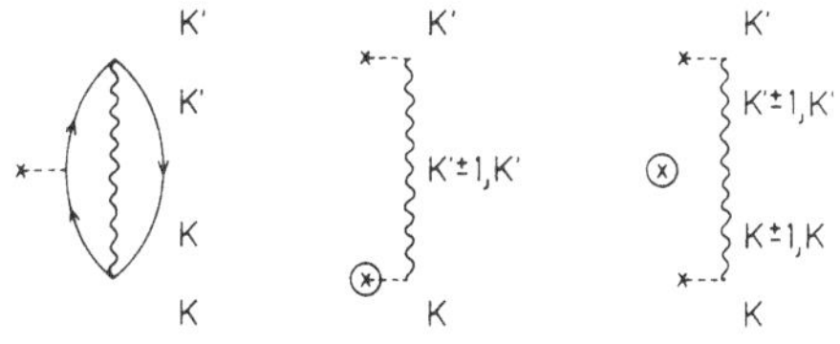

Fig.2 Diagrammatic contributions to the quadrupole matrix elements.

$\bar{Q}_\gamma(O(G))$ are corrected through diagrams of Fig.2. The results are

$$\text{(graph 2) (a)} \quad \Delta Q^\ell_m = D^2_{m,o}\left(-\frac{3\bar{Q}_\beta}{2WD}+3\right) + 2^{-1/2}\left(D^2_{m,2} - D^2_{m,-2}\right) \times \left(\frac{-3\bar{Q}_\gamma}{2WD} + 3^{1/2}\right)$$

$$\text{(graph 2) (b)} \quad \Delta Q^\ell_m = D^2_{m,o}\frac{3\bar{Q}_\beta}{WD} + 2^{-1/2}\left(D^2_{m,2} + D^2_{m,2}\right)\frac{3\bar{Q}_\gamma}{WD} \tag{18}$$

$$\text{(graph 2) (c)} \quad \Delta Q^\ell_m = D_{m,o}\left(-\frac{3\bar{Q}_\beta}{2WD}\right) + 2^{-1/2}\left(D^2_{m,2} + D^2_{m,-2}\right) \times \left(-\frac{3\bar{Q}_\gamma}{2WD}\right)$$

Again, the divergent terms cancel in the final physical result. The net result is that $\bar{Q}_\beta$ is replaced by $\bar{Q}_\beta + 3$ and $\bar{Q}_\gamma$, by $\bar{Q}_\gamma + 3^{1/2}$. These corrections can be verified, for instance in the value of the ground state energy (Casimir operator C_2)

$$-\frac{V}{2}\left((\bar{Q}_\beta+3)^2 + (\bar{Q}_\gamma + 3^{1/2})^2\right) = -\frac{V}{2}\left(\bar{Q}^2_\beta + \bar{Q}^2_\gamma + 6\bar{Q}_\beta + 2.3^{1/2}\,\bar{Q}_\gamma + O(1)\right) \tag{19}$$

IV. CONCLUSION

It appears possible to construct a perturbation theory in a deformed basis, in spite of the singularities that are mentioned in I. Our solution implies introducing explicitely divergent terms in the effective Hamiltonian, which are originated in the constraints. The singularities introduced by the new term cancel the infrared catastrophes inherent to the problem.

The method has been thoroughly verified for two-dimensional rotation in refs.3. This is the first presentation of the extension to three dimensional rotations. A large fraction of the terms appearing in refs.1 has been checked in the calculation presented in section III, but not all of them. For instance, the last term in eqs.1 does not contribute either to the moment of inertia or to the quadrupole matrix elements. However, this term appears in higher order calculations which are in progress.

ACKNOWLEDGEMENT.

The present investigation has been carried in collaboration with Drs. O. Civitarese and H. Sofia from our theory group in Buenos Aires.

REFERENCES

1. V. Alessandrini, D.R. Bes and B. Machet; Nucl. Phys. B142 (1978) 489.
2. J.P. Elliot, in Selected Topics in Nuclear Theory, ed. F. Janouch (Intern. At. Energy Agency, Vienna, 1963).
3. V. Alessandrini, D.R. Bes and B. Máchet; Phys. Lett. 80B (1978) 9; D.R. Bes, G.G. Dussel and R.P.J. Perazzo; Nucl. Phys. A340 (1980) 197.

THE BOSON FERMION-HYBRID REPRESENTATION AND THE NUCLEAR FIELD THEORY*

Cheng-Li Wu

Department of Physics
Jilin University
Changchun, Jilin, People's Republic of China

M.W. Guidry

Department of Physics and Astronomy
University of Tennessee
Knoxville, Tennessee 37916

Jin-Quan Chen[1] and Da Hsuan Feng

Department of Physics and Atmospheric Science
Drexel University
Philadelphia, Pennsylvania 19104

I. INTRODUCTION

Mottelson in 1968[1] proposed the ideas of a nuclear field theory (NFT) as a theoretical scheme to treat the collective (Phonon pr particle-hole) and single-particle modes (fields) coupling. This theory was subsequently proliferated by the Copenhagen - Buenos Aires group into a diagrammatic perturbation theory (DPT)[2]. The central theme of the DPT is to introduce a set of "empirical" rules for the evaluation of the NFT diagrams.

---

*Work partially supported by the National Science Foundation, the Department of Energy and the Oak Ridge Associated Universities.

[1]Permanent Address: Department of Physics, Nanjing University
Nanjing, The People's Republic of China

In order to validate this theory, summing schemes were developed so that exact NFT results (i.e., summing up all the allowed diagrams to infinite order) can be obtained. The first was by Liotta and Silvestre-Brac[3] who introduced an infinite order summing procedure which is applicable for the case of three particles outside closed shells. Wu[4], in a later paper, proposed a general summing scheme and applied it to the case of four particles outside closed shells. In both cases, the exact NFT and the shell model give identical predictions of the energy spectrum and other observables. Fig.1 is an example of such a comparison between the results of these two seemingly different theories. It should be noted that the calculations presented in ref.3 and ref.4 do not place any restriction as to the type of interactions used.

It is of course well known in quantum mechanics that only equivalent representations can produce identical informations about a physical system[5]. The shell model is based entirely on a fermion representation. Thus, in view of refs.3 and 4, it is only natural for one to seek a transformation between the usual fermion representation and the representation which would explicitly contain both the fermion and the boson degrees of freedom. What we have in mind is to find a representation which is behind the NFT. In this talk, we shall first generalize the representation theory in quantum mechanics and establish the criteria for an equivalent transformation. We will then introduce a representation which we shall term the Boson-Fermion-Hybrid-Representation (BFHR). We will show that this representation is equivalent to the fermion representation. Furthermore, we will demonstrate that the NFT is but an "effective" way, albeit ingenious, to solve the BFH Schrodinger equation and the aforementioned "empirical" rules can emerge naturally. We recently learned* that Girardeau has for sometimes been working on the Fock-Tani representation which may bear resemblence to our BFHR.

II. THE USUAL UNITARY TRANSFORMATION BETWEEN EQUIVALENT SPACES

The usual representation transformation in standard quantum mechanics[5] deals with unitary transformation; the spaces concerned are of equal dimensions. In this section, we shall recast this "common" knowledge into a slightly different form; this form will be more convenient for the next section when we generalize the transformation theory.

Suppose representation G_a is defined by n single (fermion) operators $\{a^\dagger_{\mu i}\}$ i = 1 ... n where μ_i denotes the necessary quantum numbers necessary to fully specify a single particle state. On the

*We are grateful to Dr. A. Klein who brought Dr. Girardeau's work to our attention.

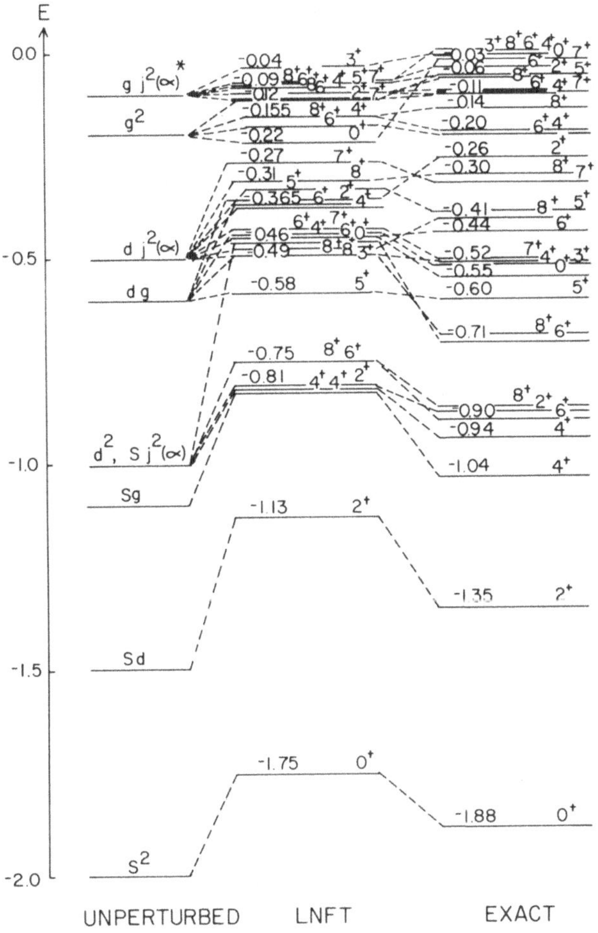

Fig.1 The spectra of four particles moving in a single-j shell (j = 15/2).
UNPERTURBED - The free boson energies.
LNFT - The results of the lowest order NFT diagrams.
EXACT - The results of the exact NFT and the shell model. These two calculations show no differences.

other hand, suppose G_b is another representation $\left\{b^{\dagger}_{\mu_i}\right\}_{i=1\ldots n}$ where again, there are other n single particle levels μ_i. If G_a and G_b equivalent representations, then there must exist a unitary transformation between $\{a^\dagger\}$ and $\{b^\dagger\}$ such that

$$a^{\dagger}_{\lambda_i} = \sum_{\mu_i} U_{\mu_i\lambda_i} b^{\dagger}_{\mu_i} \qquad (1)$$

In eq.1, the matrix U is unitary. For simplicity, we shall hereafter refer to eq.1 symbolically as a=Ub.

Let

$$\psi_k(a) = \sum_{\lambda_1\ldots\lambda_N} C^k_{\lambda_1\ldots\lambda_N} \left(a^{\dagger}_{\lambda_1} \cdots a^{\dagger}_{\lambda_N}\right)_k |0\rangle \qquad (2)$$

and

$$\Psi_k(b) = \psi_k(Ub)$$

$$= \sum_{\mu_1\ldots\mu_N} D^k_{\mu_1\ldots\mu_N} \left\{b^{\dagger}_{\mu_1} \cdots b^{\dagger}_{\mu_N}\right\}_k |0\rangle \qquad (3a)$$

$$D^k_{\mu_1\ldots\mu_N} = \sum_{\lambda_1\ldots\lambda_N} C^k_{\lambda_1\ldots\lambda_N} U_{\mu_1\lambda_1} \cdots U_{\mu_N\lambda_N} \qquad (3b)$$

The wavefunction $\psi_k(a)$ and any operator $\hat{L}(a)$ in G_a must be equivalent to the corresponding wavefunction $\Psi_k(b)$ and operator

$$\hat{\mathscr{L}}(b) = \hat{L}(Ub) \qquad (4)$$

The usual definition of equivalence, is,

$$\langle\psi_k(a)|\hat{L}(a)|\psi_{k'}(a)\rangle = \langle\Psi_k(b)|\hat{\mathscr{L}}(b)|\Psi_{k'}(g)\rangle \qquad (5)$$

An important point to note here is that if $\psi_k(a)$ is a solution of the Schrodinger equation in the representation G_a then $\Psi_k(b)$ must necessarily be the solution of the Schrodinger equation in the representation G_b. Thus

$$H(a)\psi_k(a) = E_k\psi_k(a) \qquad (6a)$$

$$\mathscr{H}(b)\Psi_k(b) = E_k\Psi_k(b) \tag{6b}$$

To summarize, the usual unitary transformation can be succintly written in the following way:

$$\begin{aligned} a &\rightarrow a(b) = Ub \\ \Psi_k(a) &\rightarrow \Psi_k(b) = \psi_k(Ub) \\ L(a) &\rightarrow \mathscr{L}(b) = L(Ub) \end{aligned} \tag{7}$$

III. GENERALIZED REPRESENTATION TRANSFORMATION

In G_a, let $A^\dagger_{J\alpha}(a)$ be the creation operator of certain fermion groups. As an example, for fermion pairs

$$A^\dagger_{J\alpha}(a) = A^\dagger_{J\lambda_1\lambda_2}(a) = \left(a^\dagger_{\lambda_1}\ a^\dagger_{\lambda_2}\right)_J \tag{8}$$

and N (a) be the number operators of the fermions in the state

$$N_{\lambda_i}(a) = a^\dagger_{\lambda_i}\ a_{\lambda_i} \tag{9}$$

Clearly, any wavefunction $\psi_K(a)$ and operator L(a) can be written as a function of a_{λ_i}, $A_{J\alpha}(a)$ and $N_{\lambda_i}(a)$

$$\phi_K(a, A) = \psi_K(a) \tag{10a}$$

$$L(a, A, N) = L(a) \tag{10b}$$

The single particle operators a on the l.h.s. of eq.10, is necessary only if the number of nucleons in question is odd.

Under the transformation

$$a \rightarrow a(b) = Ub \tag{11a}$$

$$A_{J\alpha}(a) \rightarrow A'_{J\alpha}(b, B) \tag{11b}$$

$$N(a) \rightarrow N'(b, B) \tag{11c}$$

the wavefunction and operators in the representation G_a go over as

$$\phi_K(a, A) \rightarrow \Phi_K(b, B) \equiv \phi_K(a, A') \tag{12a}$$

$$L(a, A, N) \rightarrow \mathscr{L}(b, B) = L(a, A', N') \tag{12b}$$

It can be readily proved that if the set of operators

$$\left\{a_{\lambda_i}(b)\right\}, \quad \left\{A'_{J_\alpha}(b, B)\right\} \quad \text{and} \quad \left\{N'_{\lambda_i}(b, B)\right\}$$

in G_b obey the same algebraic (i.e. commutation) relation of operators $\left\{a_{\lambda_i}\right\}$, $\left\{A_{J_\alpha}(a)\right\}$, and $\left\{N_{\lambda_i}(a)\right\}$ in G_a, then the representation G_a and G_b are equivalent. The proof is as follows.

Obvioulsy we have

$$\begin{aligned}&\langle\phi_K(a, A)|L(a, A, N)|\phi_{K'}(a, A)\rangle \\ &\qquad = \langle\phi_K(a, A')\,|L(a, A', N')|\phi_{K'}(a, A')\rangle\end{aligned} \tag{13}$$

Since the evaluation of the matrix elements depends solely on the commutation relations between the operators. While according to eq.12, eq.13 is equivalent to

$$\begin{aligned}&\langle\phi_K(a, A)|L(a, A, N)|\phi_{K'}(a, A)\rangle \\ &\qquad = \langle\Phi_K(b, B)|\,\mathscr{L}(b, B)|\Phi_{K'}(b, B)\rangle\end{aligned}$$

Therefore the representation G_a and G_b are equivalent.

In summary, eqs.11 and 12, instead of eq.7 define a more general transformation; i.e., transformation involving two representations wth different dimensions. Clearly, if the "additional" degrees of freedom $\{B\alpha\}$ is absent, then the

the transformation of eqs.11 and 12 is just the standard unitary transformation.

There is a rather subtle point associated with the general equivalence relations of eq.13. Such an equivalence can only ensure that the two representations have identical observables. However, it does not mean that $\Phi_K(b, B)$ is a solution of the Schrodinger equation in the representation G_B, even though $\phi_K(a,A)$ is a solution of the G_a Schrodinger equation. Needless to say, this situation is different from the usual unitary transformation. We mentioned that if eq.5 is satisfied, then eq.6 must automaticall be true. However, for the generalized case, where one is not dealing with representations of the same dimensions, it can in fact be shown that the wavefunctions in G_b are highly degenerate. The reason for this degeneracy can best be seen from the following simple example: Consider the 0^+ g.s. state of four particles in a single j shell, denoted as $\psi_K(a) = |j^2, 0^+\rangle$. If $j \leq 7/2$, it is well known[6] that such a state can be expressed as $\phi_K(a,\bar{A}) = |j^2(\lambda), j^2(\lambda), 0^+\rangle$ where λ is any permissible value (0,2,4, .. etc.). In other words, eq.10a can be written generally as

$$\psi_K(a) = \phi_K^{(\alpha)}(a, A) \qquad (\alpha = 1 \ldots f) \tag{15}$$

where f is the total number of different but equivalent ways of casting $\psi_K(a)$ in the form of $\phi_K(a A)$. Of course, all $\phi_K^{(\alpha)}(a, A)$ are equivalent in G_a; however, their counterpart, obtained via transformation due to eqs.11 and 12 in G_b are different. In fact, even though they all can satisfy the equivalence condition of eq.13, not all will satisfy Schrodinger equation in the representation G_b. It is sufficient to point out here that for each state $\psi_K(a)$ in G_a, there is a unique state $\Psi_K(a, B)$ in G_b which will satisfy the Schrodinger equation

$$\mathscr{H}(a, B)\Psi_K(a, B) = E_K\Psi_K(a, B) \tag{16}$$

The solution will, of course, satisfy the equivalence relations:

$$\langle\psi_K(a)|L(a)|\psi_{K'}(a)\rangle = \langle\Psi_K(b, B)|\mathscr{L}(b, B)|\Psi_{K'}(b, B)\rangle \tag{17}$$

From the above discussion, it is clear that a generalized representation transformation can exist and should be carried out by the following algorithm: First, the transformations of operators (eqs.11 and 12) be carried out; second, solve the Schrodinger equation in the G_b representation (eq.16); third, retain only those solutions of eq.16 which can satisfy the equivalence relations of eq.17. The details of this proof is beyond the scope of this talk, and will not be presented here.

IV. THE BFH REPRESENTATION

We now return to the original problem of seeking the BFHR and its transformation properties.

In the last section, we stated that in a generalized transformation, the key issue is to seek the operator set $\{b, A', N'\}$ in G_b. In order to do so, we must first define precisely what we mean by G_{BFH}, and to do this would require some physical considerations.

Suppose that we have only the now famous Interacting Boson Model (IBM)[8] in mind (however, it must be stressed that this assumption is only for simplicity). To this end, we need to consider only the building blocks of the IBM, i.e., the identical fermion pairs. We shall further assume that the fermion pair structure given by the appropriate two body Schrodinger equation, is known. This means that the wavefunction of the fermion pair is

$$\psi_F^{(\alpha J)}(a) = \psi_F^{(\alpha J)^\dagger}(a)\,|0\rangle = \sum_{\lambda_1 \leq \lambda_2} C^{\alpha J}_{\lambda_1\lambda_2} A^\dagger_{J\lambda_1\lambda_2}(a)\,|0\rangle \qquad (18)$$

which is of course the solution of the two body Schrodinger equation:

$$H(a)\psi_F^{\alpha J}(a) = \omega_{\alpha J}\psi_F^{\alpha J}(a) \qquad (19)$$

In eq.19, $\omega_{\alpha J}$ is the energy eigenvalue. In eq.18, the field operator $\psi_F^{(\alpha J)^\dagger}(a)$ satisfies the following commutation relation:

$$\left[\psi_F^{(\alpha J)}, \psi_F^{(\beta J')^\dagger}\right]_- = \delta_{\alpha\beta}\delta_{JJ'} + \text{partial contracted terms} \qquad (20)$$

So far we are still discussing a fermion pair. Now we direct our attention to the boson problem. A "real" boson, which is created by the boson operator $B^\dagger_{\alpha J}$ and has the same energy $\omega_{\alpha J}$ as the fermion pair, must, be definition, obey the bosonic commutation relations

$$\left[B_{\alpha J}\ ,\ B^{\dagger}_{\beta J'}\right] = \delta_{\alpha\beta}\delta_{JJ'} \tag{21}$$

A glance at eqs.20 and 21 will suggest that a fermion pair in the BFH representation (where one must include the boson degrees of freedom as an inherent part) should have two components: A "bosonic" and a "nonbosonic" components. Clearly the bosonic component must have the structure of $B_{\alpha J}$ and will contribute solely to the $\delta_{\alpha\beta}\delta_{JJ'}$ term in eq.20 while the non-bosonic part will contribute to the partially contracted terms in the same equation. In other words, if $\psi_{\alpha J}$ is the corresponding wavefunction in the BFH representation, then it should be

$$\psi'_{\alpha J} = \psi_F^{\alpha J}(A')|0\rangle \equiv (B^{\dagger}_{\alpha J} + \tilde{\psi}^{\dagger}_{\alpha J})|0\rangle \tag{22}$$

So, the question here is to find the "non-bosonic" part $\tilde{\psi}^{\dagger}_{\alpha J}$.

Intuitively, the nonbosonic part should just be the original fermion part with the boson part "substracted". A straightforward (and totally trivial) subtraction, i.e. $\tilde{\psi}^{\dagger}_{\alpha J} = \psi_F^{(\alpha J)\dagger} - B^{\dagger}_{\alpha J}$ cannot of course serve our purpose here since one will immediately go back to the original fermion space. This is obvious from eq.22. Thus, what we need to do here is to seek a nontrivial subtraction in order that the boson information can be extracted. To this end, we shall define the BFH representation denoted as G_{BFH} as follows: The G_{BFH} consists of three independent (i.e. commuting) components: $\{a^{\dagger}_{\lambda}\}_{\lambda_1=1\ldots n}$, $\{B^{\dagger}_{\alpha J}\}$ and $\{\bar{B}^{\dagger}_{\alpha J}\}$. The operator set $\{\bar{B}^{\dagger}_{\alpha J}\}$, which we shall term the "negative" boson operators has the following algebraic relations:

$$\left[\bar{B}_{\alpha J'}, \bar{B}^{\dagger}_{\beta J'}\right] = -\delta_{\alpha\beta}\delta_{JJ'} \tag{23a}$$

$$\left[\bar{B}_{\alpha J'}, \bar{B}_{\beta J'}\right] = \left[\bar{B}^{\dagger}_{\alpha J'}, \bar{B}^{\dagger}_{\beta J'}\right] = 0 \tag{23b}$$

$$\bar{B}_{\alpha J}|0\rangle = 0 \tag{23c}$$

The significance of the negative sign of eq.23a can be seen in the following discussion.

We have indicated that the game here is to find the "non-bosonic" part of the pair wavefunction $\psi'_{\alpha\tau}$ (eq. 22). Any expectation value evaluated in the BFH representation must of course be the same as that evaluated in the fermion space. Therefore, in the BFHR, the bosonic contribution must be subtracted away. With the minus sign in eq.23a, one can shown that any expectation value evaluated with $\{\bar{B}_{\alpha\bar{J}}\}$ must be equal in magnitude and opposite in sign to the expectation value evaluated by $\{B_{\alpha J}\}$. Therefore in this context, $\{\bar{B}^{\dagger}_{\alpha J}\}$ merely plays the role of "subtracting" the bosonic part from a fermionic pair and it should not be interpreted as an operator which will generate physical states.

V. THE BFHR AND THE NFT

Having defined the BFHR, we are now in a position to construct operators $\{A'_{J\lambda_1\lambda_2}\}$ and $\{N'_{\lambda_i}\}$ which are demanded by the equivalence conditions previously discussed. These operators are

$$A^{\dagger}_{J\lambda_1\lambda_2} = \tilde{A}^{\dagger}_{J\lambda_1\lambda_2} + \sum_{\alpha} C^{\alpha J}_{\lambda_1\lambda_2} B^{\dagger}_{\alpha J} \tag{24a}$$

where

$$\tilde{A}^{\dagger}_{J\lambda_1\lambda_2} = A^{\dagger}_{J\lambda_1\lambda_2} + \sum_{\alpha} C^{\alpha J}_{\lambda_1\lambda_2} \bar{B}^{\dagger}_{\alpha J} \tag{24b}$$

Similarly, the number operator in the BFHR is

$$N'_{\lambda_i} = \tilde{N}_{\lambda_i} + \sum_{\substack{\alpha\beta JJ' \\ \lambda_1\leq\lambda_2}} C^{\alpha J}_{\lambda_1\lambda_2} C^{\beta J'}_{\lambda_1\lambda_2} \left(\delta_{\lambda_i\lambda_1} + \delta_{\lambda_i\lambda_2}\right) B^{\dagger}_{\alpha J} B_{\alpha J} \tag{25a}$$

$$\tilde{N}_{\lambda_i} = N_{\lambda_i} - \sum_{\substack{\alpha\beta JJ' \\ \lambda_1\leq\lambda_2}} C^{\alpha J}_{\lambda_1\lambda_2} C^{\beta J'}_{\lambda_1\lambda_2} \left(\delta_{\lambda_i\lambda_1} + \delta_{\lambda_i\lambda_2}\right) \bar{B}^{\dagger}_{\alpha J} B_{\beta J'} \tag{25b}$$

It can be shown that $\{a, A', N'\}$ in the BFHR obeys the same algebraic structure as $\{a, A, N\}$ in the fermion representation. Thus, according to Section III, the transformation from $\{a, A, N\}$ to $\{a, A', N'\}$ will define an equivalent representation transformation. This transformation is what we call the BFH transformation. Under this transformation, any operator $L(a, A, N)$ in the fermion space will be transformed into the BFHR as

$$L(a, A'N') = \mathscr{L}(a, B, \bar{B}) \tag{26}$$

For example, for the even A nuclei, the fermionic Hamiltonian is

$$\begin{aligned} H_F(a,A,N) &= \sum_{\lambda} \varepsilon_{\lambda} N_{\lambda} \\ &+ \sum_{\substack{\lambda_1 \leq \lambda_2 \\ \lambda_3 \leq \lambda_4}} \langle \lambda_1 \lambda_2; J | V_F | \lambda_3 \lambda_4; J \rangle A^{\dagger}_{J\lambda_1\lambda_2}(a) A_{J\lambda_3\lambda_4}(a) \end{aligned} \tag{27}$$

while in the BFHR, the (transformed) Hamiltonian is

$$\begin{aligned} \mathscr{H}_{BFH} &= H(a,A',N') \\ &= \sum_{\lambda} \varepsilon_{\lambda} \tilde{N}_{\lambda} + \sum_{\alpha J} \omega_{\alpha J} B^{\dagger}_{\alpha J} B_{\alpha J} \\ &+ \sum_{\lambda_1 \leq \lambda_2, \lambda_3 \leq \lambda_4} \langle \lambda_1 \lambda_2; J | V_F | \lambda_3 \lambda_4; J \rangle \tilde{A}^{\dagger}_{J\lambda_1\lambda_2} \tilde{A}_{J\lambda_3\lambda_4} \\ &+ \sum_{\lambda_1 \leq \lambda_2 \alpha J} Z_{\alpha J}(\lambda_1 \lambda_2) \left\{ A^{\dagger}_{J\lambda_1\lambda_2} B_{\alpha J} + B^{\dagger}_{\alpha J} A_{J\lambda_1\lambda_2} \right\} \end{aligned}$$

where

$$Z_{\alpha J}(\lambda_1 \lambda_2) = \langle \lambda_1 \lambda_2; J | V_F | \psi_F^{\alpha J} \rangle \tag{29}$$

is the boson-fermion vertices.

As we have previously mentioned in Sec.III, the wavefunction in the BFHR cannot be obtained by the straightforward transformation. One has to solve the BFH Schrodinger equation

$$\mathcal{H}_{BFH}\,\psi_{BFH} = E\,\psi_{BFH} \tag{30}$$

It is known that[4] the NFT Hamiltonian has, of course, just the same form as eq.28, except that all the $\tilde{N}_\lambda$ and $\tilde{A}_{J\lambda_1\lambda_2}$ in eq.28 must be replaced by N_λ and $A_{J\lambda_1\lambda_2}$. On the other hand, the NFT requires that (i) a "no-bubble" rule be imposed and (ii) only boson space be used as the initial and final (P space) space. In the following, we shall demonstrate that the NFT Hamiltonian with rules (i) and (ii) is in fact an effective way to "eliminate" the negative bosons and solve the problem in a perturbative manner.

To begin with, it is easy to check that $\{a_{\lambda_i}\}$, $\{\hat{N}_{\lambda_i}\}$ and $\{\tilde{A}_{J\lambda_1\lambda_2}\}$ must satisfy the same algebraic relations as $\{a_{\lambda_i}\}$, $\{N_{\lambda_i}\}$ and $\{A_{J\lambda_1\lambda_2}\}$. The only difference is

$$\left[A_{J\lambda_1\lambda_2}\ A_{J\lambda_1'\lambda_2'}\right] = \delta_{JJ'}\delta_{\lambda_1\lambda_1'}\delta_{\lambda_2\lambda_2'}$$

$$+ \text{ partially contracted terms}$$

while

$$\left[\tilde{A}_{J\lambda_1\lambda_2},\ A_{J\lambda_1'\lambda_2'}\right] = \text{Partially contracted terms.} \tag{31b}$$

It is easy to visualize that the complete contracted terms $\delta_{JJ'}\delta_{\lambda_1\lambda_1'}\delta_{\lambda_2\lambda_2'}$ is precisely, in the diagrammatic language, a bubble. Therefore, if the boson space is the P space and the "no-bubble" rule is active, than $\{\tilde{N}_\lambda\}$ and $\{\tilde{A}_{J\lambda_1\lambda_2}\}$ in eq.28 can be replaced by $\{N_\lambda\}$ and $\{A_{J\lambda_1\lambda_2}\}$. In this way, $\mathcal{H}_{BFH}$ becomes H_{NFT}.

Why must one use the boson space as the P space in the NFT? If one were to use $\{B^\dagger_{\alpha J}\}$ as the P space, then the Q space must be either $\{A^\dagger_{J\lambda_1\lambda_2}\ B^\dagger_{\alpha J}\}$ or equivalently $\{A^\dagger_{J\lambda_1\lambda_2}\ \tilde{A}_{J\lambda_1\lambda_2}\}$. However, $\{A^\dagger_{J\lambda_1\lambda_2},\ \tilde{A}_{J\lambda_1\lambda_2}\}$ is just $\{A^\dagger_{J\lambda_1\lambda_2}\}$ with the "no-bubble" rule. On the other hand, if one were to use $\{A^\dagger_{J\lambda_1\lambda_2}\}$ as the P space, then the Q space must then be $\{B^{\dagger}_{\alpha\tau}, \bar{B}^{\dagger}_{\alpha\tau}\}$. In this case, it is not possible to solve the BFHR Schrodinger equation (28) without explicitly taking into account the "negative" bosonic degrees of freedom. Of course, if there is no attempt to avoid the "negative" bosons, then one is free to choose any P space as one may wish.

In conclusion, we have now shown that the BFHR is the representation which the NFT is built upon. In fact, we have demonstrated that the NFT is an effective, albeit clever, way of treating the full BFHR problem.

ACKNOWLEDGEMENT

We like to thank Professors Igal Talmi, A. Klein, W.T. Pinkston, Chia-Chang Shih, V. Oberaker, M. Vallieres and J.M. Yuan for useful discussions. CLW is grateful to the Physics Departments of Drexel and Vanderbilt University for providing the necessary financial assistance as well as a good atmosphere for collaborative research. Two of us (DHF and CLW) are grateful to the Physics Department of the University of Tennessee for its warm hospitality.

REFERENCES

1. B.R. Mottelson, J. Phys. Soc. Japan (supplement), 24, 87 (1968); A. Bohr and B.R. Mottelson, Nuclear Structure, Vol.2 (Addison-Wesley, New York, 1975).
2. P.F. Bortignon, R.A. Broglia, D.R. Bes and R. Liotta, Phys. Rep. Phys. Rep. 30C, 305 (1977).
3. R.J. Liotta and B. Silvestre-Brac, Nucl. Phys. A309, 301 (1978).
4. C.L. Wu, Nucl. Phys. A349, 114 (1980); C.L. Wu and D.H. Feng, Phys. Lett. 96B, 243 (1980); Phys. Rev. C.(in press); Ann. of Phys. (in press).
5. A. Messiah, Quantum Mechanics, (John Wiley and Sons, Inc. 1968).
6. A. de-Shalit and Igal Talmi, Nuclear Shell Theory.

MICROSCOPIC CALCULATIONS OF THE FISSION BARRIER OF SOME ACTINIDE NUCLEI WITH SKYRME-TYPE INTERACTION USING A TWO-STEP ITERATIVE METHOD

A.K. Dutta

Department of Physics
University of Saskatchewan
Saskatoon, Sasketchewan
Canada, S7N 0W0

Michael Vallières

Department of Physics & Atmospheric Science
Drexel University
Philadelphia, Pennsylvania 19104

R.K. Bhaduri, I. Easson and M. Kohno

Department of Physics
McMaster University
Hamilton, Ontario
Canada, L8S 4M1

I. ABSTRACT

For a given two-body force, we formulate a method for estimating the energy in a constrained Hartree-Fock calculation of a deformed nucleus. This is done by using the information from the first two iterations of the self-consistency cycle. After checking the accuracy of the method for spherical nuclei, deformation energy curves for ^{168}Yb and ^{240}Pu are obtained using the Skyrme III interaction and compared with the corresponding CHF calculations. Further, the method has been used to test the sensitivity of the fission barrier height to the density dependence of the interaction. Calculations were performed with five Skyrme-Type forces and dependence of second barrier height in ^{240}Pu on certain physical quantities is investigated. The interaction which gives better agreement with experimental results for ^{240}Pu has been used to trace deformation energy curves for some other actinides nuclei. The results are found to be in good agreement with other theoretical and experimental results.

II. INTRODUCTION

The aim of this paper is to present a simple and accurate method of estimating the results of a static Hartree-Fock (HF) calculation using the information from the first two iterations of the self-consistency cycle . The method, called the two-step iterative method (TSIM) will be shown to be capable of tracing the energy versus deformation curve in close correspondence with constrained Hartree-Fock (CHF) results . Since CHF calculations take up excessive computer time even with Skyrme-type interactions, the TSIM may be very useful, particularly for obtaining fission barriers of actinides where the dependence of barrier height on the details of the effective interaction is worth investigating.

III. THE METHOD

The HF energy for a system of N fermions of one type, interacting via a two-body interaction v_{12} is

$$E(\rho) = tr_1 \hat{t}_1 \rho_1 + \frac{1}{2} tr_1 tr_2 \rho_1 \rho_2 v_{12} \tag{1}$$

where $\hat{t}_1$ is the kinetic energy operator and ρ is the one body density matrix normalised to N. The self-consistent one-body potential is $\hat{u}_1 = tr_2(\rho_2 v_{12})$. Choosing a trial density $\rho^{(o)}$ to start the self-consistency cycle, Eqn.(1) can be written as

$$E(\rho) = E^{(o)} + tr_1[(\hat{t}_1 + u_1^{(o)})\delta\rho_1] + \frac{1}{2} tr_1 tr_2 (\delta\rho_1 \delta\rho_2 v_{12})$$

where $E^{(o)} = E(\rho^{(o)})$, $u_1^{(o)} = tr_2(\rho_2^{(o)} v_{12})$ and $\delta\rho_1 = \rho_1 - \rho_1^{(o)}$. If $\rho^{(1)}$ is the density obtained by solving in the one-body field $u_1^{(o)}$, then $E(\rho)$ can be rewritten as

$$\begin{aligned} E(\rho) = {} & E^{(o)} + \hat{tr}_1[(\hat{t}_1 + u_1^{(o)})(\rho_1^{(1)} - \rho_1^{(o)})] \\ & + tr_1[(t_1 + u_1^{(o)})(\rho_1 - \rho_1^{(1)})] + \frac{1}{2} tr_1 tr_2 (\delta\rho_1 \delta\rho_2 v_{12}) \end{aligned} \tag{2}$$

The approximation consists of dropping the last two terms in Eqn.(2), so that the HF energy is approximated by

$$E(\rho) \approx E^{(o)} + tr_1 \, (\hat{t}_1 + u_1^{(o)})(\rho_1^{(1)} - \rho_1^{(o)}) \tag{3}$$

Although we cannot justify the approximation analytically, it has been seen that the two neglected terms nearly cancel each other. The method is first tested in spherical nuclei where the trial density $\rho^{(o)}$ was chosen in two different ways: first through the harmonic oscillator wavefunctions and second through the Woods-Saxon wavefunctions. Binding energies are displayed in Table I. A poorer choice of $\rho^{(o)}$ leads to a less accurate $E^{(o)}$, but the correction term $\Delta E = tr_1[(\hat{t}_1 + u_1^{(o)})(\rho_1^{(1)} - \rho_1^{(o)})]$ is correspondingly larger. This makes the final estimate for the energy to be nearly independent of the choice of $\rho^{(o)}$. Note that our method is microscopic but not variational and can therefore overestimate the HF energy. Further, although the correction term ΔE is reminiscent of the Strutinsky theorem[3], the connection between the two is remote. Our $E^{(o)}$ already contains shell-effects and no "smoothing" of any kind is involved. For the harmonic oscillator $\rho^{(o)}$, the neglected terms in Eqn.(2) were numerically found to be of the order of 90 MeV in ^{208}Pb but opposite in sign! In the case of Woods-Saxon $\rho^{(o)}$, these terms were much smaller ($\sim$5 MeV). Depths of 44 and 40 MeV were chosen for neutron and proton fields with equal radii 1.27 $A^{1/3}$ fm and a diffuseness parameter of 0.67 fm.

For deformed nuclei the method has to be slightly modified to implement a constrained calculation and to include pairing. In order to study the variation of total energy with deformation of the system, an external driving field must be added to the Hamiltonian. We use a linear constraint on the quadrupole moment <Q>. The constrained energy to be varied is

$$E(\lambda) = tr_1\hat{t}_1\rho_1 + \frac{1}{2}\, tr_1 tr_2 \rho_1\rho_2 v_{12} + \lambda tr_1 \hat{Q}_1\rho_1$$

where λ is the Lagrange multiplier to be determined by the constraint on <Q>. Starting with a trial density $\rho^{(o)}$, we obtain

$$E(\rho) = E^{(o)}(\lambda) + tr_1(\hat{t}_1 + u_1^{(o)})\delta\rho_1 + \frac{1}{2}\, tr_1 tr_2 \delta\rho_1\delta\rho_2 \, v_{12} \, ,$$

where $E^{(o)}(\lambda) = E^{(o)}(\lambda, \rho^{(o)})$ and $u_1^{(o)}(\lambda) = tr_2\rho_2^{(o)}v_{12}+\lambda\hat{Q}_1$. If $\rho^{(1)}$ is the density generated by $u_1^{(o)}(\lambda)$, then after making our usual approximation, the CHF energy at a given deformation is given by

$$E_{CHF} \approx E^{(o)} + tr_1[(t_1+u_1^{(o)}(\lambda))(\rho_1^{(1)} - \rho_1^{(o)})] - \lambda tr_1\hat{Q}_1\rho_1^{(o)} \tag{4}$$

TABLE I

NUCLEUS	$E^{(o)}$	E	$E^{(o)} + \Delta E$	E(HF)
^{40}Ca	-310.85	-8.67	-319.52	-318.00
	-304.17	(-14.72)	(-318.89)	
^{90}Zr	-834.58	-17.75	-752.33	-749.74
	(-676.12	(-85.54)	(-751.66)	
^{208}Pb	-1558.02	-30.82	-15.88.55	-1587.44
	(-1372.48)	(-213.10)	(-15.86.58)	

Energies (in MeV) of closed shell nuclei. The numbers in parenthesis are results with harmonic oscillator $\rho^{(o)}$, while the unparenthesized numbers are obtained by using $\rho^{(o)}$ obtained from a Woods-Saxon potential.

The value of λ is so adjusted that the quadrupole moment obtained from $\rho^{(1)}$ is the same as that obtained from $\rho^{(o)}$. The trial density $\rho^{(o)}$ is obtained from a deformed Woods-Saxon potential. The deformation is introduced through the (c, h) parameterization[4].

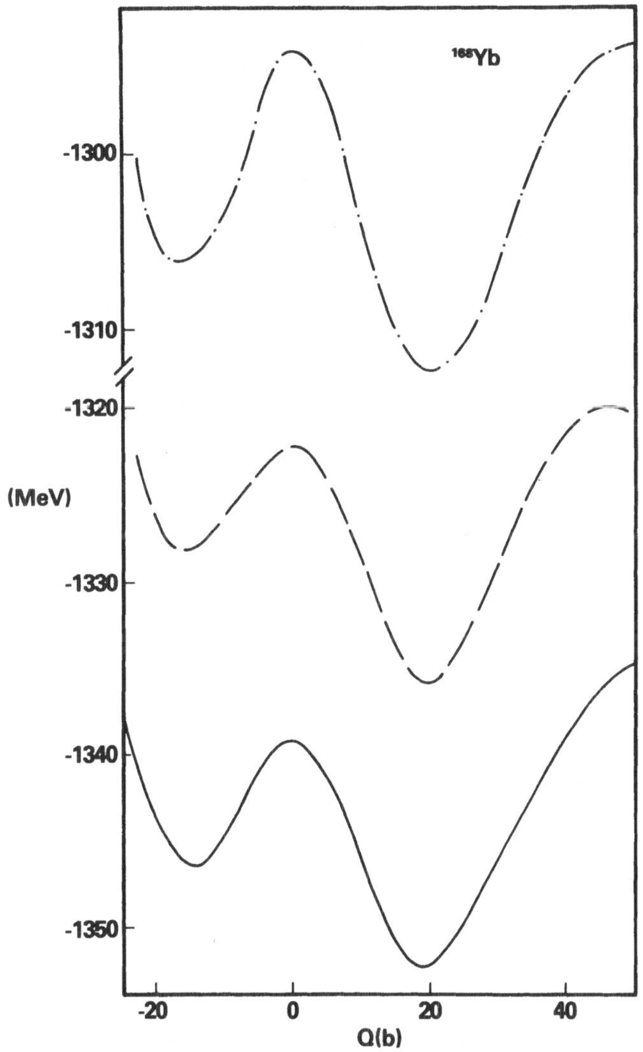

Fig.1 The binding energy of ^{168}Yb as a function of the quadrupole moment; the dashed-dot line is $E^{(o)}$, the dashed line is the result of TSIM and the solid line is the exact HF result.

Pairing effects can be easily incorporated into our approximation. We use the HF + BCS scheme formulated by Vautherin[5] and choose a constant gap $\Delta = 12A^{-1/2}$ for all deformations. Such a choice seems to be in fair agreement with the HFB calculations by Gogny[6]. The condensation energy $E_p = -\frac{\Delta}{2}\Sigma_i u_i v_i$ where v_i^2 is the occupational probability of the ith state and $u_i^2+v_i^2 = 1$. To start with, the gap Δ is related to the pairing strength $G^{(o)}$ by $\Delta = G^{(o)}\Sigma_i u_i^{(o)} v_i^{(o)}$, where $u_i^{(o)}$ and $v_i^{(o)}$ are the solutions of BCS equations obtained by the starting Woods-Saxon potential. In the next iteration, Δ is kept fixed by letting $G^{(o)}$ change to $G^{(1)}$; that is, a new set of $u_i^{(1)}$ and $v_i^{(1)}$ gives $\Delta = G^{(1)}\Sigma_i u_i^{(1)} v_i^{(1)}$. This results in a change in condensation energy

$$E_p^{(1)} - E_p^{(o)} \simeq E_p^{(o)} \left[\frac{G^{(o)} - G^{(1)}}{G^{(1)}}\right]$$

This term is added to Eqn.4, keeping in mind that $E^{(o)}(\lambda)$ already contains $E_p^{(o)}$ and the densities are calculated with proper occupational probabilities.

We use Skyrme III interaction to compare our results with the corresponding CHF results[7] for ^{168}Yb and ^{240}Pu. The binding energy as a function of the quadrupole moment is plotted in Figs.1 and 2; the zeroth order estimate $E^{(o)}$ is also plotted. The difference in absolute energies between TSIM and CHF calculation is nearly equal to the centre-of-mass correction which is not included in TSIM results. Otherwise, the overall shape is reproduced reasonably well, as seen in Figs.1 and 2. For ^{240}Pu, the CHF second barrier beight is underestimated by about 6 MeV. Incidently, it may be noted that our final results are very similar to the "expectation value method" of Brack[7]. The latter is simply the evaluation of $E^{(o)}$, but with a carefully chosen Woods-Saxon potential. It should also be mentioned that although the TSIM estimates the binding energy quite well, the method is not meant to provide us with a good density[1].

IV. FISSION BARRIER OF ^{240}Pu

CHF calculations with the Vautherin-Brink (VB) version of Skyrme forces and using a constant gap, yield a second barrier much higher than the experimental value for ^{240}Pu, even after including estimated corrections[8] for spurious rotational energy, mass- and γ-asymmetry, truncation of basis etc. Although the barrier height can be lowered if a surface dependent pairing strength is used[9], such a choice is not favoured by the HFB calculations[6]. Indeed the VB type Skyrme forces imply a very high incompressibility K, which in turn can incluence the deformation energy. Zamick[10] has shown that there exists a strong correlation between K and the exponent of the density dependence in the effective two body interaction. Several Skyrme-type forces have been parameterized with density-dependence ρ^{α} where α is less than one. We have chosen Skyrme III (SIII), interactions due to Köhler (SKa)[11] the Orsay group (SKM)[12] and Kyoto group (SMK)[13], with corresponding values α = 1, 1/3, 1/6 and 1/6 ; their respective compressibility K in nuclear matter is 356, 263, 217 and 226 MeV. It should be noted that the GO force[14] also has α = 1/6. For interest, we have also included Skyrme V (SV) in our TSIM calculations. This force has no density dependence and yields K = 306 MeV.

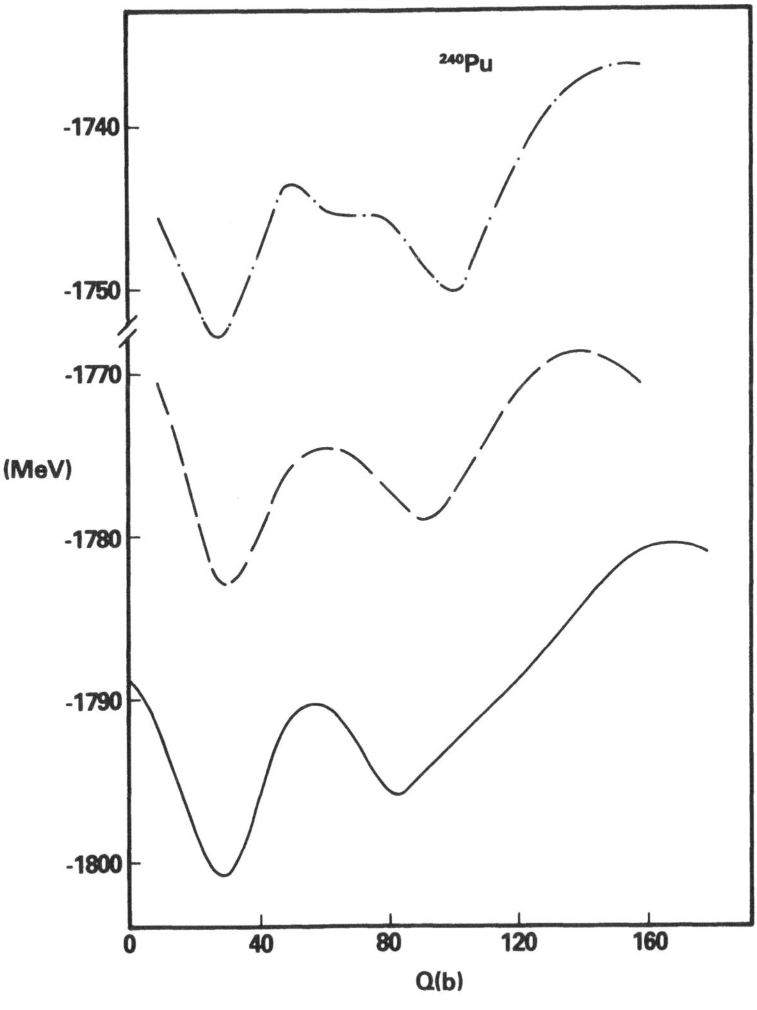

Fig.2 Same as in Fig.1, but for ^{240}Pu.

Although the TSIM underestimates the CHF second barrier, this underestimation is not expected to depend on the interaction used and therefore some quantitative conclusions can be obtained.

The height of the second barrier is found to decrease in the following order: SV, SIII, Ska, SMK, and SKM. The barrier height hus decreases with decreasing K, SV being an exception, We have also looked at the effective surface energy a' of Pu due to these interactions. The values of $a_s' = a_s(1 + k\ I\ )$, where $I = (N - Z)/A$ are obtained after fitting the liquid-drop mass formula through the parameterization

$$E(N, Z) = a_v(1 + k_v I^2)A + a_s(1 + k_s I^2)A^{2/3}$$

$$+ a_c(1 + k_c I^2)A^{1/3} + a_o$$

Values of a and k are chosen to be those of nuclear matter and the other five parameters are determined from the energies of spherical nuclei ^{16}O, ^{40}Ca, ^{48}Ca, ^{90}Zr and ^{208}Pb. We find that the second barrier height decreases as a_s' decreases. Again, SV is an exception. An unrealistic value of effective mass (0.38) in SV may be the reason for the high barrier. Similar conclusion for SIV was made by Flocard[16] who found that the interaction gave a higher barrier than SIII in ^{240}Pu. The other four interactions we considered have nearly the same effective mass and spin-orbit strength. Hence we can expect that the difference of the second barrier height reveals the difference in the bulk properties of these interactions. We find that the barrier height by the SMK interaction is in close agreement with the experimental estimates[9]. It should be noted that the value of K = 20 MeV given by the SMK interaction is close to that obtained from giant monopole excitation calculations[17].

V. FISSION BARRIER OF OTHER ACTINIDES

We have used the SMK interaction to trace the deformation energy curves of ^{232}Th, ^{252}Fm and ^{258}Fm. It is well known that theoretical and experimental values for the inner barrier of the Th-isotopes are far from being in agreement. This discrepancy increases up to 4 MeV for lighter isotopes of Thorium. Uncertainty due to the lack of self-consistency in the shell-correction method and errors in the liquid-drop parameters are not enough to explain this "Th-anomaly". We find that in ^{232}Th the inner barrier is 3-4 MeV compared to the experimental value of 5 $\sim$ 6 MeV[18], thus verifying the conclusions of the shell-correction approach.

Experiments on fission of Fm-isotopes have revealed that a

transition from mass-asymmetric to mass-symmetric fission occurs between ^{256}Fm and ^{258}Fm. This is interpreted as a consequence of the vanishing of the second barrier. A macroscopic-microscopic type calculation[19] does explain this transition. Using the TSIM, we have found that, indeed in ^{258}Fm the SMK interaction does not exhibit a second barrier.

V. CONCLUSION

The TSIM allows an accurate estimate to the binding energy of a system describable by HF equations to be obtained in two(2) iterations. The method was shown to be very accurate for both spherical and deformed nuclei.

The TSIM allows an easy study of the fission barriers. The second barrier height was found to be sensitive on the density dependence of the effective interaction used. In particular, using the SMK force, we have shown how the second fission barrier disappear in going from ^{256}Fm to ^{258}Fm, corresponding to the mass-asymmetric to mass-symmetric transition. It is clear that the fission barrier calculation ought to be taken into account in searching for the best effective interaction; the study of the ground state properties of spherical and deformed nuclei is not enough in this task

REFERENCES

1. A.K. Dutta, R.K. Bhaduri, M.K. Srivastava and M. Vallieres, Phys. Lett. 84B (1979) 17;
   A.K. Dutta, M. Vallieres, R.K. Bhaduri and I. Easson, Nucl. Phys. A341 (1980) 461.
2. H. Flocard, P. Quentin, A.K. Kerman and D. Vautherin, Nucl. Phys. A203 (1973) 433.
3. V.M. Strutinsky, Nucl. Phys. A95 (1967) 420; A122 (1968) 1.
4. M. Brack, J. Damgaard, A.S. Jensen, H.C. Pauli, V.M. Strutinsky and C.Y. Wong, Rev. Mod. Phys. 44 (1972) 320.
5. D. Vautherin, Phys. Rev. C7 (1973) 296.
6. D. Gogny, Nuclear Self-consistent Fields, ICTP Conf. Trieste, 1965, ed. G. Ripka and M. Porneuf (North-Holland, Amsterdam 1975) p.333.
7. M. Brack, Phys. Lett. 71B (1977) 239.
8. M. Brack, "Static Deformation Energy Calculations: from Microscopic to Semiclassical Theories", Int. Sym. Phys. and Chem. of Fission, Jiilich, 1979, paper IAEA-SM.241/C1, to be published by IAEA (Vienna).
9. H. Flocard, P. Quentin, D. Vautherin, M. Veneroni and A.K. Kerman, Nucl. Phys. A231 (1974) 176.
10. L. Zamick, Phys. Lett. 45B (1973) 313.

11. H.S. Kohler, Nucl. Phys. A258 ·(1976) 301.
12. H. Krivine, J. Treiner and O. Bohigas, Nuc. Phys. A336, (1980) 155.
13. S. Nishizaki, M. Kohno and K. Ando, Contribution paper in 1980 RCNP Int. Sym. on Highly Excited States in Nuclear Reactions (Osaka, Japan).
14. D.W.L. Sprung and P.K. Banerjee, Nuc. Phys. A168 (1971) 273. X. Campi and D.W.L. Sprung, Nucl. Phys. A194 (1972) 401.
15. Y.H. Chu, B.K. Jennings and M. Brack, Phys. Lett. 68B (1977) (1977) 407.
16. H. Flocard, These d'Etat, Orsay (1975).
17. O. Bohigas, A.M. Lane and J. Martorell, Phys. Rep. 51 (1979) 267.
18. H.C. Britt, review paper IAEA-SM-241/A1, Int. Sym. Phys. and Chem. of Fission, Julich, 1979.
19. M.G. Mustafa and R.L. Ferguson, Phys. Rev. C18 (1978) 301.

LINEAR RESPONSE RPA CALCULATIONS TO SPHERICAL OPEN-SHELL NUCLEI

A. Moalem and J. Bar-Touv

Department of Physics
Ben-Gurion University
Beer-Sheva, Israel

ABSTRACT

We describe a simple extension of the Linear Response RPA model to spherical open-shell nuclei. The model allows replacing the uncorrelated p-h vacuum of the standard RPA by a correlated ground state wave function from preliminary model calculations. The model is exactly soluable in the sense that it avoids the necessity to truncate the vector space and thus allows the inclusion of the entire continuum spectrum.

There has been much interest recently in the nuclear region above particle threshold emission. Data from various photonuclear reactions, electron and hadron scattering from nuclei[1] provide evidence for new giant multipole resonances (GMR) besides the well-known giant dipole resonance (GDR). Unlike other collective states, such as the first quadrupole (2_1^+) and octupole (3_1^-) states in even-even nuclei, the GMR are mostly governed by the average nuclear properties. However, there exists certain isotopic dependence, particularly in light nuclei, where the addition of one or two nucleon may induce significant differences in the strength distributions.[2]

The systematic locations of the electric multipole resonances and their contributions to the corresponding energy weighted sum rules are well reproduced using random phase approximation (RPA) theories which are based on the Hartree-Fock (HF) ground states of Skyrme-type zero range interactions.[3-5] Yet the application of these theories is limited to closed shell nuclei. It is the purpose of the present note to outline a simple extension of the linear response RPA model of Bertsch and collaborators[3,4] to spherical open shell nuclei. Such an extension is most desirable not only for evaluating isotopic effects as mentioned above but also to treat more realistically the so-called "closed shell nuclei" (e.g. ^{16}O, ^{42}Ca, ^{90}Zr and ^{208}Pb). The extension is based on the introduction of occupation parameters $\theta_{\ell j}$ in the early stage of determining the single particle basis and single particle Hamiltonian (H_o) which are needed for the 1p-1h RPA calculations. These occupation parameters are common to other methods which can also treat open-shell nuclei, such as the Tensor Open-Shell RPA[6] and can be derived from shell model calculations or otherwise from particle transfer data. The modifications imposed by the extension of the model concern the exact form of the unperturbed Green's function[7] only and will be discussed in some detail. Other than that calculations of the transition strength follow exactly a track similar to that of ref.4. The RPA Green's function is calculated directly in co-ordinate space by matrix inversion. The dimensions of the matrix are determined by the mesh size in coordinate space and thus avoids the necessity to truncate the 1p-1h configurational space as is the case in the standard RPA calculations.

In the LRM the transition strength is expressed by[3]

$$\sum_n |\langle n|F|o\rangle|^2 \, \delta(E_n - E_o - E) = \frac{1}{\pi} \text{Im} \int dr\, dr'\, F(r) \cdot G(r,r',E)\, F(r') \tag{1}$$

where F is a single-particle operator representing an external field and G(r,r',E) is the RPA Green's function given by[2]

$$G(r, r', E) = G^{(0)} \left[1 - \frac{\delta^2 E}{\delta\rho^2} G^{(0)}\right]^{-1} \tag{2}$$

Here E is the total HF energy which has functional dependence on the density ρ and $G^{(0)}$ is the unperturbed particle-particle Green's function.

For a closed shell nuclei and in a discrete spectral decomposition $G^{(0)}$ is given by

$$G^{(0)}(r, r', \omega) = -\sum_p \sum_h \Phi_h^*(r)\, \Phi_p^*(r) \cdot \left(\frac{1}{\varepsilon_p - \varepsilon_h - \omega} + \frac{1}{\varepsilon_p - \varepsilon_h + \omega}\right) \phi_h(r')\, \phi_p(r') \tag{3}$$

where the sum is over unoccupied orbits p (particle) and occupied orbit h (hole), with the HF orbital and energies denoted by Φ_λ- and ε_λ. The integral on the right side of eq.1 is the response function which represents the density oscillations induced in a nucleus that is subjected to the external field F. The poles of G determine the energies of the excited states of the system.

As a starting point we derive an intrinsic single particle field of the nucleus using the static HF theory with zero-range forces.[8] Thus we search for eigenstates of a single particle Hamiltonian which is a functional of its own eigen functions, i.e.

$$h[\phi_j]\ \phi_i = \varepsilon_i\ \phi_i \tag{4}$$

where $h[\phi_j]$, the HF single particle Hamiltonian is determined by a variational condition on the total energy. A relevant point to the present application of the RPA to open-shell nuclei is the introduction of occupation parameters, θ_λ, in the early stage of determining the single particle basis. The Hamiltonian density is expressed in terms of three local densities: the nucleon density

$$\rho(\bar{r}) \equiv \sum_\lambda \theta_\lambda\ |\phi_\lambda\ (\bar{r})|^2 \tag{5}$$

the kinetic energy density

$$\tau(\bar{r}) \equiv \sum_\lambda \theta_\lambda\ |\nabla\phi_\lambda\ (\bar{r})|^2 \tag{6}$$

and the spin density

$$J(\bar{r}) \equiv -\, i \sum_\lambda \theta_\lambda\ \phi^*_\lambda(\bar{r}) \cdot [\nabla\phi_\lambda \times \sigma] \tag{7}$$

λ represents a set of the usual quantum number $n\ell jt_3$ so that spin and charge states are included in the summation. In the case of spherically symmetric system the occupation numbers can be taken as $\theta_\lambda = N_\lambda/(2j_\lambda+1)$, where N_λ is the number of protons or neutrons that occupy the orbital λ. For an open shell nucleus we distinguish between three groups of particle states: (a) fully occupied with $\theta_a=1$, (b) partially occupied with $0<\theta_b<1$ and (c) unoccupied with $\theta_c=0$. The 1p-1h excitations which contribute to eq.3 fall now into four categories: (i) $h \in a \rightarrow p \in c$, (ii) $h \in a \rightarrow p \in b$, (iii) $h \in b \rightarrow p \in c$ and (iv) $h \in b \rightarrow p \in b$. The contribution from the first two categories is given by

$$- \sum_{h \subset a} \sum_{p \subset b,c} \theta_h (1 - \theta_p) \phi_h^*(r) \phi_p^*(r)$$

$$\cdot \left(\frac{1}{\varepsilon_p - \varepsilon_h - \omega} + \frac{1}{\varepsilon_p - \varepsilon_h + \omega} \right) \phi_h(r') \phi_p(r') \qquad (8)$$

The factor $(1-\theta_p)$ accounts for the blocking of the partially occup- pied orbits (b) to unoccupied orbits (c)

$$- \sum_{h \subset b} \sum_{p \subset c} \theta_h (1 - \theta_p) \phi_h^*(r) \phi_p^*(r)$$

$$\cdot \left(\frac{1}{\varepsilon_p - \varepsilon_h - \omega} + \frac{1}{\varepsilon_p - \varepsilon_h + \omega} \right) \phi_h(r') \phi_p(r') \qquad (9)$$

and the 1p-1h excitations among the partially occupied orbits (b)

$$- \sum_{h \subset b} \sum_{p \subset b} \theta_h (1 - \theta_p) \phi_h^*(r) \phi_p^*(r)$$

$$\cdot \left(\frac{1}{\varepsilon_p - \varepsilon_h - \omega} + \frac{1}{\varepsilon_p - \varepsilon_h + \omega} \right) \phi_h(r') \phi_p(r') \qquad (10)$$

The last expression may include 1p-1h excitations from a higher partially occupied single particle orbit to a lower one. These contributions should be avoided, i.e., one has to add the sum

$$\Delta = \sum_{h \subset b} \sum_{\substack{h \subset b \\ \varepsilon_p < \varepsilon_h}} \theta_h (1 - \theta_p) \phi_h^*(r) \phi_p^*(r)$$

$$\cdot \left(\frac{1}{\varepsilon_p - \varepsilon_h - \omega} + \frac{1}{\varepsilon_p - \varepsilon_h + \omega} \right) \phi_h(r') \phi_p(r') \qquad (11)$$

It is straightforward now to realize that the sum of expressions (7) - (9) amounts to

$$\tilde{G}^{(0)} = - \sum_{hp} \theta_h (1 - \theta_p) \phi_h^*(r) \phi_p^*(r)$$

$$\cdot \left(\frac{1}{\varepsilon_p - \varepsilon_h - \omega} + \frac{1}{\varepsilon_p - \varepsilon_h + \omega} \right) \phi_h(r') \phi_p(r') \qquad (12)$$

where now h and p run over all orbits. To treat the continuum exactly, we introduce now the single particle Green's function with the well-knonw representation

$$g_{\ell j}(r, r', E) = \frac{1}{H_o - E} = \sum_{p\ a,b,c} \phi_p^*(r) \left(\frac{1}{\varepsilon_p - E}\right) \phi_p(r) \quad (13)$$

Since the term in the square bracket of eq.(12) is antisymmetric with respect to the exchange of p and h, the sum with $\theta_h\theta_p$ vanishes identically. We thus can omit the factor θ_p ub eq.(11) and replace the sum over p by the single-particle Green's functions. The true unperturbed Green's function is then given by

$$G^{(0)} = -\sum_h \theta_h \phi_h^*(r) [g(r,r', \varepsilon_h + \omega) + g(r,r', \varepsilon_h - \omega)] \phi_h(r') + \Delta \quad (14)$$

Both berms in the last expression involve occupied states only, and thus can be calculated exactly once the ground state is specified. Obviously the correction term Δ should be included only if two or more partially occupied orbits are involved. For the simple case of ^{12}C, 1p-1h excitations among the $p_{3/2}$ and $p_{1/2}$ do not contribute to $J^{\pi}=1^-$ and $J^{\pi}=3^-$ excitations, while contribution to $J^{\pi}=2^+$ is negligible and does not affect much the detailed strength distribution.

We now turn to describe the calculational procedure and the results for ^{18}O. The self consistent field and the RPA response are calculated using Skyrme type forces[10]. Following refs.3,4 we adjust the strength of the p-h interaction to place the spurious 1^- state at zero energy. In order to take full advantage of the simple manner by which the LRM accounts for the complete 1p-1h continuum, we preserve the spherical symmetry of the HF single-partilce field by employing uncoupled scheme to represent the ground state wave function of ^{18}O. The strength distributions are then calculated for each multipole using two different ground state wave functions. The first consists of a spherical ^{16}O core plus two neutrons equally distributed among the various m states of the $1d_{5/2}$ orbit, i.e. $\theta(1d_{5/2})=1/3$. The second is the intermediate coupling ground state wave function of refs.11, 12. The occupation parameters extracted out from this wave function are as follows: for neutron orbits;

$$\theta(1p_{1/2}) = 0.864;\ \theta(2s_{1/2}) = 0.579;\ \theta(1d_{5/2}) = 0.186$$

and for proton orbits

$$\theta(1d_{5/2}) = 0.045;\ \theta(1p_{1/2}) = 0.864;\ \theta(2s_{1/2}) = 0.$$

The calculated photonuclear cross section for $J^{\pi}=1^{-}$ and T=1 states in ^{18}O are presented in fig.1. The comparison we also included in fig.1 our results for ^{16}O. We refer first to the cross sections obtained with the simple shell model. In the region up to 15 MeV the isovector dipole strength is distributed among seven narrow

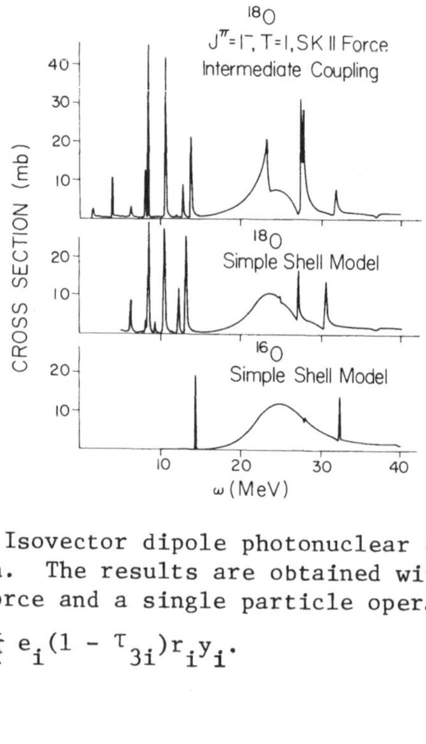

Fig.1 Isovector dipole photonuclear cross section. The results are obtained with a SKII force and a single particle operator $F_{-}= \sum_i \frac{1}{2} e_i(1 - \tau_{3i})r_i y_i$.

resonances at energies 6.4, 8.3, 8.6, 9.4, 10.6, 12.4 and 13.3 MeV which exhaust a total of ∿15% of the energy weighted dipole sum (EWDS). In the experimental photoabsorption cross section of ^{18}O 2 one also observes seven narrow resonances at 5.1, 10.3, 11.1, 13.1, 13.8, 14.7 and 15.8 MeV which also exhaust ∿15% of the EWDS. These resonances which are commonly referred to as the pygmy resonance are related to excitations of the two valence neutrons. Thus apart from an overall shift of 2-3 MeV towards the lower energies the model accounts for the observed pygmy resonance. This discrepancy is mainly due to the defficiencies of the calculated HF single particle energies, which yield a neutron threshold at ∿5 MeV compared to an experimental value of 8.05 MeV. In the energy region of 15-30 MeV we observe the broad and smooth peak of the GDR. The locations (23. MeV), the width and the fact that the cross sections remain substantial up to 40 MeV are in agreement with experiment.

Using now the intermediate coupling wave function of ref.11, 12, one notices that the smooth distribution of the simple shell model GDR becomes structured with an extra peak at about 23 MeV and a peak at 27 MeV being enhanced. These differences are obviously structural effects and can be understood in terms of the wave functions we used. There are 2p-0h components (two particles in the s-d shell and zero hole in the p shell) common to both ground state wave functions of ^{18}O, but the intermediate coupling wave function includes in addition 4p-4h terms. Such terms lead to 5p-3h $J^{\pi} = 1^{-}$ states which cannot decay by the emission of two neutrons to the ground states of ^{16}O. They can decay however to excited states such as the first monopole or quadrupole states in ^{16}O which contain substantial 4p-4h admixtures. It is worth mentioning that the main peak at 23.7 MeV is more prominent in the (γ, 2n) cross section than in the single photoneutron cross sections[2]. Clearly the example above pleads for more attention to the approximation made for the ground state wave function. The application of the model to ^{12}C and ^{16}O using realistic ground state wave functions yields also a better agreement with experiment for the electric $J^{\pi} = 1^{-}$, 2^{+} and 3^{-} strength distributions. In particular the model improves on properties of the 2_1^+ and 3_1^- states.

ACKNOWLEDGEMENT

One of us (A.M.) acknowledges the hospitality extened to him during his stay at the Tandem Accelerator Laboratory of the University of Pennsylvania.

REFERENCES

1. F.E. Bertrand, Ann. Rev. Nucl. Sci. 26, 457 (1975); and references therein.
2. See for example B.L. Berman, Atomic Data and Nucl. Date Tables 15, 1319 (1975);J.G. Woodworth et al., Phys. Rev. C19, 1667 (1979); B.L. Berman et al., Phys. Rev. Lett. 36, 1441 (1976); A. Moalem et al., Phys. Rev. C20, 1593 (1979).
3. G. Bertsch and S.F. Tsai, Phys. Reports 18C, 125 (1975).
4. S. Shlomo and G. Bertsch, Nucl. Phys. A243, 507 (1975).
5. P. Ring and J. Speth, Phys. Lett. 44B, 477 (1973); S. Krewald and J. Speth, Phys. Lett. 52B, 295 (1974); S. Krewald et al.,

Phys. Rev. Lett. 33, 1386 (1974); K.F. Liu and G.E. Brown, Nucl. Phys. A265, 385 (1976).

6. C. Ngo-Trong et al., Nucl. Phys. A313, 15 (1979); and references therein.
7. J. Bar-Touv et al., Nucl. Phys. A339, 303 (1980); A. Moalem and J. Bar-Touv, submitted for publication in Phys. Rev. C, J. Bar-Touv and A. Moalem, submitted for publication in Nucl. Phys. A.
8. D. Vautherin and D.M. Brink, Phys. Rev. C5, 626 (1972).
9. C. Mahaux and H.A. Weidenmuller, Shell-model Approach to Nuclear Reactions (North-Holland, Amsterdam, 1969).
10. M. Beiner et al., Nucl. Phys. A238, 29 (1975); and references therein.
11. A.P. Zuker, Phys. Rev. Lett. 23, 983 (1963).
12. B. Buck and J.B. McGrory, Phys. Rev. Lett. 21, 39 (1968).

COLLECTIVE GYROMAGNETIC RATIO FROM DENSITY DEPENDENT HARTREE FOCK (DDHF) CALCULATIONS

I. EVEN NUCLEI

II: ODD NUCLEI

D.W.L. Sprung and S.G. Lie

Department of Physics
McMaster University

And

M. Vallières

Department of Physics and Atmospheric Sciences
Drexel University

The collective gyromagnetic ratio and moment of inertia of deformed even, odd-proton and odd-neutron axially symmetric nuclei have been calculated in the cranking approximation using wave functions obtained with the Skyrme force S-III. Good agreement with experiment is found for g_R, while the moment of inertia is about 20% too small due to the too great spread of the single particle energies produced by the S-III force near the Fermi surface. The cranking formula leads to better results than the projection method (in which one simply takes the expectation value of the relevant operator in the deformed HF ground state, neglecting corrections of relative order $1/\langle \tilde{J}^2 \rangle$). In particular, the cranking results follow nicely the exceptionally large/small g_R for the odd proton/neutron nuclei around mass 153-167.

I. INTRODUCTION

It was shown by Milsson and Prior[1], and also by Griffin and Rich[2], that with the inclusion of pairing forces, the cranking model could give a good account of moments of inertia of deformed rare earth nuclei, and also of their collective gyromagnetic ratio.

Since these early calculations used Nilsson model wave functions, they contain many parameters and some judicious assumptions must be made to secure good results. In recent years, Hartree-Fock (HF) calculations using density-dependent forces have had considerable success in explaining many nuclear ground state properties: binding energy, size, deformation, and electron scattering form factors among others. It is a more severe challenge to the DDHF calculations to reproduce non-static properties such as the moment of inertia and the collective gyromagnetic ratio g_R, to which we turn our attention here.

In a series of papers, Villars[3-6] has discussed the relationship between the collective model and the Hartree-Fock theory. In Villars' theory, states of good angular momentum are projected from a deformed HF intrinsic state. This projection is carried out approximately, with successive corrections proportional to higher powers of $1/\langle J^2\rangle$, where the mean value is taken in the HF ground state. Recently the projection method has been applied by Zaringhalam and Negele[7]) to the calculation of inelastic form factors for electron scattering, taking corrections to first order in $1/\langle J^2\rangle$. Villars[4] showed that the projected Hartree-Fock (PHF) value for the gyromagnetic ratio is

$$g_R^o = \langle \underset{\sim}{M} \cdot \underset{\sim}{J} \rangle / \langle \underset{\sim}{J}^2 \rangle \qquad (1)$$

where

$$\underset{\sim}{M} = \sum_i \left(g_\ell \underset{\sim}{\ell}_i + g_s \underset{\sim}{S}_i \right) \qquad (2)$$

is the magnetic moment operator.

Subsequently Villars and Copper[6] attempted a unified theory of nuclear rotations, based on a transformation to intrinsic and collective coordinates. They were able to show that g_R and the moment of inertia should be calculated from the self-consistent cranking formula of Thouless and Valatin[8]. They argue[6] that eq.(1) corresponds to using the rigid-body value for the moment of inertia, which is known to be quite incorrect for nuclear rotations.

With the aim of seeing how well DDHF calculations can reproduce collective nuclear properties, we have carried out calculations of g_R and $\mathcal{J}$ in both the PHF and cranking approximations.

II. THEORY

In the cranking model[1], one has

$$_{cr}g_R^{cr} = \sum_{\beta \neq \alpha} \langle \Phi_\alpha | M_x | \Phi_\beta \rangle \frac{1}{E_\beta - E_\alpha} \langle \Phi_\beta | J_x | \Phi_\alpha \rangle \tag{3}$$

where $|\Phi_\beta\rangle$ are a complete set of states. In our work, $|\Phi_\alpha\rangle$ is taken to be a Hartree-Fock ground state with a BCS pairing interaction included following Vautherin[9], so each of the self consistent orbitals $|k\rangle$ has an occupation v_k^2. The excited states $|\Phi_\beta\rangle$ will then be particle-hole excitations because J_x is a one-body operator. The cranking moment of inertia is given by a similar expression in which M_x is replaced by J_x. For this reason, we will not write out the parallel formulae for $\mathscr{I}_{cr}$.

Hartree Fock calculations for even-even deformed rare earth nuclei were carried out by Flocard, Quentin and Vautherin[10]. The wave function was required to be invariant under time reversal, and axial symmetry was assumed. The single particle orbitals then occur in pairs with values of $j_z = +\kappa, -\kappa$ to which one assigns the same occupation v_{k}^2. In fact, only the states with $\kappa > 0$ need be calculated, the "time reversed" states with $\kappa < 0$ making an equal contribution to the density. In the case of an odd nucleus, the wave function will not be invariant under time reversal, as the above symmetry between orbital pairs need not hold. It is very convenient however, to preserve it, and we do this by means of the "filling approximation", according to which the odd nucleon is put half into the orbital with $\kappa = \mu > 0$, and half into its conjugate $\bar{\mu}$. The particular orbital μ is selected by looking up the ground state spin and parity of the odd nucleus, and selecting the relevant orbital just above or below the Fermi surface of an adjacent even nucleus.

The wave function of an odd nucleus, in our approximation, is then

$$|\mu\rangle = \frac{1}{\sqrt{2}} (\alpha_\mu^\dagger - \alpha_{\bar{\mu}}^\dagger) |BCS\rangle \tag{4}$$

where $|BCS\rangle$ is the HF-BCS ground state for the even core, but computed self-consistently in the presence of the odd particle. We employ the "blocking" prescription, so that $v_\mu^2 = 0$ when the pairing equations are solved. Since

$$\underset{\sim}{M} = \sum_i [g_\ell \underset{\sim}{j}_i + (g_s - g_\ell) \underset{\sim}{S}_i] \tag{5}$$

($g_\ell = 1, 0$ for protons, neutrons respectively and $g_s = 5.58269$, -3.82630 n.m.), one can rearrange eq.(3) into the frame

$$g_R^{cr} = \frac{\mathcal{J}_{cr}^{p}}{\mathcal{J}_{cr}} + (g_s^p - 1)\frac{W_p}{\mathcal{J}_{cr}} + g_s^n \frac{W_n}{\mathcal{J}_{cr}} \tag{6}$$

where $\mathcal{J}_{cr}^{p}$ is just the proton contribution to $\mathcal{J}_{cr}$ and W_p, W_n are given by similar expressions, but with S_x replacing M_x, and the sum restricted to proton or neutron orbits only. Expressing J_x and S_x in the Hartree-Fock basis leads to[11]

$$W_q = \hat{\sum_{k\ell}}' \langle k|S_+|\ell\rangle^* \langle k|j_+|\ell\rangle (u_k v_\ell - u_\ell v_k)^2/(E_k + E_\ell) \tag{7.a}$$

$$+ \frac{1}{2}\hat{\sum_{k\ell}}'' \langle k|S_+|\bar{\ell}\rangle^* \langle k|j_+|\bar{\ell}\rangle (u_k v_\ell - u_\ell v_k)^2/(E_k + E_\ell) \tag{7.b}$$

$$+ \frac{1}{2}\sum_{k}' \langle k|S_+|\mu\rangle^* \langle k|j_+|\mu\rangle (u_k u_\mu + v_k v_\mu)^2/(E_k - E_\mu) \tag{7.c}$$

$$+ \frac{1}{2}\sum_{\ell}' \langle \mu|S_+|\ell\rangle^* \langle \mu|j_+|\ell\rangle (u_\ell u_\mu + v_\ell v_\mu)^2/(E_\ell - E_\mu) \tag{7.d}$$

$$+ \frac{1}{2}\sum_{k}'' \langle k|S_+|\bar{\mu}\rangle^* \langle k|j_+|\bar{\mu}\rangle (u_k u_\mu + v_k v_\mu)^2/(E_k - E_\mu) \tag{7.e}$$

Here, $\hat{\Sigma}$ means that the sum is to be carried out once including $k,\ell = \mu$, then once omitting $k,\ell = \mu$, and then the results averaged. The primed sum is carried out only over those orbitals with $\kappa > 0$, while the doubly primed sum (7b to 7e) is only for orbitals with $\kappa = \frac{1}{2}$ (because $\bar{\kappa}$ can then be connected to a $\kappa = \frac{1}{2}$ state via j_+).

Here, lines (a,b) represent particle-hole excitations of the even core. Lines (c,d,e) are terms corresponding to the odd particle being excited from the state $|\mu\rangle$ to another state. In the case of an even nucleus, these terms will be absent, and there will be no reference to $|\mu\rangle$ in lines (a,b). In the "projection" method, one arrives at a similar expression

$$g_R^o = \frac{\langle J^2 \rangle_p}{\langle J^2 \rangle} + (g_s^p - 1)\frac{W_p^o}{\langle J^2 \rangle} + g_s^n \frac{W_n^o}{\langle J^2 \rangle} \qquad (8)$$

Here, W_q^o is given by an expression similar to (7) only omitting the energy denominators. Similarly, $\langle J^2 \rangle$ is given by an expression differing from that for $\mathcal{J}_{cr}$ only by the omission of the energy denominators.

III. CALCULATIONS

We first carried out calculations for a number of even rare earth and trans-uranic nuclei[12]. A selection of results are shown in Table 1. From eq. (6) and (7), we expect that $g_R \approx Z/A$, on the grounds that in the leading term, the proton and neutron contributions to $\mathcal{J}$ (or $\langle J^2 \rangle$) should be roughly in the ratio Z:N. What we see is that g_R^o is systematically closer to Z/A than is g_R^{cr}, and that g_R^{cr} lies closer to the experimental value(s) in the majority of cases. At the same time, we noted that $\mathcal{J}_{cr}$ is systematically smaller than the experimental moment of inertia by about 20%. We believe that this is because the Skyrme force S-III gives single particle levels too spread out near the Fermi surface. If we arbitrarily compress the single particle spectrum to correspond to an effective mass $m^* \approx 1$, instead of $m^* \approx 0.75$, then increases to the experimental value[12]. All of the nuclei included in Table 1 are good rotators, as evidenced by $E_{4^+}/E_{2^+} \approx 3.0$-$3.3$, and it is only for such cases that we have found reasonable results.

More recently we have extended our calculations to a number of

TABLE I

Gyromagnetic Ratio Calculated Using Eq.(8) For g_R^{cr} And Eq.(1) For g_R^o

	Z/A	g_R^o	g_R^{cr}	Experimental					
^{152}Sm	0.408	0.431	0.446	0.416(25)	0.277(28)	0.35 (3)	0.33 (6)	0.27 (7)	0.31 (6)
^{158}Gd	0.405	0.402	0.376	0.315(25)	0.385(22)				
^{162}Dy	0.407	0.399	0.370	0.37 (4)	0.362(25)	0.363(24)			
^{166}Er	0.410	0.392	0.338	0.329(27)	0.312(6)	0.305(15)			
^{174}Yb	0.402	0.378	0.306	0.337(7)	0.247(13)	0.338(15)			
^{178}Hf	0.404	0.367	0.299	0.237(14)	0.300(35				
^{182}W	0.407	0.355	0.267	0.266(9)	0.249(24)	0.233(27)			
^{184}W	0.402	0.355	0.271	0.295(10)	0.280(18)	0.275(25)			
^{188}Os	0.404	0.370	0.332	0.305(14)	0.280(21)	0.310(27)			
^{190}Os	0.400	0.372	0.341	0.39 (4)	0.355(29)	0.34 (3)	0.21 (2)	0.331(16)	
^{192}Os	0.396	0.385	0.378	0.41 (6)	0.20 (31)	0.349(18)	0.278(19)	0.39 (3)	0.382(39)
				0.399(18)	0.280(10)				

odd rare earth nuclei[11].. The results are depicted in Fig.1. As before, we find that g_R^{cr} differs more from Z/A than does g_R^o, and agrees much better with experiment. The cases of ^{163}Dy, ^{165}Ho and ^{167}Er are interesting because of the big fluctuations seen both in g_R^{cr} and in experiment.

In comparing neighbouring even and odd nuclei, what we find is that the core terms (7.a,b) change very little. The main difference comes from (7.c,d,e), and that in these sums it is invariably only two states with $\kappa = \mu \pm 1$ which make a significant contribution. Also, it is the change in $\mathscr{I}_{cr}$ (or $<J^2>$) and not W_q (or W_q^o) which is most important, For an odd proton nucleus, a big increase in $\mathscr{I}_{cr}^p$ can raise g_R^{cr} closer to unity, as seen in ^{165}Ho (0.526) compared to ^{166}Er (0.338). For an odd neutron nucleus, a big increase in $\mathscr{I}_{cr}^n$ in the denominator of (6) can similarly lower g_R^{cr} as seen in ^{167}Er (0.090) compared again to ^{166}Er. In order to

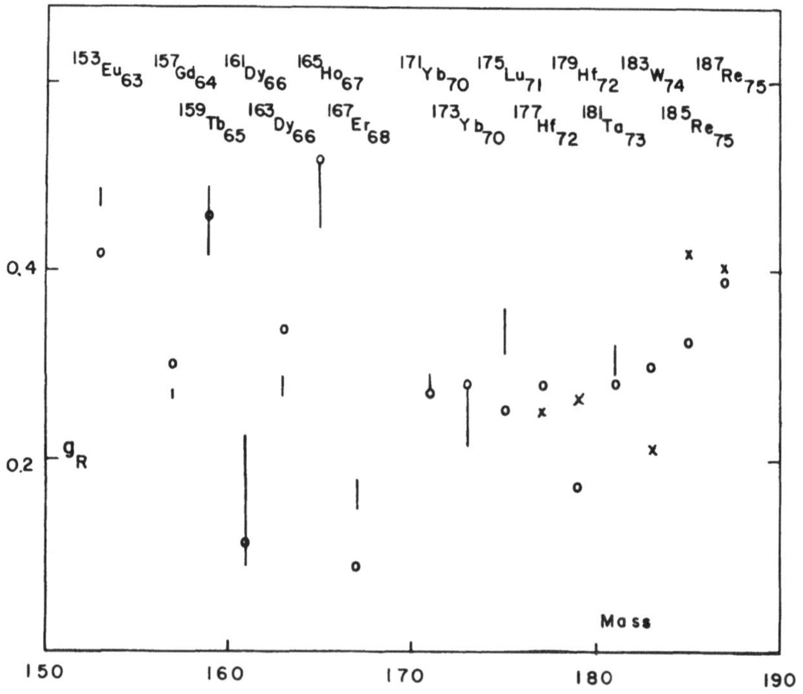

Fig.1. Calculated g_R^{cr} (small o) compared to experiment (small x). Where there are several experimental values, these are joined by a solid bar. No experimental errors are indicated in this figure.

have these large effects, one also requires a small energy difference $E_k - E_\mu$ in (7.c). Typically the important energy differences in this region are of order 2 MeV, but in the two cases cited, one of the strongly coupled states happens to lie within 1 MeV of the odd orbit, which amplifies the contribution to $\mathcal{J}_{cr}$. In calculations of g_R^o, one does not have energy denominators, so this particular enhancing effect on the odd particle contribution cannot occur.

At the same time, we note that $<\underset{\sim}{J}^2>/2_{cr} \sim 2.5$ MeV, so numerically $<J^2>$ is about five times larger than $\mathcal{J}_{cr}$. The odd particle contribution is therefore much less significant when added to $<J^2>$, for g_R^o, than it is added to $\mathcal{J}_{cr}$ for g_R^{cr}. Thus, the projection method is simply unable to reproduce the big fluctuations in g_R which are seen experimentally, but the cranking formula can.

IV. CONCLUSION

HF calculations using the Skyrme force S-III give very good agreement for the collective gyromagnetic ration of well deformed rare earth nuclei. The moments of inertia are about 20% small, which we ascribe to the single particle energy spectrum being too diffuse when the S-III force is used. The agreement here is comparable to or better than in the classic work of Nilsson and Prior, but the present calculation is parameter free in the sense that all the force parameters were adjusted previously to secure agreement of other nuclear properties. The cranking formula employed in this work leads to better agreement than the use of the projection formula g_R^o. Pairing reduces both $\mathcal{J}_{cr}$ and g_R^{cr} from the projected values and is necessary to secure agreement with experiment.

V. ACKNOWLEDGEMENT

The authors are grateful to the Natural Sciences and Engineering Research Council of Canada for continued support under operating grant A-3198.

REFERENCES

1. S.G. Nilsson and O. Prior, Mat. Fys. Medd. Dan. Vid. Selsk. 32, No. 16 (1960).
O. Prior, F. Boehm and S.G. Nilsson, Nucl. Phys. A110 (1968) 257.
2. J.J. Griffin and M. Rich, Phys. Rev. 118 (1960) 850.
3. F.M.H. Villars, in Proc. Int. School of Physics Enrico Fermi, Course 23 (Academic Press, N.Y. 1963) pp.1-47.
4. F.M.H. Villars, Proc. Int. School of Physics Enrico Fermi, Course 36 (Academic Press, N.Y. 1966) pp.1-13; pp.14-42.
5. F.M.H. Villars and N. Schmeing-Rogerson, Ann. of Phys. 63 (1971) 443.
6. F.M.H. Villars and G. Cooper, Ann. of Phys. 56 (1970) 224.

7. A. Zaringhalam and J.W. Negele, Nucl. Phys. A288 (1977) 417.
8. D.J. Thouless and J.G. Valatin, Nucl. Phys. 31 (1962) 211.
9. D. Vautherin, Phys. Rev. C 7 (1973) 296.
10. H. Flocard, P. Quentin and D. Vautherin, Phys. Lett. 46B (1973) 304.
11. D.W.L. Sprung, S.G. Lie and M. Vallieres, Nuclear Physics, A 352 (1981) 19-29.
12. D.W.L. Sprung, S.G. Lie, M. Vallieres and P. Quentin, Nucl. Phys. A326 (1979) 37.

A PHENOMENOLOGICAL STUDY OF NUCLEAR CURRENTS

Ching-Liang Lin

Department of Physics and Astronomy
University of Massachusetts
Amherst, MA 01003

I. ABSTRACT

Nuclear currents are derived by assuming charge conservation and additive property for the nucleon-nucleon interaction inside the nucleus, which are then applied to calculate the magnetic form factor of the plane wave Born approximation. Good agreement with experiment is obtained.

II. INTRODUCTION

In the last two decades, both electromagnetic properties and structure of nuclei have been extensively investigated by elastic and inelastic electron scattering from nuclei [1,2] [see also ref.5]. With the recent availability of high resolution and intensity facilities precise experiemntal information have become obtainable, and mesonic exchange currents are found to play a more significant role in electromagnetic transitions than ever assumed before. However, for incident energy less than 300 MeV, the most important contribution from the nuclear currents is of one-body currents which consist of convection and magnetization currents [1]. We find that those currents can respectively be classified into irrotational current $\vec{J}^{(1)}_{irot.}(X)$ and rotational current $\vec{J}^{(1)}_{rot.}(X)$ defined as follows:

$$\vec{\nabla} \times \vec{J}^{(1)}_{irot.}(X) = 0, \tag{1}$$

$$\vec{\nabla} \cdot \vec{J}^{(1)}_{rot.}(X) = 0, \tag{2}$$

where the superscript means one-body. The source of $\vec{J}^{(1)}_{rot.}(X)$ is clearly the intrinsic spin of the nucleon owing to the fact that the nucleon has its size finite. The existence of that intrinsic freedom of the nucleon and Pauli exclusion principle result in the $(\hat{\ell}.\hat{S})$ - force and exchange character of nucleon-nucleon interaction. For the long-range part of that interaction we can have nuclear potentials in which all models of low-energy nuclear physics are based and therewith the non-relativistic treatments based upon the Schrodinger equation become available. Our main concern is whether those model Hamiltonians used in low-energy nuclear Physics can simultaneously explain the nuclear currents in the long-range interaction.

III. PHENOMENOLOGICAL APPROACH FOR NUCLEAR CURRENTS

We restrict the scattering to discrete nuclear levels so that the Hamiltonian which is applicable to explain wider ranges of experimental results of low-energy nuclear physics are more useful for our purpose to obtain nuclear currents. Furthermore, we assume that the charge conservation exists in nucleon-nucleon interaction currents and the additive property holds out for nucleon-nucleon interaction inside the nucleus. These two assumptions can be combined into one formula:

$$\vec{\nabla} \cdot \hat{\underset{\sim}{J}}^{(n)} + \frac{i}{h} [\hat{H}^{(n)}, \rho] = 0, \tag{3}$$

where n=1,2,3... and $\hat{J}^{(n)}$ is the n-body current operator corresponding to n-body correlation Hamiltonian $\tilde{H}^{(n)}$, thus the currents considered here are purely due to the correlations among nucleons, in spite of the fact that there are currents induced by charged projectile. For the one-body current the central force Hamiltonian, which includes the most useful harmonic oscillator Hamiltonian, is used:

$$\hat{H}^{(1)} = - \frac{h^2}{2M} \nabla^2 + (\text{central force}) + (\hat{\ell}.\hat{S} \text{ force}), \tag{4}$$

where M is the nucleon mass. For simplicity, we can use the product of $(\hat{\ell}.\hat{S})$ operator and $(\hat{\ell}.\hat{S})$-force strength $V_{\ell s}$ as our $\hat{\ell}.\hat{S}$ force, then from eqs.(3) and (4) the obe-body current

$$\hat{J}^{(1)}(\vec{r}) = \rho_o [\frac{ih}{M} \vec{\nabla} + \frac{1}{\hbar} V_{\ell s}(\vec{r} \times \vec{\sigma})]\rho(\vec{r}) \tag{5}$$

is obtained, which includes the Pauli spin operator $\vec{\sigma}$, the charge density $\rho(\vec{r})$ of the nucleons and its normalization constant ρ_o. On the right side of eq.(5), the first term is the conventional convection current which is an irrotational one-body current. The second term is different from the traditional magnetization current, nevertheless it is also originated from the intrinsic behaviour of the nucleon and reflects its rotational character. Furthermore, the anomalous property of nucleon magnetic moment, in other words the pionic exchange effect, as well as the nuclear many-body correlation effect can be incorporated into our $\hat{\underset{\sim}{J}}^{(1)}(\vec{r})$ through a reliable Hartree-Fock potential $V_{\ell s}$.

The two-body currents which are mainly originated from the still ambiguous nuclear forces even at large-range distance, can be obtained from the commonly used exchange forces based upon Pauli exclusion principle. Though the stability of nuclei comes from either the Heisenberg or Majorana forces, the latter is more important than the former for the most stable nucleus ^{4}He [3]. Accordingly we consider only the Majorana force V_{ij} in the derivation of our exchange currents, and further assume that: nucleons are point particles; the superposition principle holds for electromagnetic fields inside the nucleus; and there is no zero-range interaction between two of the same kind of nucleons because of the existence of the hard core, that is:

$$[\hat{V}_{ij}, \rho]\,|P,P\rangle = 0 \tag{6a}$$

$$[\hat{V}_{ij}, \rho]\,|n,n\rangle = 0 \tag{6b}$$

Therefore exchange currents exist only when nucleons exchange charged pions, that is:

$$[\hat{V}_{ij}, \rho]\,|n,P\rangle \neq 0. \tag{6c}$$

The exchange currents are thus derived and can be expressed:

$$\vec{\nabla} \cdot \hat{\underset{\sim}{J}}^{(2)} = -\frac{i\rho_o}{\hbar} \sum_{i>j} V(\vec{r}_{ij};\, \vec{\sigma}_i \cdot \vec{\sigma}) \times \left(\hat{\tau}_+^{(i)}\hat{\tau}_-^{(j)} - \hat{\tau}_-^{(i)}\hat{\tau}_+^{(j)}\right)\left(\delta(\vec{r}-\vec{r}_i) - \delta(\vec{r}-\vec{r}_j)\right) \tag{7a}$$

or

$$\hat{J}^{(2)}\left(\vec{r};\vec{r}_i,\vec{r}_j\right) = \frac{i}{4\pi\hbar}\rho_o \sum_{i>j} V\left(r_{ij};\vec{\sigma}_i\cdot\vec{\sigma}_j\right)$$
$$\times\left(\hat{\tau}_+^{(i)}\hat{\tau}_-^{(j)} - \hat{\tau}_-^{(i)}\hat{\tau}_+^{(j)}\right)$$
$$\times\vec{\nabla}\left(\frac{1}{|\vec{r}-\vec{r}_i|} - \frac{1}{|\vec{r}-\vec{r}_j|}\right)$$

with $\vec{\nabla}V(r_{ij};\vec{\sigma}_i\cdot\vec{\sigma}_j) = 0$, where clearly the $\hat{J}^{(2)}(\vec{r};\vec{r}_i,\vec{r}_j)$ is an irrotational current, and $r_{ij}=|\vec{r}_i-\vec{r}_j|$.

IV. PLAVE WAVE BORN APPROXIMATION

With Lorentz guage and c.g.s. Gaussian unit the transverse electromagnetic multipole operators in the plane wave Born approximation are [1]:

$$\hat{T}^{mag.}_{J\lambda}(|\vec{q}|) = i^J \int d^3X j_J(|\vec{q}||x|)\vec{Y}^{\lambda}_{JJ1}(\Omega_{\vec{x}})\cdot\hat{\underset{\sim}{J}}(\vec{x}) \tag{8}$$

$$\hat{T}^{e\ell.}_{J\lambda}(|\vec{q}|) = \frac{i^J}{|\vec{q}|}\int d^3X[\vec{\nabla}X(j_J(|\vec{q}|\,|\vec{x}|)\vec{Y}^{\lambda}_{JJ1}(\Omega_{\vec{x}}))]\cdot\hat{\underset{\sim}{J}}(\vec{x}) \tag{9}$$

then the form factors related to the differential cross section of a point electron scattered by an unpolarized nucleus to discrete nuclear states are given as follows:

$$\left|F^{mag.}_J(|q|)\right|^2 = \frac{\left|\langle I_f|\,|\frac{1}{C}\hat{T}^{Mag.}_J(|\vec{q}|)|\,|I_i\rangle\right|^2}{2\,I_i + 1} \tag{10}$$

$$\left|F^{e\ell.}_J(|\vec{q}|)\right|^2 = \frac{|\langle I_f|\,|\frac{1}{C}\hat{T}^{e}_J\cdot(|\vec{q}|)|\,|I_i\rangle|^2}{2\,I_i + i} \tag{11}$$

where C is the light speed and all notations in Eqs. (8) - (11) are as usual. It is easy to show that the irrotational current $\underset{\sim}{J}^{(2)}(\vec{r};\vec{r}_i,\vec{r}_j)$ has no contribution either to $F^{e\ell.}_J$ or to $F^{mag.}_J$ because the dynamical part of $\hat{T}^{e\ell.}_{J\lambda}$ and the kinematic part of $\hat{T}^{mag.}_J$ become zero respectively.

V. NUMERICAL CALCULATION FOR M4 TRANSITION TO 19.0 MeV LEVEL OF OF ^{16}O

For simplicity, we use point charge model as usual for $\rho(\vec{r})$ in Eq.(5), and take into account the size effects with the convolution theory. Therefore, we merely multiply the resultant reduced matrix element by $\exp\left(-\frac{|\vec{q}|^2}{4a^2}\right)$ factor with the use of the Gaussian distribution $\rho_{od} \exp[-a^2(\vec{r}-\vec{r}')^2]$ for the charged particle, in the calculation of the form factor for M4 transition to 19.0 MeV state of ^{16}O using stretch model. Since the ground state spin and the 19.0 MeV state spin are respectively 0^+ and 4^+, the main term of the wave function can be generated by 1 $P_{3/2}$-hole and 1 $d_{5/2}$-particle such that we have for reduced matrix element:

$$\begin{aligned} &<[(1d_{5/2})^1(1P_{3/2})^{-1}]4^-| |\hat{T}^{mag.}_{J=4}(|\vec{q}|)| |0^+> \\ &= - <(1d_{5/2})^1| |\hat{T}^{mag.}_4(|\vec{q}|)| |(1P_{3/2})^1>. \end{aligned} \tag{12}$$

If the harmonic oscillator wave function [1] with the size parameter b is used for the single-particle wave function, the above reduced matrix element becomes:

$$\begin{aligned} &<(1d_{5/2})^1| |\hat{T}^{mag.}_4(|\vec{q}|)| |(1P_{3/2})^1> \\ &= \frac{\rho_o V_{\ell s}}{\hbar} \frac{3+\sqrt{2}}{120\sqrt{7\pi}} b^5 q^4 \exp\left[-\left(\frac{b^2}{4}+\frac{1}{4a^2}\right)q^2\right] \end{aligned} \tag{13}$$

where the irrotational current does not contribute because the stretch model entails vanishing kinematic factors. For the comparison of the form factor calculated with that of the experiment [4], the conventional center of mass correction factor [5] is used, and the form factor becomes:

$$\begin{aligned} \left|F^{mag.}_{F=4}(|\vec{q}|)\right|^2 &= \left(\frac{3+\sqrt{2}}{120\sqrt{7\pi}} \frac{\rho_o V_{\ell s}}{\hbar C}\right)^2 \\ &\times b^{10} q^8 \exp\left\{-\frac{1}{2}\left[\left(1-\frac{1}{A}\right)b^2+\frac{1}{a^2}\right]q^2\right\} \end{aligned} \tag{14}$$

where the strength $V_{\ell s}$ = 10.9 MeV is taken from charge density analysis for elastic electron scattering [6], the parameter $[(1-1/A)b^2+1/a^2]=(1.8\ \text{fm})^2$ with $1/a$=0.59 fm [7] is used, and the normalization constant ρ_o is normalized to the total charge of ^{16}O divided by its whole volume. We show the result in the figure.

Figure 1.

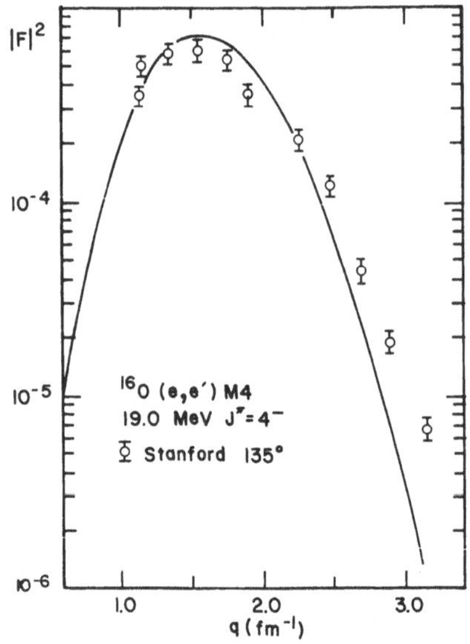

VI. CONCLUSION AND DISCUSSION

Throughout the numerical calculations, we have not adjusted any of the parameters in eq.(14). As shown in the figure the calculated values and experimental results are in good agreement in both magnitude and shape of the form factor except a slight discrepancy in high momentum transfer region. It is our belief that the discrepancy could be due to the existence of a two-body rotational current in that region. Conversely, if the magnetic form factor is known in terms of our model, the $\hat{\ell}.\hat{S}$-force strength can be obtained from electron scattering data.

ACKNOWLEDGEMENT

The author would like to thank the nuclear physics group of the Department of Physics and Astronomy of the University of Massachusetts/ Amherst for providing the library and office support for the work; and to Mrs. Dorthy Pascoe for her excellent typing of this paper.

REFERENCES

+Research supported by National Science Council of Taiwan, China.
++On leave from Department of Physics, National Taiwan University, Taiwan, China.

1. T. deForest and J.D. Walecka; Adv. in Phys. 57 (15) (1966) 1.
2. H. Baier; Fortschritte der Physik 27 (1979) 209
3. E. Segre; Nuclei and Particles (W.A. Benjamin, Inc.) (1977).
4. I. Sick, E.B. Hughes, T.W. Donnelly, J.D. Walecka and G.E. Walker; Phys. Rev. Lett. 23 (1969) 1117.
5. H. Uberall; Electron Scattering from Complex Nuclei, Part A (Academic Press, 1971).
6. W.J. Gerace and G.C. Hamilton; Phys. Lett. 39B(4) (1972) 481.
7. R.S. Hicks; private communication (1980).

STUDY OF DELTA - LOG INTERACTION*

Afsar Abbas

Serin Physics Laboratory, Rutgers University

Piscataway, N.J. 08854

I. ABSTRACT

It is found that small values of σ in zero range Skyrme interaction $-t_o\delta(\vec{r}_1-\vec{r}_2) + t_3\delta^\sigma(R)\delta(\vec{r}_1-\vec{r}_2)$ lead to smaller values for the compression modulus. It would therefore be of interest to study the limiting case $\sigma \to 0$; this leads to a delta - log interaction for binding energy, effective two body interaction for single particle energies and the effective particle-hole interaction. This interaction is used to study the isotope shift.

II. INTRODUCTION

Various values of $\sigma(\leq 1)$ have been used in the zero range Skyrme interaction

$$V = -t_o\ \delta(\vec{r}_1 - \vec{r}_2) + t_3\ \delta^\sigma(R)\ \delta(\vec{r}_1 - \vec{r}_2) \tag{1}$$

The more popular ones being $\sigma = 1$[1] and $\sigma = 2/3$[2]. It has however been shown that there exists a stong correlation between compression modulus in nuclear matter and σ[3]. L. Zamick[4] found that small values of σ lead to better values for the compression modulus. In fact recent experimental[5] results seem to indicate that σ should be significantly smaller than one.

It would therefore be of some interest to study the limiting case $\sigma \to 0$. This when applied to the above interaction leads to

*Supported by the National Science Foundation.

a delta-log interaction, wherein we find that the parameters which give attraction and saturation are explicit functions of average kinetic energy, binding energy and integrations involving densities for a particular nuclei. We study the isotope shifts for ^{16}O and ^{40}Ca using zero range Skyrme interaction with σ going from 1 to 0.

III. ZERO RANGE SKYRME INTERACTION IN THE LIMIT $\sigma \to 0$

When harmonic oscillator wavefunctions are used the total energy for zero range Skyrme interaction is given as

$$E = \frac{A}{b^2} - \frac{B}{b^3} + \frac{C}{b^{3+3\sigma}} \tag{2}$$

where $b = \sqrt{h/m\omega}$ and A, B, C are functions independent of b. The first term is <T>, the mean kinetic energy. The total energy for N = Z nucleus is also given as

$$E = \frac{h^2}{2m}\int \tau d\vec{r} - \frac{3}{8} t_o \int \rho^2 d\vec{r} + \frac{3}{8} t_3 \int \rho^{2+\sigma} d\vec{r} \tag{3}$$

On putting $dE/db = 0$ and comparing one gets t_o and t_3 as explicit functions of <T>, E, σ, $\int \rho^2 d\vec{r}$ and $\int \delta^{2+\sigma} d\vec{r}$.

Changing the density ρ to $\rho + \delta\rho$ in E above and sorting out the terms for single particle energies[6] one gets functionals like $\int \rho^{1+\sigma}\delta\rho d\vec{r} / \int \rho^{2+\sigma} d\vec{r}$ whose limit $\sigma \to 0$ is taken by expanding them around $\sigma = 0$ in a Taylor series. From here one finds that potential for single particle energy looks like

$$U(r) = T + A\rho(R) + B\rho(R)\ell n \frac{\rho(R)}{\rho(0)} \tag{4}$$

where A and B are functions of <T>, E and integrals $\int \rho^2 \ell_n\{\rho(R)/\rho(o)\} d\vec{R}$ and $\int \rho^2(R) d\vec{R}$. Sor far we have just tried to show that in the limit $\sigma \to 0$ the potential looks like a delta-log interaction. To get the parameters for the repulsive and the attractive parts explicitly let us write the interaction for determining the binding energy in the form

$$V = - J\delta(\vec{r}_1 - \vec{r}_2) + K\delta(\vec{r}_1 - \vec{r}_2)\ell n\{\rho(R)/\rho(0)\} \tag{5}$$

Self consistency with the binding energy gives J and K. When ρ is replaced by $\rho + \delta\rho$ in the total energy and terms which correspond to single particle energies and particle-hole interactions are collected together, one finds that the effective two body interaction for single particle energies V_{SP} and the effective particle-hole interaction V_{PH} becomes

$$V_{SP} = -(J - K/2)\delta(\vec{r}_1 - \vec{r}_2) + K\rho(R)\ell n\{\rho(R)/\rho(0)\} \tag{6}$$

$$V_{PH} = -(J - 3K/2)\delta(\vec{r}_1 - \vec{r}_2) + K\delta(\vec{r}_1 - \vec{r}_2)\ell n\{\rho(R)/\rho(0) \tag{7}$$

where

$$K = (8/9)(<T> - 3E)/\int \rho^2(R)d\vec{R} \tag{8}$$

$$J = (8/3)(<T> - E)/\int \rho^2(R) + KC\int \rho^2(R)\ell n\{\rho(R)/\rho(0)\}d\vec{R}/\int \rho^2(R)d\vec{R} \tag{9}$$

IV. RESULTS

The parameters J and K which determine the repulsive and attractive parts of the delat-log interaction are shown in Table 1.

Calculations have been done in the R.P.A. formalism with two different choices of single particle energies SP I and SP II [7]

SP I: Single particle potential energies are calculated using the particular zero range Skyrme interaction or the delta-log interaction while the single particle kinetic energies are given by the harmonic oscillator values.

SP II: Single particle energies are those of the degenerate harmonic oscillator.

The compression modulus κ is given by the expression[4]

$$\kappa = \frac{1}{A}[<T> + 9E_B + \sigma(3<T> + 9E_B)] . \tag{10}$$

We take $<T>/A = 18$ MeV and $E/A = 8$ MeV as the mean kinetic energy and the binding energy per particle respectively. We get

$$\kappa = 90 + 126\sigma \text{ MeV}$$

TABLE I

J And K For A Few Nuclei

A	ν	J(MeV-fm^3)	K(Mev-fm^3)
4	0.450	423.891	423.481
16	0.323	626.474	405.448
40	0.256	415.518	414.851
80	0.215	541.437	413.505

And Isotope TABLE II O And Ca

κ And Isotope Shifts For ^{16}O And ^{40}Ca

σ	κ	ISOTOPE SHIFT			
		^{16}O		^{40}Ca	
		SP I	SP II	SP I	SP II
0(delta log)	90	0.249	0.249	0.319	0.365
0.2	115.2	0.229	0.231	0.297	0.339
0.4	140.4	0.214	0.219	0.273	0.315
0.6	165.6	0.202	0.208	0.254	0.294
0.8	190.8	0.192	0.199	0.238	0.275
1.0	216.0	0.183	0.191	0.221	0.256

and isotope shifts for various values of σ and delta-log interaction are given for two nuclei ^{16}O and ^{40}Ca in Table II.

The isotope shift for a pure delta function has the wrong sing. Density dependence leads to a correct sign for the isotope shift. We found the correct sign for the isotope shift with our delta-log interaction too. However one notes the undesirable feature of the magnitude of the isotope shift increasing as one goes to lower values of σ. By and large it appears that there is no drastic change as one goes from the zero range Skyrme interaction to the delta-log interaction.

V. ACKNOWLEDGEMENT

I would like to thank Dr. Larry Zamick for stimulating conversation and helpful comments.

REFERENCES

1. D. Vautherim and D.M. Brink, Phys. Rev., C5 (1972) 626.
2. S.A. Moszkowski, Phys. Rev., C2 (1970) 402.
3. X. Campi and D.W.L. Sprung, Nucl. Phys., A194 (1972) 401.
4. L. Zamick, Phys. Lett., 45B (1973) 313,
   L. Zamick, Nucl. Phys. A249 (1975) 63.
5. P. Quentin and H. Flocard, Ann. Rev. of Nucl. Part. Sc,, 28 (1978) 523.
6. R. Sharp and L. Zamick, Nucl. Phys., A208 (1973) 130.
7. A. Abbas and L. Zamick, Phys. Rev., C (1980).

OVERVIEW OF EXPERIMENTAL TESTS OF THE INTERACTING BOSON APPROXIMATION (IBA)

R.F. Casten

Brookhaven National Laboratory
Upton, New York 11973

I. ABSTRACT

A broad overview is presented of the principal tests to date of the IBA, for both even and odd nuclei, and including the evidence concerning the possible existence of the recently proposed supersymmetries. Although some details are presented, the aim is rather to survey the principal characteristics of the model, to assess the overall quality and extent of the agreement with experiment, and to indicate where further testing would be most useful.

II. INTRODUCTION

Within the last year, the author has presented two reviews of the experimental status of the IBA. The first surveyed both even and odd mass nuclei. The second was centered on a more thorough look at even nuclei. Although written only a few months after the first, its content was substantially different, including, for example, discussions of the greatly expanded testing of the model, in the interim, in the Pt-Os-W region, in particular with IBA-2 calculations, the first consequential tests involving high spin states and quadrupole moments, and the emergence and implications of the first significant discrepancies with model predictions. The present review will not repeat these detailed discussions except insofar as they are useful for establishing an overview of where the IBA currently stands vis a vis a confrontation with experiment. The emphasis here will be on the relation between the symmetry structure of the IBA and experimental tests of the model, on the overall quality of agreement, on the parameters of the model and their variations, and on those areas where the potential for future tests is abundant.

III. THE IBA AND OTHER COLLECTIVE MODELS

The shell model, with residual interactions, is generally considered, if not an ideal, at least the best existing approach to the nuclear structure problem. Unfortunately, as Talmi[3] has repeatedly emphasized, it is an impossible approach for most nuclei where the shell model space is simply too large for practical calculations. Furthermore, even if such calculations were feasible, it would be virtually impossible either to identify the special states that are collective or to perceive simplicities and regularities in the enormously large number of wave function components. Therefore, much of the history of nuclear structure physics, especially in heavy nuclei, can be viewed as an attempt to simplify the problem by truncating the shell model space or by constructing alternate schemes that emphasize specific degrees of freedom.

Naturally, each such approach has limitations, most obviously in that each selects out a subset of states and/or nuclei for emphasis and therefore those that are not so encompassed are treated poorly if at all. The most well known, extensively developed, and highly successful scheme has clearly been the collective model of Bohr and Mottelson[4] along with its numerous offshoots, including the extension to odd-mass nuclei in the Nilsson[5] model. Developed as an attempt to interpret the collective excitations of deformed nuclei, it has had enormous success both in correlating empirical information and as a predictive tool as well and has become the paradigm against which new models are judged. Though originally proposed as a phenomenological model, it has a substantial microscopic foundation, primarily centering on RPA-type calculations in the framework of a pairing plus quadrupole Hamiltonian. The most successful microscopic application of the Bohr-Mottelson Hamiltonian has probably been the calculations of Kumar and Baranger[6] or the related outgrowth of these, the Dynamic Deformation Theory[6] which succeeded in predicting many properties of the lowest collective states in extended series of nuclei, including both the difficult Sm and Pt-Os transition regions. Another recent microscopic approach to collective excitations has been the boson expansion calculations of Tamura and co-workers[7].

Despite the considerable successes of collective models such as these, there has nevertheless remained a somewhat ad hoc character in their use: that is, for each different type of nucleus (vibrational, transitional, quadrupole symmetric, triaxial, etc.) a different phenomenological model is applied,or, if a microscopic approach is used, its success has generally not been uniform across broad series of nuclei: for example, the boson expansion models often converge much more poorly for deformed than for spherical nuclei[7].

It was in the spirit of attempting to overcome this limitation that the IBA was developed[8-13] and it is in this respect that it has achieved its greatest successes. It is an attempt to obtain an extreme simplification of the nuclear structive problem by grossly truncating the shell model space so as to lead to a simple boson Hamiltonian that is applicable with comparable quality to broad regions of the periodic table, encompassing nuclei of widely different types, and which automatically yields a comprehensive set of low lying collective states. The model is phrased in a rather abstract boson space and involves rather sophisticated group theoretic ideas. As a result the simplicity and broadness of coverage is gained at the expense of a simple geometrical interpretation of the resultant predictions, although recent work,[14-16] has clarified the correspondence between the IBA Hamiltonian and that of the Bohr-Mottelson collective model. This point is raised because it highlights the fact that the IBA should not be considered as a replacement for geometrical models but rather as complementary to them. That is, the primary appeal of the IBA is often not to be found in its predictions for a given nucleus but in its systematic treatment of extensive sequences of nuclei within a single unified framework. Indeed, since any nucleus must have an approximate classical shape, a geometrical model incorporating that shape should account at least for the broad characteristics of the collective excitations. There are two respects, however, in which the IBA can be preferable even in this respect. First, it automatically includes the fact that these are a finite number of particles outside closed shells, whereas geometrical models implicitly assume an infinite dimensional space. Secondly, the IBA automatically includes full sequences of intrinsic excitations, including some of their mutual interactions, whereas the higher lying intrinsic collective excitations in geometrical models often have to be introudced explicitly and their interactions parameterized (e.g., band mixing).

IV. THE STRUCTURE AND SYMMETRIES OF THE IBA

The following discussion is based on an extensive series of theoretical papers on the IBA cited in refs.8-13 as well as others referenced in refs.1 and 2. The basic structure of the IBA is obtained by writing a boson Hamiltonian involving two types of bosons carrying angular momentum zero (s bosons) and two (d bosons). In a given nucleus there is a fixed total number of bosons, N, equal to the sum of the number of s bosons (n_s) and d bosons (n_d). N is simply given by one half the number of protons and neutrons outside their respective nearest closed shells. In the earlier and conceptually simpler of the two IBA formulations (IBA-1), no distinction is made between proton and neutron bosons or between particle or hole bosons.

In somewhat schematic form the IBA-1 Hamiltonian may be

$$H = \varepsilon \sum_m d_m^+ d_m + \frac{1}{2} \sum_{L=0,2,4} \sqrt{2L+1}\, C_L \left[(d^+ d^+)^L (\tilde{d}\tilde{d})^L \right]^0 \qquad (1)$$

$$+ \text{(terms that change } n_d)$$

where d^+ (or d) creates (or destroys) a d-boson, $\varepsilon = \varepsilon_d - \varepsilon_s$ is the d-boson energy relative to that of an s-boson (which latter is aribtrarily set to zero), and the C_L and the (unwritten) coefficients of the d boson number changing terms are other parameters. The superscripts denote angular momentum coupling.

In IBA-1, the basis states are those of the extreme vibrational limit and are characterized as

$$|n_d, n_\beta, n_\Delta, L>$$

where $n_\beta(n_\Delta)$ is the number of pairs (triplets) of d-bosons coupled to zero spin and L is the total angular momentum. In the vibrational limit the ground state quantum numbers $(n_d, n_\beta, n_\Delta) = (0,0,0)$, the first excited, 2^+ state, is (1,0,0), the 4^+ and 2^+ members of the "two phonon" triplet are characterized by identical quantum numbers (2,0,0) but with the two d-bosons coupled to spins 4 and 2, respectively. The 0^+ member of this triplet, of course, has $n_d=2$ and thus $(n_d, n_\beta, n_\Delta) = (2,1,0)$. The construction of the higher states is similar.

In Eq.(1), the first term simply counts the number of d-bosons and multiplies by the energy of each. It alone would produce a harmonic vibrator spectrum. The second term does not affect the wave functions (the operator is diagonal) but gives an anharmonic vibrator with broken degeneracies. The last terms of Eq.(1) can be either one (e.g., d^+d^+ds) or two (e.g., d^+d^+ss) d-boson number changing; they greatly complicate the wave functions and are essential in all but the simplest of nuclear species. Different relative sizes of the different terms lead to radically different spectra. While the detailed structure of the last term in Eq.(1) is not particularly complicated, it is relatively unenlightening. It is possible, and generally preferable, to rewrite Eq.(1) in terms of specific combinations of these terms that correspond to characteristic boson multipole operators. Thus, one can write a completely equivalent Hamiltonian as[12,17]

$$H = \varepsilon n_d + \kappa Q\cdot Q + \kappa' L\cdot L + \kappa'' P\cdot P + T_3[(d^+\tilde{d})^3 (d^+\tilde{d})^3]_0^0$$
$$+ T_4[(d^+\tilde{d})^4 (d^+\tilde{d})^4]_0^0 \qquad (2)$$

The coefficient ε, κ, κ', κ'', T_3 and T_4 are parameters. Here the boson operators Q, L, and P correspond to quadrupole, spin-dependent and pairing interboson interactions. For example, and for the future discussion, the expression for Q is

$$Q = (d^+s + s^+\tilde{d})^{(2)} + \chi(d^+\tilde{d})^{(2)} \quad (3)$$

The coefficient χ is important and will be discussed below. In cartain cases it has well defined values, in others it can be treated as a parameter.

The structure of Eq.(2) is particularly instructive and closely related to one of the most important characteristics of the IBA-1, its predictions of specific limiting coupling schemes corresponding to characteristic symmetries wherein level energies and transition rates (reaction or electromagnetic) can be described by analytic expressions and obey specific quantum number selection rules. The symmetries arise from the inherent group structure associated with the six dimensional space spanned by the five components of the d-boson and the single s-boson. The group decomposition of this space has three non-trivial subgroups, called SU(5), SU(3) and O(6). The first two are already well known and correspond to the familiar vibrator and symmetric rotor limits, respectively. The O(6) limit was previously unrecognized, and indeed the almost simultaneous prediction of it and its empirical discovery[18] was one of the earliest triumphs of the IBA. It has been shown[18,19] to correspond to the γ-unstable picture. The SU(5) limit is best thought of in terms of Eq.(1). In its harmonic form, with degenerate sets of levels, it corresponds to the first, or ε, term. Anharmonicities can be introduced, within the SU(5) limit, by non-zero C_L coefficients. The SU(3) or rotor limit is characterized in Eq.(2), not surprisingly, by the Q•Q term although κ', the coefficient of L•L, can also be non-zero. The O(6) limit corresponds to finite values of the T_3 and κ'' terms in Eq.(2); it is the former which gives the limit its characteristic energy spacings and wave function regularities within each collective family of states and the latter which determines the relative energies of each family. Thus, to recapitulate each symmetry corresponds to the dominance of certain characteristic terms in Eqs.(1) and (2). These are: ε:SU(5), κ:SU(3), T_3:O(6). Each of the symmetries has been discussed in detail in the literature. In Fig.1 we summarize the low lying levels and the strong transitions in each. Several points are worth noting. First, the SU(3) limit corresponds to a very specific type of rotor, namely one in which levels of the same spin in the β and γ bands are degenerate. Higher bands also have characteristic degeneracies. Thus, it tends to correspond more to the early part of a deformed region than to the middle where the deformations might actually be larger. Secondly, the E2 transition rates and transfer

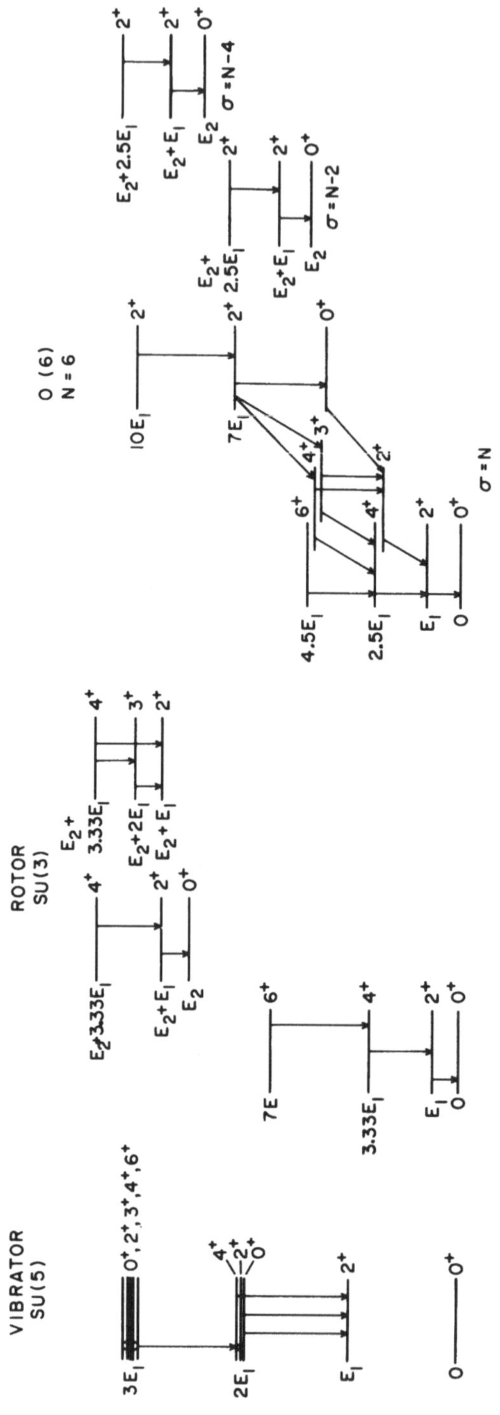

Fig.1 Low lying levels and connecting strong E2 transitions for the three limiting symmetries.

reaction cross sections are not identical in the IBA and geometrical descriptions. For transitions allowed in both treatments, the transition rates in the IBA approach those of the geometrical models in the infinite dimensional limit: for finite N, they are generally given by the geometrical value modulated by a function of the boson number N. There are, however, some actual and highly important differences in the selection rules. The most notable occurs in the SU(3) limit: in the simplest phenomenological geometrical description the levels of the β and γ bands deexcite by moderately enhanced transitions to the ground band but transitions between β and γ bands are forbidden since they involve a change of two vibrational quanta. In the IBA, on the other hand, the states in each limit are associated into families according to their group representations. A fundamental E2 selection rule is that transitions (induced by operators which are generators of the group) between representations are forbidden. Thus, in SU(3) for example, the β and γ bands belong to the same representation but to a different one than the ground band. Thus there are allowed (albeit typically weak) transitions (not shown in Fig.1) between β and γ bands but in the strict SU(3) limit not from either band to the ground band.

The third limit in Fig.1 is the O(6) limit. It corresponds, in the infinite N limit, to the γ-unstable model of Wilets and Jean. The levels are grouped into principal families (representations) according to a quantum number σ. Within a representation the levels are classified by quantum numbers τ (similar to a phonon quantum number) and ν_Δ (which is related to the number of triplets of d-bosons coupled to zero spin). As with SU(3), the inter-representation transitions are forbidden: this differs from the γ-unstable model with β vibrational excitations but, again, as with SU(3), this selection rule can be broken by small perturbations to the limit.

It is interesting to note that in both SU(3) and O(6) the forbiddenness of inter-representation ($\Delta(\lambda, \mu)$ or $\Delta\sigma \neq 0$) transitions arises from a subtle cancellation. The operator Q in Eq.(2) (which is the same as the E2 transition operator) is taken to be composed of generators of the relevant group. For each limit the proper choice of χ assures this. Thus $(d^+\tilde{d})$ is not a generator of O(6), and so $\chi = 0$ for this limit. In SU(3) $\chi = -\sqrt{35}/2 = -2.958$. In O(6), the $\Delta\sigma \neq 0$ transition rates are zero due to a cancellation in the terms arising from the different wave function amplitudes. Therefore, a miniscule change of these amplitudes, without introducing d-boson number changing perturbations, will destroy the cancellation. In SU(3) the forbiddenness arises from the perfect cancellation of the contributions arising from each of the two terms in the transition operator Q. Therefore $\beta \to g$ or $\gamma \to g$ transitions can arise either by perturbing the wave functions or the operator. In actual calculations, however, it turns out that relatively large symmetry breaking in the Hamiltonian has much less effect on the transition rates than changes in the value of χ. Due

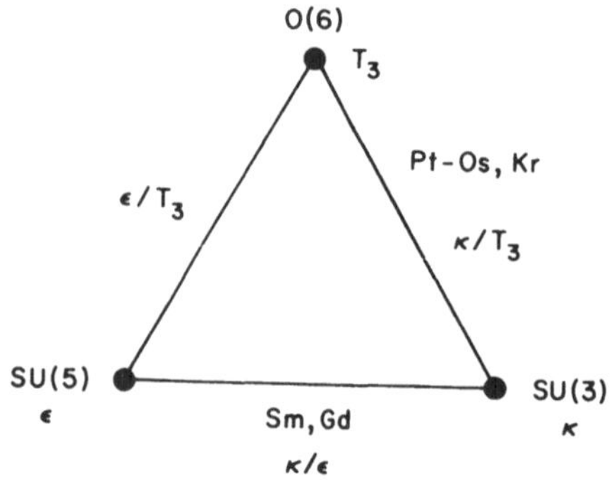

Fig.2 Symmetry triangle representing the three IBA symmetries, the transition legs between them the corresponding coefficients of the IBA-1 Hamiltonian of Eq.(2), and some nuclei representing different cases.

to this and to the fact that χ is only rigorously defined in the limiting symmetries, an interesting and as yet incompletely resolved question is the bahavior of χ between symmetries. Clearly, from the above comments, its value will have a dominant effect on the calculated electromagnetic properties of a given nucleus. Alternatively, it can easily be fixed from the ratio of inter- to intra-representation transition rates.

Much of the simplicity of the IBA can be illustrated by the symmetry triangle pictured in Fig.2. Each vertex corresponds, as indicated, to a given symmetry and the relevant coefficient is indicated. One of the most powerful and appealing aspects of the IBA is the simplicity with which it can treat what have historically been among the most theoretically difficult and complex sequences of nuclei, the transition regions between the extreme limiting coupling schemes. In the IBA, these correspond to the legs of the triangle and can be specified in terms of a single parameter, namely the ratio of the coefficients describing the two related vertices. The ratio, in effect, specifies the relative position along the leg. Thus, for example, an O(6) $\rightarrow$ rotor (SU(3)) transition region (e.g., W-Os-Pt) can be specified by a monotonically changing value of κ/T_3 which can be considered as specifying, in a crudely phrased sense, the relative "SU(3)-ness" and "O(6)-ness" of a nucleus. The SU(5) $\rightarrow$ SU(3) transition (e.g., Sm isotopes) would be described in terms of the parameter κ/ε. It is important to remark also here that the generality of the IBA, that is, its potential applicability to any nucleus not at or adjacent to closed shells, means that one can also calculate IBA predictions for any other point in our out of the triangle and that a series of nuclei with changing properties need not follow a path along one of the legs. Further, it is significant that a sequence of nuclei with varying properties may also arise at a constant point in the triangle by virtue of the changes, with nucleus, of boson number N. It might be noted that while the symmetry triangle was originally introduced for pedagogical reasons it has recently been treated more formally, by Gilmore and Feng[16], who has constructed various surfaces, inside a three dimensional extension of it, which specify the regions of dominance of each symmetry and help to elucidate the character of the phase transitions between them.

The discussion so far has concerned the so-called IBA-1. In an attempt to refine the model and to bring it closer to a microscopic basis, the IBA-2 was developed. It differs principally in that it explicitly distinguishes proton and neutron bosons and thus leads to the possibility that the parameters describing each (neutron and proton) part of the Hamiltonian will vary in a systematic and a priori predictable way throughout a (neutron or proton) shell. It therefore implicitly enables a distinction between particle and hole bosons as well.

The Hamiltonian for the IBA-2 is simply

$$H = H_\pi + H_\nu + h_{\pi\nu}$$

where the notation is obvious. Clearly, this can, in general, be rather complicated. A considerable simplification has been proposed[11-13] and used in nearly all tests of the IBA-2 and is, in practical terms, what is usually considered to be the IBA-2. In this approximation

$$H = \varepsilon(n_{d_\pi} + N_{d_\nu}) - \kappa Q_\pi \cdot Q_\nu + V \qquad (4)$$

where V is a Majorana term and where it has been assumed that $\varepsilon_\pi = \varepsilon_\nu$ and the $Q_\pi(Q_\nu)$ operators are defined analogously to the operator Q of Eq.(3) with coefficient $\chi_\pi(\chi_\nu)$. The only modification of Eq.(4) that is occasionally used is the addition of a diagonal term of the form $\frac{1}{2}\sum_L \sqrt{2L+1}\, C_L (d^+d^+)^L (\tilde{d}\tilde{d})^L$ for either neutron or protons which is sometimes useful for tuning the level energies. The Majorana term has little effect on existing calculations and will not be discussed furhter.

It is clear from Eq.(4) and the discussion above therefore that, that, except for the fine tuning obtainable by introducing the C_L parameters, there are four parameters in the model: ε, κ, χ_π and χ_ν. If one assumes that a given shell can be approximated by a single j shell of appropriate (large) degeneracy, then it is possible to derive the expected behavior of these parameters (see Eqs.(3.32), (3.33) of ref.13). It turns out that ε and κ are approximately constant within a major shell and that χ_π (or χ_ν) varies in a smooth, nearly linear, way from negative to positive values, passing through zero at mid-shell.

The IBA-2, therefore, though more microscopic and presumably superior to IBA-1, actually contains fewer free parameters: once a couple of nuclei within a major shell are treated, and values of ε and κ are determined, they are esentially fixed for the entire shell. Once the $\chi_\pi(\chi_\nu)$ values are found for a given series of isotopes (isotones) they are applicable to any other nuclei with the same valence number of neutrons (protons).

V. TESTS OF THE IBA

1. General Survey

The experimental tests of the IBA to date have thus far centered in several well defined areas, the symmetries, the transition legs between them, high spin states, two nucleon transfer reactions, and odd mass nuclei and the question of supersymmetries. Many of the most extensive tests are schematically summarized in Table I and, where applicable, labelled according to the relevant aspect of the symmetry triangle involved. A quick scan of the Table reveals some interesting aspects. First, only one of the three symmetries is listed, the O(6) limit, although vibrator and rotor nuclei are more familiar. The reason is partly an accident of historical timing (detailed calculations for near SU(3) nuclei are underway) and partly due to basic differences between the limits. For the vibrator, the differences between the geometrical model and the IBA are, except for the aforementioned allowed (but low energy) β-γ transitions in the IBA (which for a near SU(3) nucleus would be almost impossible to detect empirically), the differences with the IBA do not become substantial until multi-β and multi-γ vibrational levels are reached. Unfortunately, in most cases, either these have hardly been studied in sufficient detail or they occur near or above the pairing gap where admixtures with quasiparticle levels are likely and may prevent meaningful comparisons. In contrast, in the O(6) limit, the states belonging to higher representations may occur in principle at any energy and, in practice, in the Pt isotopes, happen to occur well below the pairing gap so that a detailed study has been possible. Furthermore, to carry out a detailed and comprehensive test of a model like the IBA that aims at an interpretation of complete sequences of families of levels requires a correspondingly thorough experimental technique that discloses all low lying levels and their decay properties. In practice, the (n, γ) reaction is perhaps the only one that can assure such generality

TABEL I

Survey Of Some Tests Of The IBA

Topic: General Category	Topic: Specific Type	Nuclei	References
Symmetry	O(6)	^{196}Pt	18,19,20
Transition Legs	O(6)→SU(3)	Pt-Os-W	21-23
		Kr	24
	SU(5)→SU(3)	Sm	9,13
	(O(6)→SU(5))	Ba,Ce,Xe	13,25
Deformed (near-SU(3)) nuclei	Detailed	^{168}Er	26,27
	Collective vibrational B(E2) values	several Gd, Dy nuclei	28
EO transitions	Survey	Many	29
	O(6) nuclei	Pt,Os,W	22,23,30
Transfer Reactions	(p,t), (t,p)	Os,Pt	31,32
		W	33
		Sm	9,34
High Spin	Large N Deformed Nuclei	Dy	35
		Actinides	36
	Small N (near-O(6)) nuclei	Kr	37,38
		Pt	20,39
Other		Pd,Ru,Se, (p,p'), etc.	Many (see refs. 1,2)
Odd Mass Nuclei	single-j	Pd	40
		Rb	41
		Eu	42
		Ag, Tc	43
	multi-j	Rb	44
		Eu	13,42,45,46
	supersymmetry	Ir,Au	47,48

and it is only recently that thorough studies of near-SU(3) nuclei have been completed.

Another interesting aspect of the Table is the preponderance of tests in the W-Os-Pt region. Again, there are several reasons. For one, the vibrator-rotor transition in the N = 90 region has long been considered relatively well understood, in contrast to the extremely complex Os-Pt region. Secondly, the IBA description of SU(5) → SU(3) transition is reasonably close to a geometrical one. Thirdly, the discovery of the O(6) limit in Pt immediately provided[18] a new benchmark for this heretofore difficult region, that is, a clear cut limiting symmetry at one end and the possibility of treating the region more simply in terms of relatively small and smoothly changing perturbations to this newly recognized limit.

Finally, the O(6) → SU(5) transition is noted in parentheses, for two reasons. First, the transition seems less clearcut theoretically than the others. Some of the Ba, Ca and Xe nuclei appear to approach an SU(3)-like structure enroute. Secondly, the data are extremely sparce and so, at present, the study referenced is less an experimental test of the IBA than an extensive set of IBA predictions that already exist and still await confrontation with data.

The following sections will contain brief summaries of some of the main tests of the IBA. The reader is referred to refs.1 and 2 and the original literature for more details.

2. Specific Studies: Even Nuclei

a. Symmetries and transition legs

Although the IBA was proposed in 1974 its most important early achievement was the prediction of the O(6) limit as a new, hitherto unrecognized, symmetry. This occurred essentially simultaneously with its empirical discovery[18] in ^{196}Pt following a series of (n, γ) studies that disclosed all low spin levels below the pairing gap in ^{196}Pt and showed that there was a one-to-one correspondence between the predicted levels and E2 transitions and the empirical findings.

With this limit as an anchor point, it was immediately clear that a new interpretation of the Pt→Os→W region, as undergoing an O(6) → SU(3) transition, was appealing and, from a casual inspection of several branching ratios, qualitatively correct. Detailed IBA-1 calculations were carried out[21] for this leg of the symmetry triangle, as a function of essentially only the one smoothly varying parameter κ/T_3. The calculations established this conjecture and exhibited remarkable agreement with experiment, for a very large number of absolute B(E2) values and branching ratios in $^{186-194}$Os, $^{188-196}$Pt.

These were soon followed by IBA-2 calculations[22,23] of Os and Pt and of W. One expects, a priori, for near O(6) nuclei (where $\chi_\pi \simeq -\chi_\nu$), that there will be little difference between IBA-1 and IBA-2 calculations for electromagnetic transition rates but that substantial differences will occur for level energies and quadrupole moments. On account of this expectation the IBA-1 comparisons of refs.18 and 21 for the O(6) limit and O(6) $\rightarrow$ SU(3) calculations paid little attention to these latter quantities, where indeed there were substantial discrepancies, and concentrated on the E2 properties of the Pt-Os nuclei, where excellent agreement was found. Thus it is particularly satisfying to note that, in the IBA-2 calculations for Pt-Os (ref.22), the B(E2) values changed little from the IBA-1 results, that these differences did grow, as expected, with deviations from the O(6) limt, and that the level energies and quadrupole moments changed substantially and in the direction of the experimental results[49]. Overall, the agreement for both Pt-Os and W is more than satisfactory[22,23]. There are only two discrepancies of importance. One centers on ^{194}Os where the IBA-2 cannot reproduce (without a sudden change in parameters) the empirically indicated[50] break in the systematic trend toward smaller prolate deformations in Os, wherein ^{194}Os appears to jump to a larger deformation than ^{192}Os and is presumably oblate. The other discrepancy is that the IBA calculations underpredict the quadrupole moments deduced from Coulomb excitation experiments in Pt and, indeed, fare poorly in this regard compared to asymetric rotor, pairing plus quadrupole, and boson expansion technique calculations (see ref. 49).

It is instructive to consider the parameters deduced in these studies in more detail. They are plotted in Fig.3 for W along with the theoretical χ_ν values for a single j shell and the χ_ν values for Pt-Os. The striking similarity in the χ_ν values obtained in the independent Pt-Os and W studies suggest that χ_ν (and by implication, χ_π) is a global parameter which depends almost solely on neutron (proton) number. While this is theoretically satisfying, it is also of practical importance since it greatly facilitates extrapolation of parameters to unstudied (e.g., far from stability) nuclei. Secondly, the empirical χ_ν parameters are rather similar to the theoretical ones in magnitude and trend but clearly display deviations, namely a faster falloff toward midshell. This can be easily interpreted[23]. Since a given shell is actually composed of several j subshells the behavior of χ_ν should be a series of negative-to-positive trending $\chi_\nu(j)$ functions, each with a steeper slope reflecting the lower degeneracy of an individual orbit or subshell compared to the complete shell. The overall behavior of χ_ν would then be a combination of several linked S-shaped segments. This appears to be exactly what is happening in the extracted W parameters where the subgroups $f_{7/2}$-$h_{9/2}$, $i_{13/2}$, and $p_{3/2}$-$f_{5/2}$-$p_{1/2}$ are partly detached from each other.

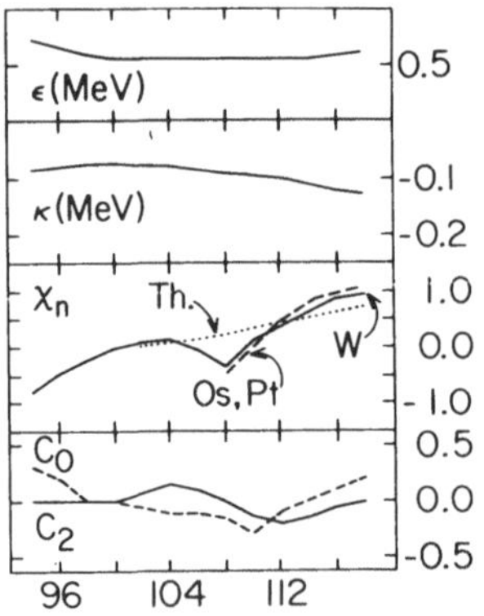

Fig.3 IBA-2 parameters for W deduced in ref.23. The χ_ν parameters for Pt-Os from ref.22 and the theoretical single j shell prediction[12,13] for χ_ν are also shown.

The discrepancies, mentioned before, concerning ^{194}Os and the quadrupole moments in Pt can now be viewed in this light. In both cases an IBA description corresponding to a larger deformation would improve the agreement. But this would imply, in the context of Fig.3, more negative values of χ_ν near neutron number 118 and therefore a substantial break in the χ_ν systematics. The authors of ref.22 chose not to permit this: if it is in fact the case, though, it could imply a greater sensitivity of χ_ν to subshell effects than hitherto believed. This is clearly an area for considerable further study which might indeed lead to a better understanding of the relationship of the IBA-2 to its underlying shell model foundations.

The Kr isotopes have been shown[24] to exhibit an O(6) → SU(3) transition as well although the data are not as extensive as for the W-Pt region. The most interesting aspect of the Kr isotopes, however, centers on the high spin yrast states and will be discussed below in that context.

The next transition leg to be studied extensively is the SU(5) → SU(3) transition that occurs in the Sm isotopes[9,13]. As with the O(6) → SU(3) transition, the calculations here were first carried out in the IBA-1 formalism with a single parameter specifying the location of a nucleus along the leg. Excellent results were obtained for energy levels, E2 transition rates, and two nucleon transfer reaction cross sections. In addition, E0 transition rates, isomer, and isotope shifts were calculated. Where data exist for these latter quantities the agreement is reasonable. However, much more data is called for and, since the calculations already exist, can be used to provide stringent tests of the model.

b. Deformed nuclei

Recently, the first extensive tests of the IBA for well deformed nuclei have appeared. In one, Coulomb excitation was used[28] to obtain B(E2) values for the lowest levels of the ground, β and γ bands for a considerable series of nuclei in the first half of the rare earth region. The results were compared with IBA calculations in the SU(3) limit and in a perturbed SU(3) scheme. Two significant results emerge. First, the SU(3) predictions are not in good agreement with the data which highlights the fact that, although these nuclei are well deformed, they are not good SU(3) nuclei (most do not have nearly degenerate β and γ bands). Secondly, the broken SU(3) calculations are in excellent agreement with the empirical B(E2) values.

The other test[26,27] of the IBA for deformed nuclei deals with ^{168}Er and presents the most detailed test to date of the IBA in a single nucleus. The calculations were carried out with the IBA-1 Hamiltonian

$$H = -\kappa Q \cdot Q - \kappa' L \cdot L + \kappa'' P \cdot P \qquad (5)$$

The constants κ and κ' of the SU(3) limit were fixed from the energies of the first two 2^+ states. The only parameter that was varied was κ". The P P term was included in Eq.(5) because this term is important in the O(6) limit and ^{168}Er deviates from the SU(3) limit in just that direction, namely with the first excited 0^+ state considerably higher in energy than the second $2^+(2^+_\gamma)$ state. The results for level energies are shown in Fig.4. Not only are the "β" and "γ" bands reproduced, but there is also agreement for the higher representations. Relative B(E2) values were also calculated, with one additional parameter (χ in the Q operator). Empirically, the most striking feature of the decay of the first 0^+ band is that transitions to the γ band dominate over those to the ground band. This feature, inherent in the SU(3) limit, persists in near-SU(3) nuclei whereas the opposite behavior is predicted by the simplest treatment of the Bohr-Mottelson approach. Clearly this is a significant success for the IBA. Since the dominance of β→γ over β→g transitions can be arranged in the geometrical collective model as well, by the explicit introduction of β→γ bandmixing, the interesting question arises as to the relation between the collective model with bandmixing and the IBA. In the former, the β→γ transitions proceed by virture of K mixing so that the transition matrix element contains, in effect, small amplitudes for intraband transitions. In the IBA, the wave functions in the SU(3) limit already contain K admixtures automatically. The β→γ matrix element proceeds via these admixtures as well as by a direct K = 2 term. The relative size of these two terms and the relation of their boson number dependence to, for example, the variations across the rare

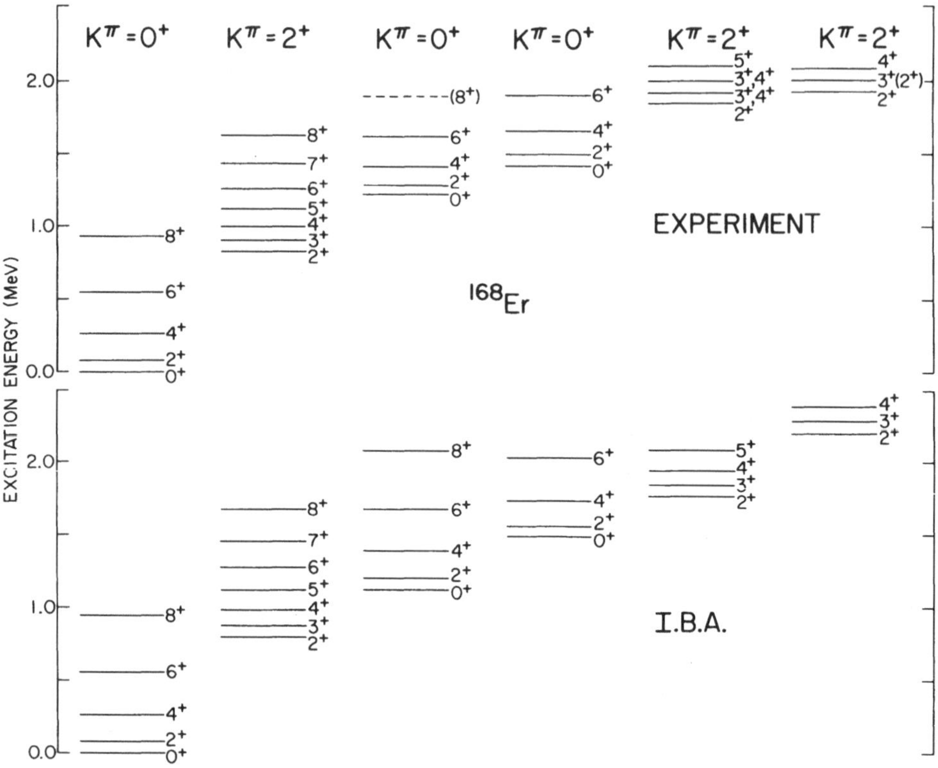

Fig.4 Comparison of the IBA with the empirical positive parity levels of ^{168}Er. Data from ref.26, calculations from ref.27.

earth region of the β→γ mixing parameter in the collective model are topics of considerable interest in understanding the relation between these two models.

c. Two nucleon transfer and E0 transitions.

The above tests of the IBA have stressed the overall quality of predicted level energies and E2 transitions and rely for their interest as much on their thoroughness and extensiveness as on the detailed results for any given quantity. As such they test the overall character of the IBA wave functions. A complementary approach is to test as directly as possible the detailed makeup of individual wave functions. For this one generally needs processes that entail specific simple operators. Two excellent examples are two nucleon transfer reactions and E0 transitions, in both cases with emphasis on transitions between 0^+ states. Both display characteristic selection rules and transition rates in each limit that lead to easily testable predictions. For example, since the ground state in the SU(5) limit has $n_d = 0$ while all excited states have $n_d \neq 0$, there can be no cross section in (t, p) or (p, t) to excited 0^+ states in SU(5) nuclei. For O(6) nuclei, the group theoretical selection rule, $\Delta\sigma = \pm 1$, predicts that, above midshell (where N decreases with increasing mass) there should be one excited 0^+ state (that with $(\sigma, \tau) = (N-2, 0)$) excited in (p, t) but none in (t, p). Further, the (p, t) cross section ratio $R = S(0^{+\prime})/S(g.s.)$ should obey a simple analytic expression, given in ref.33, which yields $R \sim 0.14$ for N = 13. As a final example, in an SU(5) → SU(3) transition there is a sudden reduction of a factor of $\tilde{\sim}3$ in the predicted (t, p) ground state transition. Many of these predictions, as well as others, have been verified. Others remain to be tested. The reader is referred to refs.1 and 2 and the original literature[9,31-34] for details of individual studies.

E0 transitions provide an elegant and even more direct test of the IBA wave functions. The E0 operator is of the form $\alpha(d^+\tilde{d})^0 + \beta(s^+s)^0$ which can be rewritten, using $N = n_s + n_d$, as $T(E0) = N_0 + \alpha' n_d$ where N_0 and α' are constant. Thus E0 transitions are directly dependent on the matrix element $\langle\Psi_i|n_d|\Psi_f\rangle$ amd therefore provide a rather direct probe of the basic s-d boson structure of the states involved. It is also apparent that they are very trivially calculable (in some cases by inspection) given a set of IBA wave functions. Thus, for example, it is clear that there can be no E0 transitions in a pure SU(5) nucleus. In an O(6) nucleus, the E0 selection rules are $\Delta\sigma = 2$, $\Delta\tau = 0$. This implies, for example, only one ground state E0 transition (from the $\sigma = N-2$, $\tau = 0$ 0^+ state), and also that certain pairs of higher lying 0^+ states are linked by E0 transitions. These predictions are currently being tested and preliminary results[30] show excellent agreement in ^{188}Os and ^{196}Pt.

d. High spin states

Considerable attention has naturally focussed on testing the IBA for high spin yrast states. There are two principal predictions of interest here, both a direct reflection of the explicit treatment of finite boson number in the IBA. First, there should be a spin cutoff at J = 2N independent of the structure of the nucleus. Secondly, the B(E2: J→J-2) value should fall off rapidly as J→2N. Most tests to date have dealt with deformed nuclei in which the second prediction is often expressed as a predicted falloff below that of the rotational model.

The spin cutoff prediction is probably of limited significance in that it occurs at the very limit of expected applicability of the model, where additional degrees of freedom, for example from a g-boson, from the need for effective N values, or for two quasiparticle states, may well appear. As for the B(E2) falloffs they

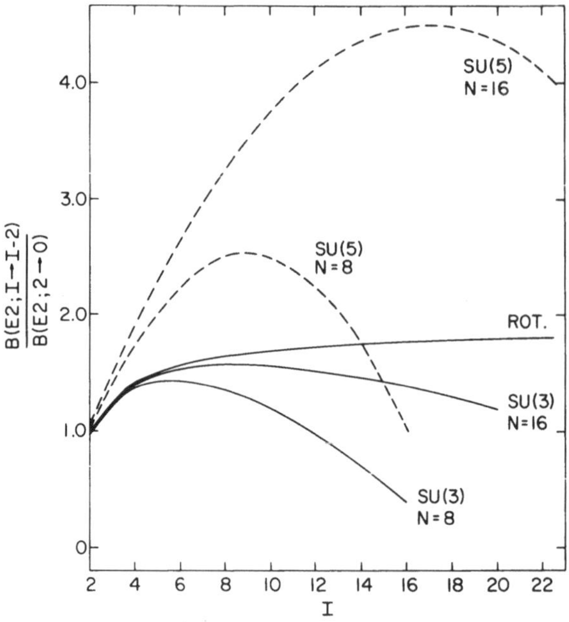

Fig.5 B(E2) values for the geometrical rotor model, the SU(5) and SU(3) limits of the IBA for the ground state band.

must be carefully assessed. Fig.5 shows the behavior of the IBA B(E2) values in the SU(5) and SU(3) limits. While both eventually fall to zero and while the IBA predictions for each case fall below the corresponding geometrical model predictions, it is clearly not sure that the B(E2) values in a deformed nuclei will necessarily fall below the rotational model predictions since small admixtures of SU(5) character can substantially raise the predicted B(E2) values except very close to J = 2N. This is not to suggest that nuclei such as the Dy isotopes, where some of the best high spin B(E2) values exist[35], deviate from SU(3) in this particular way but only to show a simple example of how a broken SU(3) scheme can alter the IBA predictions from those of the SU(3) limit. The point is simply that in order to assess the success or failure of the IBA for high spin states, one must either measure the B(E2) values very close to J = 2N or first use the low lying, low spin levels to define the actual IBA Hamiltonian and then use this to obtain the predicted high spin B(E2) values.

Data in the actinides[36] and in Dy (ref.35) are cited in Table I but to date are inconclusive primarily for lack of the requisite detailed IBA calculations. Adequate tests[37,38,20,39] do exist, however, for ^{78}Kr and ^{194,196}Pt, and yield opposite conclusions. The

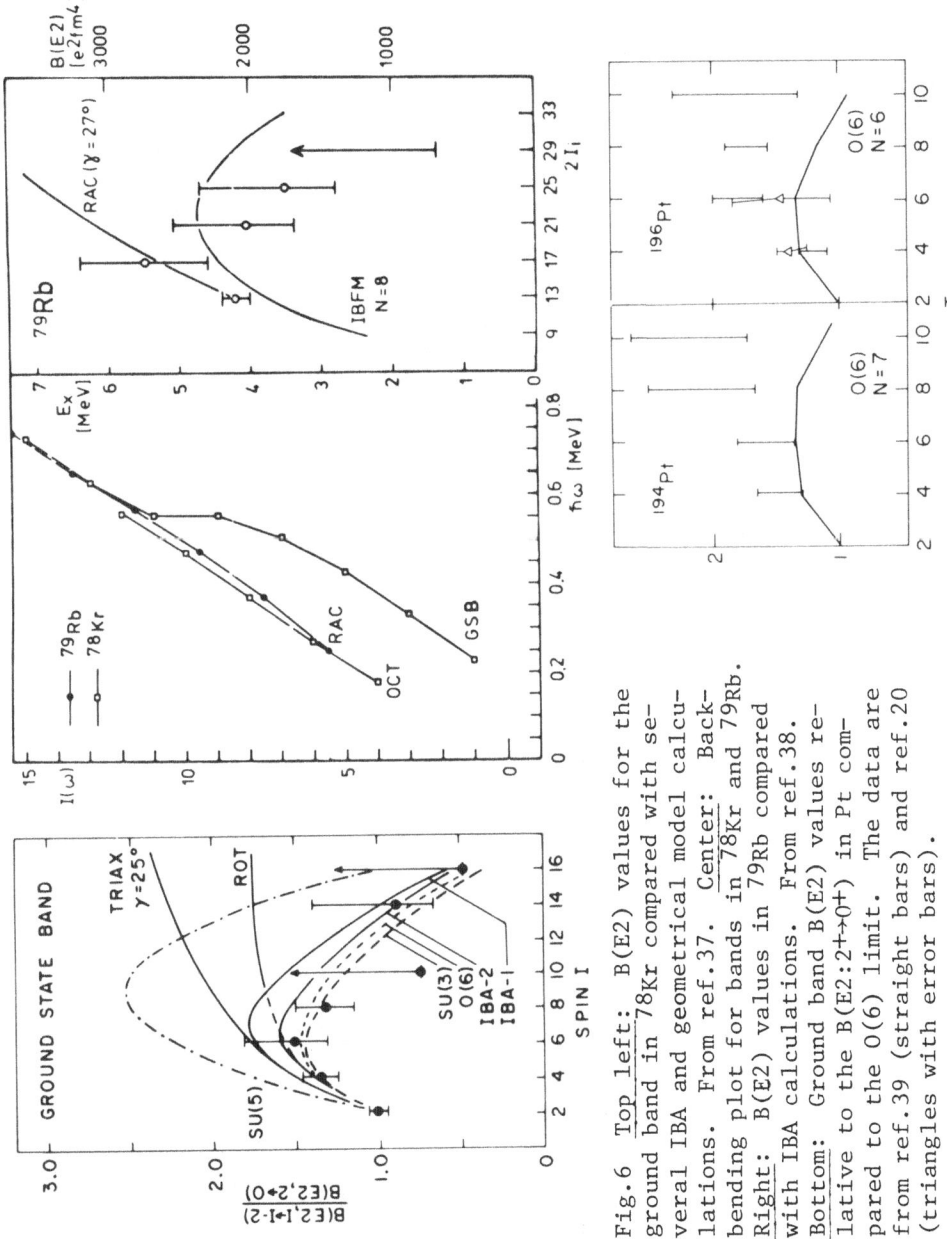

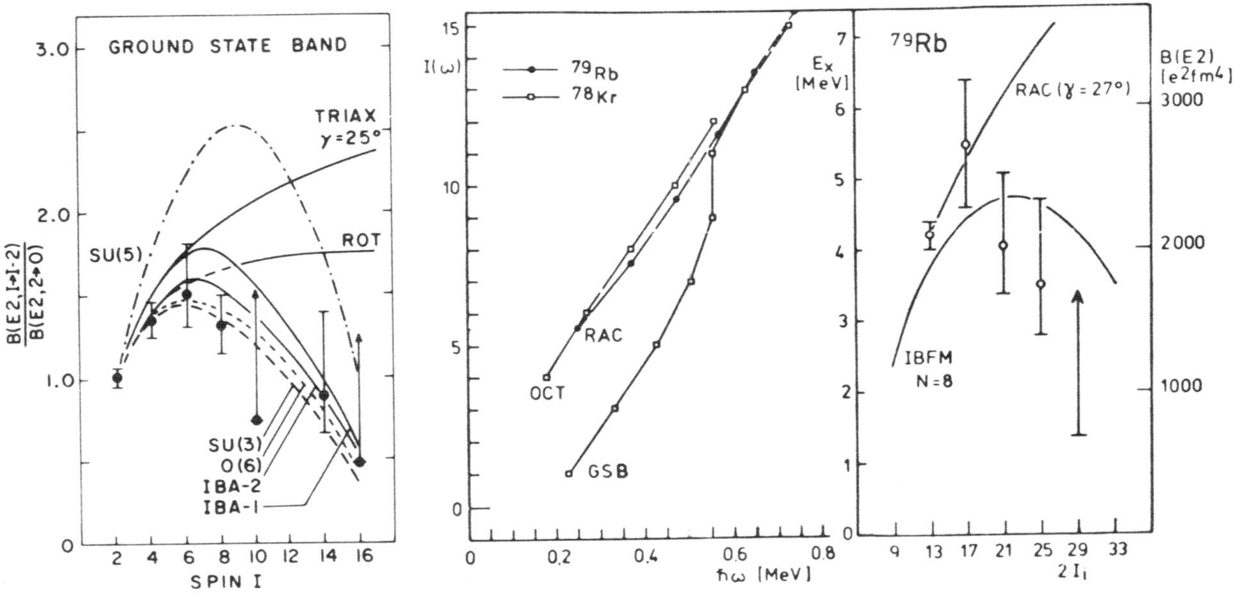

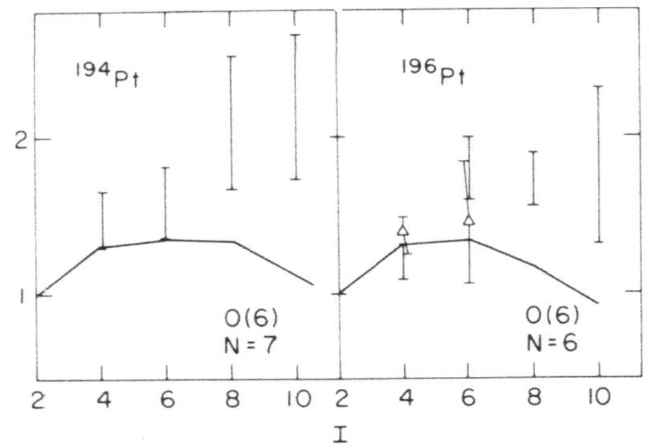

Fig.6 Top left: B(E2) values for the ground band in ^{78}Kr compared with several IBA and geometrical model calculations. From ref.37. Center: Backbending plot for bands in ^{78}Kr and ^{79}Rb. Right: B(E2) values in ^{79}Rb compared with IBA calculations. From ref.38. Bottom: Ground band B(E2) values relative to the B(E2:$2^+\to0^+$) in Pt compared to the O(6) limit. The data are from ref.39 (straight bars) and ref.20 (triangles with error bars).

Kr results[37,38] are shown in Fig.6. There is clearly a falloff below the rotational prediction and in agreement with several IBA calculations (that lavelled IBA-2 probably being the most reasonable). However, there is a known upbend in ^{78}Kr (see center top panel of Fig.6) and is possible that the B(E2) falloff could be ascribed to admixtures of an aligned two-quasiparticle state. However, this possibility can be eliminated by utilizing recent results[38] for ^{79}Rb. As seen in the top right panels there is no upbend in ^{79}Rb but there is still a falloff in B(E2) values. Lieb has discussed[38] these results in detail including arguments to show that the odd nucleon in ^{79}Rb does not alter the core structure.

In ^{194,196}Pt, the O(6) limit is applicable and any deviations therefore are in the direction of the SU(3) limit which has a similar B(E2) behavior as O(6) (see, for example, the SU(3) and O(6) curves for ^{78}Kr in Fig.6). Therefore, there is little ambiguity about the predicted B(E2) values. In addition the boson number is low enough that the B(E2) falloffs are expected at relatively low spin. The empirical and calculated B(E2) values are shown in Fig.6. Two sets of empirical values are shown for ^{196}Pt. While the more extensive data, from multiple Coulomb excitation studies[39] at GSI, indicate a clear discrepancy with the IBA calculations, the B(E2) values of ref.20, based on direct lifetime measurements, while limited to ^{196}Pt and to lower spin states, suggest a smaller discrepancy and encourage the need for refined uncertainties.

3. The IBA For Odd Mass Nuclei

The IBA can be extended to odd mass nuclei[51,52,13,45] by writing the Hamiltonian

$$H = H_B + H_F + V_{BF}$$

where B and F stand for boson and fermion and V_{BF} represents the boson fermion interaction. H_B is just the IBA Hamiltonina for the even core, H_F is that of the single particle. The essence of the model, called the IBFA, is to specify V_{BF}. Clearly this can be rather complicated but it has been suggested[51] that a good approximation is:

$$V_{BF} = A[(d^+\tilde{d})^{(o)}(a_j^+a_j)^{(o)}]^{(o)} + \Gamma[Q^{(2)}(a_j^+\tilde{a}_j)^{(2)}]$$
$$+ \Lambda(d^+\tilde{a}_j)^{(j)}(da_j^+)^{(j)}]^{(o)}$$

where we have written V_{BF} for the case of a single possible j orbit for the odd fermion. The first term is a monopole one which acts more or less as a scale factor on the energies and will not be further discussed. The second term is a direct term similar to that which gives the core-particle interaction in an intermediate coupling formalism. It serves, for example, to split the degeneracy of the five levels arising from the coupling of a particle in shell modle orbit j to the first 2^+ state of the core. The third term is an exchange term which arises from the Pauli principle since particles in a given shell model orbit may participate in the boson micro-structure. This is a blocking type of effect and clearly becomes more important the more particles there are in a given shell model orbit. Therefore, in effect, it allows one to simulate the movements of the Fermi surface across a series of nuclei. To see the effects of the exchange term, consider a single j shell such as the unique parity orbit in a prolate deformed nucleus. Then, low values of a Λ correspond to the Fermi surface near the low K orbits. The resultant spectrum resembles a decoupled band picture. For a heavier nucleus, Λ is larger, corresponding to a Fermi surface near the K orbits, and the resulting spectrum resembles a strong coupling scheme characteristic of many well deformed nuclei. Of course intermediate schemes are also possible. When the core is not that of a deformed rotor, other varities of nuclear level spectra emerge. It is clear then that, like the IBA, the IBFA is potentially a very general approach to odd mass nuclei and well worth the extensive effort required to thoroughly test it.

The model is also applicable to the multi-j shell case but rapidly becomes much more complicated. The parameters Γ, A and Λ becomes multi-dimensional arrays although model dependent prescriptions do exist to inter-relate the members of each array. Some detailed multi-j shell calculations have indeed been performed formed,[13,42-45] notably in Rb, Ag, Tc and Eu. The Rb study[44] in particular concluded that the spectra resulting from such calculations are similar to those in the single j shell case but that the parameters are strikingly different, especially if two or more of the j shells are strongly coupled by the quadrupole operator. The implication is perhaps that the parameters deduced in a single j shell case are effective parameters that to some extent mock up the effects of the neglected shells. Since the study of the multi-j shell case is only in its beginning stages, we shall not indulge in further discussion except to comment that the first results of such calculations suggest[44] that it is prudent not to attribute too much direct meaning to the parameters deduced for a single j shell case. One awaits with interest further developement of the IBA for the multi-j case.

The first IBFA calculations[40,41] were carried out for ^{105,109}Pd and ^{79,81,83}Rb. The former calculations[40] dealt with two related topics centering on the negative parity levels from the $h_{11/2}$ neutron orbit in the Pd nuclei. In ^{109}Pd, it was shown that the IBFA worked remarkably well for the favored aligned high spin unique parity (i.e., yrast) levels. The study of ^{109}Pd, however, dealth with the low spin levels of the same parentage, that is, the favored and unfavored antialigned levels. It can be shown that the particle-rotor model, with or without axial asymmetry, cannot, even in principle, reproduce the experimental levels. On the other hand, the IBFA calculations produces much better agreement. The principal origin of the improvement is the much expanded set of core states automatically incorporated in the IBFA description. Table II exhibits the probability distribution of different wave function components in comparison to the particle rotor model. While the latter model is limited to the even core states of the (quasi) ground band (or the quasi γ band if asymmetry is included), in the IBFA calculations the ground band accounts in fact for considerably less than half of the wave function probability distribution.

The Rb case[41] was the first extensive test of the IBFA for high spin states. The even core parameters were taken from a systematic study[24] of the Kr isotopes which span an O(6)→rotor transition. The fit to the levels of the Rb isotopes included both favored and unfavored levels with spins ranging from 5/2 to 21/2.

The most successful and impressive IBFA study[42] to date is probably that for the Eu isotopes which span a transition from nearly decoupled to strongly coupled with many level crossings across these isotopes. Nearly all of these have been reproduced in an extremely simple set of single j shell calculations.

TABLE II

Probability Distribution (in %) of Categories of Core States in IBFA and Particle Rotor Wave Functions for ^{105,109}Pd (from ref.40).

Category of Levels	Category of Core States			
	Quasi Ground Band		Other	
	IBFA	PR	IBFA	PR
Low spin favored (^{109}Pd)	≈36	100	≈64	0
Low spin unfavored (^{109}Pd)	≈40	100	≈60	0
High spin favored (^{105}Pd)	≈49	100	≈51	0

Since the Pd, Rb and Eu calculations have been discussed in an earlier review[1], we choose to highlight here some more recent results. The same group involved in the Rb study has followed that work up with detailed studies of Tc and Ag and again obtained good agreement with the data. For illustration, we show recent results[43] for ^{105}Ag and ^{99}Tc in Fig.7 where levels up to 3 MeV are included. Of particular interest is an extensive series of unfavored states in ^{99}Tc.

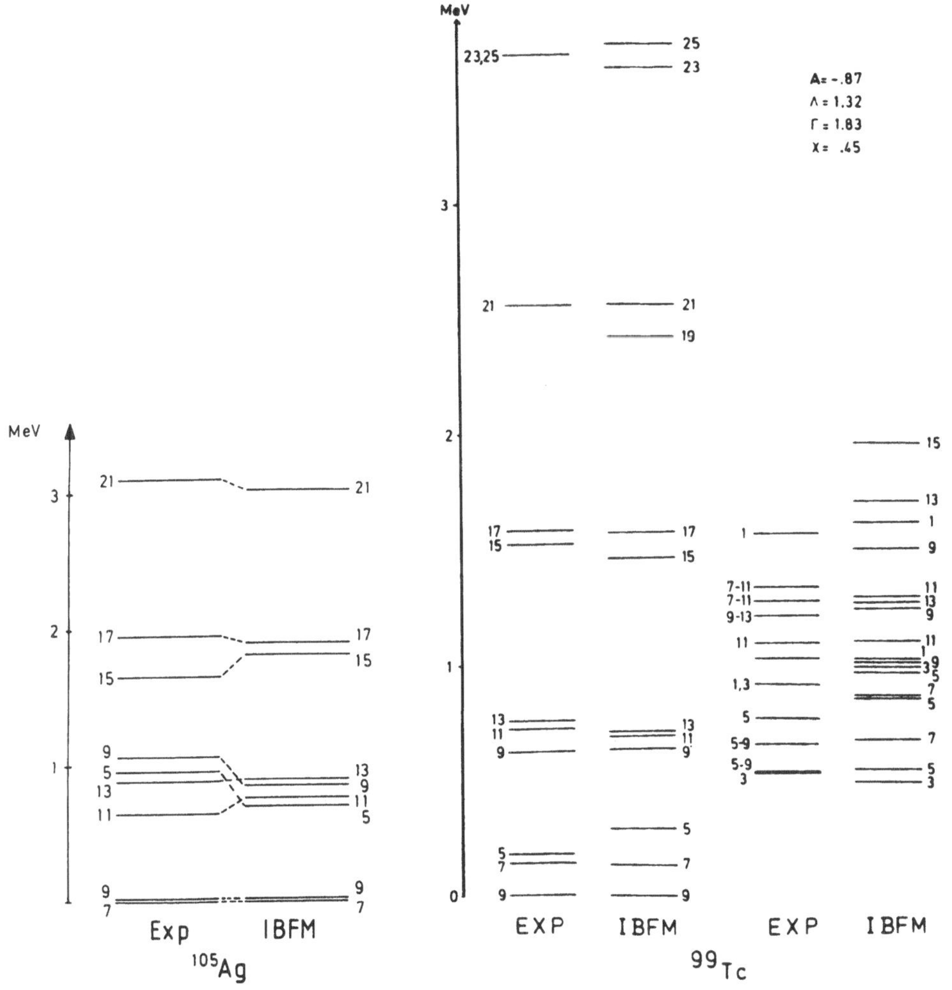

Fig.7 Levels in ^{105}Ag and ^{99}Tc compared to IBFA calculations. The numbers indicated on the energy levels are twice the spin values. From ref.43.

The application of the IBFA to nuclei such as these clearly offers a challenge both to more sophisticated extensions of the model, to the multi-j shell case, and to more accurate and expanded sets of empirically determined level spins. An entire area that has barely been touched on, with the notable exception of the multi-j shell canculations for Rb, is the calculation of B(E2) values[44] in odd mass nuclei. The relevant computer codes are still partially in the testing stage but already provide an opportunity for greatly expanded testing of a very promising model.

4. Supersymmetries

Undoubtedly the most intriguing aspect of the IBFA is the prediction[53] of super symmetries. These are similar to the symmetries of even nuclei but are more extensive, encompassing both even and odd nuclei in the same scheme wherein one set of quantum numbers, one eigenvalue equation and one set of expressions for electromagnetic transition rates and charged particle transfer amplitudes applies to both types of nuclei. The supersymmetry scheme thus offers a potential unifying concept of great promise. There are limitations, of course. While the IBFA can be calculated for nearly any odd mass nuclei, a supersymmetry arises only when the underlying even core corresponds to one of the three IBA symmetries. Secondly the supersymmetry may involve either a single j shell or certain combinations of j shells. In the latter case, however, the several j shells must be degenerate[53] or, in practice, at least considerably closer in energy than the energy separation of multiplets in the resultant spectra: this seldom will occur. Nevertheless even the possibility of the existence of supersymmetries in isolated cases has profound repercussions on our understanding of the relation between even and odd nuclei and, as did the O(6) limit in even nuclei, may provide new bench-marks that permit the reinterpretation of even and odd mass nuclear level schemes in terms of small supersymmetry breaking. Furthermore, the question of supersymmetries is of high relevance to elementary particle physics as well where the unification of bosonic and fermionic degrees of freedom is an essential goal.

So far, the only supersymmetry worked out in detail[53] is for a $j = 3/2$ particle coupled to an O(6) core. The most likely region to search for this is then in a positive parity shell where only the $s_{1/2}$ orbit is likely to interfere and where the even nuclei obey the O(6) symmetry. In practice, the Au and Ir nuclei in the $A \sim 195$ region offer the most promising nuclei.

Fig.8 shows, in simplified form, the lowest level groups of the 3/2 x O(6) super-symmetry for the odd mass nucleus incorporated in the structure. The quantum numbers are $[N, (\sigma_1\sigma_2\sigma_3), (\tau_1\tau_2), \nu_\Delta$ and J]. However, for the odd nucleus, $\sigma_2 = \sigma_3 = 1/2$ and $\tau_2 = 1/2$ for all levels and so, the essential quantum numbers are only σ ,

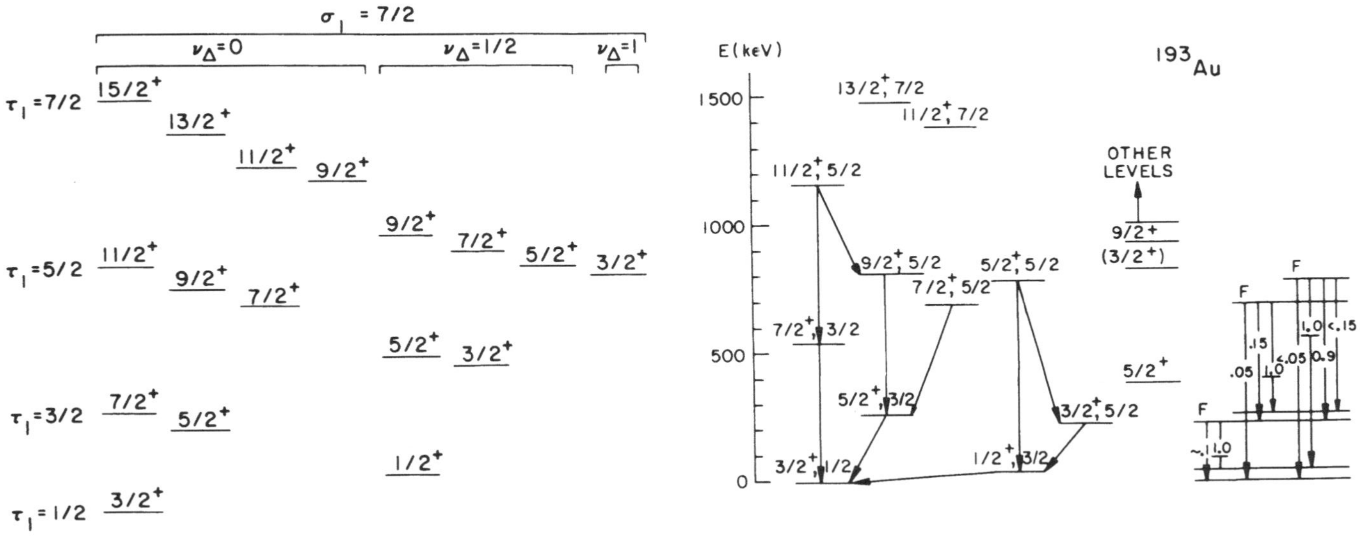

Fig.8 Left: Low lying positive parity levels from the σ_1 = N + 1/2 group of the 3/2 ⊗ O(6) supersymmetry for N = 3. Right: Positive parity levels of ^{193}Au with the principal E2 decay branches indicated. At far right are the relative reduced E2 transition rates for the τ_1 = 5.2 states with J $\leq$ 7/2, including several upper limits. The forbidden transitions in the supersymmetry are indicated by an F. Figure taken from ref.48.

which is analogous to σ in O(6), τ_1 (analogous to τ) and ν_Δ and J. The eigenvalue expression (for the odd nucleus) is

$$E(N, \sigma_1, \tau_1, J) = \frac{1}{4} A[\sigma_1(\sigma_1 + 4) + 3/2]$$

$$+ B[\tau_1(\tau_1 + 3) + 3/4] + J(J+1)$$

which is identical in structure to the O(6) eigenvalue equation with the exception of the constants added to the A and B terms. The figure shows the case for N = 3 for which the maximum $\sigma_1 = N + 1/2 = 7/2$. The E2 selection rules are $\Delta\sigma = 0$ and $\Delta\tau = 0, \pm 1$.

At first glance the scheme in Fig.8 is similar to a weak coupling picture. However, there is no $3/2^+$ level corresponding to a $3/2^+ \otimes 2^+$ multiplet. The first excited $3/2^+$ level has $\tau_1 = 5/2$ and is not this missing $3/2^+$ levels as can be seen from its decay properties: by virtue of the τ selection rule this state decays to the three $\tau_1 = 3/2$ levels but not to the ground state as would a $[3/2^+ \otimes 2^+]^{3/2}$ level. Not shown in the figure is the next higher lying $3/2^+$ state which is the sequence head of the next σ_1 group with $\sigma_1 = 5/2$.

Most of the tests[47,48] of the supersymmetry to date rely on the character of the $3/2^+$ states. One concerns the odd Au nuclei and has been discussed in detail by Wood[48]. A prerequisite to testing the supersymmetry is to veryfy that the condition of an isolated j = 3/2 particle exists. This has not been done in ^{193}Au but proton stripping reactions leading to ^{195}Au suggest[48] the main $s_{1/2}$ strength will lie at or above 1 MeV in ^{193}Au and thus at least the low lying levels are a prime candidate for the supersymmetry. The levels and strong E2 transitions are shown on the right in Fig.8 and the appropriate levels are labelled by the quantum numbers (J^π, τ_1). It is clear that there are strong perturbations to the energy level patterns predicted on the left. Nevertheless, the "missing" $3/2^+$ level indeed seems to be absent: the lowest $3/2^+$ state decays only to the $\tau_1 = 3/2^+$ J = $1/2^+$ level and not to the ground state. Also, the $\tau_1 = 5/2$, J = $5/2^+$level strongly decays by an allowed $\Delta\tau = 0$ transition to the $\tau_1 = 5/2$, J = $3/2^+$ state. Finally, two other forbidden, $\Delta\tau = 2$, transitions from the upper levels are also weak although this is perhaps of less significance since they are also forbidden in the weak coupling limit. Similar analyses[48] of ^{195,197}Au indicate a progressive breakdown of the scheme there as the forbidden ground state transition from the lowest excited $3/2^+$ state becomes stronger, with increasing mass.

The supersymmetry scheme also predicts relative charged particle transfer reaction cross sections. For (t, p) the L = 0 selection

TABLE III

(t, α) strengths to $3/2^+$ states in $^{193,195}Ir$ compared to supersymmetry predictions (from ref.47).

Final Nucleus	E_x(keV)	S_{exp}(rel)	$S_{supersymmetry}$
^{193}Ir	0	≡1.00	≡1.00
N=7	180	0.07	0
	460	0.69	0.64
^{195}Ir	0	1.00	1.00
N=6	234	0.16	0
	287	0.23	0.60

rules are $\Delta\sigma_1 = \pm 1$, $\Delta\tau = 0$ and thus, in the Pt-Ir region, only the ground state transition is allowed. L = 0 (t, p) transitions satisfying $\Delta\tau = 0$) to excited states would be $\Delta\sigma_1 \geq 2$. Analytic expressions relate the (t, p) strength in the even and odd mass systems. For the Pt and Ir nuclei near A = 195 these expressions predict nearly equal strengths. Cizewski[47] has compared the empirical and predicted results and shown excellent agreement for ground state transitions on targets of $^{194,192}Pt$ and $^{193,191}Ir$ (the Ir transitions are $3/2^+ \rightarrow 3/2^+$ and display L = 0 angular distributions). On the other hand, in the ^{191}Ir(t, p) reaction, excited $3/2^+$ states (albeit above 1 MeV) are populated and the calculations overpredict the ground state transition strengths in Os.

Finally, the supersymmetry predictions for single proton pick-up reactions have been tested in this region. The experimental studies[47] utilized the $^{194,196,198}Pt$ (t, α) $^{193,195,197}Ir$ reaction for which the supersymmetry predicts the selection rules $\Delta\sigma_1 = \pm 1/2$, $\Delta\tau = 1/2$. The target has σ = N and N increases with decreasing mass. Thus, two $3/2^+$ states should be populated, namely the ground state of the odd nucleus which has $\sigma_1 = N + 1/2$ and the $\tau = 1/2$ excited $3/2^+$ state with $\sigma_1 = \sigma_1(g.s.) - 1 = N - 1/2$. The relative cross sections should be approximately N:N+4. The $\tau = 5/2$ $3/2^+$ state of the $\sigma = N + 1/2$ group should not be populated. The results[47] are shown in Table III and show fine agreement for ^{193}Ir

and fair agreement for ^{195}Ir. However, there is also relatively large (t, p) strength to a low lying $1/2^+$ state at ~70 keV in ^{195}Ir and a suggestion of similar strength in ^{193}Ir in contrast to the prerequisites for an isolated $3/2^+$ single particle orbit.

Thus to summarize, there are several lines of evidence, mostly concerning the properties of ground and excited $3/2^+$ states, that suggest the existence of supersymmetry-like structure. At the same time, it is clear that the level energies are strongly perturbed, and that there is some low lying $1/2^+$ single particle strength that partially renders the predictions inapplicable and which must introduce some symmetry breaking. One hopes now for more thorough studies, based on these pioneering early results, to establish whether or not supersymmetries exist and the extent and nature of symmetry breaking. The question is of high importance in terms of a verification of a major prediction of the IBA, for the promised unification of odd and even nuclei it would afford, and for its significance in other areas of physics as well.

5. Conclusions

A broad survey of tests of the IBA has been carried out, with emphasis on the overall patterns of agreement and on the newest tests, such as those in ^{168}Er, the tests of the IBFA in Ag and Tc, and of the supersymmetry scheme in Ir and Au. In general the IBA has been remarkably successful both in individual nuclei and for broad sequences of nuclei, for both odd and even nuclei, for level energies, for electromagnetic transition rates and for charged particle transfer reactions. Certain discrepancies have appeared, and have been commented on, and major areas for further testing, such as deformed even nuclei, E0 transitions, multi-j shell analyses of odd nuclei and the search for supersymmetries have been cited.

ACKNOWLEDGEMENT

I wish to express appreciation to F. Iachello, A. Arima, I. Talmi, O. Scholten, R. Broglia, B. Barrett, P. von Brentano, A. Gelberg, K.P. Lieb, D.D. Warner, G. Lo Bianco, J.L. Wood and J.A. Cizewski for numerous discussions and for much help in keeping abreast of theoretical and experimental developments in the IBA. I am grateful to all the others whose unpublished experimental work is cited in the references for permission to quote their results prior to publication.

REFERENCES

1. Richard F. Casten, Nucl. Phys. A347, 173 (1980).
2. Richard F. Casten, Proceedings of the Workshop on Interacting

Boson Fermion Systems, Erice, Sicily, Italy, June, 1980 (Plenum Press, to be published).
3. I. Talmi, in Interacting Bosons in Nuclear Physics, ed. by F. Iachello (Plenum Press, New York, 1979), p. 79.
4. A. Bohr and B.R. Mottelson, Physics Scripta 22, 468 (1980) preprint No. 80/19, 1980.
5. S.G. Nilsson, Mat. Fys. Medd. Dan. Vid. Selsk. 29, No. 16 (1955).
6. K. Kumar and M. Baranger, Nucl. Phys. A122, 273 (1968) and for example, K. Kumar and J.B. Gupta, Nucl. Phys. A304, (1978) 295.
7. T. Tamura, K. Weeks and T. Kishmoto, Phys. Rev. C20 (1979) 307 (1979) 307.
8. A. Arima and F. Iachello, Phys. Rev. Lett. 35, (1975) 1069: Ann. Phys. (N.Y.) 99. (1976) 253; 111, (1978) 201; Phys. Rev. Lett. 40, (1978) 385; Phys. Rev. C16, (1977) 2085.
9. O. Scholten, F. Iachello and A. Arima, Ann. Phys. (N.Y.) 115, (1978) 366.
10. A. Arima, T. Otsuka, F. Iachello and I. Talmi, Phys. Lett. 66B, (1977) 205.
11. T. Otsuka, A. Arima, F. Iachello and I. Talmi, Phys. Lett. 76B, (1978) 139; Nucl. Phys. A309, 1 (1978).
12. F. Iachello in Interacting Bosons in Nuclear Physics, ed. by F. Iachello (Plenum Press, New York, 1979), p.1.
13. O. Scholten, in Interacting Bosons in Nuclear Physics, ed. by F. Iachello (Plenum Press, New York, 1979), p. 17 and O. Scholten, Thesis, 1980.
14. A.E.L. Dieperink and O. Scholten, Nucl. Phys. A346, 125 (1980).
15. J.N. Ginocchio and M.W. Kirson, Nucl. Phys. A350 (1980) 31.
16. R. Gilmore and Da Hsuan Feng, contribution to this conference and Da Hsuan Feng, R. Gilmore and S.R. Deans, Phys. Rev. C,
17. Manual for IBA-1 Code Package PHINT, written by O. Scholten.
18. J.A. Cizewski, R.F. Casten, G.J. Smith, M.L. Stelts, W.R. Kane, H.G. Borner, and W.F. Davidson, Phys. Rev. Lett. 40, (1978) 167; J.A. Cizewski, R.F. Casten, G.J. Smith, M.R. Macphail, M.L. Stelts, W.R. Kane, H.G. Borner, and W. W.F. Davidson, Nucl. Phys. A323, (1979) 349.
19. R. Casten, in Interacting Bosons in Nuclear Physics, ed. by F. Iachello (Plenum Press, New York, 1979), p. 37.
20. H.H. Bolotin, A.E. Stuchbery, I. Morrison, D.L. Kennedy, C.G. Ryan and S.H. Sie, preprint UM-P-80/S7, University of Melbourne, 1980.
21. R.F. Casten and J. Cizewski, Nucl. Phys. A309, (1978) 477 and, to be published; R.F. Casten, M.R. Macphail, W.R. Kane D. Breitig, K. Schreckenhach, and J.A. Cizewski, Nucl. Phys. A316, (1979) 61.
22. R. Bijker, A.E.L. Dieperink, O. Scholten and R. Spanhoff, Nucl. Phys. A344, 207 (1980).

23. P.D. Duval and B.R. Barrett, preprint and Proceedings of the Workshop on Interacting Boson Fermion Systems, Erice, Sicily, Italy, June, 1980 (Plenum Press, to be published).
24. U. Kaup and A. Gelberg, Z. Phys. A293, (1979) 311; A. Gelberg and U. Kaup, in Interacting Bosons in Nuclear Physics, ed. by F. Iachello (Plenum Press, New York, 1979), p.59.
25. G. Puddu, O. Scholten and T. Otsuka, Nucl. Phys. A348, 109 (1980).
26. W.F. Davidson, D.D. Warner, K. Schreckenbach, H.G. Borner, J. Simic, M. Stojanovic, M. Bogdanovic, S. Koicki, W. Gelletly, R.F. Casten, G. Orr and M.L. Stelts, J. Phys. G., to be pu published.
27. D.D. Warner, R.F. Casten and W.F. Davidson, Phys. Rev. Lett. 45, 176 (1980).
28. F.K. McGowan and W.T. Milner, preprint to be published.
29. Denis Hageman, preprint 1978, KVI report 54.
30. W.R. Kane, R.F. Casten, D.D. Warner, K. Schreckenbach, H. Faust, private communication, 1980.
31. J.A. Cizewski, E.R. Flynn, Ronald E. Brown, and J.W. Sunier, Phys. Lett. 88B, (1979) 207.
32. P.T. Deason, C.H. King, T.L. Khoo, J.A. Nolen, Jr., and F.M. Bernthal, Phys. Rev. C20, (1979) 927.
33. R.R. Betts and M.H. Mortensen, Phys. Rev. Lett. 43, (1979) 616
34. A. Saha, O. Scholten, D.C.J.M. Hageman and H.T. Fortune, Phys. Lett. 85B, (1979) 215.
35. H. Emling, P. Fuchs, E. Grosse, R. Kulessa D. Schwalm, R.S. Simon and H.J. Wollersheim, International Conference on Nuclear Behaviour at High Angular Momentum, Strasbourg, April 22-24, 1980.
36. H. Ower, Th. W. Elze, J. Idzko, W. Stelzer, H. Emling, P. Fuchs, E. Grosse, D. Schwalm, H.J. Wollersheim, N. Kaffrell and N. Trautmann, International Conference on Nuclear Behaviou at High Angular Momentum, Strasbourg, France, April 22-24, 198 1980.
37. H.P. Hellmeister, U. Kaup, J. Keinonen, K.P. Lieb, R. Rascher, R. Ballini, J. Delaunay and H. Dumont, Phys. Lett. 85B, (1979) 34 and Nucl. Phys. A332 (1979) 241.
38. J. Panqueva, K.P. Hellmeister, F.J. Bergmeister and K.P. Lieb, preprint and K.P. Lieb, Proceedings of the Workshop on Interacting Boson-Fermion Systems, Erice, Sicily, Italy, June 1980 (Plenum Press, to be published) and K.P. Lieb, private communication
39. J. Idzko, K. Stelzer, Th. W. Elze, H. Ower, H.J. Wollersheim, H. Emling, P. Fuchs, E. Grosse, R. Piercey, and D. Schwalm International Conference on Nuclear Behaviour at High Angular Momentum, Strasbourgh, April 22=24, 1980.
40. R.F. Casten and G.J. Smith, Phys. Rev. Lett. 43 (1979) 337.
41. A. Gelberg and U. Kaup, in Interacting Bosons in Nuclear

Physics, ed. by F. Iachello (Plenum Press, New York, 1979). p. 59; U. Kaup, A. Gelberg, W. Gast, and P. von Brentano, Proc. International Conference of Medium-Heavy Nuclei, May 1-4, 1979, Rhodos, Greece, p. 40; A. Gelberg, private communication, 1980.

42. G. LoBianco, N. Molho, A. Moroni, S. Angius, N. Blasi, and A. Ferrero, J. Phys. G: Nucl. Phys. 5 (1979) 697; N. Blasi and O. Scholten, KVI-Annual Report, 1979.

43. D. Hippe, H.W. Schuh, A. Gelberg, U. Kaup, K.O. Zell and P. von Brentano, Proceedings of the International Conference on Nuclear Physics, Berkeley, California, August 24-30, 1980, (LBL-11118), p.859., and P. von Brentano, Private communication, 1980.

44. U. Kaup, A. Gelberg, P. von Brentano, Phys. Rev. C22, 1738 (1980).

45. O. Scholten, contribution to this conference and references therein.

46. G. LoBianco, Proceedings of the Workshop on Interacting Boson-Fermion Systems, Erice, Sicily, Italy, June 1980 (plenum Press, to be published).

47. J.A. Cizewski, Proceedings of the Workshop on Interacting Boson-Fermion Systems, Erice, Sicily, Italy, June 1980 (Plenum Press, to be published).

48. J.L. Wood, Proceedings of the Workshop on Interacting Boson-Fermion Systems, Erice, Sicily, Italy, June 1980 (Plenum Press, to be published).

49. C. Baktash, J.X. Saladin, J.J. O'Brien and J.G. Alessi, Phys. Rev. C22, (1980) 2383.

50. R.F. Casten, A.I. Namenson, W.F. Davidson, D.D. Warner and H.G. Borner, Phys. Lett. 76B, (1978) 280.

51. F. Iachello and O. Scholten, Phys. Rev. Lett. 43, (1979) 679.

52. F. Iachello, Nucl. Phys. A347, (1980) 51.

53. F. Iachello, Phys. Rev. Lett. 44, (1980) 772, and private communication, 1980.

THE INTERACTING BOSON MODEL IN THE CONTINUOUS BASIS REPRESENTATION†

Robert Gilmore

Institute for Defence Analyses
400 Army-Navy Drive, Arlington, VA 22202

Da Hsuan Feng

Department of Physics and Atmospheric Science
Drexel University, Philadelphia, PA 19104

I. ABSTRACT

A number of properties of the Interacting Boson Model can only be described adequately in the SU(6) coherent state basis. Two of these properties are treated in the present work: the ground state energy phase transition and the reformulation of the Hamiltonian as a second order differential operator. These two properties are strongly evocative of analogous properties of the liquid drop model of collective nuclear motions. The two models are demonstrably not equivalent (the first has a finite spectrum, the second does not). However, their properties are so strikingly similar that a comparison of their logical foundations seems to reveal which details of each theory must be modified to bring them into equivalence. The appropriate modifications are proposed.

II. THE INTERACTING BOSON MODEL

Medium and heavy even-even nuclei have spectra containing large numbers of low-lying collective excited states. In these nuclei, identical nucleons outside closed shells (shell model cores)

†Work partially supported by the National Science Foundation under grant # Phy-7908402.

seem to exhibit strong pairing interactions. These interactions are strongest for nucleons of opposite spin, and decrease in strength as the total angular momentum of the pair increases. The interaction seems to be attractive for small J and weakly repulsive for large J.[1] The crossover point occurs in the neighborhood of J=4. As a result, identical nucleons are strongly bound in opposite spin pairs in the J=0 state (s-bosons) and somewhat less strongly bound in opposite spins pairs in the five J=2 states (d-bosons). Nucleon pairs corresponding to J=4, 6, 8 --- are either very weakly bound or unbound. As a consequence, the low lying collective excitations might be adequately explained in terms of s- and d-bosons and the interactions among them. These considerations were formalized by Arima and Iachello in the Interacting Boson Model.[2] This model not only accurately describes the low lying spectra of individual even-even nuclei, but more importantly accounts for nuclear systematics over a large portion of the periodic table of the nuclides.[3-5]

If the residual interactions between the s- and d-bosons is neglected, the d-bosons form a 5-fold degenerate excited multiplet. If the binding energies of the s- and d-bosons are assumed equal, the six bosons are energy-degenerate and form the basis for a linear representation of the group SU(6). The group SU(6) is the dynamical group (not a symmetry group) of the Interacting Boson Model (IBM). This group has previously been introduced for the description of nuclear processes, but in terms of nonlinear realizations rather than linear representations.[6]

In the IBM each boson may be in one of six possible states $|i\rangle$ (or any linear superposition of them). Here i=1 corresponds to the s-boson and $2\leq i\leq 6$ corresponds to the five d-bosons. The shift operator e^{α}_{ij} transforms the αth boson from state j to state i. Collective operators E_{ij} are defined by

$$E_{ij} = \sum_{\alpha=1}^{N} e_{ij}^{(\alpha)} \tag{1}$$

where N is the number of bosons present (i.e., half the number of nucleons outside of closed shells). These collective operators obey the standard SU(6) commutation relations

$$[E_{ij}, E_{rs}] = E_{is}\delta_{jr} - E_{rj}\delta_{si} \tag{2}$$

Intensive collective operators are defined as the mean of the corresponding collective operators: E_{ij}/N.

The IBM is strictly an algebraic model. Coordinates of nucleons or nucleon pairs do not appear in this model. Therefore, in the context of this model all physical operators are represented as functions of the intensive collective operators E_{ij}/N. This is in essence a mean field assumption. This assumption leads to the following consequences:

- All matrix elements of E_{ij}/N or any function of the generators E_{ij} of SU(6) can be explicitly computed in in any representation.

- All matrix elements of the shift operators E_{ij} are zero between different irreducible representations of SU(6).

Since SU(6) is a compact algebra, all its irreducible representations are finite dimensional. As a consequence, the spectrum of any operator (e.g., the hamiltonian $\hat{\mathscr{H}}$) is finite. No problems of ad hoc truncation of the spectrum of $\mathscr{H}$ occur in the IBM. The intensive hamiltonian $\hat{\mathscr{H}}/N$ must be a hermitian function of the intensive collective operators E_{ij}/N:

$$\hat{\mathscr{H}}/N = h(E/N) \tag{3}$$

in order for thermodynamic limits ($N\to\infty$) to exist.[7,8] Once the hamiltonian is specified and its eigenstates computed, it is a relatively straightforward matter to compute the transition rates for M1 or E2 processes between pairs of states.

The operator $\hat{\mathscr{H}}$ acts in the Hilbert space which carries the fully symmetric representation $[N,\dot{0}\}$ of SU(6) corresponding to N bosons. This space has dimension $(N+5)!/N!5!$

The Hamiltonian chosen for the IBM is a sum of Casimir invariants for the subgroups of SU(6) shown in Fig.1.

$$\begin{aligned}\hat{\mathscr{H}}/N = {} & A_0\mathcal{C}^1[U(6)]/N^1 + A_1\mathcal{C}^1[U(5)]/N^1 + A_2\mathcal{C}^2[U(3)]/N^2 + \\ & A_3\mathcal{C}^2[U(4)]/N^2 + A_4\mathcal{C}^2[O(5)]/N^2 + A_5\mathcal{C}^2[U(2)]/N^2\end{aligned} \tag{4}$$

The subgroups of SU(6) play two roles in the IBM

The Casimir invariants provide models for residual boson-boson interactions.

The sequence of subgroups in any chain provide a set of labels for basis vectors in the Hilbert space.

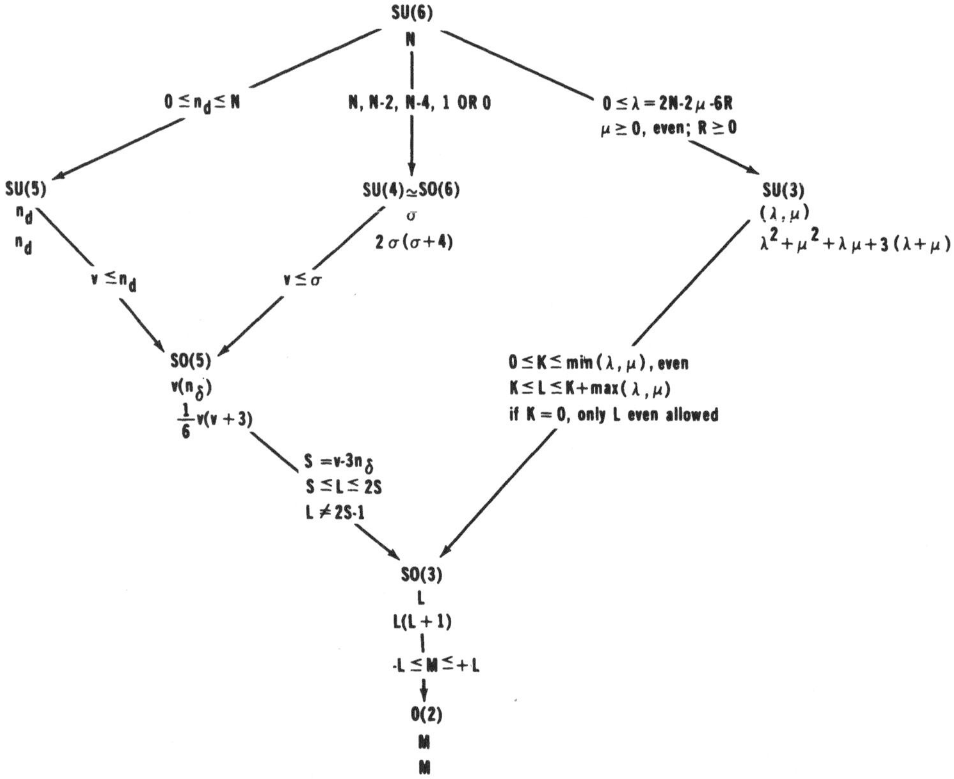

Fig.1 A complete set of basis states in each SU(6) symmetric invariant subspace is labelled by the representation labels (quantum numbers) associated with one of three chains of nested subgroups. The first line under each subgroup indicates the representation labels generated by that subgroup. The second line provides the value of the Casimir invariant of the subgroup which appears in the IBM. The reducibility conditions (splitting rules) for representation of a group on restriction to a subgroup are indicated along the downward pointing arrow connecting a group with a subgroup.

Fig.1 summarizes the IBM hamiltonian (4). The first line under each group indicates the representation labels generated by the subgroup. Labels within parenthesis (n_δ, K) are required to lift degeneracies and are not actually subgroup labels. The second line under each group indicates the value of the appropriate Casimir invariant within an irreducible representation. The splitting rules for representations of a group on restriction to a subgroup are indicated along the downward pointing arrows relating a group with a subgroup.

Three chains connect the group U(6) with U(1). Within each chain the group labels (including degeneracy labels) completely characterize a set of orthonormal basis states in the U(6) Hilbert space. The spectrum of the IBM hamiltonian (4) is particularly simple to compute when $\hat{\mathscr{H}}$ is diagonal with respect to one of these three sets of states. The eigenvalues are

$$\text{I:}\quad (A_2 = A_3 = 0)\qquad E([N];\ n_d;\ \nu(n_\delta);\ L;\ M) =$$

$$A_0N + A_1n_d + A_4\nu(\nu + 3)/6 + A_5L(L + 1)$$

$$\text{II:}\quad (A_1 = A_3 = A_4 = 0)\qquad E([N];\ (\lambda,\ \mu);\ L(K);\ M) = \tag{5}$$

$$A_0N + A_2(\lambda^2 + \mu^2 + \lambda\mu + 3(\lambda + \mu)) + A_5L(L + 1)$$

$$\text{III:}\quad (A_1 = A_2 = 0)\qquad E([N];\ \sigma;\ \nu(n_\delta);\ L;\ M) =$$

$$A_0N + A_3(N - \sigma)(N + \sigma + 4)/4 + A_4\nu(\nu + 3)/6 + A_5L(L + 1)$$

The range of values of these subgroup indices and their multiplicity are summarized in Fig.1. Nuclear systematics are obtained by assuming a weak dependence of the coefficients A_i on N.

Examples of nuclei conforming to these three types of pure spectra have been found. These spectra are like the spectra of the anharmonic vibrator [U(5)],[3] the axisymmetric rotor [U(3)],[4] and the γ-unstable rotor [U(4)$\simeq$O(6)].[5] Fig.2 shows the regions in the periodic table of the nuclides where practically pure behavior according to one of these three chains is observed.[9] Those regions left unshaded correspond to nuclei for which the residual boson-boson interaction is modeled by the Casimir invariants of two or all three of the groups: U(5), U(3), O(6). In these regions, spectra must be obtained by a numerical matrix diagonalization.

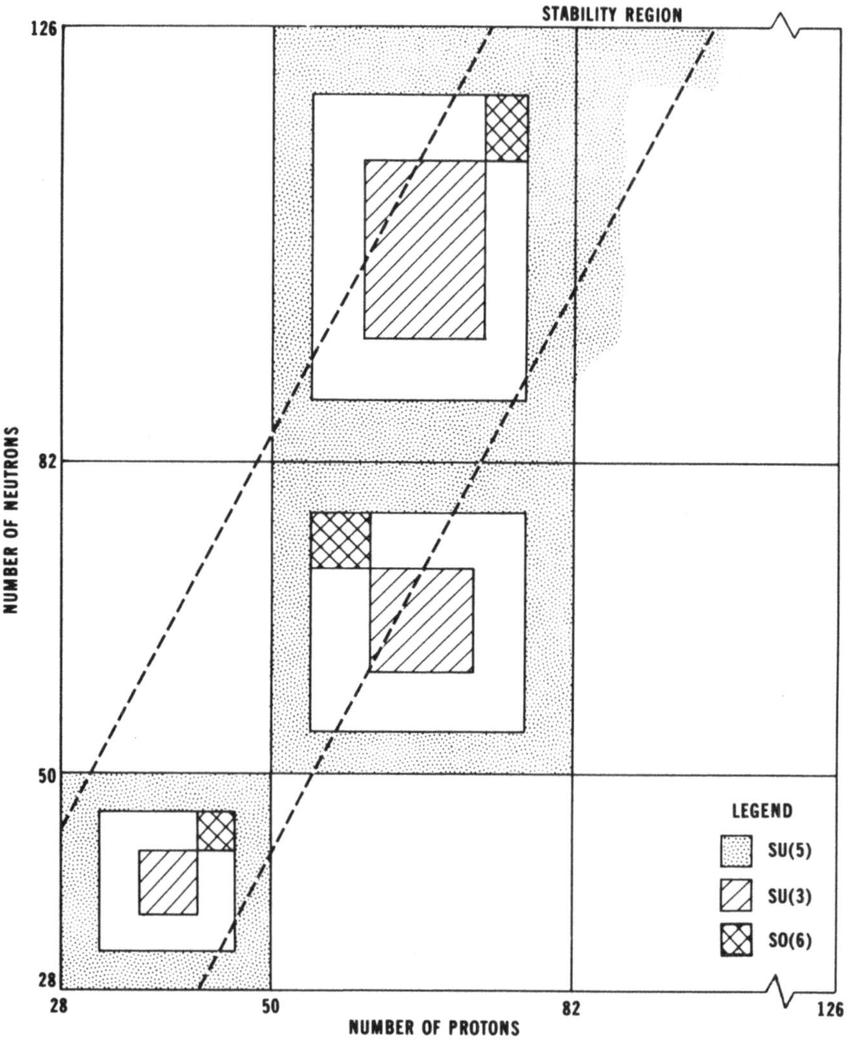

Fig.2 The periodic table of even-even nuclei is divided into regions in which nuclear properties conform closely to one of the three IBM chains. These regions are shown schematically by different shadings. Nuclei in the unshaded region are modelled by IBM hamiltonians which must be diagonalized numerically.

There is widespread (but not perfect) agreement between observed spectra and transition rates and the predictions of fitted IBM hamiltonians, both for individual nuclei and for nuclear systematics. This failure of the IBM to account for all observations must be taken in the same spirit as the failure of the band theory of solids or the Dirac theory of the hydrogen atom to account perfectly for all observations. Discrepancies between observations and IBM predictions do not invalidate the theory. Rather, they will reinforce our confidence in the model by clearly delineating its regions of validity while at the same time deepen our understanding of the residual physical mechanisms which are present (e.g.: superconductivity,[10] quantum electrodynamics[11]).

III. THE CONTINUOUS BASIS REPRESENTATION

The algebraic properties of the IBM are in principal all known since the matrix elements of the shift operators E_{ij} are known in any representation. These operators span the Lie algebra su(6) of the Lie group SU(6). While su(6) has only an algebraic structure, the group SU(6) has a topological structure as well. This topoligical structure can be used to illuminate the properties of the IBM from a completely different viewpoint. The interplay between the algebraic and topological properties of the IBM will be discussed more fully in Section 4.

The topological properties of the IBM are described most easily by introducing a continuous basis representation. Formally, the "basis" states in this representation are the coherent states for SU(6).[12] A great deal is known about these states. They are defined in terms of four mathematical structures.[13] These structures are described here in the context of the IBM.

1. A Lie group G. For the IBM, this is the dynamical group SU(6).

2. A Hilbert space V^{Λ} which carries an irreducible representation Γ^{Λ} of G. For the IBM, the invariant subspace and irreducible representation are fully symmetric and characterized by the Young tableau $\{N,\dot{0}\}$.

3. An extremal state $|\Lambda, \text{ext}>$. For the IBM this state corresponds to the presence of N s-bosons. This is the IBM ground state in the absence of residual interactions. This state is sometimes called by the fancy name "condensate."

4. A stability subgroup H. This subgroup leaves the extremal state invariant up to a phase factor. The phase factor is unimportant since projective representations of the Hilbert space carry all physical information. For the IBM the stability subgroup is U(5).

Coherent states for SU(6) are defined as follows. For an arbitrary group element $g \varepsilon SU(6)$ there is a unique coset decomposition[14]

$$g = \Omega h \tag{6}$$

with $h \varepsilon H$ (stability subgroup) and $\Omega \varepsilon G/H$ (coset representative). Then

$$g\left|{\Lambda \atop \text{ext}}\right> = \Omega h\left|{\Lambda \atop \text{ext}}\right> = \Omega\left|{\Lambda \atop \text{ext}}\right> \text{ x (phase factor)}$$

$$= \left|{\Lambda \atop \Omega}\right> \text{ x (phase factor)} \tag{7}$$

The state $|\Lambda,\Omega>$ is called a coherent state[12,13]. For the IBM, $\Omega \varepsilon SU(6)/U(5) \simeq S^{10}$, the 10-dimensional sphere.

Coherent states are labeled by group elements. It is this connection with Lie group theory that endows the IBM with a host of geometric properties. For present purposes it is sufficient to label the coherent states $|N,\Omega>$ by the six complex parameters $(z_1,z_2,z_3,z_4,z_5,z_6)$ which obey the normalization condition

$$\sum_{i=1}^{6} |z_i|^2 = 1 \tag{8}$$

These complex numbers may be interpreted as the components of a single boson in the 6-dimensional s-d boson space. An overall phase factor may be removed from this single particle wave function to make z_1 real and nonnegative. The coherent state $|N,\Omega>$ is just the fully symmetric N^{th} order tensor product based on the single boson wave function[12,15]

$$\left|{N,\dot{0} \atop \Omega}\right> = \left(\left|{1,\dot{0} \atop z_1,z_2,z_3,z_4,z_5,z_6}\right>\right)^{\otimes N} \tag{9}$$

Generalized coherent states have a large number of useful properties. One of these properties will be exploited in Section 3 and two will be exploited in Section 4.

1. Expansion of Coherent States. The coherent states may be represented as superpositions of diagonal states $|\Lambda$, "M">. Here Λ indexes the invariant hilbert space ([N] for IBM) and "M" may be chosen according to one of the three chains shown in Fig.1, as a Gel'fand-Tsetlein basis,[16] or in some other convenient way.

Then

$$\Omega | {}^{N}_{ext} >$$

$$\text{def.} \quad /\!/ \quad \backslash\!\backslash$$

$$|{}^{N}_{\Omega}> \; = \; |{}^{N}_{"M"}> \Gamma^{N}_{"M",ext}(\Omega) \tag{10}$$

The matrix elements $\Gamma^{N}_{"M",ext}(\Omega) = <N\ "M"|\Omega|N,ext>$ have been computed for the unitary groups in the Gel'fand-Tsetlein basis.[15]

2. Overcompleteness of Coherent State. The coherent states $|\Lambda,\Omega> \sim |[N],z>$ are vastly overcomplete. Any state $|\psi> \varepsilon V^{\Lambda}$ can be expanded in terms of the coherent states

$$|\psi> \; = \; \int |\Lambda,\Omega> <\Lambda,\Omega|\psi> \, d\mu(\Omega) \tag{11}$$

The measure $d\mu(\Omega)$ is determined from the Lie groups G and H.[13] For the IBM, $G/H = SU(6)/U(5) \sim S^{10}$ and $d\mu(\Omega) \sim |z_1|^{-1}$. In addition, any operator $\hat{\theta}$ mapping V^{Λ} into itself can be expressed as a superposition of coherent state "projectors"

$$\hat{\theta} \; = \; \int |{}^{\Lambda}_{\Omega}> F_{\hat{\theta}}(\Omega) <{}^{\Lambda}_{\Omega}| \, d\mu(\Omega) \tag{12}$$

In both (11) and (12) the weight functions are nonunique.

3. Expectation Values. Expectation values in the coherent state representation are rather simply expressed:[17]

$$\begin{aligned} <{}^{N}_{z}| E_{ij} |{}^{N}_{z}> \; &= \; N z_i^* z_j \\ <{}^{N}_{z}| E_{ij}^2 |{}^{N}_{z}> \; &= \; N(N-1)(z_i^* z_j)^2 \end{aligned} \tag{13}$$

In particular, the classical limit can be written down at sight:

$$\begin{aligned} \lim_{N\to\infty} <{}^{N}_{z}| (E_{ij}/N)^2 |{}^{N}_{z}> \; &= \; (z_i^* z_j)^2 \\ \lim_{N\to\infty} <{}^{N}_{z}| f(E_{ij}/N) |{}^{N}_{z}> \; &= \; f(z_i^* z_j) \end{aligned} \tag{14}$$

TABLE I

Classical Limits of the Casimir Operators In the Interacting Boson Model

$$\langle {}^{N}_{Z} | \mathcal{H}/N | {}^{N}_{Z} \rangle = \Sigma \text{ coefficient} \times \text{Tr } M^{+} M$$

GROUP	Dim M	M	CONDITIONS
U(5)	5×5	$M_{ij} = Z_i \delta_{ij}$	$2 \leq i, j \leq 6$
SU(4) ≃ SO(6)	6×6	$M_{ij} = \text{Im } Z_i^* Z_j$	$1 \leq i, j \leq 6$
SO(5)	5×5	$M_{ij} = \text{Im } Z_i^* Z_j$	$2 \leq i, j \leq 8$
SU(3)	$2(11) + (44) + (55)$	$(56) + \sqrt{2}[(14) + (42)]$	$(64) + \sqrt{2}[(35) + (51)]$
		$2(22) + (44) + (66)$	$(54) + \sqrt{2}[(36) + (62)]$
			$2(33) + (55) + (66)$
	$(ij) = Z_i^* Z_j$		
SO(3)	3×3	$M(SO(3)) = \text{Im } M(SU(3))$	

TABLE II

Phase Transitions That May Occur in the Interacting Boson Model

PHASE TRANSITION TYPE	OPERATOR TO BE MINIMIZED	EXPERIMENTAL PARAMETERS	ORDER PARAMETERS
Ground State Energy Phase Transition	$\mathcal{H}$	IBM PARAMETERS $A_1 \rightarrow A_5$	SU(6) Coherent State Parameters
Thermodynamic Phase Transition	$\mathcal{F}$	Temperature T	SU(6) Coherent State Parameters or off diagonal matrix elements of single particle density operator ρ
High Spin Phase Transition	$\mathcal{H}$ $\mathcal{F}$	$A_1 \rightarrow A_5$ T	SU(6) Coherent State Parameters " , ρ
Dynamical Phase Transition	$S = \int \mathcal{L}\, dt$	$A_1 \rightarrow A_5$ (T)	TDHF orbital invariants

where f is a polynomial in the shift operators E_{ij}/N. The classical limits of the Casimir operators which appear in the IBM (1.4) are simple to compute. The results are summarized in Table 1.

The algebraic studies of the IBM described in Section 1 can now be complemented by a number of geometric studies. The geometric studies initiated so far fall into six broad categories:

1. Phase Transitions
2. Transformation Theory
3. Dynamical Equations
4. Quantum-Classical Correspondence
5. Transition Operator Expansions
6. Phenomenological Properties.

Space considerations limit us to a description of only the first two categories, presented in the following two sections.

IV. PHASE TRANSITIONS

Phase transitions are qualitative changes in the properties of a system. The properties of a system are generally characterized by order parameters which are introduced to facilitate the description of the system. In many instances the state of the system is determined by minimizing a suitable variational function (Hamiltonian, free energy, action integral) with respect to the order parameters. The resulting minimized function is then independent of the system order parameters, and depends only on the experimental parameters which constrain the system. This function is generally not analytic. Points in the experimental parameter space at which the minimized function is not smooth are called phase transition points. The order of the phase transition is the degree of the derivative of the minimized function (with respect to the experimental parameters) at which a discontinuity first occurs. Table 2 summarizes the types of phase transitions that may be studied for the IBM. We will be concerned entirely with ground state energy phase transitions.

Phase transitions are more or less sharply defined depending on whether the number of particles present is large or small. However, for mean field models of the type under discussion, numerical and analytical results indicate that the concept of phase transition retains its validity even for small $N (N \sim 8)$[18]. Phase transitions in the IBM are studied by computing the expectation value of the IBM hamiltonian (4) in the coherent state representation. The minimum value of the resulting function provides an upper bound on the ground state energy per boson. In the classical limit ($N \to \infty$) the expectation value (4) can be written down at sight as explained in Sec.2 (cf. 14 and Table 1.). The expectation value of $\mathcal{H}/N$ for finite N is obtained simply from the classical limit by multiplying the terms in (4) by the factor[19]

- 1 for linear Casimir invariants
- (1 - 1/N) for quadratic Casimir invariants.

A lower bound on the ground state energy can also be computed. This is obtained from the classical limit by multiplying the classical limits of the terms appearing in (4) by the factors[19]

- (1 + 1/N) for linear Casimir invariants
- (1 + 1/N) (1 + 2/N) for quadratic Casimir invariants.

The three functions, the upper and lower bounds (for finite N) and the classical limit, differ by order 1/N and therefore have the same critical properties. The finite size effect merely serves to renormalize slightly the values of the experimental parameters A_i which appear in the IBM and therefore the locations, but not the types, of critical points. For this reason it is sufficient to study the critical properties of the IBM in the classical limit.

This study is accomplished by studying the competition between the three terms organizing the three chains shown in Fig.1. For this study

$$A_0 = A_4 = A_5 \tag{15}$$

and the three remaining parameters are nonnegative. The study is further simplified by observing that the Hamiltonian (4) is homogeneous in the experimental parameters A_i, so that the ground state is unchanged by the scaling transformation $A_i \rightarrow SA_i$, $S > 0$. It is therefore sufficient to study the critical properties of the IBM in the triangular region

$$A_1 + A_2 + A_3 = 1 \qquad A_i \geq 0, \quad i = 1,2,3 \tag{16}$$

The functional form of the classical limit is simplified somewhat by transforming to the intrinsic frame. In this frame the SU(6) coherent state assumes the form[20]

$$\begin{aligned}
&(z_1, z_2, z_3, z_4, z_5, z_6) = \\
&= (\sin\tfrac{\theta}{2}, \cos\tfrac{\theta}{2}\sin\gamma/\sqrt{2}, 0, \cos\tfrac{\theta}{2}\cos\gamma, 0, \cos\tfrac{\theta}{2}\sin\gamma/\sqrt{2}) \\
&= (1+\beta^2)^{-\frac{1}{2}}(1, \beta\sin\gamma/\sqrt{2}, 0, \beta\cos\gamma, 0, \beta\sin\gamma/\sqrt{2})
\end{aligned} \tag{17}$$

TABLE III

Selected Properties of the Three Limits in the Interacting Boson Model

GROUP	CHAIN	GEOMETRICAL PICTURE	FUNCTIONAL FORM OF CASIMIR INVARIANT		LOCATION OF MINIMUM		CORRESPONDING PSEUDOSPIN OPERATOR
			PROJECTIVE COORDINATES	SPHERICAL COORDINATES	θ	φ	
U(5)	I	A_1 Anharmonic Vibrator	$\frac{\beta^2}{1+\beta^2}$	$\frac{1}{2}(1+\cos\theta)$	180°	-	J_z/N
U(3)	II	A_2 Axisymmetric Rotor	$\frac{1+\frac{3}{4}\beta^4-\sqrt{2}\,\beta^3\cos3\gamma}{(1+\beta^2)^2}$	$\frac{7}{16}+\frac{2}{16}\cos\theta+\frac{7}{16}\cos^2\theta$ $-\frac{2}{4}\sin\theta(1+\cos\theta)\cos3\gamma$	70°	0	$\frac{1}{4}\frac{J_z}{N}-\frac{7}{8}\left\{\frac{J_+}{4},\frac{J_-}{N}\right\}$ $-\left\{\frac{J_++J_-}{N}\right\}\left\{\frac{1}{2}+\frac{J_z}{N}\right\}$
SO(6)	III	A_3 γ - Unstable Rotor	$\left(\frac{1-\beta^2}{1+\beta^2}\right)^2$	$\cos^2\theta$	90°	-	$-2\left\{\frac{J_+}{N_1},\frac{J_-}{N}\right\}$

The expectation value of the Casimir invariants of U(5), U(3), and SO(6) in this coordinate system are summarized in Table 3.

The ground state critical properties of the IBM were studied by computing the expectation value $<\hat{\mathscr{H}}/N>$ at each point in the triangular region (16), minimizing with respect to the order parameters β, γ, and then searching for discontinities. Fig. 3 summarizes the results of this search. A curve of first order phase transitions extends across the triangular region (16) from the edge $A_3=0$ to the edge $A_2=0$, where it terminates in a second order phase transi-

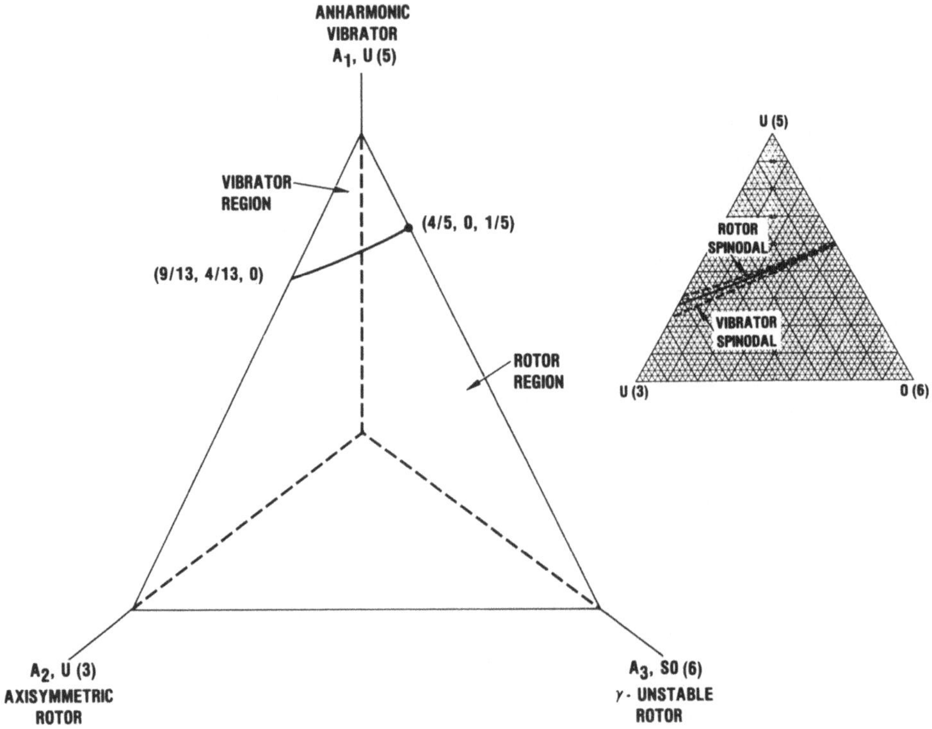

Fig. 3 A line of first-order ground state energy phase transitions extends from the point $(A_1, A_2, A_3) = (9/13, 4/13, 0)$ across the triangular region to the point $(4/5, 0, 1/5)$, where it terminates in a second order phase transition. This line of first order phase transitions is surrounded by spinodal lines marking the limits of metastable vibrator and rotor states (inset). No other islands of metastability exist in the A_1, A_2, A_3 orthant.

tion. The boundaries of this curve, the first order phase transition at $(A_1, A_2, A_3) = (9/13, 4/13, 0)$ and the second order phase transition at $(4/5, 0, 1/5)$, have previously been determined by Dieperink, Scholten, and Iachello[20]. This first order phase transition line divides the parameter space into two regions, one characterized by vibrator-like behavior, the other by rotor-like behavior. A small region of vibrator metastability extends below the first order phase transition line. It is bounded by the spinodal line obeying the linear equation

$$A_1 - 2A_2 - 4A_3 = 0 \tag{18}$$

Similarly, a region of rotor metastability extends into the vibrator regime, where it is bounded by a spinodal line. The spinodals obey the 2/3 power law dependence characteristic of the cusp catastrophe and the van der Waals fluid. Indeed, a close analogy between either and the IBM is possible. The regions between the two spinodal lines parameterize IBM hamiltonians which exhibit pronounced nuclear isomerism. There are no other regions or islands of metastability in the triangular region (16).

Nuclear phase transitions will occur in the complement to the shaded regions shown in Fig.2.

The grounded state critical properties of the IBM are very closely related to the ground state critical properties of a pseudospin model Hamiltonian. The pseudospin operator counterparts of the Casimir operators of U(5), U(3) and SO(6) are indicated in Table 3. The expectation values of these three pseudospin operators are given in the SU(2) coherent state representation. It is clear that there is a homomorphism between the IBM expectation values in the intrinsic frame (θ,γ) and the pseudospin model expectation values in the SU(2) coherent state representation (θ,ϕ) under the identification

$$(\theta,\ 3\gamma) = (\theta,\ \phi) \tag{19}$$

The relation between the IBM and the pseudospin model is summarized in Fig.4. Transformation from the full SU(6) coherent state representation to the intrinsic coordinate system effects a projection from the sphere S^{10} to the sphere S^2. Transformation from the IBM in the intrinsic frame to the pseudospin model in the SU(2) continuous basis representation is accomplished by a 3-1 wrapping of the sphere S^2 onto itself. It is clear that some information is lost in projecting the IBM in the intrinsic frame down to the pseudospin model because of the 3-1 homomorphism $S^2 \rightarrow S^2$ which is involved. It is not known how much information is lost in the projection from

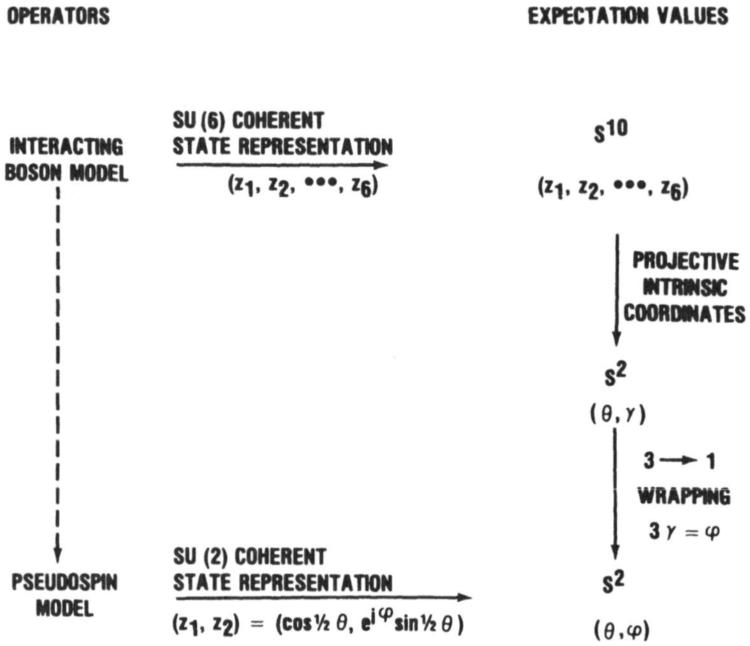

Fig.4 The classical limit of the IBM is defined over a 10-dimensional sphere S^{10}. Computations are facilitated by transforming to the intrinsic reference frame (β,γ), diffeomorphic to the 2-sphere $S^2(\theta,\gamma)$. The classical limit of any pseudospin model is defined over the 2-dimensional Bloch[21] sphere S^2 (θ,ϕ). With the identifications provided in Table 3, the 3-1 homomorphism $(\theta, 3\gamma) = (\theta,\phi)$ provides a mapping between the ground state critical properties of the IBM and those of the corresponding pseudospin model.

S^{10} down to S^2 involved in the restriction from the generalized SU(6) coherent states to the intrinsic frame states.

V. CHANGE OF BASIS TRANSFORMATIONS

A complete orthonormal system of basis vectors $|\Lambda, "M">$ in V^Λ and the overcomplete set of coherent states $|\Lambda, \Omega>$ are related by

$$\left|{\Lambda \atop \Omega}\right> = \sum_{"M"} \left|{\Lambda \atop "M"}\right>\left<{\Lambda \atop "M"}\right|{\Lambda \atop \Omega}\right> \tag{20}$$

$$|{}^{\Lambda}_{"M"}\rangle = \int |{}^{\Lambda}_{\Omega}\rangle \langle{}^{\Lambda}_{\Omega}|{}^{\Lambda}_{"M"}\rangle \, d\mu(\Omega) \tag{21}$$

The functions $\langle"M"|\Omega\rangle$ have one foot in each field: they are half algebraic, half geometric. They are in fact the matrix elements of the unitary transformation between the algebraic and the geometric representations of the model hamiltonian.

Functions such as this with one foot in each field are actually familiar from the elementary theory of angular momentum. The functions $\langle\theta\phi|{}^{\ell}_{m}\rangle \sim \Gamma^{\ell}_{\underline{m}o}(\theta, \phi) \sim Y^{\ell *}_{\underline{m}}(\theta, \phi)$ are creatures of this type. Matrix elements in a mixed basis representation can be taken in two ways using resolutions of the identity in either the discrete or the continuous basis:

$$\begin{array}{c} \langle\theta\phi|L_{\pm}|{}^{\ell}_{m}\rangle \\ /\!/ \qquad \backslash\!\backslash \\ \langle\theta\phi|L_{\pm}|\theta'\phi'\rangle\langle\theta'\phi'|{}^{\ell}_{m}\rangle = \langle\theta\phi|{}^{\ell}_{m'}\rangle\langle{}^{\ell}_{m'}|L_{\pm}|{}^{\ell}_{m}\rangle \end{array} \tag{22}$$

The matrix elements of the shift operators in these two representations are

$$L_{\pm}|\theta'\phi'\rangle = e^{\pm i\phi}\left(\pm\frac{\partial}{\partial\theta} + i\cot\theta\frac{\partial}{\partial\phi}\right) \times$$

$$\delta(\cos\theta - \cos\theta')\delta(\phi - \phi') \tag{23}$$

$$\langle{}^{\ell}_{m'}|L_{\pm}|{}^{\ell}_{m}\rangle = \sqrt{(\ell \mp m)(\ell \pm m + 1)}\,\delta_{m',m+1}$$

In exactly the same way, matrix elements of the shift operators E_{ij} may be taken in either a discrete or a continuous basis

$$\begin{array}{c} \langle"M"|E_{ij}|\Omega\rangle \\ /\!/ \qquad \backslash\!\backslash \\ \langle"M"|E_{ij}|"M'"\rangle\langle"M'"|\Omega\rangle = \langle"M"|\Omega'\rangle\langle\Omega'E_{ij}|\Omega\rangle \end{array} \tag{24}$$

The matrix elements in the algebraic representation are particularly easy to take in the Gelfand-Tsetlein basis $|n_6, n_5, n_4, n_3, n_2, n_1\rangle$, where $n_6 \geq n_5 \geq n_4 > n_3 \geq n_2 \geq n_1 \geq 0$. The matrix elements are

$$\langle \underset{\sim}{n}' | E_{ij} | \underset{\sim}{n} \rangle = (n_{i+1} - n_i)^{1/2} (n_j - n_{j-1})^{1/2} \delta_{n_i', n_i+1} \times \\ \times \delta_{n_j', n_j-1} \prod_{k \neq i,j} \delta_{n_k', n_k} \tag{25}$$

In the continuous representation these matrix elements are[22]

$$\langle \Omega' | E_{ij} | \Omega \rangle = z_i \partial_i + z_j z_i^* [\tfrac{1}{2} N - \tfrac{1}{2}(z \cdot \nabla + z^* \cdot \nabla^*)] \\ \times \delta(\Omega' - \Omega) \tag{26}$$

where $$z_1 = [1 - z^* \cdot z]^{1/2}$$

$$\partial_1 = \frac{\partial}{\partial z_1} \rightarrow \frac{1}{2} z_1 [N - (z \cdot \nabla + z^* \cdot \nabla^*)]$$

The transformation of the shift operators E_{ij} from the algebraic to the geometric representation is effected by the unitary transformation $\langle "M" | \Omega \rangle$. The representation of shift operators E_{ij} by first order differential operators is called a $\mathscr{D}$-algebra mapping; the corresponding operator is denoted $\mathscr{D}(E_{ij})$.

By the same means the IBM hamiltonian can be transformed from the algebraic to the geometric representation. In the algebraic representation $\hat{\mathscr{H}}$ is simply a hermitian matrix whose dimensionality corresponds to that of the space $V^{\{N,0\}}$ on which it acts. This matrix can be diagonalized by numerical techniques. In the geometric representation $\hat{\mathscr{H}}$ is a partial differential operator. This operator is of second order because $\hat{\mathscr{H}}$ involves only linear and quadratic Casimir operators. In the continuous representation the energy eigenvalue equation becomes essentially a Schrödinger equation on S^{10}. As such, it is unlike any previous Schrödinger equation proposed to describe nuclear systems.

This second order differential eigenvalue equation

$$\mathscr{D}(\hat{\mathscr{H}}) \langle \Omega | \Psi \rangle = E \langle \Omega | \Psi \rangle \tag{27}$$

can in principal be solved for both the eigenvalues and the eigenfunctions. However, these can be obtained more simply by exploiting the unitary transformation and rewriting (27) as

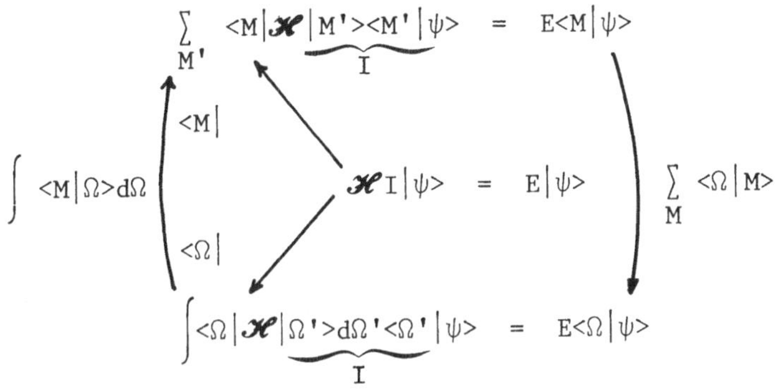

Fig.5 The IBM may be solved in either the algebraic or the goemetric representation. In the latter, a Schrodinger-like equation is involved. The relation between these two representations is summarized here.

$$\langle\Omega|\hat{\mathcal{H}}|\Omega'\rangle\ \langle\Omega'|\Psi\rangle = E\langle\Omega|\Psi\rangle$$

$$\langle\Omega|\text{"}M\text{"}\rangle\ \langle\text{"}M\text{"}|\hat{\mathcal{H}}|\text{"}M'\text{"}\rangle\ \langle\text{"}M'\text{"}|\Psi\rangle = E\langle\Omega|\text{"}M\text{"}\rangle\ \langle\text{"}M\text{"}|\Psi\rangle \tag{28}$$

The coefficients $\langle\text{"}M\text{"}|\Psi\rangle$ are obtained by numerical diagonalization of the hamiltonian $\hat{\mathcal{H}}$ in the algebraic (orthonormal) basis. The functions $\langle\Omega|\text{"}M\text{"}\rangle$ are also known in the Gel'fand-Tseltein basis. Therefore, the eigenfunctions of the second order differential operator $\mathcal{D}(\hat{\mathcal{H}})$ can be computed for any nucleus whose spectrum has been successfully modeled by an IBM hamiltonian. The change of basis transformation between the algebraic and the geometric representations is summarized in Fig.5

VI. RELATION BETWEEN THE LIQUID DROP MODEL AND THE INTERACTING BOSON MODEL

The shell model, in its various modifications, ramifications, and equivalent representations, has accounted reasonably well for single particle nuclear properties. All further developments which attempt to account for nuclear single particle properties must at some point make contact with the shell model. Historically, there has been one other widely studied model of collective nuclear motion. This is the liquid drop model of Bohr and Mottelson[23]. It is inevitable that at some point the Interacting Boson Model of collective nuclear properties be compared with the liquid drop model of collective nuclear motion. The foundations for such a comparison are laid in the present section. In subsection a we summarize the Bohr-Mottelson liquid drop model; in subsection b we summarize the Arima-Iachello Interacting Boson Model. These synopses are in terms of the logical inputs to these models; the outputs of these models (spectra, transition rates, etc.) are adequate to justify their side-by-side consideration here. Subsection c is devoted to a brief summary of the important modifications and extensions of these two models. In Subsection d we make a comparison of the assumptions of these two models. Some assumptions are quite similar; others differ in important details. This comparison of logical inputs is crucial. There is no point in trying to demonstrate the equivalence of these two models if the underlying assumptions are not in some way equivalent. Similarly, it is not too useful to show that the models are not equivalent unless we can put our collective finger on the assumptions which differ in the two models. In Subsection e we indicate how those assumptions which differ in the two models can be modified slightly so as to give these models a congruent underlying logical structure, and where further work must be concentrated if the equivalence of (relation between) these two models is to be shown.

a. Synopsis oɩ the Bohr-Mottelson Liquid Drop Model of Collective Nuclear Motion

In this Subsection we review briefly the logical underpinnings of the classical model of Bohr and Mottelson.

1. Physical Basis. This model is based on the theory of oscillations of a liquid drop around its equilibrium shape.

2. Physical Bias. The starting point for the model is classical continuum mechanics.

3. Mathematical Bias. The model is palpably geometric.

4. Shell Model Inputs: None

5. Starting Point. The radius $R(\theta, \phi)$ is a single valued function of the angular coordinates

$$R(\theta, \phi) = \sum_{L,M} a^L_M Y^L_M (\theta, \phi) > 0 \tag{29}$$

6. Approximations. The following series of approximations reduce the complexity from that of continuum mechanics to that of the classical mechanics of a system with a small number of degrees of freedom

- odd L terms represent a displacement of the drop, and are neglected.
- L = 0 term is constant
- L = 2 terms describe the quadrupole oscillations, which are important
- L = 4 terms "can be shown to be of much less importance"

than quadrupole deformations, and are neglected. $L \geq 6$ terms are also neglected.

7. Technical Condition. The L=2 terms eventually become the dynamical variables of the liquid drop model. These five complex parameters obey a reality constraint

$$a_M^{L^*} = (-)^M a_{-M}^L \tag{30}$$

There are thus five real coordinates in this model.

8. Dynamics. In lowest order approximation, the oscillation energy can be written as the sum of a quadratic form in the coordinates a_M^L and a quadratic form in the velocities $\dot{a}_M^L$ or the associated momenta π_M^L:

$$E \simeq \frac{1}{2}B \sum_M |\dot{a}_M^L|^2 + \frac{1}{2}C \sum_M |a_M^L|^2 = \frac{B}{2} \sum_M |\pi_M^L|^2 + \frac{C}{2} \sum |a_M^L|^2 \tag{31}$$

9. Dynamical Variables. The five real variables a_M^L and the canonically conjugate momenta π_M^L constitute the dynamical variables of the theory.

10. Configuration Space. This is $\tilde{\mathbb{R}}^5$, spanned by the five real variables a_M^L. (We remark here that this is not quite correct, for the a_M^L must obey the positivity constraint $R(\theta, \phi) > 0$.)

11. Phase Space. This is $\tilde{\mathbb{R}}^5 \otimes \mathbb{R}^5$, spanned by a_M^L and π_M^L. Both copies of $\mathbb{R}^5$ are Euclidean. This can be determined from the metric tensor $G_{\mu\nu}$ which is determined from the expression for the kinetic energy: $T = 1/2[G_{\mu\nu}\pi^\mu\pi^\nu]$.

12. Underlying Hilbert Space. Before the transformation from the classical to the quantum theory can be carried out, the Hilbert space for the quantum theory must be established. This is the space of square-integrable functions defined over configuration space $\mathscr{L}^2(\mathbb{R}^5, \sqrt{|G|})$.

13. Quantization. This is carried out by means of the usual Dirac prescription, following Pauli:

$$[\pi^\mu, a_\nu] = \frac{\hbar}{i}\left\{\pi^\mu, a_\nu\right\} = \frac{\hbar}{i}\delta_\nu^\mu \tag{32}$$

14. Method of Solution. Quantization transforms the classical Hamiltonian function into a second order differential operator.

TABLE IV

Logical Inputs To The
Bohr-Mottelson Liquid Drop Model Of Collective Nuclear Motion
And The
Arima-Iachello Interacting Boson Model Of Collective Nuclear States

NUMBER	AUTHORS MODEL COLLECTIVE NUCLEAR	BOHR-MOTTELSON LIQUID DROP MOTION	ARIMA - IACHELLO INTERACTING BOSON STATE
1.	Physical Basis	Liquid Drop Motion	Cooper Pairing
2.	Physical Bias	Classical Mechanics	Quantum Mechanics
3.	Mathematical Bias	Geometry	Algebra
4.	Shell Model	- -	a. Inert Closed Core b. Residual Nucleon-Nucleon Interaction
5.	Starting Point	$R(\theta,\phi) = \sum_{L,M} a_M^L Y_M^L(\theta,\phi) > 0$	$\|J\rangle = \sum_{j,j'} \alpha_{jj'} \|(jj')J\rangle$
6.	Approximations		
	J odd	Represent Nuclear Displacement Ignored	Require Higher Harmonic Oscillator Shells, Ignored
	J = 0	Constant	Strong Interaction
	J = 2	Principal Vibrational Mode	Slightly Weaker Interactions
	$J \geq 4$	Much Less Importance Than L = 0, 2 Terms	Much Less Importance Than J = 0, 2 Terms
7.	Technical Condition	$a_M^L = (-)^M a_M^{L*}$	$s^+s + d^+d = N$
8.	Dynamics	$E \simeq \frac{1}{2} B \sum \|a_M^L\|^2 + \frac{1}{2} C \sum \|\dot{a}_M^L\|^2$	$\mathscr{H} \simeq f(E_{ij})$
9.	Dynamical Variables	a_M^L , π_M^L	z, z^*, SU(6) Coherent State Parameters (not a direct input).
10.	Configuration Space	$\tilde{\mathbb{R}}^5$ Euclidean 5 dim.	SU(6)/U(5) compact 10 dimensional
11.	Phase Space	$\tilde{\mathbb{R}}^5 \otimes \mathbb{R}^5$ Euclidean 10 dim.	SU(6)/U(5), compact 10 dimensional
12.	Underlying Hilbert Space	$\mathscr{L}^2(\mathbb{R}^5, \sqrt{\|G\|})$	$\mathscr{L}^2[SU(6)/U(5);\ d\mu;\ \{N,\mathring{0}\}]$
13.	Quantum Classical Mapping	$[\pi^\mu, a_\nu] = \frac{\hbar}{i}\{\pi^\mu, a_\nu\} = \frac{h}{i}\delta_\nu^\mu$	$f(z) = \langle {}^N_z \| \mathscr{H} \| {}^N_z \rangle$
14.	Method of Solution Solution	Transfrom to Intrinsic Coordinates Solve Schrodinger-like Equation	Matrix Diagonalization

This operator is simplified by transformations to the intrinsic coordinate system (β, γ). The resulting Schrodinger equation must then be solved for energy eigenvalues and eigenfunctions.

b. Synopsis of the Arima-Iachello Interacting Boson Model of Collective Nuclear States

This subsection is written and numbered in parallel with the preceeding subsection. The paralles is made manifest in Table 4.

1. Physical Basis. This model is based on the theory that similar nucleons obey Cooper pairing to a greater or lesser extent.

2. Physical Bias. The starting point for this model is the independent particle picture of the nucleus as expressed in the nuclear shell model.

3. Mathematical Bias. This model is unabashedly algebraic

4. Shell Model Inputs:
   a. Inert closed core
   b. Basis states are drawn from valence orbitals of next shell

5. Starting Point. Nucleons of the same isotopic spin and opposite spin form Cooper pairs

$$|J\rangle = \sum_{jj'} \alpha_{jj'} |(jj')J\rangle \tag{33}$$

Here nucleons with angular momenta j and j' couple to form a boson with angular momentum J. The sum extends over identical nucleons of opposite spin in valence orbitals outside of closed shells, and $\alpha_{jj'}$ are suitable coupling coefficients.

6. Approximations. Only nucleons in valence orbitals outside of closed shells are considered. The interactions between nucleons of opposite isospin are ignored. The following series of further approximations reduce the complexity of the problem from a field theory to a quantum theory

- odd J terms can arise only from nucleons in adjacent opposite parity harmonic oscillator levels and therefore are absent.
- J = 0 term is present. In fact, this s boson is strongly bound.
- J = 2 terms (d-bosons) are important, but not as strongly bound as the s-boson.

- $J = 4$ terms are weakly attractive or repulsive, and are therefore neglected.

- $J \geq 6$ terms are also neglected.

7. Technical Condition. The s- and d-bosons formed from a single nucleon pair span a six-dimensional Hilbert space. These wave functions obey a unitarity constraint, so that only five complex amplitudes are independent. The operators which effect transitions from one boson state to another span the Lie algebra **u**(6) and generate the dynamical group U(6).

8. Dynamics. The residual interaction after formation of the shell model core and the strongly bound nucleon pairs can be expressed entirely as a function of the u(6) shift operators. In particular, the Hamiltonian can be approximated by a linear superposition of linear and quadratic Casimir invariants of the three naturally occurring subgroup chains illustrated in Fig. 1. All physical operators (e.g., E2, M1 operators) can also be expressed as a function of the **u**(6) shift operators.

9. Dynamical Variables. The Interacting Boson Model, in its original formulation, possesses no dynamical variables at all. These are introduced by means of the continuous basis representation. In particular, they are essentially the SU(6) coherent state parameters z SU(6)/U(5).

10. Configuration Space. This is the coset space SU(6)/U(5), a complex 5-dimension (real 10-dimensional) compact space.

11. Phase Space. Normally the dimension of phase space is twice that of the basic configurations space. However, we have shown that in the classical limit of quantum mechanics, phase space may be identified with configuration space. Therefore phase space is also SU(6)/U(5).

12. Underlying Hilbert Space. The Hilbert space on SU(6)/U(5) is reducible, and decomposes into a number of invariant subspaces. Each is indexed uniquely by an integer N = 0, 1, , ... This integer characterizes

- Mathematically, the order of the symmetric tensor representation of SU(6) carried by that space. The dimension of the space and size of the representation matrix is $(N + 6 - 1)!/N!(6 - 1)!$.

- Physically, the number of nucleon pairs (bosons) outside the shell model core.

These Hilbert spaces are denoted $V^N = \mathscr{L}^2[SU(6)/U(5);\ d\mu;\ \{N, \dot{0}\}]$.

The measure $d\mu$ is the invariant measure on the coset SU(6)/U(5).

13. Classical Mapping. This is carried out by taking the expectation value of the IBM Hamiltonian in the coherent state basis.

$$f_N(z) = \langle {}^N_z | \hat{\mathcal{H}} | {}^N_z \rangle \tag{34}$$

14. Method of Solution. The IBM Hamiltonian is a D x D matrix, where $D = \dim V^N = (N+5)!/N!5!$. Numerical diagonalization of this matrix determines energy eigenvalues and the corresponding D-dimensional eigenvectors.

c. Evolution of the Collective Models.

The successes of the liquid drop model and the interacting boson model have been adequate to encourage attempts to extend their ranges of validity. The evolution of these models, though done independently, proceed along remarkably paralle lines. This parallel development is summarized in Table 5 and summarized below.

The initial models, the liquid drop model and the IBM - IBA - 1 (interacting boson approximation - 1) make no distinction between protons and neutrons.

Attempts to distinguish between protons and neutrons led to one class of modification of the original collective models. On the

TABLE V

The evolution of collective models has proceeded in a series of parallel developments

LIQUID DROP MODEL	INTERACTING BOSON MODEL
Liquid Drop	IBA - 1 = IBM
Two Fluid Model	IBA - 2
Nucleon Coupled to Surface Oscillations	IBFA
-	Supermultiplet Model

the liquid drop side, this took the form of two fluid models[24]. On the IBM side, this took the form IBA - 2[25-28].

Another generalization of the initial collective models involved attempts to couple a single nucleon to the collective motion/states. On the liquid drop side, this took the form of coupling the single nucleon motion to the nuclear surface oscillations. On the IBM side, this took the form of the interacting boson-fermion approximation (IBFA)

One further modification, peculiar to the IBM side, has been the attempt to include nuclei with both even and odd numbers of valence nucleons into a single supermultiplet[31]. This model not only has no counterpart on the liquid drop side, but also represents the only successful attempt so far to find a supersymmetry in nature.

d. Strategic and Tactical Assumptions.

In Subsections a and b we have provided a list of the logical inputs to the two collective nuclear models. These inputs are summarized and compared in Table 4. A glance at this Table will show an almost direct parallel in the assumptions intrinsic to the two theories. The only point at which the parallel breaks down is at point # 13. It is clear that the parallelism cannot be perfect; one theory has a classical starting point, so must be quantized, while the other theory has a quantum origin, so its classical limit can be taken. Therefore, the point at which the parallel breaks down is the necessary dual between these two theories.

In order for a comparison such as that summarized in Table 4 to be possible, it is clear that the two theories must live in the same ballpark. It is not so clear whether these two theories are on the same team or opposing teams.

To resolve this question, it is useful to break down the logical inputs to these two models into two subsets, the "strategic" and the "tactical" inputs. The first three inputs are strategic, the remaining eleven are tactical.

It is clear from Table 4 that the two theories cannot be equivalent, while at the same time they cannot be very different either. It is therefore useful to see if minor modifications can be made to the logical inputs of these theories to bring them to a state in which they are equivalent.

In searching for modifications of these input assumptions, it is only necessary to concentrate on those inputs which are "tactical" (i.e., technical). For if the technical considerations of these two theories can be brought into conformity and it can be shown that the modified theories are equivalent, then the "strategic" considerations will simply be a manifestitation of the representation chosen (modified liquid drop, modified IBM) and nothing more.

This leaves open the question of how these two (modified) theories are to be compared. We confront this question now. In principle this comparison may be carried out in two ways.

i. The liquid drop model is transformed to an algebraic representation, and compared with the IBM in its original algebraic formulation.

ii. The IBM is transformed to a geometric representation, and compared with the liquid drop model in its original geometric formulation.

Either method can be used for the comparison of the two theories. For if the theories are in fact equivalent, the success of either comparison mechanism guarantees the success of the other.

We will concentrate on the second method of comparison. The continuous basis representation provides a transformation mechanism between the algebraic and the geometric representations of the IBM. At the present time no such luxury exists with respect to the liquid drop model. In fact, one of the major consequences of a successful modification and comparison of these two theories will be an algebraic formulation of the liquid drop model, together with a unitary transformation between its algebraic and geometric representations.

e. Critical Comparison of Tactical Assumptions

We proceed now to analyze the 10 "tactical" assumptions. For each of these inputs the similarities in the assumptions for the two models will be pointed out and the differences stressed. A specific suggestion will then be made for modifying the input assumption for the liquid drop model and/or the interacting boson model so that the assumptions for the two models at each input level are essentially equivalent.

4. Shell Model Input. The IBM has two shell model inputs, the liquid drop model has none. The first shell model input to the IBM is the assumption of an inert closed core. This can be visualized geometrically as a spherical core inside a liquid drop. It is therefore useful to modify the liquid drop model by considering a "tidal planet model". In this model a hard spherical core of radius R_c is embedded within the liquid drop. Oscillations of the surface are studied, as in the liquid drop model. Now, however, the radial equation obeys the constraint

$$R(\theta, \phi) \geq R_c \tag{35}$$

The second shell model input to the IBM is used primarily as a crutch to visualize the occurrence of bosons in terms of shell model Cooper pairs.

5. Starting Point. It is convenient to take as the starting point of the liquid drop model the following expression for the nuclear radius $R(\theta, \phi)$:

$$R(\theta, \phi) = \text{Constant} - s\, Y_0^0 + \sum_{\mu=-2}^{+2} q_\mu\, Y_\mu^2(\theta, \phi) \geq R_c \qquad (36)$$

The determination of the constant and the interpretation of the six couple parameters s, q_μ will be discussed shortly. The starting point of the IBM remains unchanged.

6. Approximations. In both models odd J excitations are excluded by intrinsic reasons, and excitations with $J \geq 4$ are neglected as being unimportant. The $J = 2$ excitations are important in both models, but the $J = 0$ excitation is included in one model (IBM) and excluded in the other. It is easily seen that oscillations involving only the quadrupole mode are not volume preserving. To maintain a constant nuclear density (a lower energy requirement than compressive oscillations) it is necessary to include also the $J = 0$ mode. This has been done in the expression (36) for $R(\theta, \phi)$. In the standard liquid drop model s is a constant, but the incompressibility requirement means that s now becomes a (dynamical) variable.

Remark: The $J = 2$ (surface quadrupole oscillations) and the $J = 0$ (compressional) modes have been considered separately in the liquid drop model. We are here proposing to consider the two modes, with dynamical variables (s, q_μ), together.

7. Technical conditions. The reality condition on the radius $R(\theta, \phi)$ leads to the conditions

$$s^* = s, \qquad q_\mu^* = (-)^\mu q_{-\mu} \qquad (37)$$

The modified liquid drop model is described by six real variables. There is one constraint among these six variables. The origin of this constraint is the incompressibility requirement. It is convenient to choose the constant appearing in (36) in the following way. The maximum allowed value of $\beta^2 = \sum q_\mu^* q_\mu$ occurs when min $R(\theta, \phi) = R_c$, the shell model core redius. For this value of β, the constant is chosen so that $s = 0$. Then as β^2 decreases s^2 increases. Then the amplitude s, q_μ obey a relationship

$$f(s,\ q_\mu) = s^2 + \sum_\mu |q_\mu|^2 \ \underset{-}{\sim}\ \text{Constant}$$

which may be diffeomorphic to the unitarity condition obeyed by the s- and d- bosons of the IBM.

In the IBM, the d-boson amplitudes are complex and independent while the s-boson amplitude is a real nonnegative function of d_μ: $s = (1 - \Sigma|d_\mu|^2)^{\frac{1}{2}}$. In the liquid drop model the quadrupole oscillation amplitudes obey the reality conditions (37), and the s-wave amplitude is a nonnegative function of the q : $s \sim (1 - \Sigma|q_\mu|)^{\frac{1}{2}}$.

8. Dynamics. Attempts to compare these two models must first be concerned with demonstrating the equivalence of the underlying kinematics. Only then can the secondary issue of detailed dynamics be considered. If there is in fact an equivalence between suitably modified models, then each assumed liquid drop hamiltonian will have as an image an IBM hamiltonian, and vice versa. The detailed study of dynamics is a secondary issue in the present study, where the primary issue is the equivalence of the kinematics.

9. Dynamical variables. In the liquid drop model the dynamical variables are the six real amplitudes $(s,\ q_\mu)$. The incompressibility requirement reduces this to five independent real variables q_μ, which form a coordinate system for a space Q^5 which is not flat. In the IBM the dynamical variables are the amplitudes $(s,\ d_\mu)$. The unitarity condition reduces this number to five complex variables d_μ, which provide a coordinate system on the coset SU(6)/U(5)

10. Configuration Space. These are the curved spaces Q^5 and SU(6)/U(5) of real dimensions 5 and 10 for the liquid drop model and the IBM, respectively.

11. Phase Space. For the liquid drop model this is a curved 10-dimensional space determined from the prescriptions of constrained hamiltonian dynamics. In the case of the IBM, it would normally be expected that the dimension of the phase space would be twice that of the configuration space. However, we have previously shown that for hamiltonian systems derived as the classical limit of a quantum mechanical system, phase space may be identified with configuration space. Therefore, for the IBM, phase space is also the coset SU(6)/U(5).

12. Underlying Hilbert Space. For the modified liquid drop model this is the space of square integrable functions defined on Q^5:

$\mathcal{L}^2(Q^2; \sqrt{|G|})$. The metric tensor G is determined from the kinetic energy expression, taking account of the incompressibility constraint. This Hilbert space is reducible, and should be stratified by a parameter, ν, which measures the volume of liquid surrounding the solid shell model core.

The Hilbert space underlying the IBM is $\mathcal{L}^2[SU(6)/U(5);\ d\mu;\ N]$ This space is already stratified by the integer N (number of bosons), but it is too big. It is a Hilbert space defined over five complex variables, whereas the liquid drop hilbert space is defined over five real variables. We have already pointed out that the weight functions in $\mathcal{L}^2[SU(6)/U(5);\ d\mu;\ N]$ needed to construct wave functions in (11) are nonunique. It may be true that these weight functions are unique if we demand that they are distributions with nonzero values on a real 5-dimensional subspace of SU(6)/U(5). If so, we could choose this subspace to correspond to the classical configuration space defined by $d_\mu^* = (-)^\mu d_{-\mu}$. In this way the hilbert space of the liquid drop model and the IBM become more nearly equal. Equality is:

- The liquid drop configuration space is diffeomorphic with the 5 dimensional subspace of SU(6)/U(5) described above.
- the index ν which stratifies the liquid drop hilbert space is 1 - 1 with N, the index which describes fully symmetric representations of SU(6)

13. Classical-Quantum Correspondence. The classical mapping of the IBM is simply constructed using the continuous basis representtation. The hard part is quantizing the liquid drop model and requantizing the classical image of the IBM. The starting point for this quantization scheme is a compact manifold of real dimension 10 which possesses the following three structures:

- metric
- symplectic
- imaginary

The quantized form of the classical energy function will then become an elliptic differential operator depending on 5 real variables (those of configuration space). For a large class of model energy functions this differential operator will be of second order, and contain a term independent of differentials, the effective potential. The operators so obtained may be similar to, but cannot be identical to, the Bohr-Mottelson equation. The latter is defined on $\mathbb{R}^5$ and has an unbounded spectrum. The former are defined on compact Q^5 and, being elliptic operators, must have a finite spectrum (for fixed ν or N).

Remark: The differential operators obtained by this method will not be directly comparable with the differential operator obtained from the IBM by change of basis from the algebraic to the geometric representation (cf. Fig.5). The latter is a differential operator depending on 10 real variables (the phase space coordinates and momenta), the former depend only on the 5 real coordinates parameterizing configuration space.

14. Method of Solution. The Schrödinger equations obtained by (re)quantizing the classical function defined over phase space may be solved directly (a losing proposition), or else the algebraic eigenvalue equation may be solved numerically. The algebraic expressions for the eigenvalues can then be transformed to the geometric expressions (wave functions) by a suitable unitary transformation.

SUMMARY

The SU(6) coherent state representation gives the Interacting Boson Model an extra degree of freedom. This degree of freedom greatly extends the types of studies which can be carried out on this model. Two types of studies facilitated by the continuous basis representation are described in the present work. These are the study of phase transitions and the determination of the differential operator structure of the IBM hamiltonian. The phase transition studies are generally reflective of properties of nuclear models based on geometric insights. We have determined a separatrix in the IBM parameter space between those models exhibiting predominately rotor-like behavior and those exhibiting vibrator-like behavior. These results are complemented by the transformation theory results. We have shown how to transform the IBM hamiltonian from a matrix to a differential operator resembling a Schrödinger operator. We have also shown how the eigenfunctions of this operator can be constructed from outputs of the computer codes developed to diagonalize the original (algebraic) IBM hamiltonian. The resulting eigenfunctions may then be compared directly with eigenfunctions of geometric nuclear models.

This similarity of results for the IBM in the geometric (coherent state) representation and the geometric (liquid drop) model of the nucleus is both provocative and suggestive of a deeper underlying relation between these two approaches to the study of collective nuclear properties. To be sure, the theories are not equivalent in their present forms, as is most easily seen from their spectral properties (finite for IBM, unbounded or truncated by ad hoc means for liquid drop models). The input biases, predilections and assumptions of these two theories have been compared and have been found to be strikingly similar. This comparison has also revealed which assumptions of each model must be modified if it is desired to make both models equivalent representations of a common underlying physical theory. The benefits of such an effort, though clear, have been stated specifically. In addition, the specific modifications required to provide this equivalence have been prepared.

REFERENCES

1. J.P. Schiffer and W.W. True, Rev. Mod. Phys., 48, 191(1976).
2. A. Arima and F. Iachello, Phys. Rev. Lett. 35, 1069(1975).
3. A. Arima and F. Iachello, Ann. Phys. (NY) 99, 253(1976).
4. A. Arima and F. Iachello, Ann. Phys. (NY) 111, 201(1978).
5. A. Arima and F. Iachello, Ann. Phys. (NY) 123, 468(1979).
6. D. Janssen, R.V. Jolos, and F. Donau, Nucl. Phys. A 224, 93(1974).
7. D. Ruelle, Statistical Mechanics, NY: Benjamin, 1969.
8. R. Gilmore, Physica 86A, 137(1977).
9. I Bars, Proceedings of the IX th International Colloquium on Group Theoretical Methods in Physics, Cocoyoe, Mexico, June 1980 (to be published).
10. J. Bardeen, L.N. Cooper and R. Schrieffer, Phys. Rev. 108, 1175(1957).
11. W.E. Lamb and R. Retherford, Phys. Rev. 79, 549(1950); 81, 222(1951); 86, 1014(1951). W.E. Lamb, Rep. Progr. Phys. 14, 19(1951); Phys. Rev. 85, 259(1952).
12. R. Gilmore, Ann. Phys. (NY) 74, 391(1972).
13. R. Gilmore, Rev. Mex. de Fisica 23, 143(1974).
14. R. Gilmore, Lie Groups, Lie Algebras and Some of Their Applications, NY: Wiley, 1974.
15. F.T. Arecchi, R. Gilmore, and D.M. Kim, Lett. Nuovo Cimento 6, 219(1973).
16. I.M. Gel'fand and M.L. Tsetlein, Dokl. Adad. Nauk. SSSR 71, 840(1950).
17. R. Gilmore, J. Math Phys. 20, 891(1979).
18. R. Gilmore and D.H. Feng, Phys. Lett. 76B, 26(1978).
19. R. Gilmore, J. Phys. A9, L65(1976).
20. A.E.L. Dieperink, O. Scholten, and F. Iachello, Phys. Rev. Lett. 44, 1747(1980).
21 F.T. Arecchi, E. Courtens, R. Gilmore, and H. Thomas, Phys. Rev. A6, 2211(1972).
22. R. Gilmore, C.M. Bowden, and L.M. Narducci, Phys. Rev. A12, 1019(1975).
23 A. Bohr and B.R. Mottelson, Mat. Fys. Medd. Dan. Vid. Selsk. 27 (no.16) (1953).
24. A. Bohr and B.R. Mottelson, Nuclear Structure. Vol.II: Nuclear Deformations, Reading, Mass: Addison-Wesley, 1975.
25. A. Arima, T. Otsuka, F. Iachello and I. Talmi, Phys. Lett. 66B, 205(1977).
26. T. Otsuka, A. Arima, F. Iachello and I. Talmi, Phys. Lett. 76B, 139(1978).
27. T. Otsuka, A. Arima, and F. Iachello, Nucl. Phys. A309, 1(1978).

27. T. Otsuka, A. Arima, and F. Iachello, Nucl. Phys. A309, 1(1978).
28. T. Otsuka, Ph.D. Thesis, University of Tokyo, 1979.
29. F. Iachello and O. Scholten, Phys. Rev. Lett. 43, 679(1979).
30. O. Scholten, Ph.D. Thesis, Rijks University Gröningen, 1980.
31. F. Iachello, Phys. Rev. Lett. 44, 772(1980).

SHELL MODEL STRUCTURE OF THE INTERACTING BOSON MODEL AND THE INTERACTING BOSON FERMION MODEL

Igal Talmi

The Weizmann Institute of Science
Rehovot, Israel

Nuclei are very complex systems composed of a large number of strongly interacting protons and neutrons. One could expect that such systems will have extremely complicated and unwieldy level spectra. However, in spite of their complexity, nuclei exhibit remarkably simple regularities which make their spectra interesting objects of research. Many of these regularities are associated with individual nucleon motion and find their expression in the shell model[1]. Nuclei with closed shell or with a few nucleons away can be very well described in terms of simple shell model configurations and effective two-body interactions between nucleons[2]. As we move away from closed shells, the spectra become more complex. Yet in spectra of nuclei with several protons and neutrons outside closed shells, other remarkable regularities emerge. The low-lying levels can be grouped into bands whose energies in many cases correspond to those of a symmetric rotor. In other cases the bands are more vibrational but they are all characterized by strong intra-band E2 transitions. Such rotational or quasi-rotational spectra found a simple and highly successful description in terms of the collective model[3].

The natural question arises whether collective states could be understood within the framework of the (spherical) shell model. Certainly this is possible in principle since the shell model wave functions form a complete set of states. Still, a very complicated expansion involving perhaps billions of shell model states is not what should be looked for. We would like to understand in a simple way, be it even approximate, the emergence of collective features from the shell model description and to learn which ingredients of the nuclear interaction give rise to one collective behavior or another. This problem has occupied nuclear theorists for many years without too much progress towards its solution.

In the last few years a new model has been developed. The interacting boson model gives a very good description of the various limits of collective spectra of even-even nuclei, like vibrational and rotational ones[4]. By varying the few parameters of a simple boson Hamiltonian a smooth transition is obtained from one limit to another[5]. In its initial formulation (IBA)[4] there are two kinds of bosons, s-bosons with $\ell=0$ and d-bosons with $\ell=2$. The IBA Hamiltonian contains single boson energies and boson-boson interactions. It is very, very different from Hamiltonians obtained by various boson expansions which led to infinite series of higher and higher orders of boson operators.

The relationship of bosons to the Hamiltonian of the collective model is an old one. If the collective coordinates and momenta in the Bohr Hamiltonian are expressed in terms of creation and annihilation operators, the quantum states can be built from phonons of quadrupole surface vibrations. Thus, the second quantized version of the collective Hamiltonian includes d-bosons. This approach has been successfully applied to small vibrations around a state of spherical equilibrium. It was not easy to extend it to cases with a deformed state of minimum energy. A few years ago an important step was made by the construction of quadrupole operators which obeyed the commutation relations of a U(6) Lie algebra[6]. This is no mean feat since the underlying space has only 5 dimensions (components of a quadrupole operator). The eigenstates of a Hamiltonian constructed with such operators can be characterized by the irreducible representations of U(6). The fully symmetric representations of U(6) can be characterized by Young diagrams having one row only of length N. The nature of this quantum number N could not have been made clear. Unlike the previous treatments of d-bosons (with the U(5) symmetry group) this number was not the number of d-bosons. The latter was not generally conserved.

The nature of this number N was elucidated by Arima and Iachello who independently introduced a much simpler version of U(6) symmetry[4]. In their IBA model, s-bosons were introduced along with the d-bosons forming together a basis 6-dimensional space. As a result, the conserved number N turned out to be the total number of d- and s-bosons. Arima and Iachello recognized that starting with this version of U(6), Hamiltonians could be written down whose eigenstates could be characterized by the irreducible representation of SU(3). A rotational structure could thus emerge for many even-even nuclei in the same way that it was obtained in the SU(3) model of Elliott in the fermion s,d-shell. This was realized[7], also by the authors of ref.6.

So far, IBA can be considered as an elegant way to obtain properties of solutions of collective Hamiltonians. From this point of view the s-boson could be viewed as a mathematical gimmick to simplify the formulation and solution of the problem and the number

N would not have a clear physical meaning. There is, however, a much more promising approach which can lead to a shell-model, or microscopic, description of the interacting boson model and therefore also a nuclear collective motion. In this approach, the s-boson is considered to have as much physical meaning as the d-boson. In fact, in the rotational case, states of s-bosons and d-bosons are thoroughly admixed in the SU(3) wave functions. Taking seriously the s-bosons, as well as the conservation of N leads to a simple interpretation in terms of shell model states[8,9].

The simplest fermion systems that have integral spin and could correspond to bosons, are fermion pairs. Such a fermion pair could be a pair of valence nucleons or a particle-hole pair. If we have to construct a particle-hole pair with J=0 and positive parity the particle must be raised across two major shells from the hole. Such high excitations cannot be expected to play an important role in low-lying states. We are led to choose for the fermion pairs particle-particle pairs with J=0 and J=2. With this interpretation the conservation of N finds a natural explanation. Its conservation (of boson number) is simply due to the conservation of the fermion number in nuclei. We can now ask whether the particle pairs should be proton-proton and neutron-neutron pairs or perhaps proton-neutron pairs. The considerations above exclude also proton-neutron pairs. In heavier nuclei, where most collective states appear, protons and neutrons are in different major shells. It would not be possible to form a pair of valence proton and a valence neutron with J=0 and positive parity. The restriction to pairs of identical nucleons does not mean that we ignore the interaction between protons and neutrons. It has been long argued that the strong and attractive proton-neutron interaction gives rise to collective excitations and even to the shell model itself[3]. As we shall see later, this interaction is a major ingredient of the boson model. It will be introduced, however, at the proper stage and in the proper fashion.

We are led to consider pairs of identical nucleons which can be best studied in semi-magic nuclei. Such pairing into J=0 states has a long history in nuclear physics. The idea of pairing was introduced by Racah for atomic spectra through his seniority scheme as far back as 1943[10]. It has been applied by several authors to nuclear spectra where it plays a much more important role. Since the appearance of the Bardeen, Cooper and Schrieffer (BCS) theory of superconductivity it has been applied in many papers to nuclear spectra. Although perhaps familiar to many readers, I will not use the BCS version of pairing theory because of three reasons. The first is that it (approximately) diagonalizes the pairing interaction which is not a good approximation to the effective interaction between nucleons in nuclei[3]. The second reason is that the BCS wavefunction does not have a definite number of fermions. This is no difficulty when applied to electrons in metals but it may be a poor approximation for a few valence nucleons. The third reason is related to the present context. The BCS theory deals with J=0 pairing only, J=2 pairs are treated in a different manner. An alternative approach, that of generalized seniority[11], strictly conserves nucleon number and can accommodate a wide cross of effective interactions. It also treats J=0 and J=2 pairs in a similar fashion.

Consider the operator which creates a state of two identical nucleons in the same orbit j coupled to J=0,

$$S_j^+ = \frac{1}{2}\sum (-1)^{j-m} a_{jm}^+ a_{j-m}^+ \;. \tag{1}$$

The normalization of the operator (1) is chosen so that it satisfies with its hermitean conjugate and their commutator the standard commutation relations of the Lie algebra of SU(2). The seniority scheme can be defined and all its properties derived from this algebra of quasi-spin operators[2]. We will not discuss it here but go immediately to a generalization. If we consider several j-orbits in the same major shell we can construct a creation operator of a correlated J=0 pair by writing[11]

$$S^+ = \sum \alpha_j S_j^+ \tag{2}$$

If all α_j are equal the operator (2) is a component of a total quasi-spin operator and the seniority scheme is easily extended to several j-orbits.[12] In actual nuclei, however, having equal α_j is too restrictive and leads to erroneous results. We must live with unequal α_j loosing the nice properties of SU(2). Still, as we shall presently see, some important properties of the seniority scheme survive even in this generalization.

We next consider a shell model Hamiltonian H for the valence nucleons in a major shell and normalize it by $H|o\rangle=0$. The state $|o\rangle$ is the J=0 ground state of a nucleus without the valence nucleons, with closed shells only. We now ask under which conditions a "condensate" of pairs created by (2) will be an eigenstate of H (containing single nucleon energies and two-body effective interactions only). More precisely, we impose the condition

$$H(S^+)^n|o\rangle = E_n(S^+)^n|o\rangle \tag{3}$$

for any n and look for Hamiltonians which will satisfy it. We first derive from (3) by putting n=1 the condition

$$HS^+|o\rangle = V_o S^+|o\rangle \tag{4}$$

Putting now n=2 we can derive another condition

$$[[H, S^+], S^+] = \Delta(S^+)^2|o\rangle \tag{5}$$

Where Δ is a number. If conditions (4) and (5) are satisfied than it can be shown[11] that (3) holds for <u>any</u> value of n and that the eigenvalues are given by the simple expression

$$E_n = nVo + \frac{n(n-1)}{2}\Delta \tag{6}$$

Thus, binding energies of even-even nuclei are exactly linear and quadratic functions of n. Nucleon pair separation energies are linear functions of n. An important point is that no sub-shell effects can appear. No matter what the values of α_j are or what are the actual values of single nucleon energies, the prescription (3) for the ground states guarantees that no breaks will occur in the binding energy or separation energy curves. These features agree very well with the experimental binding energies.

Let us ask at this stage what is the connection between these fermion states (3) and boson states. We can first check whether the commutation relations between the operator S^+ and its hermitean conjugate $(S^+)^+$ are similar to those between creation and annihilation operators of s-bosons, namely

$$[s, s^+] = 1 \tag{7}$$

The commutation relations between the fermion pair operators are much more involved. They are given by

$$[(S^+)^+, S^+] = \sum_j \frac{1}{2}(2j+1)\alpha_j^2 - \sum \alpha_j^2 \sum_m a^+_{jm} a_{jm} \tag{8}$$

From (8) it is seen that if the second term on the r.h.s. of (8) can be neglected in comparison with the first term, fermion pairs do behave like bosons. The relation (8) could then be the same as (7) if S^+ will be properly normalized. This, however, is a trivial result which is a good approximation only if the number of nucleons is very small compared to the number of states in the major shell,

$2\Omega=\Sigma(2j+1)$. In actual cases this is not the case and we must either abandon the naive correspondence of operators or go into the wilderness of boson expansions.

There is, however, a direct and simple way to express the results obtained above in terms of bosons. This method gives up the ambitious program of expressing fermion pair operators in terms of boson operators. Instead, we make a correspondence of mapping between the fermion pair states (3) and boson states $(s^+)^n|o\rangle$. We can now construct a boson Hamiltonian which will have the same eigenvalues (6) for these boson states. Such a Hamiltonian is

$$V_o s^+ s + \frac{1}{2}\Delta (s^+)^2 s^2 \tag{9}$$

The boson Hamiltonian (9) obviously cannot replace the fermion Hamiltonian in all of its millions of states. It has the same eigenvalues only for one of the many thousands of J=0 states. We also see from (9) that $s^\dagger$ is not a replacement for the $S^\dagger$ pair creation operator nor is it intended to be a first term in an expansion. If that were the case, the second term in (9) would have the form of a four fermion interaction. Yet both terms in (9), a single boson term and a boson-boson interaction, are equivalent to a shell model Hamiltonian with at most two fermion interactions Another aspect of the nature of the correspondence between fermion and boson states is seen when n increases beyond Ω. For $n>\Omega$ all fermion states (3) vanish because of the Pauli principle. Boson states, however, are well defined and have the eigenvalues (6). Thus, there is no singularity in the boson Hamiltonian or boson states. Beyond $n=\Omega$ they simply do not correspond to fermion states.

Let us consider now pair states with J=2 which have (generalized) seniority v=2 (the states (3) have v=0). We define an operator which creates a correlated J=2 pair including amplitudes of nucleon orbits in a major shell by

$$D_M^+ = \sum_{j\leq j'} \beta_{jj'} \frac{1}{\sqrt{1+\delta_{jj'}}} \sum_{mm'} (jmj'm'|jj'2M)\, a_{jm}^+ a_{j'm'}^+ \tag{10}$$

We ask again under which conditions on the shell model Hamiltonian will the states $(S^+)^{n-1}D^+|o\rangle$ be eigenstates. We derive, from n=1 the condition

$$HD_M^+|o\rangle = V_2 D_M^+|o\rangle \tag{11}$$

Putting n=2 we obtain, in any case of interest, the condition

$$[[H, S^+], D_M^+]|o\rangle = \Delta S^+ D_M^+|o\rangle \tag{12}$$

From (11) and (12) follows for <u>any</u> value of n

$$H(S^+)^{n-1} D_M^+|o\rangle = (E_n + V_2 - V_o)(S^+)^{n-1} D_M^+|o\rangle \tag{13}$$

Thus the structure of these J=2 states, through the conditions (11) and (12), completely determines their eigenvalues. The positions of these J=2 levels above the J=0 ground state are independent of n and are given by $V_2 - V_o$. No subshell effects are seen. These features agree very well with the experimental data, for example in all Sn isotopes with neutron numbers between 52 and 80.

Also here we can obtain the eigenvalues in (13) from a boson Hamiltonian if we map the fermion states in (13) on boson states $(s^+)^{n-1} d_\mu^+|o\rangle$. These boson states, along with boson states $(s^+)^n|o\rangle$ are eigenstates, with eigenvalues (13) and (6) respectively of the boson Hamiltonian

$$V_o s^+ s + V_2 \sum d_\mu^+ d_\mu + \tfrac{1}{2}\Delta (s^+)^2 s^2 + \Delta s^+ s \sum d_\mu^+ d_\mu \tag{14}$$

The same comments made about the boson Hamiltonian (9) are valid also for (14). The boson Hamiltonian (14) has simple eigenvalues also for states with more d-bosons. In fact all states with n_d d-bosons are degenerate and their eigenevalues are given by

$$V_o n_s + \Delta \tfrac{1}{2} n_s(n_s - 1) + V_2 n_d + \Delta n_s n_d \tag{15}$$

To which fermion states do these boson states correspond? There are actual shell model Hamiltonians that have eigenstates of the form (3) with unequal α_j values, and eigenstates of the form (13)[11]. Such Hamiltonians give good agreement with experimental data, for instance in the Ni isotopes. It has not been checked whether the application of two (or more) D^+ operators leads to eigenstates of actual shell model Hamiltonians. In simple models such states may be eigenstates.

There is a technical point which should be clarified. States constructed with two or more D^+ operators may not be orthogonal to states with smaller numbers of D^+ operators. For example, the state

$(D^+.D^+)|o\rangle$ is, in general, not orthogonal to the v=0 state $(S^+)^2|o\rangle$. This difficulty occurs also for equal α_j values and for a single j-orbit. In order to make the correspondence between these fermion states and the orthogonal boson states we must project out from them amplitudes of states with lower (generalized) seniority. From states constructed with n_d fermion pair operators D^+ coupled in a certain order and n_s operators S^+ we first project out all components which can be constructed with n_s+1 operators S^+. The resulting states are mapped onto boson states of the form $(s^+)^{n_s}(d^+)^{n_d}_{\gamma JM}|o\rangle$ where γ stands for the quantum number which characterizes orthogonal states with the same J and M. In certain models, developed by Ginocchio[13], this correspondence can be shown to be exact.

It should be noted that the definition made above limits thc the number of d-bosons in any given state to at most Ω. Beyond the middle of the shell, if more D^+ operators are used, the resulting states will have only lower seniorities and will not survive the projection. To make sense of the mapping procedure we must consider beyond the middle of the (major) shell, states of fermion holes, rather than fermion states. Thus, beyond the middle of the shell the boson number n_s+n_d will be equal to the number of holes in the completed valence shell.

As we saw, for certain states of semi-magic nuclei, boson states and Hamiltonians can be used rather than the fermion description. The states for which this can be done are the low-lying states of interest. These include the J=0 ground states with very simple binding energy formulae and the first J=2 states with constant spacings between them and ground states. Still, the shell model description is very simple and using bosons does not lead to actual simplifications. It is only when we consider nuclei with valence protons and valence neutrons that the boson model reveals its full power.

Spectra of nuclei with both valence protons and neutrons contain an incredibly large number of levels. Most of them are of no interest but we would like to obtain a simple description of the low-lying levels, those which form the various collective bands. For semi-magic nuclei a simple description of the interesting states was based on generalized seniority. Could seniority be a guide also when the proton-neutron interaction is considered? There are nuclea configurations with $1f_{7/2}$ and with $1g_{9/2}$ protons and neutrons and we can check whether seniority, properly defined, is a good scheme in those cases. Whereas the T=1 interaction between $g_{9/2}$ nucleons is diagonal in the seniority scheme, the T=0 interaction strongly violates it (and so it is for $f_{7/2}$ nucleons). The simplest interaction that isolates seniority so badly is the quadrupole-quadrupole interaction. Since the "quadrupole degree of freedom" is so important in deformed nuclei[3], we shall adopt the quadrupole interaction between protons and neutrons. For simplicity we deal with nuclei where valence protons and neutrons occupy different major shells. All such states, in which valence protons are in states fully occupied by neutrons, have good isospin $T=(N+Z)/2$. We can therefore consider, in spite of charge independence, an interaction between protons and neutrons which is different from that between identical nucleons.

It is perhaps worth while to put stress on the difference between the T=1 and T=0 interactions. The quadrupole-quadrupole interaction which appears in the j^n configurations of protons and neutrons is strong, attractive and breaks seniority in a major way. How could seniority be a good scheme for states of j^n configurations with maximum isospin $T=n/2$? The quadrupole interaction may well contribute to $j^2T=1$ states but its seniority breaking non-diagonal matrix elements may be cancelled by interactions containing other multipoles. In fact, the expansion of a two-body interaction into scalar products of irreducible tensor operators is not unique if limited to $j^2T=1$ states. Since such states have even J-values, only the expansion coefficients are not uniquely determined. For example, any two-body interaction which is diagonal in the seniority scheme can be expanded in terms of odd tensors and a k=0 tensor (scaler) only[14]. The situation for several j-orbits is more complicated but in all known cases, generalized seniority is an adequate scheme for identical nucleons.

The Hamiltonian of the combined system of protons and neutrons is thus taken to be

$$H = H_p + H_n + Q_p \cdot Q_n \tag{16}$$

All states of nuclei with valence protons and neutrons can be constructed by coupling all states of the protons to all states of the neutrons. The states in this scheme are defined by

$$|\alpha_p J_p \alpha_n J_n JM> \tag{17}$$

where α_p enumerates all orthogonal states of the proton group with the same values of J and M and α_n is similarly defined for the neutrons. All billions of states of the system are obtained by coupling the millions of states of the protons to the millions of states of neutrons. The matrix of the quadrupole interaction is given in the scheme (17) in a simple way as follows

$$\langle \alpha_p J_p \alpha_n J_n JM | Q_p \cdot Q_n | \alpha'_p J'_p \alpha'_n J'_n JM \rangle$$

$$= (-1)^{J_n + J'_p + J} \begin{Bmatrix} J_p & J_n & J \\ J'_n & J'_p & 2 \end{Bmatrix} (\alpha_p J_p || Q_p || \alpha'_p J'_p) \times \tag{18}$$

$$\times (\alpha_n J_n || Q_n || \alpha'_n J'_n)$$

The dimensions of the matrix (18) for a given J (and M) are prohibitively large. Still, the form (18) of the matrix suggests a possible truncation scheme[8,9]. We should keep in (18) only those states which have large matrix elements of the quadrupole interaction. Those are the states constructed by proton states with large matrix elements of the proton quadrupole operator and neutron states with large matrix elements of the neutron quadrupole operator. It can be shown[15] that the states of generalized seniority considered above for protons (and for neutrons) have this property. Furthermore, H_p (and H_n) in (16) are diagonal in this set of states. We thus truncate the set of proton states to those constructed with S_p^+ and D_p^+ operators and that of the neutron states to those constructed with S_n^+ and D_n^+. We assume that this is a good approximation to states included in the low-lying collective bands.

This truncation reduces drastically the dimensions of the Hamiltonian matrices for the various J-values. It still implies the use of multifermion wave functions which is difficult. The Pauli principle complicates the actual calculation of matrix elements and due to the detailed structure of S^+ and D^+ we should have very detailed information on the matrix elements of the quadrupole operators. These difficulties can be avoided by making another drastic approximation[8,9]. We replace proton and neutron states (in the truncated space) by states constructed with s_p, d_p proton bosons and s_n, d_n neutron bosons. We saw above how H_p and H_n can be replaced by boson Hamiltonians. We must now find boson quadrupole operators which will have the same matrix elements between boson states as the fermion quadrupole operator between corresponding fermion states. Such boson operator can be defined in some cases where the α_j are equal[8,9,16]. In the case of the Ginocchio model[13] it is possible, in certain cases, to carry out calculations in the corresponding boson space which exactly reproduce the results of the fermion calculations[13,17]. In general, this is very difficult to do. Still, we can learn from the simple cases how to parametrize the boson quadrupole operators to a good approximation. The actual values of the parameters, which depend on the number of proton and neutron s- and d-bosons can be determined from the experimental spectra. Such detailed calculations have been carried out successfully in several regions of nuclei[18,19]. Good agreement is obtained not only for energy levels but also for electromagnetic E2 transitions. In such calculations, effective charges were introduced for proton and neutron bosons. The actual values are quite reasonable. Moreover, the effective charges are kept constant throughout the shell and fit also transitions in semi-magic nuclei. This indicates that the effective charges can be ascribed to polarization of the closed shells by the valence nucleons. The polarizations by individual bosons (or nucleons) are independent of each other and can be reproduced by a change in the single boson (or nucleon) quadrupole moment. The very large enhancement of E2 transition rates in the rotational regions is due only to the coherence effects of the SU(3) type admixtures of s- and d-boson states.

We started with a boson model with unspecified s- and d-bosons (IBA)[4]. We reached now an interacting boson model (IBM) which has s_p and s_n bosons as well as d_p and d_n bosons[8,9]. It is a more complicated model in which the various symmetries are not manifest. Still, it is a more detailed model and gives a much better description of the experimental data. In certain cases the more detailed IBM model can be replaced by the simpler one (IBA). If in the boson Hamiltonian the interaction between proton bosons and between neutron bosons is taken to be the same as between proton and neutron bosons, the lowest eigenstates are fully symmetric in the proton and neutron bosons. In this case, eigenstates can be characterized by eigenvalues of a certain spin vector F-spin ($F=1/2$, $M_F=1/2$ for proton bosons and $F=1/2$, $M_F=-1/2$ for neutron bosons)[8]. symmetric states have $F=n/2$ where n is the total number of proton and neutron bosons. Such states can be constructed by using one kind of s- and d- bosons only.

An interesting extension of the boson model is to couple to the system of bosons other degrees of freedom which are not included in the model. Such degrees of freedom could be negative parity states giving rise to negative parity bands and high spin states of a nucleon pair giving rise to a "super-band" and "backbending". Of special interest is the extension of the model to odd nuclei by coupling a single fermion to the system of bosons[20,21]. If a quadrupole interaction between the single fermion and the bosons is introduced, collective bands are obtained also for the odd nucleus. If the even-even core has a rotational spectrum, such a spectrum emerges also for the odd nucleus. The situation is similar to that of the Nilsson scheme but there is an important difference. If more nucleons identical to the single nucleon are added, the Nilsson orbits are filled and the positions of the various bands change. The ground state band will have the K value of the highest orbit not completely filled. In the boson fermion model with only a quadrupole interaction the positions of the various bands are fairly independent of the number of bosons. In the Nilsson scheme the change of order of the bands is due to the Pauli principle. It stands to reason that the Pauli principle should be introduced also

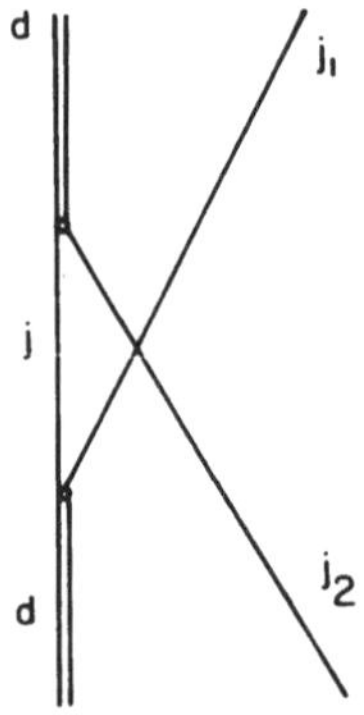

Fig.1. Diagram corresponding to the exchange term in the boson-fermion interaction.

into the boson fermion model. In the microscopic description of the model adopted above, the bosons represent pairs of fermions partially in the orbit occupied by the single fermion. It is thus natural to take seriously corrections due to the Pauli principle.

An exchange term was introduced in ref.20 in the form

$$(\tilde{d}\times a^{+}_{j_1})^{(j)}\cdot(d^{+}\times\tilde{a}_{j_2})^{(j)} \tag{19}$$

This exchange term, which corresponds to the diagram in fig.1, can simulate the desired effects of the Pauli principle. By adjusting the strength of this term it was possible to change the positions of the various bands in the odd nucleus[20]. In the rotational region these changes are similar to the results of using the Nilsson scheme. How can we understand the introduction of the operator (19)? The calculations for odd nuclei have been carried out with one kind of s- and d-bosons (IBA). Therefore the diagram in fig.1 does not have a clear meaning. Surely, the exchanged fermions must be of the same kind as the fermion to which they are coupled within the boson. The Pauli principle applies only to identical nucleons. Yet the vertices in Fig.1 would in this case represent the interaction binding two identical nucleons to a state with J=2. This interaction is, however, not very strong and vanishes in the case of the pairing interaction. In our model the strong and attractive interaction is between protons and neutrons in which case however the Pauli principle has no effect.

According to our picture, the Pauli corrections could only effect the matrix elements of the quadrupole operator of the odd group of nucleons. It is not just the sum of a single fermion operator and the operator of the even group. Due to the antisymmetry

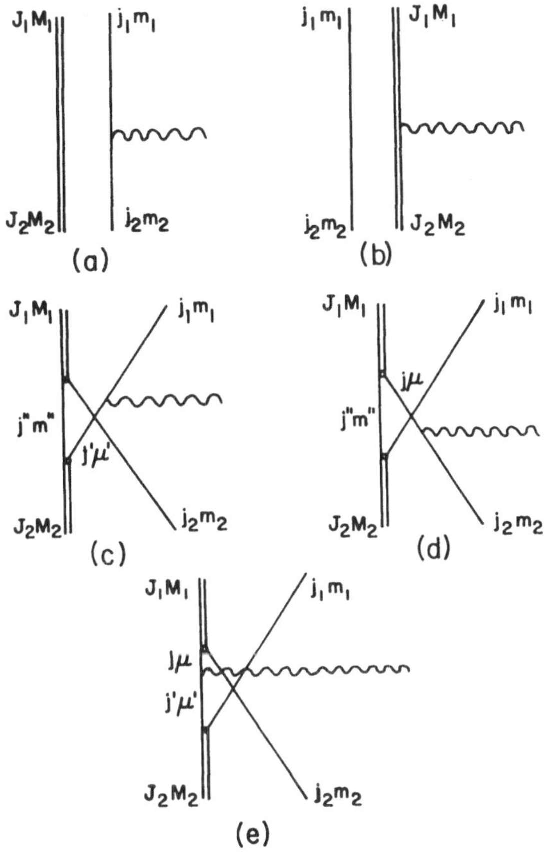

Fig.2. Diagrams corresponding to the matrix elements of the quadrupole operator, including exchange terms, for three identical nucleons.

of the wave functions there are also exchange contributions which we should calculate. The simplest possible calculation is of the quadrupole operator of three identical fermions between the states $B^{+}_{J_1M_1}a_{j_1m_1}|o\rangle$ and $B^{+}_{J_2M_2}a^{+}_{j_2m_2}|o\rangle$ [22,23]. The direct evaluation leads to the sum of three terms. These are matrix elements of the single fermion operator, matrix elements of the quadrupole operator of the even group (to be replaced by the appropriate boson operator) and the last is an exchange term. These terms correspond to diagram (a), diagram (b) and the three diagrams (c), (d), (f) of fig.2. We now construct a boson-fermion operator which has the same exchange matrix elements when calculated between the states $b^{+}_{J_1M_1}a^{+}_{j_1m_1}|o\rangle$

and $b^{+}_{J_2M_2}a^{+}_{j_2m_2}|o\rangle$ where b^{+} are boson creation operators which commute with the a^{+}_{jm}, a_{jm} operators. In the exact step we consider the scalar product of this operator and the boson quadrupole operator of the even group of nucleons of the other kind. The result is a boson-boson-fermion interaction. It can, however, be reduced to a boson-fermion interaction if we go over to the description with one kind of bosons (IBA). Using F-spin symmetric states we finally arrive at the form (19) arising from the exchange terms. The coefficients of the exchange terms correspond to having in the vertices of fig.1 the quadrupole interaction between protons and neutrons. It is interesting, even though not surprising, that fig.1 can be used as derived from identical nucleons but with quadrupole interactions acting between unlike ones.

In summary, the interacting boson model is very useful because of its simplicity. It offers a good approximation to the collective model with all its symmetries and at the same time can be viewed as a good approximation to full scale shell model calculations.

REFERENCES

1. M.G. Mayer and J.H.D. Jensen, Elementary Theory of Nuclear Shell Structure, Academic Press, New York (1955).
2. I. Talmi, Rev. Mod. Phys. 34 (1962) 704.
3. The work carried out by A. Bohr and B.R. Mottelson on the collective model is described in their book Nuclear Structure Vol.II Nuclear Deformations, W.A. Benjamin, Reading MA (1975).
4. A. Arima and F. Iachello, Phys. Rev. Lett. 35 (1975) 1069, Ann. Phys. (N.Y.) 99 (1976) 254; 111 (1978) 201; 123 (1979) 468.
5. O. Scholten, F. Iachello and A. Arima, Ann. Phys. (N.Y.) 115 (1978) 325.
6. D. Janssen, R.V. Jolos and F. Donau, Nucl. Phys. A224 (1974) 93.
7. R.V. Jolos, F. Donau and D. Janssen, Yad. Fiz. 22 (1975) 965.
8. A. Arima, T. Otsuka, F. Iachello and I. Talmi, Phys. Lett. 66B (1977) 205.
9. T. Otsuka, A. Arima, F. Iachello and I. Talmi, Phys. Lett. 76B (1978) 141.
10. G. Racah, Phys. Rev. 63 (1943) 367.
11. I. Talmi, Nucl. Phys. 172A (1971) 1.
    S. Shlomo and I. Talmi, Nucl. Phys. A198 (1972) 81.
    I. Talmi, Riv. Nuovo Cimento 3 (1973) 85.
12. A.K. Kerman, Ann. Phys. (N.Y.) 12 (1961) 300.
    A.K. Kerman, R.D. Lawson and M.H. MacFarlane, Phys. Rev. 124 (1961) 162.
    R.D. Lawson and M.H. MacFarlane, Nucl. Phys. 66 (1965) 80.
13. The complete summary of these models is given in J.N. Ginocchio, Ann. Phys. (N.Y.) 126 (1980) 234.

14. A full discussion of the seniority scheme is found in A. de Shalit and I. Talmi, Nuclear Shell Theory, Academic Press, New York (1963).
15. I. Talmi, Simple Models of Complex Nuclei, Lecture Notes, Verenna Summer School 1980 (to be published).
16. T. Otsuka, A. Arima and F. Iachello, Nucl. Phys. A309 (1978) 1.
17. J.N. Ginocchio and I. Talmi, Nucl. Phys. A337 (1980) 431.
18. F. Iachello, G. Puddu, O. Scholten, A. Arima and T. Otsuka, Phys. Lett. 89B (1979) 1.
G. Puddu, O. Scholten and T. Otsuka (to be published).
19. O. Scholten in Interacting Bosons in Nuclear Physics, F. Iachello, Editor, Plenum Press, New York and London (1979).
20. F. Iachello and O. Scholten, Phys. Rev. Lett. 43 (1979) 679.
21. F. Iachello and O. Scholten, Phys. Lett. 91B (1980) 189.
F. Iachello, Phys. Rev. Lett. 33 (1980) 772.
22. I. Talmi in Interacting Bose-Fermi systems in Nuclei, F. Iachello, Editor, Plenum Press, New York and London (to be published).
23. Equivalent results were obtained by O. Scholten and A.E.L. Dieperink in Interacting Bose-Fermi systems in Nuclei, F. Iachello, Editor, Plenum Press, New York and London (to be published).

TESTING THE INTERACTING-BOSON AND INTERACTING BOSON-FERMION APPROXIMATIONS BY THE STUDY OF RADIOACTIVE DECAY SCHEMES WITH THE UNISOR ON-LINE ISOTOPE SEPARATOR

John L. Wood

School of Physics
Georgia Institute of Technology
Atlanta, GA 30332

I. ABSTRACT

Detailed studies of low-spin states are considered from the viewpoint of radioactive decay scheme studies and the description of such states in terms of the interacting-boson approximation (IBA) and interacting boson-fermion approximation (IBFA). The essential points of the IBA and IBFA are outlined. A discussion of excited 0^+ states in even nuclei is given, with particular emphasis on those kinds of 0^+ states that lie outside of the IBA model space. The techniques and criteria currently employed in radioactive decay scheme studies, as for example done at UNISOR, are summarized. The structure of the 0^+ states in the neutron-deficient Pt and Hg isotopes is discussed. Evidence is presented which suggests that proton intruder orbitals, such as $\pi\ h_{9/2}$, play a significant role in the structure of the 0^+ states in this region.

II. INTRODUCTION

In the last few years the study of low-energy nuclear structure has reached a significant level of maturity due to two major factors. First, it is now possible experimentally to determine very completely the low-energy structure of whole regions of nuclei. Second, for the first time, a truly unified description of collective and single-particle degrees of freedom is available that is directly connected with the shell model. This theoretical description, called the interacting boson approximation (IBA), and the interacting boson-fermion approximation (IBFA) for the case of an unpaired nucleon, is unique among theories of low-energy nuclear structure in that it provides simple unambiguous prescriptions for the collective behavior expected in any given nucleus, based upon the shell model configuration of the nucleus. In other words, it is a model with a unique predictive power, qualitatively connected with the shell model, and not open to the arbitrary appendage of extra degrees of freedom to the model, or the artificial enhancement of terms in the model Hamiltonian in the pursuit of "a good fit". (A comprehensive outline of the IBA and IBFA is given below.)

Our experimental program at UNISOR has two primary goals. First, it is aimed at providing a very comprehensive data base in selected regions of the mass surface so that the IBA and IBFA models can be tested adequately to determine if they provide a complete predictive power, i.e. is there a one-to-one correspondence between theory and experiment? Second, it is aimed at elucidating the properties of shell model intruder states (shell model states that appear on the "wrong side" closed shells) and hence, provide a unified description of the intruder state degree of freedom within the IBA and IBFA models.

A crucial ingredient of this work is the systematic investigation of low-energy structural features of sequences of isotopes (and isotones) over many adjacent mass numbers. This has a two-fold purpose. First, it enables us to build up a very complete picture of nuclear structure that connects the regions of stable nuclei (where e.g., particle transfer reactions on, and Coulomb excitation of, stable targets have permitted the measurement of detailed spectroscopic properties of individual nuclear states) with regions far from stability, where detailed spectroscopic information is very limited. Second, it provides a map of the excitation degrees of freedom as a function of the changing proton and neutron number over broad mass regions. We find that many low-energy phenomena in nuclei can be understood adequately only by obtaining information over a wide range of mass numbers. Thus, the experimental investigations are conducted simultaneously over quite a number of isotopes, since it has been found that the construction of level schemes is greatly facilitated by a knowledge of neighboring isotopes (both closer to and further from the stability line).

The use of radioactive decay as the means to populate excited states of nuclei makes information on low-spin low-energy states especially accessible. Thus, such a program complements the only other experimental means of studying excited states of (neutron-deficient) nuclei far from stability, namely, in-beam reaction γ-ray and conversion electron spectroscopy. In-beam reaction spectroscopy is confined to a narrow band of (mainly) high-spin states in the region of the yrast levels, and hence, radioactive decay scheme studies are the only means one has of exploring the low-spin structure of nuclei far from stability. The high predominance of low-spin states in the model spaces of the IBA and IBFA thus makes radioactive decay scheme studies far from β stability uniquely

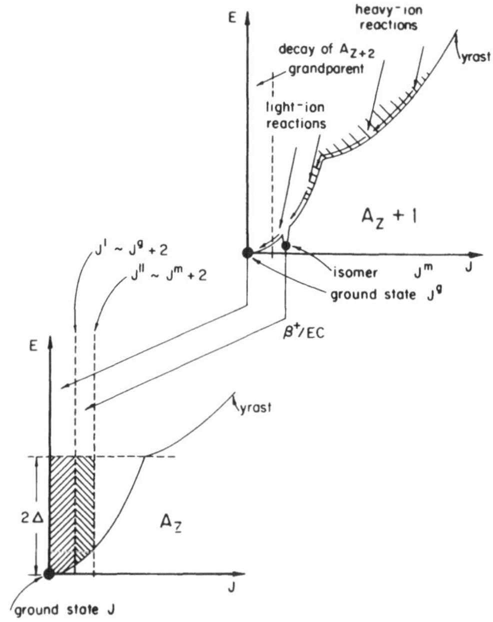

Fig.1 A schematic representation of the β decay of a ground state and isomeric state in an yrast diagram. The line marked yrast represents the envelope of the lowest lying states of a given spin. The discontinuity is intended to depict the discontinuity (often called backbending) observed above the pairing gap (2Δ). The shaded areas represent the ranges of spins and energies of states populated in β decay and charged particle induced reaction spectroscopy. One can thus view a β-decay state in neutron-deficient nuclei as a kind of "spin-filter" that provides a unique path to the population of low-spin states.

suited to test the predictions of these theories. In fact, it is quite generally found that with the various competing nuclear models developed in recent years, only a knowledge of the low-spin states can provide a test that is able to distinguish between them. Fig.1 dramatizes the role played by radioactive decay in the study of low-spin states in nuclei far from β stability.

It is clear that a primary requirement for following systematic trends through a sequence of nuclei is completeness, i.e., all states up to a given spin-parity and energy must be found. This demands that very detailed decay scheme studies must be made. The experimental criteria for completeness in decay scheme studies far from β stability have been discussed by the author in ref.1. In addition to the requirement of completeness, some rules for using nuclear systematics are necessary; a tentative set of rules has been proposed in ref.1. This set of rules has been developed and has proved to be a vital tool for the interpretation of the structure of nuclei in the $180<A<210$ region. This has already beeen discussed in some detail in refs.1, 2. Important points contained in refs.1, 2 that are relevant to the present discussion are emphasized below.

The present discussion focuses on the low-lying excited 0^+ states in the neutron-deficient Pt and Hg isotopes and what can be learned about them; and, their description in an IBA framework, through particle-core coupling, by the study of excited states in neighboring odd-mass Au and Tℓ isotopes. Particular emphasis is placed on the role of the $h_{9/2}$ proton shell model intruder orbital both in its contribution to core structures and in its use as a probe of the core.

III. THE INTERACTING BOSON AND INTERACTING BOSON-FERMION APPROXIMATIONS

At the heart of the interacting boson approximation, in its most up-to-date form (see e.g., ref.3), is the proposition that the low-energy collective behavior of nuclei can be approximated by proton and neutron bosons with angular momentum 0 and 2 (s and d bosons). These interact weakly through a pairing force between two proton bosons or two neutron bosons, and through a quadrupole force between a proton boson and a neutron boson; with the conservation of the number of proton bosons and the number of neutron bosons. Under certain circumstances[4,5] the distinction between protons and neutrons can be neglected and the Hamiltonian for the approximation then possesses an SU(6) dynamical symmetry. This Hamiltonian (often referred to as IBA1) has three limiting cases that possess exact solutions[6,7,8], all of which are realized in nuclei. The power of the formalism is that it can provide a simple phenomenological description of nearly all the known collective behavior in heavy ($A \gtrsim 60$) nuclei in terms of a few parameters (the boson-boson interactions). This description includes the large number of cases (formerly referred to as "transitional nuclei") intermediate between these three limits[9,10]. (The case intermediate between SU(5) and O(6) has not been studied yet.) The three cases are defined in terms of the three sub-group chains[11,12]

$$SU(6) \supset SU(5) \supset O(5) \supset O(3)$$
$$SU(6) \supset O(6) \supset O(5) \supset O(3)$$
$$SU(6) \supset SU(3) \supset O(3)$$

which exhaust the number of possibilities for SU(6). These limits are now known to possess well-defined geometrical analogs[13,14] which can be identified with simple cases of the geometrical collective or Bohr model (now often referred to as the Frankfurt model see e.g. Refs.15,16); and, which is based on a quantized liquid drop description of the nuclear collective motion. The three geometrical limits are: the anharmonic quadrupole vibrator - SU(5), the gamma-soft axially asymmetric vibrator - O(6), and a particular case (degenerate β and γ bands) of the axially symmetric rotor - SU(3).

The interacting boson approximation in the form IBA1 does not possess a predictive power. Howerver, when the distinction between protons and neutrons is retained, the boson-boson interactions exhibit a smooth universal behavior as revealed by a few empirical fits in each open shell region[5,17,18]. This universal behavior permits the unambiguous prediction of the collective properties of any nucleus within the given shell region[18,19], the only limitation being that the collective degrees of freedom are confined to the s, d-boson model space.

The bosons are believed to be correlated pairs of protons and neutrons coupled to L=0 and L=2 and moving in shell model orbitals of the specific shell region[4]. This picture is consistent with the unknown behavior of nucleon correlations in nuclei near closed shells shells, which are clearly dominated by a pairing force between like nucleons and a quadrupole force between unlike nucleons[20]. The success of the interacting boson approximation in describing so many low-lying states in even nuclei suggests that it is a good approximation to the residual interaction between the nucleons in the valence proton and neutron shells. A rigorous justification for such a boson approximation of the nuclear many fermion system is being sought by a number of investigators (see e.g. Ref.21,22,23).

The coupling of an unpaired fermion to the system of s and d bosons in IBA1 has been carried out in a straightforward manner with the inclusion of an "exchange term" that effectively takes into account the Pauli exclusion principle for the odd nucleon[5,24]. The Hamiltonian (referred to as IBFA) has three limiting cases that correspond with the three boson limits of IBA1[5,24,25]. These limits can be associated with geometrical descriptions of particle-core coupling which include particle-vibrator coupling and particle-rotor coupling (Nilsson model). Perhaps most dramatically of all, it has been suggested that for specific values of the boson-boson and boson-fermion interaction strength, it is possible to classify the the boson-fermion spectrum in terms of certain dynamical supersymmetries[26]. There is experimental evidence that this classification scheme is applicable to nuclei in the region of ^{191}Ir [26] and ^{193}Au [27]. The above hierarchy of theories is summarized schematically in Fig.2.

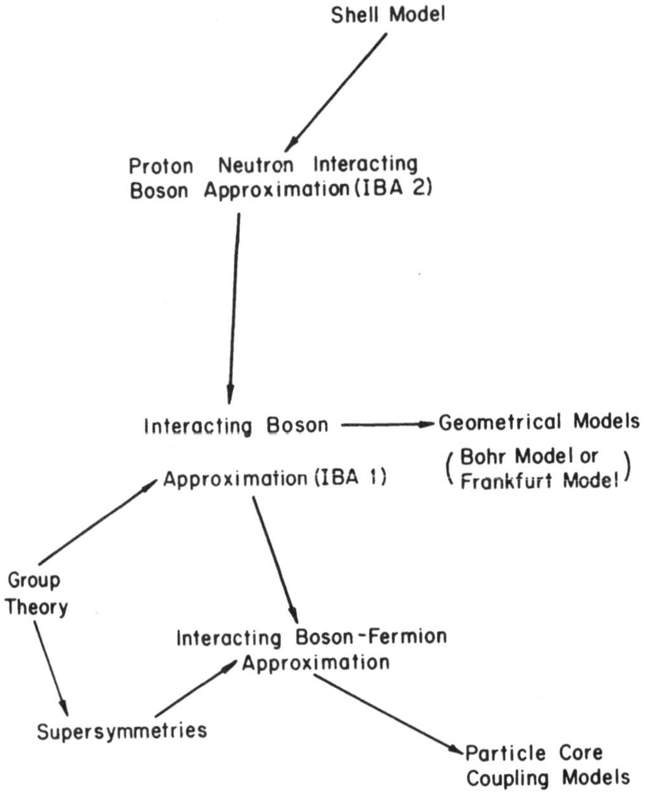

Fig.2 A diagrammatic illustration of the important branches of the IBA and related nuclear models. The arrows represent the direction in which either development has occurred or the general to the particular case has been recognized.

A major question remaining for the spectroscopist is, what about degrees of freedom outside of the IBA1, IBA2, and IBFA model spaces? Some work has been done on the description of the octupole degree of freedom by the introduction of an L-3 or f boson[6,7,9,28,24]. This description has been simplified by confining the discussion to only a single f-boson in view of the fact that two-phonon octupole states have probably not been observed yet. There is also some discussion of the coupling of s and d bosons to non-collective states in even nuclei[6,7,30], where only a single "boson" is involved. (Such modes appear to dominate the yrast structure of nuclei at high spin.) The importance of an L=4 or g boson has received only limited attention[31], and it is not clear if nucleons correlate strongly with J=4, although it has been suggested that it is important for high angular momentum states in deformed nuclei[32]. There remains the question of L=0 correlations that lie outside of the s, d-bosons space. This is discussed in some detail below.

Finally, it is tempting to consider the possibility that, having built up a description of collective motion in complex nuclei in terms of L=0 and L=2 boson "building blocks", the formalism may enable us to decompose complex nuclear excitation spectra into constituent building blocks or elementary bosons[33]. This would reveal fermion correlations that are not evident in the simple spectra near closed shells.

IV. LOW-ENERGY EXCITED 0^+ STATES

Excited 0^+ states are probably the most poorly understood modes of excitation in even nuclei. This was illustrated dramatically by the recent discovery that the lowest excited 0^+ state in $^{112-118}$Sn is strongly deformed with a well-defined rotational band built on it[34]. Consequently, it is now believed that the excitation of proton pairs across closed shells can give rise to fairly low-lying excited 0^+ states which are much more deformed than the ground state. The fact that such an excitation mode does not occur for neutrons is attributed to the Coulomb interaction[35]. Although less dramatic, the lowest excited 0^+ state in a number of the Th, U and Pu isotopes is still the subject of considerable debate (see e.g. Refs.36,37,38 and references therein). In contrast, for example, the lowest excited 0^+ state in each of the Sm isotopes[9,39] and the various excited 0^+states in ^{196}Pt (ref.40) and ^{168}Er (ref.41,42) are well described by the IBA. Since the s, d-boson space of the IBA gives rise to a significant number of 0^+ states, it is crucial that 0^+ states which lie outside of the s, d-model space are identified.

Despite the generally widespread lack of understanding of excited 0^+ states, there are some cases which appear to be at least partially understood. These cases can be characterized roughly into (a) excitations of pairs of nucleons across subshell gaps, (b) excitation of pairs of protons across closed shells, and (c) coupling of s and d-bosons in the IBA. The best example of (a) is probably the first excited 0^+ state in ^{90}Zr, of (b) are the first excited 0^+ states in the Sn isotopes, and of (c) are the first excited 0^+ states in the Sm isotopes. Although there are many examples of low-lying excited 0^+ states throughout the mass surface, with an accompanying variety of theoretical interpretations, a comprehensive cataloging of 0^+ states is considered premature at present. Thus, this discussion is confined to a few examples of cases (a) and (b).

Two further general points need to be made with regard to excited 0^+ states. The first is that two-nucleon transfer is ideally suited to the probing of the pairing collectivity of these states. The second is that the coupling of an odd-nucleon to a 0^+ state can probe its microscopic structure through the blocking effect due to the Pauli principle. The more collective (the stronger) the correlations of nucleon pairs, the more the two-nucleon transfer strength is concentrated between a single initial and a single final state, and the less the blocking effect of an unpaired nucleon. (The blocking effect is most easily understood in terms of uncorrelated nucleon pair configuration which would be completely blocked by an unpaired nucleon.)

The first excited 0^+ state in ^{90}Zr lies at 1761 keV and is, in fact, the first excited state in this nucleus. Its population in (p,t)[43], (^3He, n)[44] and (d, ^6Li)[45] transfer reactions is most readily explained as arising from proton pairing correlations: these act in the $p_{1/2}$ and $g_{9/2}$ subshells and give rise to proton configurations for the ground state and excited state of the form $\alpha(\pi\ p_{1/2})^2 + \beta(\pi\ g_{9/2})^2$ and $\alpha(\pi\ p_{1/2})^2 - \beta(\pi\ g_{9/2})$, respectively. The ground state configuration is supported by single-proton pickup

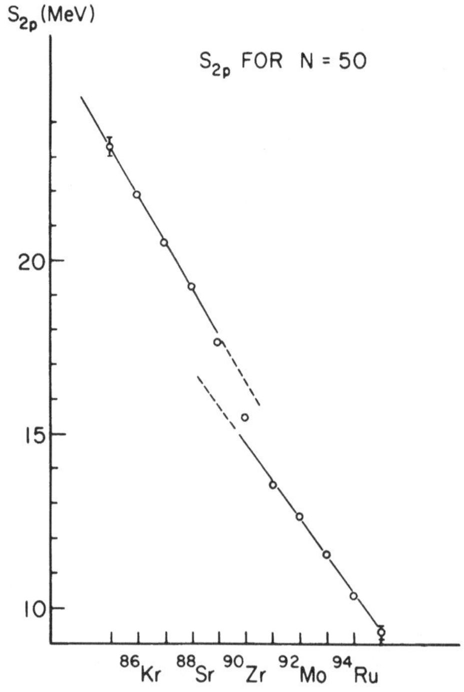

Fig.3 The two-proton separation energies (S_{2p}) for the N=50 isotopes. Note the break between Z=38 and 41, indicating a proton-subshell gap.

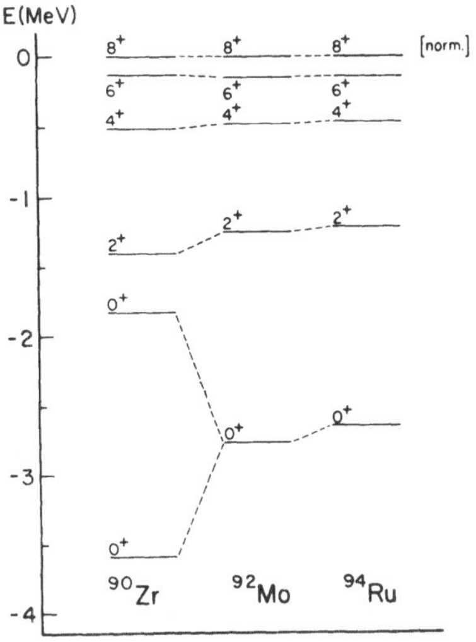

Fig.4 The energy systematics of the $(\pi g_{9/2})^{2n}$ configuration in the N=50 isotones: The systematics of the 0_1^+, 2_1^+, 4_1^+, 6_1^+, and 8_1^+ states in ^{90}Zr, ^{92}Mo and ^{94}Ru, and the 0_2^+ state in ^{90}Zr. Energies are plotted relative to the 8_1^+ state which is probably a very pure $(g_{9/2})^2$ configuration. Evidently the 0_1^+ and 0_2^+ states in ^{90}Zr mix and repel, showing that the ground state is a mixed configuration.

and stripping[47] reactions. The simplest manner in which to dramatize the occurrence of a subshell gap is to plot the two-proton separation energy (S_{2p}) against mass number for the N=50 isotones (see Fig.3). The mixing of $(\pi\ p_{1/2})^2$ and $(\pi\ g_{9/2})^2$ configurations can be visually demonstrated by plotting the states in the N=50 isotones relative to the 8^+ $\{\pi g_{9/2}^{2n}\}$ configuration (see Fig.4) as has been emphasized by Talmi[20]. It is interesting to note that this structure appears to persist into the neutron-rich Zr isotopes[45]. It has further been suggested[48] that it plays a role in the sudden appearance of deformation between ^{98}Zr and ^{100}Zr. (It should also be noted that the description[48] emphasizes the importance of proton-neutron interactions in producing deformation, which concurs with the proton-neutron IBA.) With regard to the neutron-rich Zr isotopes, it should be stressed that this degree of freedom, which is relatively well isolated at the N=50 shell closure, becomes mixed in a complicated manner, with other degrees of freedom at ^{96}Zr and behond. It is in the sense of recognizing the importance of the role of this degree of freedom in the structure of the neutron-rich Zr isotopes (and neighboring isotopes) that we aim to decompose complex spectra into simple "building blocks". The possibility should also be borne in mind that a subshell gap may only arise away from closed shells; and thus, it could be substantially masked by the other degrees of freedom (s-, d-bosons), and may only be identifiable if a reliable decomposition into building blocks is available.

The first excited 0^+ state in $^{114, 116, 118}Sn$ lies below 2MeV and is the second excited state (above the 2^+ state) in all cases. These states are strongly populated by the (^3He, n)[49] transfer reaction and only weakly populated by the (p,t)[50] and (t,p)[51] transfer reactions. In the case of ^{118}Sn, the first excited 0^+ state is also significantly populated in the (d, ^6Li) transfer reaction[52]. These data are well accounted for by interpreting these 0^+ states as proton 2p-2h excitations, resulting from the excitation of a proton pair across the Z=50 shell closure. Although the low energy of this degree of freedom was not anticipated, it can be explained by a particle-hole interaction which is large because of the Coulomb

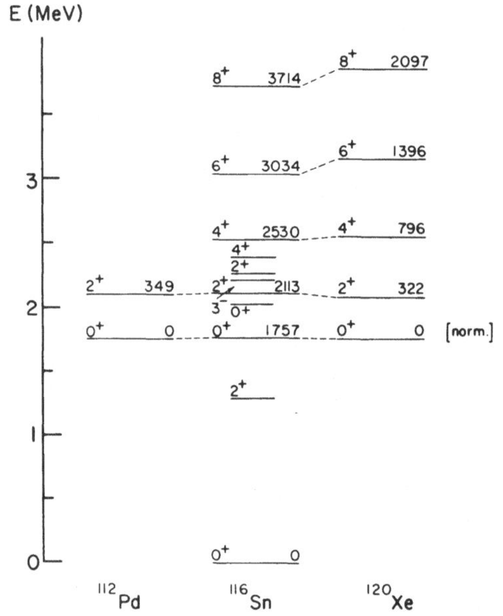

Fig.5 A comparison of the proton 2p-2h states in ^{116}Sn with the proton 4h states in ^{112}Pd and the proton 4p states in ^{120}Xe. They are normalized at the 0^+ configuration. The low-lying collective degrees of freedom due to neutron correlations alone are shown for ^{116}Sn

contribution[35]. The observation[34] of well-defined rotational bands built on these states shows that they are deformed. In fact, the rotational spacings exhibit a remarkable similarity to the Pd and Xe isotones (see Fig.5) which have 4 proton holes and 4 proton particles, respectively. This structure evidently persists into the Cd isotopes, as revealed by (^{3}He, n) transfer reaction studies[49]. There is also some evidence for rotational bands built on these states in the Cd isotopes (see e.g. Ref.53). Electromagnetic transitions between these states and other states in the Sn and Cd isotopes (see e.g. Refs. 54, 55) reveal that substantial mixing is occurring. This mixing has been investigated recently by Heyde et al.[56].

There is no information on the blocking of the first excited 0^+ states in ^{90}Zr and $^{114, 116,, 118}$Sn to the best of the author's knowledge. However, for example, the structures of the first excited 0^+ states in ^{72}Ge and $^{232, 234}$U have been probed both by two-nucleon transfer and the blocking effect. In the case of ^{72}Ge and ^{73}As (see e.g. Refs. 57, 58), the odd proton evidently blocks the excited 0^+ configuration strongly and it has not been observed as a core configuration in ^{73}As. In the case of $^{232, 233, 234}$U [59], the odd neutron partially blocks the ground-state 0^+ configuration and does not significantly block the excited 0^+ configuration. These examples are given because they illustrate strong and weak blocking of the ground and excited 0^+ state configurations. In studying nuclei far from the line of β stability, there are no stable (or long-lived) targets available for two-nucleon transfer reaction studies; and thus, the probing of 0^+ states through the blocking effect is the only method we have for investigating their structure. (At present, the systematic behavior of EO transition probabilities[60] is not understood well enough to be of use.)

V. THE STUDY OF RADIOACTIVE DECAY SCHEMES AT UNISOR: THE DECAYS ^{187m,g}Hg → ^{187}Tℓ, ^{189m}Pb → ^{189}Tℓ

The UNISOR facility operates on-line to the Oak-Ridge Isochronous Cyclotron (and will soon operate on-line to the 25MV Tandem) at the Holifield Heavy-Ion Research Facility. Isotopes far from the β-stability line are produced by heavy-ion induced compound nuclear reactions on thin targets. The isotopes so produced recoil out of the target into a catcher which slows them and thermally re-emits them into an ion-source[61,62]. Various ion-source designs[63,64] are employed, depending on the elemental species of interest. The ions are extracted from the ion-source and pass through the isotope separator to be focused and deposited on moving tape transport systems[65,66] that bring the activities to the detection systems. These consist mainly of semiconductor photon and electron detectors coupled to computer based data acquisition systems[63,67]. (However, a recent major addition to the UNISOR facility is a turnable dye laser in a collinear geometry which will be used to determine spins and static moments of long-lived ($\gtrsim$ 1 sec) states in nuclei by means of optical hyperfine spectroscopy.) The source collection time (which is equal to the source counting time) and the tape transport time are optimized for the half life of the radio-isotope under study. Although sources of activity are typically > 99% pure in mass, the production reactions generally produce a spread in Z for the mass number A of interest due to charged particle evaporation. In addition, far from β stability the Q-value for electron capture/positron decay becomes very large. The result is that extremely complicated γ-ray and coversion-electron spectra are observed. This necessitates coincidence spectroscopy with characteristic x-rays in order to make Z assignments; and, detailed γ-γ-t and γ-e-t coincidence studies to build the decay schemes. These coincidence data are accumulated event-by-event in list mode on magnetic tape. After the experiment the list data is prescanned, packed at a high density and sorted for selected energy and time gates on an IBM 370/3033. To assign weak γ-ray and electron lines, it is necessary to accumulate very good statistics ($\gtrsim 5 \times 10^6$ coincidence events) and to have excellent true-to-chance coincidence ratios ($2\tau \lesssim 10$ ns). The γ-ray and conversion-electron singles data are taken simultaneously in a multiscaling mode with time planes chosen to suit the half life of the radioisotope being studied (generally ten planes per source counting period). Our criteria for data quality are discussed further in Ref.1.

Full details of the study of the $^{187m,g}Hg \rightarrow {}^{187}Au$ decay schemes appear in Ref.68. It is sufficient to note here that sources of ^{187m}Hg (T = 2.4m) were produced by the reaction $^{180}W(^{14}N,\ p\ 6n)^{187}Hg$ and sources of ^{187}Hg (T = 2.2m) were produced by the reaction and decay $^{180}W\ (^{14}N,\ 7n)\ ^{187}T\ell\ (\beta^+)\ ^{187}Hg$. The β decay of these two states with spins of 3/2 [69] and 13/2 [70] permitted the study of essentially all states in ^{187}Au up to J=17/2 and an energy of approximately 1.2 MeV. (The occurrence of more than one β-decaying state is of immense value to detailed studies of low-spin nuclear structure, because of the range of spin values populated in the daughter, and the indication of the spins of levels populated.) The ^{187}Au level scheme was constructed from coincidence data, together with clues provided by the systematics of the odd-Au isotopes[1,2]. Spin-parities were assigned to the levels on the basis of multipolarities deduced from conversion electron data, the known ^{187}Au ground-state spin of ½ [2], in-beam reaction studies of ^{187}Au [2], and the relative population of the levels from the high and low-spin Hg β decay. Our results confirm the work of Bourgeois et al.[71] and greatly extend their scheme. (Most notably, we have observed transitions with E0 components that were not found in the previous work[71].)

A greatly simplified level scheme for ^{187}Au is shown in Fig.6. The eight positive parity states on the left are a well-defined feature throughout the odd-Au isotopes[1,2], and are due to shell

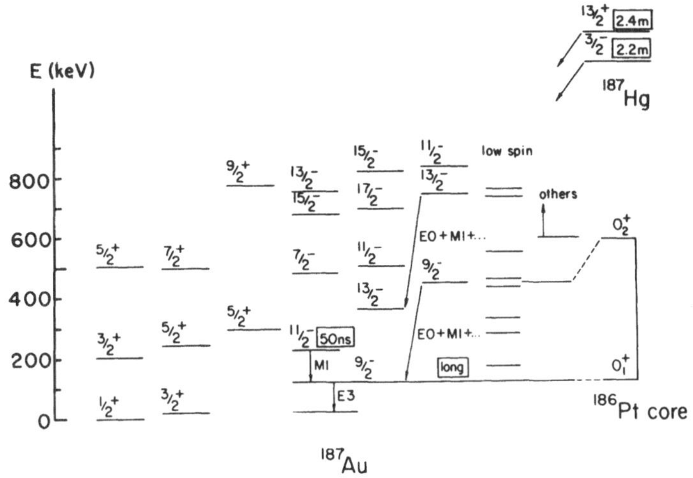

Fig.6 A schematic representation of the excited states in ^{187}Au populated in the β decay of ^{187}Hg. Many of the levels are discussed in the text. The series of levels marked "low-spin" belong to the $h_{9/2}$ band. There are many other levels above 590 keV that are not shown. The half life of the $9/2^-$ isomer is probably between 0.1 and 1 sec.

model hole states with positive parity and associated collective modes. The band of states built on the 50 ns isomer results from the $h_{11/2}$ shell model hole state. The $h_{11/2}$ band is also a well-defined feature throughout the odd-Au isotopes[1,2]. The states on the right-hand side of Fig.6, based on the $9/2^-$ long-lived isomer, result from the $h_{9/2}$ shell model intruder state and associated collective modes. This state has the properties of a particle state, as has been demonstrated in ^{189}Au [2]; and thus, couples to an effective core of ^{186}Pt. Consequently, the $h_{9/2}$ band in ^{187}Au serves as a particle-core coupling probe of the collective modes in ^{186}Pt. Most notably, the coupling of the $h_{9/2}$ particle to the excited 0^+ core state in ^{186}Pt and its associated collective modes[72,73] is observed. These are shown in Fig.6 as a band of states, two of which decay by E0+M1 (+E2) transitions. This coupling of the $h_{9/2}$ proton to the ^{186}Pt 0_1^+ (ground state) and 0_2^+ core states is central to the discussion below.

The study of the ^{189}Pb $\rightarrow$ ^{189}Tl decay scheme is in progress: This investigation is a collaboration between UNISOR and the on-line isotope separator group at GSI, Darmstadt. Sources of ^{189}Pb (50s) were produced by the reaction ^{180}W (^{16}O, 7n)^{189}Pb at UNISOR, and by the reaction ^{146}Nd(^{48}Ti, 5n)^{189}Pb at GSI. Only one half-life

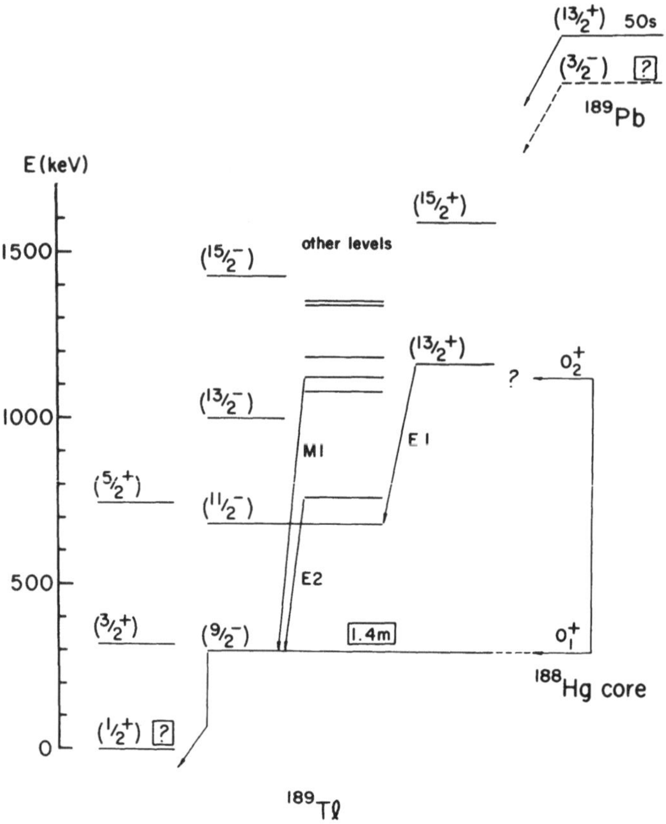

Fig.7 A schematic representation of the excited states in ^{189}Tℓ populated in the β decay of ^{189}Pb. The 1.4 min isomer is placed at ∿ 300 keV above the ground state on the basis of of systematics. The parentheses indicate that the spin assignments rely on systematics.

has been observed in the decay of ^{189}Pb; and, a comparison of the strong γ-rays seen in our work with in-beam reaction studies of ^{189}Tℓ [74] shows that we are observing the β decay of a high-spin state. This is probably the $i_{13/2}$ state; and thus, the β decay of ^{189}Pb is analogous to the heavier Pb β decays observed in other UNISOR studies (see e.g. Refs. 74, 75). The ^{189}Tℓ level scheme was constructed from the coincidence data, in-beam reaction data[74], and clues provided by the systematics of the odd-Tℓ isotopes[2]. Spin-parities are assigned on the basis of multipolarities deduced from conversion-electron data, and systematics. Because there is no direct spin information on the ^{189}Tℓ ground state and isomer (the existence of which is proposed on the basis of this work and the ^{189}Tℓ → ^{189}Hg decay scheme[1,76], these assignments must be regarded with caution.

A preliminary level scheme for ^{189}Tℓ is shown in Fig.7. The band of states build on the 1.4 min isomer is intepreted as resulting from the $h_{9/2}$ shell model intruder state and associated collective modes. The effective core for this band is believed to be the neighboring even-Hg nucleus[2], ^{188}Hg in this cae. Of particular interest to the present discussion is the search for the coupling of the $h_{9/2}$ proton to the strongly deformed band of states in ^{188}Hg. This deformed band in ^{188}Hg is built on a 0^+ state at 825 Kev[73,77], and the coupling of the $h_{9/2}$ proton to this state might be expected to give rise to a $9/2^-$ state that de-excites by an E0(+M1+E2) transition to the $9/2^-$ band head. We have not yet found any evidence for a transition with an E0 component in the decay of ^{189}Pb, and can almost certainly rule out any of the levels shown as candidates for the $h_{9/2} \otimes {}^{188}$Hg (0_1^+) coupling. The apparent non-existence of this state, within 1050 keV of excitation from the $9/2^-$ band head, is discussed below.

VI. THE LOW-LYING EXCITED 0^+ STATES IN THE NEUTRON-DEFICIENT Pt AND Hg ISOTOPES AND THE IBA

The existence of low-lying excited 0^+ states in the neutron-deficient even-Pt isotopes has been known for some time[72]. The

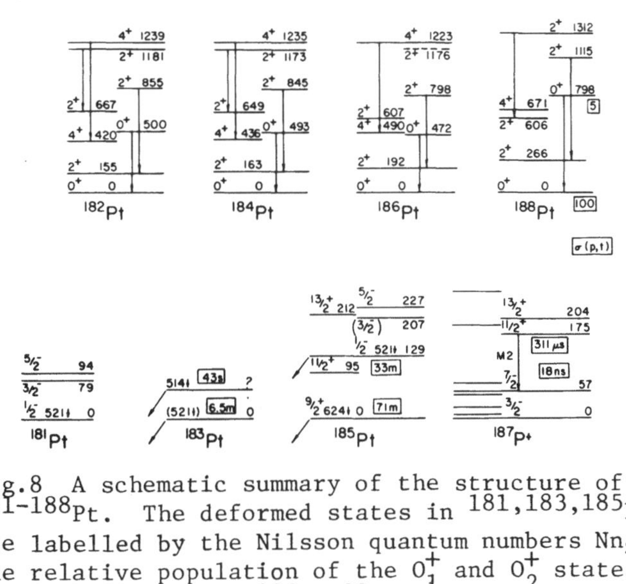

Fig.8 A schematic summary of the structure of $^{181-188}$Pt. The deformed states in ^{181,183,185}Pt are labelled by the Nilsson quantum numbers $Nn_z\Lambda\Sigma$. The relative population of the 0_1^+ and 0_2^+ states in the (p, t) study of ^{188}Pt [92] is shown.

early studies of the even-Au decay schemes (see Ref. 72 and references therein) have been followed by the more detailed investigations of Caillau, et al.[78] and Cole[73]. Various attempts have been made to interpret the 0^+ states in the Pt isotopes (see, e.g. Ref.79 and references therein). However, the subsequent discovery of rotational bands in ^{185}Pt [80] and ^{177,179,181}Pt [81], with rotational parameters differing by a factor of two from those deduced using $E(2_1^+)$ in the even neighbors, showed that an exotic structure existed at low energy in the neutron-deficient Pt isotopes. These rotational bands have not been observed yet in ^{187}Pt [82]. The essential details of these findings in the neutron-deficient Pt isotopes are summarized in Fig.8. Independently of the work on the Pt isotopes, optical pumping experiments on the neutron-deficient odd-Hg isotopes revealed a sudden increase in the mean-square charge radius between ^{187}Hg and ^{185}Hg [83]. Further optical measurements established spins, magnetic moments and spectroscopic quadrupole mements of the (ground) states in the odd-Hg isotopes and suggested the existence of the strongly deformed states[69]. A variety of investigations were directed at the neutron-deficient even-Hg isotopes (see i.g. Refs. 73, 77,84,85,86 and references therein) before the experimental picture became reasonably complete. Most recently, optical hyperfine spectroscopy with a laser has been used to directly demonstrate the shape isomerism in ^{185}Hg [70] and the weak deformation of the ^{184,186}Hg ground states[87]. Essential details of the neutron-deficient Hg isotopes are shown in Fig.9. The attempts to explain the shape isomerism in the neutron-deficient Hg isotopes have been

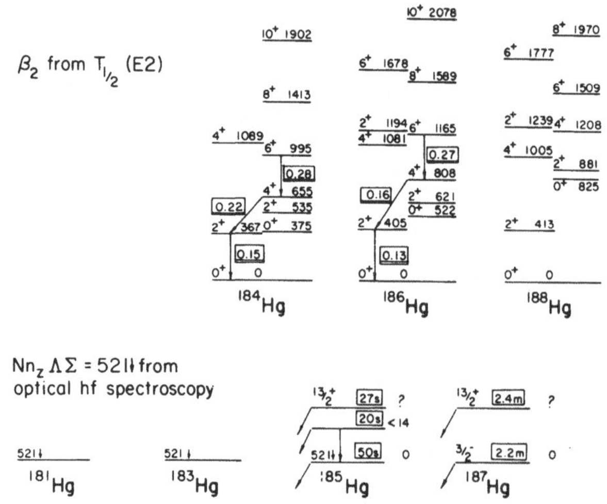

Fig.9 A schematic summery of the structure of $^{181,183-188}$Hg. The β_2 values are deduced from $T_{1/2}$(E2) measurements .

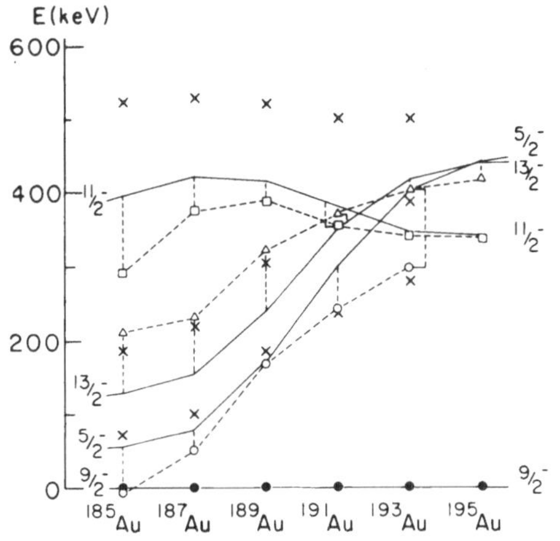

Fig.10 A comparison between the experimental energies of the lowest $9/2^-$, $5/2^-$, $13/2^-$ and $11/2^-$ members of the $h_{9/2}$ collective bands in the odd-Au isotopes, the IBFA calculations of Refs.95, 96, and calculations 95, 96 done with the Meyer ter Vehn model. Experimental energies are marked by open points, the energies calculated with the IBFA are shown as solid lines, and the Meyer ter Vehn model calculations are depicted by crosses. The dotted lines are to guide the eye.

numerous. They can be subdivided roughly into prescriptions that invoke near degenerate prolate and oblate minima in the potential energy surface; and, such minima in combination with a strong pairing correlation effect for the neutrons: the theoretical situation has been discussed recently at the Nashville Conference[88,89]. Although these features of the neutron-deficient Pt and Hg isotopes were initially unconnected, the rotational bands in $^{177,179,181}Pt$ have been identified on the basis of favored α-decay of the deformed states in $^{181,183,185}Hg$ [81]. Other recent findings that add to this picture include the observation[90] of an excitation spectrum in ^{180}Pt (following ^{180}Au decay) very similar to ^{182}Pt; a fast β-decaying isomer in ^{183}Pt [91] that supports a strong coupling scheme (and hence large deformation); a (p, t) study of levels in ^{188}Pt [92]; and, evidence for shape isomerism in ^{187}Hg seen in the β decay of ^{187}Tl [93].

Very recently, a fairly successful description of the neutron-deficient even - Pt isotopes was achieved within the framework of IBA2[94]. A study of the $h_{9/2}$ bands in the odd-Au isotopes using the

IBFA[94] suggested that this description of the collective degrees of freedom in the even-Pt isotopes[94] was satisfactory for the $h_{9/2} \otimes$ Pt collective coupling also. Some results of these latter calculations are shown in Fig.10, and some details are also given in Ref.96. However, a very important feature that is evidently missing is: where are the strongly deformed states in the even-Pt isotopes that are the analogs of the deformed states in the odd-Pt and even- and odd-Hg isotopes? A valuable clue to this is revealed in Fig.11, which shows the systematics of the yrast bands in the N=106 isotones and the deformed band in ^{186}Hg. Evidently the yrast band in ^{184}Pt increases in deformation with increasing spin, as seen from the rotational parameters extracted from the transition energies. A consideration of the low-spin excited states in ^{184}Pt (see Fig.8) strongly suggests that the 0^+ ground state and 0^+ excited state mix strongly and the ground state energy is lowered. To a lesser extent, the 2_1^+ and 2_3^+ states similarly must mix and repel. The E0 transition between these two states supports this. A similar systematic behavior is found for the other even-Pt isotopes. Thus the ^{184}Pt ground state contains the major component of a strongly deformed 0^+ configuration.

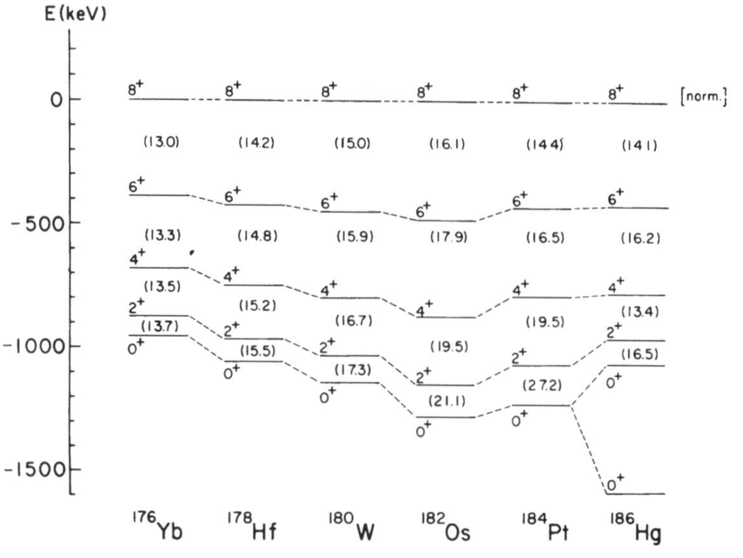

Fig.11 The 0_1^+, 2_1^+, 4_1^+, 6_1^+ and 8_1^+ states in the N=106 isotones and the 0_1^+, 0_2^+, 2_2^+, 4_1^+, 6_1^+ and 8_1 states in ^{186}Hg (see also Fig.9). The rotational parameter $h^2/2\mathcal{J} = (E_J E_{J-2})/(4J-2)$ is shown in parentheses between the levels. Energies are shown relative to the 8_1^+ states.

We now consider the 0^+ configurations in the neutron-deficient Pt and Hg isotopes in the light of the discussion above of 0^+ states and certain features of IBA2. The microscopic potential energy surface calculations suggest a situation not unlike the Sm isotopes, whereas the microscopic potential energy surface calculations that emphasize pairing suggest a neutron subshell closure also plays a role. The IBA2 calculations for the even-Pt isotopes suggest that a sufficient condition to produce the observed behavior is a situation analogous to the Sm isotopes; albeit, with a rather strong neutron-neutron quadrupole force. However, when one remembers that in IBA2 the dominant deformation producing term is the proton-neutron quadrupole force (a feature of the prescription that is re-emphasized here to be consistent with the picture of Federman and Pittel[48], an interesting and completely new possibility arises. If the only deformation producing term in IBA2 is $Q_\pi \cdot Q_\nu$, then the Pt isotopes would not be expected to have a large deformation since there are only four valence proton holes, i.e. $N_\pi=2$. Moreover, deformation would be completely ruled out for the Hg isotopes, since $N_\pi=1$. (As noted above, this was circumvented in the even-Pt isotopes with a large $Q_\nu \cdot Q_\nu$). It is asserted that if the proton intruder orbitals (such as $h_{9/2}$ which is known to lie near the Fermi energy in the odd-proton nuclei in this region) are accessible to build extra proton bosons (i.e. increase N_π), then the $Q_\pi \cdot Q_\nu$ interaction is sufficient to explain the even-Pt and -Hg isotopes. This situation would be analogous to the deformed 0^+ states in the even-Sn and -Cd isotopes.

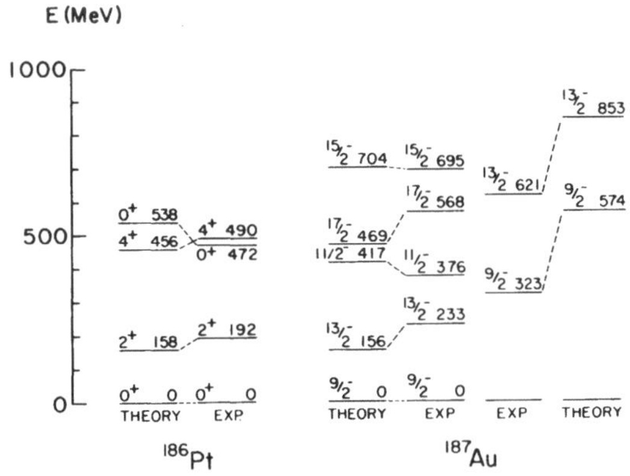

Fig.12 IBA1 calculations for ^{186}Pt and IBFA calculations for ^{187}Au. The same interacting boson parameters were used for both calculations and were derived from IBA2 parameters, as noted in the text.

The most unusual feature of this assertion is that proton pairs in the intruder orbitals play a major role in the ground state of the neutron-deficient Pt isotopes; this is believed to be unprecedented in the heavy (A > 60) nuclei. Making the reasonable assumption that the most important proton intruder orbital will be the one closest to the Fermi energy, namely $h_{9/2}$, then a simple consequence of this assertion is that the coupling of an $h_{9/2}$ proton to a Pt core in the neutron-deficient Au isotopes will block the 0_1^+ core state more than the 0_2^+ core state. The converse will hold for Hg $\otimes$ $\pi h_{9/2}$ coupling in the neutron-deficient Tℓ isotopes. This effect apparently occurring in ^{187}Au and we deduce that the blocking of the ground-state proton pairing correlations by the $h_{9/2}$ proton is costing 149 keV. The possibility that this shift is due to the quadrupole part of V_{BF} can be dismissed when one considers Fig.12. This shows the IBFA calculation for ^{187}Au [94] with core parameters taken from the ^{186}Pt calculation[93] by a projection technique[5]. A ^{186}Pt calculation using IBA1 and the projected parameters is also shown in Fig.12. The IBFA calculations demonstrates that the coupling of the $h_{9/2}$ proton to the boson quadrupole field results in a 9/2 energy of 574 keV. The non-observation of an E0 transition in ^{189}Tℓ is also consistent with this picture, since the $h_{9/2}$ blocking will effect the excited 0^+ state in ^{188}Hg and result in a raising of the E0 transition energy.

VII. CONCLUSIONS

The above discussion suggests that in the pursuit of model building (and boson building), we must proceed with caution and carefully consider a broad range of spectroscopic and structural information. The specific structure that is argued herein to be important in the neutron-deficient Pt and Hg isotopes (namely, proton intruder orbitals from the Z = 82 shell) can be incorporated readily into the IBA framework and preliminary details of the technique have been reported by van Isacker[97]. The most attractive aspect of the IBA (and IBFA) formalism is that it can incorporate such building blocks into its framework in an explicit manner. This results in predictions that are unambiguous and readily tested by experiment.

It is evident from this discussion that a large number of experimental studies are needed in this region. Apart from a few obvious spectroscopic studies such as the two-proton transfer reaction ^{184}Os(^{3}He, n)^{186}Pt, most of these studies will have to be done by radioactive decay. Many of the possibilities are in progress or planned for future investigation at UNISOR.

VIII. ACKNOWLEDGEMENTS

The experimental studies of levels in ^{187}Au and ^{189}Tℓ were done at UNISOR in collaboration with M.A. Grimm, E.F. Zganjar and J.D. Cole; and at GSI, Darmstadt in collaboration with G.M. Gowdy, L.L. Riedinger, and E. Roeckl and cowerkers. The IBA and IBFA calculations were done in collaboration with R.A. Braga. UNISOR is a consortium of thirteen institutions and is supported by them and by the U.S. Dept. of Energy, contract No. DE-AC05-760R00033. This work was supported in part by the U.S. Dept. of Energy, Contract No. DE-AS05-80ER10599.

REFERENCES

1. J.L. Wood, in Future Directions in Studies of Nuclei Far From Stability, eds. J.H. Hamilton, et al., North-Holland Publishing Co., Amsterdam (1980), p.37.
2. E.F. Zganjar, ibid., p.49.
3. Interacting Bose-Fermi Systems in Nuclei, ed. F. Iachello, Ettore Majorana International Science Series, Plenum Press, to be published.
4. A. Arima, T. Otsuka, F. Iachello and I. Talmi, Phys. Lett. 66B, 205 (1977).
5. O. Scholten, thesis, Univ. of Groningen (1980).
6. A. Arima and F. Iachello, Ann. of Phys. 99, 253 (1976).
7. A. Arima and F. Iachello, ibid. 111, 201 (1978).
8. A. Arima and F. Iachello, ibid. 123, 468 (1979).
9. O. Scholten, F. Iachello and A. Arima, ibid. 115, 325 (1978).
10. R.F. Casten and J.A. Cizewski, Nucl. Phys. A309, 447 (1978).
11. F. Iachello, in Lecture Notes in Physics #119, eds. G.F. Bertsch and D. Kurath, Springer-Verlag, Berlin and Heiderberg (1980), p.140.
12. O. Castanos, E. Chacon, A. Frank and M. Moshinskly, J. Math. Phys. 20, 35 (1979); M. Moshinsky, Notas de Fisica 2, 255 (1979).
13. J.N. Ginocchio and M.W. Kirson, Phys. Rev. Lett. 44, 1744 (1980).
14. A.E.L. Dieperink, O. Scholten and F. Iachello, ibid., 1747 (1980).
15. P.O. Hess, M. Seiwert, J. Maruhn and W. Greiner, Z. Phys. A296, 147 (1980); P.O. Hess, J. Maruhn and W. Greiner, in Future Directions in Studies of Nuclei Far From Stability, eds. J.H. Hamilton et al., North-Holland Publishing Co., Amsterdam (1980, p.151.
16. M. Moshinsky, Nucl. Phys. A338, 156 (1980).
17. O. Scholten, in Interacting Bosons in Nuclear Physics, ed. F. Iachello, Ettore Majorana International Science Series, Plenum Press, New York (1979), p.17.
18. F. Iachello, G. Puddu, O. Scholten, A. Arima and T. Otsuka, Phys. Lett. 89B, 1 (1979).
19. F. Iachello, in Future Directions in Studies of Nuclei Far From Stability, eds. J.H. Hamilton, et al., North-Holland Publishing Co., Amsterdam (1980), p.281.

20. I. Talmi, Riv. Nuovo Cimento, 3, 85 (1973); in Proc. Int. School of Physics "Enrico Fermi", Course LXIX, Varenna, 1977, North-Holland Publishing Co., Amsterdam (1977), p.352; Varenna lectures, 1980.
21. T. Otsuka, A. Arima, F. Iachello and I. Talmi, Phys. Lett. 76B, 139 (1978); T. Otsuka, A. Arima and F. Iachello, Nucl. Phys. A309, 1 (1978).
22. J.N. Ginocchio, Ann. of Phys. 126, 234 (1980).
23. J.N. Ginocchio and I. Talmi, Nucl. Phys. A337, 431 (1980).
24. F. Iachello and O. Scholten, Phys. Rev. Lett. 43, 679 (1979).
25. F. Iachello and O. Scholten, Phys. Lett. 91B, 189 (1980).
26. F. Iachello, Phys. Rev. Lett. 44, 772 (1980).
27. J.L. Wood, preprint (1980); and to be published in Ref.3.
28. A. Arima and F. Iachello, Phys. Lett 57B, 39 (1975).
29. Z. Sujkowski, D. Chmielewska, M.J.A. de Voigt, J.F.W.Janssen and O. Scholten, Nucl. Phys. A291, 365 (1977).
30. A. Gelberg and A. Zemel, Phys. Rev. C22, 936 (1980).
31. K.A. Sage and B.R. Barrett, Phys. Rev. C22, 1765 (1980).
32. A. Bohr and B.R. Mottelson, Preprint (1980).
33. F. Iachello, in Structure of Medium - Heavy Nuclei, The The Institute of Physics (London) Conference Series, No.49 (1980), p.161.
34. J. Bron, et al., Nucl. Phys. A318, 335 (1979).
35 E.R. Flynn and P.D. Kunz, Phys. Lett. 68B, 40 (1977).
36. A.M. Friedman, K. Katori, D. Albright and J.P. Schiffer, Phys. Rev. C9, 760 (1974).
37. B. Sorensen, Nucl. Phys. A236, 29 (1974).
38. R.R. Chasman, Phys. Rev. Lett. 42, 630 (1979).
39. O. Castanos, A. Frank and P. Federman, Phys. Lett. 88B, 203 (1979).
40. J.A. Cizewski, et al., Nucl. Phys. A323, 349 (1979).
41. W.F. Davidson, et al., submitted to J. Phys. G.
42. D.D. Warner, R.F. Casten and W.F. Davidson, submitted to Phys. Rev. Lett.
43. J.B. Ball, R.L. Auble and P.G. Roos, Phys. Rev., C4, 196 (1971).
44. H.W. Fielding, R.E. Anderson, D.A. Lind, C.D. Zafiratos and W.P. Alford, Nucl. Phys. A269 125 (1976).
45. A. Saha, G.D. Jones, L.W. Put and R.H. Siemssen Phys. Phys. Lett. 82B, 208 (1979).
46. B.M. Preedom, E. Newman and J.C. Hiebert, Phys. Rev. 166, 1156 1156 (1968).
47. M.R. Cates, J.B. Ball and E. Newman, Phys. Rev. 187, 1682 (197 1682 (1969).
48. P. Federman and S. Pittel, Phys. Rev. C20, 820 (1979).
49. H.W. Fielding R.E. Anderson, C.D. Zafiratos, D.A. Lind, F.E. Cecil, H.H. Wieman and W.P. Alford, Nucl. Phys. A281, 389 (1977).
50 D.G. Fleming, M. Blann, H.W. Fulbright and J.A. Robbins, Nucl. Phys. A157, 1 (1970).

51. J.H. Bjerregaard, O. Hansen, O. Nathan, R. Chapman and S. Hinds, Nucl. Phys. A131, 481 (1969); J.H. Bjerregaard, O. Hansen, O. Nathan, L. Vistisen, R. Chapman and S. Hinds, Nucl. Phys. A110, 1 (1968).
52. J. Janecke, F.D. Becchetti and C.E. Thorn, Nucl. Phys. A325, 337 (1979).
53. R.A. Meyer and L. Peker, Z. Phys. A283, 379 (1977).
54. J. Kantele, R. Julin, M. Luontama, A. Passoja, T. Poikolainen, A. Backlin and N.G. Jonsson, Z. Phys. A289, 157 (1979).
55. R. Julin, J. Kantele M. Luontama, A. Passoja, T. Poikolainen, A. Backlin and N.G. Jonsson, Z. Phys. A296, 315 (1980).
56. K. Heyde, M. Waroquier, P. van Isacker, H. Vincx and G. Wenes, preprint (1980).
57. D. Ardouin, B. Remaud, K. Kumar, F. Guilbault, P. Arignon, R. Seltz, M. Vergnes and; G.Rotbard, Phys. Rev. C18, 2739 (1978). 2739 (1978).
58. M.N. Vergnes, G. Rotbard, R. Seltz, F. Guilbault, D. Ardouin, R. Tamisier, and P. Avignon, Phys. Rev. C14, 58 (1976).
59. A.M. Friedman K. Katori, D. Albright and J.P. Schiffer, Phys. Rev. C9, 760 (1974).
60. N.A. Voinova, Izv. Akad. Nauk SSSR (ser. fiz.) 41, 1558 (1977); [transl. Bull. Acad. Sci. USSR (phys. ser.) 41 # 8, 19 (1977)].
61. R.L. Mlekodaj. E.H. Spejewski, H.K. Carter and A.G. Schmidt, Nucl. Instr. Meth. 139, 299 (1976).
62. E.H. Spejewski, R.L. Mlekodaj and H.K. Carter, Contribution to the Tenth International EMIS Conference, to be published in Nucl. Instr. Meth.
63. E.H. Spejewski, ibid.
64. R.L. Mlekodaj. ibid.
65. R.L. Mlekodaj, E.F. Zganjar and J.D. Cole, ibid.
66. H.K. Carter and R.L. Mlekodaj, Nucl. Instr. Meth. 128, 611 (1975).
67. H.K. Carter, et al., Nucl. Instr. Meth. 139, 349 (1976).
68. M.A. Grimm, thesis, Geortia Tech. (1978); and M.A. Grimm, J.L. Wood and E.F. Zganjar, to be published.
69. J. Bonn, G. Huber, H.J. Kluge and E.W. Otten, Z. Phys. A276, 203 (1976).
70. P. Dabkiewicz, F. Buchinger, H. Fischer, H.J. Kluge, H. Kremmling, T. Kuhl, A.C. Muller and H.A. Schuessler, Phys. Lett. 82B, 199 (1979).
71. C. Bourgeois, P. Kilcher, J. Letessier, V. Berg and M.G. M.G. Desthuilleirs, Nucl. Phys. A295, 424 (1978).
72. M. Finger, et al., Nucl. Phys. A188, 369 (1972).
73. J.D. Cole, thesis, Univ. of Delft (1978).
74. A.C. Kahler, thesis, Univ. of Tennessee (1978).
75. L.L. Collins, thesis, Univ. of Tennessee (1978).
76. G.M. Gowdy, thesis, Georgia Tech. (1976).
77. J.H. Hamilton, et al., Phys. Rev. Lett. 35, 562 (1975); C. Bourgeois, et al., J. de Phys. 37, 49 (1976).

78. M. Caillau, R. Foucher, J.P. Husson and J. Letessier, J. de Phys. 35, 469 (1974), and J. de Phys. Lett. 35, L-233 (1974).
79. V. Berg, et al., Proc. 34d Int. Conf. on Nuclei Far From Stability, Cargese, Corisca 1976, CERN 76-13, p.367.
80. C. Bourgeois, et al., ibid. p.456.
81. E. Hagberg, P.G. Hansen, P. Hornshøj, B. Jonson, S. Mattson and P. Tidemand-Petersson, Phys. Lett. 78B, 44 (1978), and Nucl. Phys. A318, 29 (1979).
82. A. Ben Braham, et al., Nucl. Phys. A332, 397 (1979); M. Piiparinen, S.K. Saha, P.J. Daly, C.L. Dors, F.M. Bernthal and T.L. Khoo, Phys. Rev. C13, 2208 (1976); P.J. Daly, et al., Ann. Rept. MSUCL 1977/78.
83. J. Bonn, G. Huber, H.J. Kluge, L. Kugler and E.W. Otten, Phys. Lett. 38B, 308 (1972).
84. J.D. Cole, et al., Phys. Rev. Lett. 37, 1185 (1976).
85. R. Beraud, M. Meyer, M.G. Desthuilliers, C. Bourgeois, P. P. Kilcher and J. Letessier, Nucl. Phys. A284, 221 (1977); J.D. Cole, et al., Phys. Rev. C16, 2010 (1977).
86. N. Rud, D. Ward, H.R. Andrews, R.L. Graham and J.S. Geiger, Phys. Rev. Lett. 31, 1421 (1973); D. Proetel, R.M. Diamond and F.S. Stephens Phys. Lett. 48B, 102 (1974).
87. T. Kuhl, P. Dabkiewicz, C. Duke, H. Fischer, H.J. Kluge, H. Kremmling and E.W. Otten, Phys. Rev. Lett. 39, 180 (1977).
88. S. Frauendorf, F.R. May and V.V. Pashkevich, in Future Directions in Studies of Nuclei Far From Stability, eds. J.H. eds. J.H. Hamilton, et al., North-Holland Publishing Co., Amsterdam (1980), p.133.
89. I. Ragnarsson, ibid. p.367.
90. J.P. Husson, C.F. Liang and C. Richard - Serre, J. de Phys. Lett. 38, L-245 (1977).
91. A. Visvanathan, E.F. Zganjar, J.L. Wood, R.W. Fink, L.L. L.L. Riedinger and F.E. Turner, Phys. Rev. C19, 282 (1979).
92. M. Vergnes, G. Rotbard, J. Kalifa, G. Berrier, J. Vernotte, Y. Deschamps and R. Seltz, J. de Phys. Lett. 39, L-291 (1978).
93. J.L. Wood, E.F. Zganjar and J.D. Cole, Bull. Am. Phys. Soc. 25, 739 (1980).
94. R. Bijker, A.E.L. Dieperink, O. Scholten and R. Spanhoff, Nucl. Phys. A344, 207 (1980).
95. R.A. Braga and J.L. Wood, Bull. Am. Phys. Soc. 25, 740 (1980); R.A. Braga, J.L. Wood, M.A. Grimm and E.F. Zganjar, to be published.
96. J.L. Wood, preprint (1980); and, to be published in Ref.3.
97. P. van Isacker, preprint (1980); and to be published in Ref.3.

STUDIES OF ISOTOPE SERIES WITH EFFECTIVE BOSON HAMILTONIANS

P. Federman

Instituto de Fisica, UNAM
Mexico 20, D.F.

O. Castaños and A. Frank

Centro de Estudios Nucleares, UNAM
Mexico 20, D.F.

I. ABSTRACT

We discuss the possibility of describing series of isotopes using a single IBA-1 hamiltonian. The results of detailed calculations for several regions of heavy nuclei are presented. The usefulness of the present approach to identify "intruder" states is pointed out.

II. INTRODUCTORY REMARKS

The description of atomic nuclei has presented nuclear theorists with exciting challenges now for many years. Yet, there is no model capable of describing the properties of nuclei throughout the Periodic Table.

This is due to the inherent difficulties in solving the nuclear many-body problem. An exact solutions is beyond our means and will remain so for the foreseeable future, and in any case it would be only of limited interest to physicists.

Such a situation has led naturally to the construction of nuclear models of limited scope that take into account only a few degrees of freedom, considered to be important for some specific properties. Even then, drastic and often not obvious "a priori" assumptions had to be made.

Although today the Shell Model seems very natural to us, and is probably the most powerful microscopical tool to describe nuclei, it took a lot of soul-searching to accept the model. It was not clear how the strong nucleon-nucleon interactions could lead to a central potential. In spite of it, the impressive success of the model in describing ground state spins and magnetic moments of light nuclei made the concept of a single-particle orbit meaningful and respectable[1]. With the years the model was refined and extended to treat nuclei with several nucleons outside closed shells, including the residual interactions among them[2]. Eventually, good reasons as to why the model should work in the first place were also provided[3]. Still, the usefulness of the Shell Model remains limited to nuclei with only a few valence nucleons. The descrption of heavy nuclei far from closed shells had to wait for the Collective Model, in which many-nucleon degrees of freedom are drastically replaced by collective coordinates. This is achieved by visualizing the nucleus as a liquid drop, and quantizing the movement of its surface[4]. Although the nuclear surface is far from being well defined, the above treatment is applied to the nucleus. Nevertheless, the wealth of experimental data of heavy nuclei described and correlated by the Collective Model made it into the probably most successful and widely accepted nuclear model of collective motion.

Recently, a new model has been introduced that seems to represent an intermediate step between the extreme descriptions of the Shell and Collective Models. The Interacting Boson Approximation Model (IBA) makes the assumption that the properties of low-lying states of even-even heavy nuclei away from closed shells can be described to a very good approximation with only six degrees of freedom, represented by bosons with angular momenta L=0 (s-boson) and L=2 (d-boson)[5]. The number of bosons in each case depends on the number of valence nucleons. The IBA presents features resembling a hybrid between the Shell and Collective Models. Like in the Collective Model the building blocks are bosons, but not of a phonon nature. In this case they are particle (hole)-like as in the Shell Model, and are assumed to represent in some way correlated pairs of identical valence nucleons (nucleon holes)[6]. On the other hand, the only bosons considered have angular momenta privileged by the Pairing (L=0) and Quadrupole (L=2) interactions of the Unified Model[7].

This new model has been very successful in describing the spectroscopic properties of even-even nuclei. In order to address the important question of the physical interpretation of the model parameters, Otsuka, Arima, Iachello and Talmi[8] proposed a way to deduce them microscopically from the Shell Model, by means of mapping from the fermion to the boson space.

The neutron and proton degrees of freedom are treated independently, giving rise to a generalized version of the IBA-model known as IBA-2. (The original model is now refered to as IBA-1). The IBA-2 model has been subsequently applied to several regions of the nuclear table[9], and it has been shown that a good fit to the data can usually be accomplished with parameters that follow approximately the N, Z dependence predicted by a microscopic theory which assumes a single large valence orbit or several degenerate orbits.

The main difference between IBA-1 and IBA-2 is that IBA-1 states correspond to the fully symmetric representation [N] of the group U(6), whereas in IBA-2 all two-row representations are allowed. However, from the spectra of even-even nuclei there is considerable evidence that states not fully symmetric in neutron and proton degrees of freedom do not play a significant role at the low excitation energies of interest. This fact is taken into account in IBA-2 by introducing a Majorana force which pushes up these states with respect to the symmetric ones. Thus in actual calculations one can work in many cases with the simpler IBA-1 model.

The parameters of the IBA-2 model can be related to those in IBA-1 by a projection technique[10]. In fact it has been shown for several chains of isotopes that full IBA-2 and projected IBA-1 calculations agree quite well[10], indicating that the mixture of the two-row states is generally small. This projection technique only works when dealing with a separate calculation for each nucleus.

In most calculations a nucleus by nucleus fit is carried out for series of isotopes, and thus the number of experimental levels in each fit is quite small, so it turns out that different sets of parameters can reproduce the energies equally well. It is remarkable that a set of parameters can be found which follow the systematics expected from the simple degenerate single-particle theory. Nevertheless, it seems to us that it is worthwhile to attempt a different approach to the problem of finding physically meaningful parameters. We thus propose the possibility of defining "effective" Hamiltonians, where the <u>same</u> Hamiltonian is used to describe the energy spectra of whole series of isotopes or isotones and only the total number of bosons N changes from nucleus to nucleus. In this way one expects to establish a relationship or "kinship" between the levels pertaining to these nuclei and also to be able to identify intruder states properly. Also one avoids the problem of indetermination of the parameters to a large extent, since the ratio of experimental levels to number of parameters grows by an order of magnitude. Once the effective Hamiltonian is found, one can try small variations in a nucleus by nucleus fit, (which would reflect the effect of Paluli Principle in the interactions) not allowing large fluctuations from it. As a final stage the parameters of each nucleus can be compared with the predictions of the microscopic theories.

We have performed calculations in several regions. The results indicate that in some cases it is indeed possible to find such effective Hamiltonians, but also some limitations of the model become apparent.

In the next section the IBA-1 formalism is presented, while in Section IV we test the ability of the model to describe the energy level structure and systematics of series of even-even isotopes. We also discuss the present results and speculate on possibilities to include degrees of freedom not taken into account by the model.

III. DESCRIPTION OF THE MODEL.

The interacting boson approximation model (IBA) was formulated by Arima and Iachello from a purely phenomenological point of view[5], to describe in a unified way the collective states observed in medium and heavy even-even nuclei. In this approach, collective states are constructed as states of a system of N bosons which can occupy two levels, a level with L=0 (s-boson) and another with L=2 (d-boson).

The total number of bosons (N) is fixed in each nucleus by the number of nucleon pairs outside closed shells, so in this picture the valence nucleons are paired together with resultant angular momentum either L=0 or L=2. For example, for ^{154}Sm, there are 12 protons and 10 neutrons outside the closed shells at nucleon numbers 50 and 82, giving rise to N=11 for this nucleus.

Introducing creation (d^+, s^+) and annihilation (d, s) boson operators satisfying the commutation relations

$$[d^{m'}, d^+_m] = \delta^{m'}_m , \quad [s, s^+] = 1$$
$$[d^+, d^+] = [s^+, s^+] = [d^+, s^+] = 0 \tag{1}$$

the most general rotational invariant Hamiltonian containing one and two-body interactions can be written in the second quantized form:

$$H = \varepsilon_s \eta_{oo} \xi_{oo} + \varepsilon_d \sum_m \eta_{2m} \xi^{2m} + \frac{1}{2} \sum_{\ell_1 \ell_2} \sum_{\ell_1' \ell_2'} \sum_L (-1)^{\ell_1' + \ell_2'}$$
$$\times (2L+1)^{1/2} \langle 2\ell_1, 2\ell_2\, L | V_{12} | 2\ell_1', 2\ell_2'\, L \rangle$$
$$\times \left[\left(\eta_{\ell_1} \times \eta_{\ell_2} \right)^L \times \left(\xi_{\ell_1'} \times \xi_{\ell_2'} \right)^L \right]^o_o$$

where we have introduced the notation $s^{+}\equiv\eta_{oo}$, $s\equiv\xi_{oo}$, $d^{+}_{m}\equiv\eta_{2m}$ and $d_m\equiv\xi_{2m}$, and $[\eta_{\ell_1} \times \eta_{\ell_2}]^L$ denotes angular momentum coupling of ℓ_1 and ℓ_2 to a total L.

An examination of Eq.(2) indicates that there are only seven types of Hermitian two body interactions. All of them can be expressed[11] in terms of the Casimir operators of the three chains of groups

$$\begin{array}{cccc} U(6) \supset & U(5) \supset & O(5) \supset & O(3) \\ \hat{N} & \hat{\eta}_d & \hat{\Lambda}^2 & \hat{L}^2 \end{array} \tag{3a}$$

$$\begin{array}{ccc} U(6) \supset & SU(3) \supset & O(3) \\ \hat{N} & \hat{G} & \hat{L}^2 \end{array} \tag{3b}$$

$$\begin{array}{cccc} U(6) \supset & O(6) \supset & O(5) & O(3) \\ \hat{N} & \mathscr{L}^2 & \hat{\Lambda}^2 & \hat{L}^2 \end{array} \tag{3c}$$

where underneath each group we have written the corresponding Casimir operator.

When the Hamiltonian (2) can be expressed in terms of the operators associated with one of any of the above chains, it is said to exhibit a dynamical symmetry. For these cases the eigenvalues and other spectroscopic quantities are then straightforward to derive[12]. It is remarkable that the three limiting situations, U(5), SU(3) and O(6) dynamical symmetries, have been verified experimentally to a very good approximation in particular regions of the nuclear table. Using the Casimir operators, the Hamiltonian (2), except for terms contributing to binding energies, can be written in the form:

$$H = k_1\hat{\eta}_d + k_2\hat{\eta}_d\hat{N} + k_3\hat{\eta}_d^2 + k_4\hat{L}^2 + k_5\hat{\Lambda}^2 + k_6\hat{P}^2 + k_7\hat{Q}^2 \tag{4a}$$

where $\hat{Q}^2$ and $\hat{P}^2$ are given by

$$\hat{Q}^2 = \hat{G} - \frac{1}{2}\hat{L}^2 \tag{4b}$$

and

$$\hat{P}^2 = \frac{1}{4}\hat{N}(\hat{N}+4) - \frac{1}{4}\hat{\mathscr{P}}^2 \tag{4c}$$

respectively.

The question arises, however, whether the Hamiltonian (4) can successfully describe the properties of nuclei away from those exhibiting dynamical symmetries. Generally the above Hamiltonian is strongly truncated for the sake of simplicity. This can be sometimes justified beforehand on the basis of the symmetries exhibited by the experimental spectra, but this is not the case in general.

To investigate this point we make use of analytical expressions for the states, obtained through group theoretical techniques[11,13], that carry the irreducible representations of the chain of groups (3a).

The matrix elements for the operators $\hat{Q}^2$ and $\hat{P}^2$ were constructed in this basis. A computer code was written to calculate and diagonalize the energy matrices, and perform a least square search of the parameters k_i that fit the energy levels best[14].

IV. APPLICATIONS AND DISCUSSION

The first application in the framework described in the above section was carried out for the Sm isotopes[15]. Only the lower lying levels were included in the fitting procedure, in order to compare with previous calculations. The procedure converged and four of the seven Hamiltonian parameters k_i were found not to play any significant role. The results are compared with experiment in Fig.1. The agreement is good, and the parameters obtained are comparable to the ones used previously. However, it is worth mentioning that convergence problems arise when considering all experimental levels possible. The same is true of other series of isotopes in the same region.

The next application was carried out for nuclei at the beginning of the Z = 50-82 shell, namely for the Xe and Ba isotopes. All 70 experimental levels are reasonably well reproduced for the Xe isotopes with only three parameters, while for the Ba isotopes only one experimental level seems to fall outside the model space. The results are presented in Tables 1 and 2. A direct comparison of the parameters obtained in this case with the ones of previous calculations[16] is hindered by the fact that the later ones were performed in the IBA-2 framework. As discussed in the introduction we plan at a later stage to extract the IBA-1 parameters for each isotope starting from the effective Hamiltonian, in order to confront these calculations. Nevertheless, one difference is striking. Unlike the n-p IBA-2 results[16] (in the case of the Ba isotopes), here the quadrupole interaction is not found to play a significant role.

TABLE 1

A	J	E_{exp}	E_{th}	A	J	E_{exp}	E_{th}	A	J	E_{exp}	E_{th}
116	0		.80	120	6	1.99	1.95	126	3		2.60
	2	.39	.36		6		2.39		4	.94	1.03
	2		.70		7		2.53		4	1.49	1.49
	3		1.12		8	2.05	2.25		4		1.96
	4	.92	.83		8		2.69		5	1.90	2.05
	4		1.20		9		3.33		5		2.54
	5		1.68		10	2.87	3.06		6	1.64	1.70
	6	1.53	1.41		10		3.52		6	2.21	2.18
	6		1.80	122	0		.89		6		2.66
	7		2.35		2	.33	.43		7		2.80
	8	2.21	2.10		2	.84	.82		8	2.44	2.47
	8		2.50		2		1.41		8		2.96
	9		3.11		3	1.21	1.29		9		3.64
	10	2.96	2.88		3		2.43		10	3.32	3.33
	10		3.31		4	.83	.95		10		3.84
118	0	.83	.83		4	1.40	1.37	128	0		.99
	0		1.06		4		1.81		2	.44	.48
	2	.34	.38		5	1.77	1.91		2	.97	.94
	2	.93	.74		5		2.37		2		1.56
	2	1.23	1.31		6	1.46	1.59		3		1.43
	2		1.52		6	2.06	2.03		4	1.03	1.07
	3	1.37	1.18		6		2.49		4		1.54
	3		2.27		7		2.62		5		2.13
	4	.81	.88		8	2.22	2.32		6		1.76
	4	1.44	1.26		8		2.78		7		2.89
	4		1.66		9		3.43		8		2.54
	5	1.52	1.76		10	3.03	3.13		9		3.74
	5		2.18		10		3.63		10		3.42
	6	1.40	1.47	124	0		.93	130	0		1.02
	6	2.00	1.88		2	.35	.44		2	.54	.51
	6		2.30		2	.85	.86		2	1.12	.97
	7		2.44		2		1.46		2		1.60
	8	2.07	2.17		3	1.25	1.35		3		1.52
	8		2.60		3		2.52		4	1.21	1.11
	9		3.22		4	.88	.99		4		1.60
	10	2.81	2.97		4	1.44	1.43		5		2.20
	10		3.42		4		1.89		6	1.94	1.81
120	0	.91	.86		5	1.84	1.98		6		2.32
	0		1.12		5		2.46		7		2.97
	2	.32	.40		6	1.55	1.64		8		2.61
	2	.88	.78		6		2.10		10		3.50
	2	1.27	1.36		7		2.71	132	0		1.05
	2		1.60		8	2.33	2.40		2	.67	.52
	3	1.27	1.24		8		2.87		2	1.30	1.01
	3		2.35		9		2.54		2		1.65
	4	.80	.92		10	3.23	3.24		3	1.80	1.57
	4	1.40	1.32		10		3.73		4	1.44	1.15
	4	1.71	1.74	126	0		.96		4		1.65
	4		1.96		2	.39	.47		5		2.27
	5	1.82	1.84		2	.88	.90		6		1.87
	5		2.28		2		1.51		8		2.68
	6	1.40	1.53		3	1.32	1.41				

TABLE 2

A	J	E_{exp}	E_{th}	A	J	E_{exp}	E_{th}	A	J	E_{exp}	E_{th}
134	0	1.76	1.70	130	6	1.59	1.63	124	2	.23	.26
	0	2.16	1.94		6		2.24		2		.57
	0		4.29		7		3.02		3		1.00
	2	.61	.51		8		2.46		4	.65	.67
	2	1.17	1.08		9		4.01		4		1.06
	2		2.47		10		3.42		5		1.60
	3	1.64	1.78	128	0		1.16		6	1.23	1.22
	3		4.64		2	.28	0.33		6		1.69
	4	1.40	1.19		2	.89	.72		7		2.32
	4	1.97	1.84		2		1.76		8	1.92	1.91
	4		2.57		3	1.33	1.25		8		2.44
	5		2.65		3		3.12		9		3.17
	6		2.01		4	.76	.83		10	2.69	2.73
	7		3.64		4	1.37	1.31		10		3.32
	8		2.96		4		1.86	122	0		.82
	9		3.76		5	1.93	1.94		2	.20	.23
	10		4.04		5		2.57		2		.51
132	0	1.50	1.50		6	1.41	1.47		3		.91
	0		1.96		6		2.03		4	.57	.61
	2	.47	.44		7		2.76		4		.97
	2	1.03	.95		8	2.19	2.25		5		1.47
	2		2.21		8		2.88		6	1.08	1.13
	3	1.51	1.58		9		3.70		6		1.55
	3		3.78		10	3.08	3.16		7		2.15
	4	1.13	1.05		10		3.85		8	1.70	1.78
	4	1.73	1.64	126	0		1.02		8		2.27
	4		2.31		2	.26	.29		9		2.95
	5		2.39		2	.59	.64		10	2.40	2.55
	6	1.93	1.81		2		1.58		10		3.10
	6		2.48		3		1.11	120	0		.74
	7		3.32		4	.71	.74		2	.18	.21
	8		2.70		4		1.17		2		.46
	9		4.36		5		1.76		3		.83
	10		3.72		6	1.33	1.34		4	.54	.57
130	0		1.32		6		1.85		4		.89
	2	.36	.38		7		2.53		5		1.35
	2	.91	.83		8	2.09	2.07		6	1.04	1.05
	2		1.97		8		2.65		6		1.44
	3		1.40		9		3.42		7		2.00
	4	.90	.93		10	2.94	2.93		8		1.66
	4		1.46		10		3.57		9		2.76
	5		2.15	124	0		.91		10		2.40

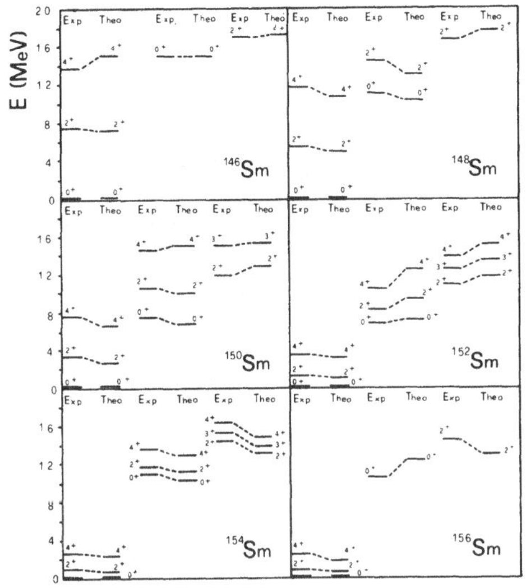

Fig.1 Comparison of experimental and theoretical J = 0^+, 2^+ and 4^+ levels of even Sm isotopes. Experimental levels are taken from ref.21.

Rather, the spectra are dominated only by the O(6) and U(5) symmetries.

Calculations were also performed at the end of the shell, both for the Os and Pt isotopes. The Os isotopes are well behaved and 76 experimental energy levels can be reproduced quite well with four parameters. But the same cannot be said about the Pt isotopes. Some 10 levels out of the 109 included in the fitting procedure are badly reproduced, and clearly fall outside the model space. They include mainly excited 0^+ states in the heavier Pt isotopes, usually predicated too low. It is striking that the main discrepancies remain even when the fit is tried isotope by isotope. No IBA calculation has been reported that reproduces all the experimentally known levels, and in particular the excited J=0^+ levels[17].

Thus, it seems clear by now that something has to be added to the IBA in order to include "intruder" states like the ones mentioned above. Some efforts have been made in this direction within the IBA framework, like induding a g-boson[18] or new s' and d' bosons[19]. Nevertheless it is clear that sometimes specific shell model orbits have to be treated explicitly if strong shell effects in the energy level systematics are to be reproduced. The IBA is supposed to represent degrees of freedom associated only with the valence shells of neutrons and protons, and indeed smooth systematics of energy levels are nicely reproduced. Nevertheless, strong shell effects in general inhibit the model from working for light and medium nuclei. The same seems to happen sometimes for heavy nuclei. In particular, strong polarization effects arising from the presence of one orbit of opposite parity among the orbits with natural parity like the Zr isotopes cannot be included in the IBA. There deformation arises through the interaction of an unnatural parity $g_{9/2}$ proton orbit with the $g_{7/2}$ neutron orbit. Some similar mechanism could explain the appearance of many excited 0^+ levels for the heavier Pt isotopes. When the proton shell is almost closed, energy may be gained by lifting the two protons into the next shell, where they can strongly interact with the valence neutrons. Indeed, the Hg isotopes (Z=80) seem to exhibit excited rotational bands, and the same has been suggested for the Pt isotopes[20]. Such a mechanism could then predict the appearance of low lying proton core-excited states near Z=82. The proton holes could be in the upcoming $11/2^-$ Nilsson proton orbit. Within the IBA framework such excitations can be described by introducing extra s' and d' proton bosons, with stronger couplings to neutron bosons. Alternatively a mixed Interacting Boson-Fermion approach might be desirable.

In conclusion, it seems that "effective" Hamiltonians can indeed be defined for several regions of heavy nuclei, which have the attractive feature of simultaneously describing the energy levels of groups of isotopes. This can be taken as a starting point for extracting physically meaningful parameters in IBA-1, which can be subsequently compared with the predictions of microscopic theories. In addition, this procedure can possibly provide a means of identifying intruder states.

We thank Stuart Pittel for fruitful discussions and for a critical reading of the manuscript.

REFERENCES

1. Maria G. Mayer, Phys. Rev. 75 (1949) 1969; Phys. Rev. 78 (1950) 16.
2. A. de Shalit and I. Talmi, Nuclear Shell Theory, Academic Press, New York, (1963).
3. L.C. Gomes, J.D. Walecka and V.F. Weisskopf, Ann. Phys. (N.Y.) 3 (1958) 241.

4. A. Bohr, Mat. Fys. Medd. Dan. Vid. Selsk. 26 (1952), 14
A. Bohr and B.R. Mottelson Mat. Fys. Medd. Dan. Vid. Selsk. 27, (1953) 16.
5. A. Arima and F. Iachello, Phys. Rev. Lett. 35 (1975) 1069.
6. A. Arima, T. Otsuka, F. Iachello and I. Talmi, Phys. Lett. 66B (1977) 205; Phys. Lett. 76B (1978) 139.
7. A. Bohr and B.R. Mottelson, Nuclear Structure, Vol. II, W.A. Benjamin, Reading, Mass., (1975).
8. T. Otsuka, Ph.D. Thesis, University of Tokyo, 1979.
9. See e.g. O. Scholten, Interacting Bosons in Nuclear Physics, ed ed. F. Iachello (Plenum Press, N.Y., 1979) p.17 and references therein.
10. O. Scholten, Ph.D. Thesis, University of Groningen, 1980.
11. O. Castanos, E. Chacon, A. Frank and M. Moshinsky, J. Math. Phys. 20 (1979) 35.
12. A. Arima and F. Iachello, Ann. Phys. (N.Y.) 99 (1976) 253; Ann. Phys. (N.Y.) 111 (1978) 201; Ann. Phys. (N.Y.) 123 (1979) 468.
13. E. Chacon, M. Moshinsky and R.T. Sharp, J. Math., Phys. 17, 668 (1976);
E. Chacon and M. Moshinsky, J. Math. Phys. 18, 870 (1977).
14. A. Frank, Ph.D. Thesis, University of Mexico, 1979.
15. O. Castanos, P. Federman and A. Frank, Phys. Lett. 88B, (1979) 203.
16. G. Puddu, O. Scholten and T. Otsuka, Preprint KVI-226, University of Groningen.
17. R. Bijker, A.E.L. Dieperink, O. Scholten and R. Spanhoff, Preprint KVI-224, University of Groningen.
18. B. Barret in "Interacting Bose-Fermi System in Nuclear Physics", Physics", ed. F. Iachello, Plenum Press 1980.
19. P. Van Isaacker in "Interacting Bose-Fermi Systems in Nuclear Physics" ed. F. Iachello, Plenum Press 1980.
20. S. Frauendorf, et al., in "Future Directions in Studies of Nuclei Far From Stability", North Holland, (1979), 133.
21. Table of Isotopes, Ed. by C.M. Lederer & V.S. Shirley, John Wiley & Sons (1978).
M. Sakai and Y. Gono, Quasi-Ground Quasi Beta and Quasi-Gamma Bands, INS-J-170. University of Tokyo (1979).

USE OF A BOSON MAPPING TO ELUCIDATE THE RELATIONSHIP BETWEEN THE IBM AND THE BOHR COLLECTIVE HAMILTONIAN AND BETWEEN A SIMPLIFIED SHELL MODEL AND THE BOHR COLLECTIVE HAMILTONIAN

Abraham Klein

Department of Physics, University of Pennsylvania
Philadelphia, Pennsylvania 19104

and

Michel Vallieres

Department of Physics and Atmospheric Science
Drexel University
Philadelphia, Pennsylvania 19104

I. ABSTRACT

The solution of two problems is outlined: (i) Given the phenomenological boson model, IBM 1, what is the equivalent Bohr Collective Hamiltonian? (ii) Given a generalized Meshkov-Glick-Lipkin model with n levels, what is the equivalent Bohr Collective Hamiltonian? The basic tool in each case is a well-known boson mapping for the symmetric representation of SU(n).

II. INTRODUCTION

In this note we outline how a well-known explicit realization of the symmetric representation of SU(n) may, in conjunction with suitable physical reasoning, be utilized to elucidate two topics of great current interest. First we solve the problem: Given a boson model belonging to the class known as IBM 1[1], i.e., a model in which we do not distinguish neutrons from protons, what is the Bohr collective Hamiltonian equivalent to this model. Second we study a class of generalized MGL[2] models with n levels and obtain its equivalent Bohr description.

III. TRANSFORMATION OF THE IBM1 MODEL

We shall not consider the most general model of the type IBM1 but only an interesting special case. Our methods, however, are applicable to the general case. The Hamiltonian is written in terms of the two basic bosons, the s boson describing a pair of nucleons coupled to angular momentum zero, and the d_μ boson, $\mu = 1...5$, describing a coupling to angular momentum two, namely

$$H = \varepsilon \sum_\mu d_\mu^\dagger d_\mu - \kappa \sum_\mu Q_\mu Q_{-\mu}(-1)^\mu \tag{1}$$

$$Q_\mu = s^\dagger \tilde{d}_\mu + d_\mu^\dagger s + \chi(d^\dagger d)_\mu^{(2)} \tag{2}$$

$$\tilde{d}_\mu = (-1)^\mu d_{-\mu} \tag{3}$$

The Hamiltonian (1) contains three parameters, ε describing the excitation energy of a d pair relative to an s pair, κ, the overall strength of the quadrupole-quadrupole force, and χ, the relative importance of the two kinds of basic quadrupole that define the general quadrupole operator (2); it also conserves the total number of bosons, N, where

$$N = s^\dagger s + \sum_\mu d_\mu^\dagger d_\mu \tag{4}$$

and N is considered to be half the number of "active fermions". The goal of the program under consideration is to replace the six degrees of freedom of the IBM by a Hamiltonian containing five quadrupole degrees of freedom and the value of N. It is understood that the equivalent Bohr-Mottelson picture emerges in the limit as N tends to infinity. Thus, if we can display a transformed Hamiltonian which admits an expansion in inverse powers of N, our aim will have been essentially achieved.

The required transformation[3] is expressed in terms of a new quadrupole boson b by means of the formulas

$$d_\mu^\dagger d_\nu = b_\mu^\dagger b_\nu \tag{5}$$

$$d_\mu^\dagger s = b_\mu^\dagger (N - h)^{\frac{1}{2}} \tag{6}$$

$$s^\dagger d_\mu = (N - h)^{\frac{1}{2}} b_\mu \tag{7}$$

$$s^{\dagger}s = N - h \tag{8}$$

where

$$h = \sum b^{\dagger}_{\mu} b_{\mu} \tag{9}$$

is clearly the number of b bosons. A similar reduction from n to n - 1 bosons can be carried out for SU(n).

It may help the reader to consider the following interpretation of Eqs.(5) - (8): In the IBM, the basis for studying a nucleus with N bosons is obtained by applying to the "vacuum" various products $(s^{+})^{n_s} (d^{+})^{n_d}$ such that $n_s + n_d = N$. An equivalent basis is obtained by applying $(b^{+})^{n}$ to the state $(s^{+})^{N}|vac\rangle$, where $n \leq N$. The b boson is thus to be considered as a multipole rather than a pairing boson, and for this reason is the object we seek ultimately to associate with the Bohr picture.

The substitution of (5 - 8) into H yields a first form of transformed Hamiltonian, which we do not record. In consequence of the square roots in (6) and (7) it is not a polynomial in the creation and annihilation operators. One might be tempted to expand in powers of (h/N), but this expansion can be justified only in the limit of weak interaction among the bosons, which coincides with the SU(5) or vibrational limit of the IBM. Since we wish a theory valid also for the other special limits of interest, the O(6) and SU(3) limits, we must proceed with greater caution.

The utilization of the transformation (5) - (8) to reach the Bohr Hamiltonian description in any of its limits involves two steps. The first is to introduce "canonical" variables x , p into the Hamiltonian by means of the formula

$$b^{\dagger}_{\mu} = \frac{1}{\sqrt{2}} (x_{\mu} - ip^{\dagger}_{\mu}) \tag{10}$$

$$b_{\mu} = \frac{1}{\sqrt{2}} (x^{\dagger}_{\mu} + ip_{\mu}) \tag{11}$$

$$[x_{\mu}, p_{\nu}] = i\delta_{\mu\nu} = [x^{\dagger}_{\mu}, p^{\dagger}_{\nu}] \tag{12}$$

The second step is to understand the relative importance of the various terms. When dealing with the oscillatory regime, small vibrations about spherical equilibrium, the commutators (12) are fulfilled by matrix elements which in order of magnitude satisfy $x \sim p \sim 1$. On the other hand, when the system undergoes a phase transition to a deformed shape, $x \sim N^{\frac{1}{2}}$, so that in order to satisfy (12), we have $p \sim N^{-\frac{1}{2}}$. This suggests that in dealing with the radicals that occur in (6) and (7), we write (scalar product notation)

$$h = b^{\dagger}b = \frac{1}{2}(x^{\dagger}x + p^{\dagger}p + 5) \qquad (13)$$

and note that if we lump the factor $x^{\dagger}x$ with the factor N and expand only in terms of the remaining quantities, $p^{\dagger}p + 5$, then successive terms will certainly diminish by reciprocal powers of N. Though this expansion is justified for large N for all regimes of the parameters and is the only one generally valid, further expansion of the coordinate dependence is justified in the spherical domain.

For full details concerning the straightforward manipulations and the results that ensue we must refer to a more detailed publication.[4] Here we include only a few remarks concerning the potential energy, which is the leading term when we carry out an expansion in powers of $(p^{\dagger}p/N)$. For example to exhibit a function, $V(\beta,\gamma)$, of the conventional shape parameters requires a rescaling

$$x \rightarrow N\,x \quad , \quad p \rightarrow N^{-\frac{1}{2}}p \qquad (14)$$

as well as the definitions

$$x^{\dagger}x = \bar{\beta}^2 \qquad (15)$$

$$\sum_{\mu} (x^{\dagger})_{\mu} (x \otimes x)^{(2)}_{\mu} = -(2/7)^{\frac{1}{2}}\bar{\beta}^3 \cos 3\gamma \qquad (16)$$

$$\bar{\beta} = \sqrt{2}\,\beta/(1+\beta^2)^{\frac{1}{2}} \qquad (17)$$

leading to the expression $(F = (\kappa N/\varepsilon))$

$$(1+\beta^2)^2[V(\beta,\gamma)/N\varepsilon] = (\tfrac{1}{2} - 2F)2\beta^2(1+\beta^2) +$$

$$4F\chi(2/7)^{\frac{1}{2}}\beta^2 \cos 3\gamma + F[1 - (\chi^2/14)]4\beta^4 \qquad (18)$$

For β^2 sufficiently small, we have a spherical nucleus. Deformed shapes are confined to oblate or prolate, $\sin 3\gamma = 0$. The γ unstable (0(6)) minimum, $\chi = 0$, occurs for $\beta^2 = (F-4)/(F+4)$ and the SU(3) limit, corresponding to $\chi = (-\sqrt{7}/2)$ and $F \to \infty$, yields $\beta = \sqrt{2}$ as the point of minimum potential energy.

IV. TRANSFORMATION OF THE GENERALIZED MGL MODEL

We study a multi-level generalization of MGL in which the Hamiltonian is a polynomial in the generators of SU(n). We consider n non-degenerate levels each with sublevel degeneracy N. Let $\alpha^\dagger_{ir}$ be the creation operator for a fermion in level r and its ith degenerate sublevel, $\alpha^\dagger_{ir}$ the corresponding destruction operator. The bilinear operators

$$A_r^s = (A_s^r)^\dagger = \sum_{i=1}^{N} \alpha^\dagger_{ir}\alpha_{is}, \qquad r, s = 1 \ldots n, \tag{19}$$

are a set of generators of the Lie Algebra, U(n). We select a Hamiltonian

$$H = H_0 + H_1 \tag{20}$$

$$H_0 = \sum_{r=1}^{n-1} \varepsilon_r J_0^{(r)} \tag{21}$$

$$H_1 = -G \sum_r [(J_+^{(r)})^2 + (J_-^{(r)})^2] \tag{22}$$

Here ε_r are the n-1 single-particle excitation energies, G is the interaction strength and

$$J_0^{(r)} = \tfrac{1}{2}(A_{r+1}^{\ r+1} - A_1^{\ 1}) \tag{23}$$

$$J_+^{(r)} = (J_-^{(r)}) = A_{r+1}^{\ 1}, \quad r=1, \ldots n-1 \tag{24}$$

are a subset of the generators of SU(n).

The semi-classical limit of the Hamiltonian is studied by

means of the equations analogous to eqs.(5) - (8). Here we write

$$A_{r+1}^{\ \ s+1} = b_r^{\dagger} b_s, \qquad r\gamma, s = 1, \ldots n-1, \tag{25}$$

$$A_{r+1}^{\ \ 1} = (A_1^{\ r+1})^{\dagger} = b_r^{\dagger}\,\theta(N) , \tag{26}$$

$$A_1^{\ 1} = \theta(N)\theta(N) \tag{27}$$

where

$$\theta(N) = \theta(N)^{\dagger} = [N - \sum_{r=1}^{n-1} b_r^{\dagger} b_r]^{\frac{1}{2}} \tag{28}$$

The passage to the large N limit proceeds in complete analogy to the study of the IBM - the introduction of canonical coordinates and expansion in powers of the kinetic energy. We are, for instance, able to define a potential energy surface and to study the phase changes of the model as a function of the parameters.

It is also of interest within the context of this model to obtain the classical limit. In contrast to the previous procedure where quantities were estimated in strict adherence to the requirements of the commutation relations, here we simply keep whatever remains when commutators are set to zero. Be examining this result we can verify the limited accuracy of methods which requantize this Hamiltonian.

Our results agree with those found by previous authors.[5] A more complete account of our work may be found elsewhere.[6]

V. CONCLUDING REMARKS

Some of the ideas outlined in this work, in particular those associated with the boson mapping of a shell model may, with advantage, also be studied[7] within the context of a model with the symmetry of 0(5), where a closed boson mapping is also known. It appears that a method used to derive this mapping is also a useful tool for studying the mapping of interest for the realistic shell model[8].

VI. ACKNOWLEDGEMENT

This work was supported in part by the U.S. Department of Energy under Contract No. EY-76-C-02-3071.

REFERENCES

1. Interacting Bosons in Nuclear Physics, ed. F. Iachello (Plenum Press, New York, 1979).
2. A.J. Lipkin, N. Meshkov, and A.J. Glick, Nucl. Phys. 62 (1965) 188.
3. D. Janssen, R.V. Jolos and F. Donan, Nucl. Phys. A224 (1974) 93; S. Okubo, J. Math. Phys. 16 (1975) 528.
4. A. Klein and M. Vallieres, U. of Pennsylvania report UPR-0156T (1980).
5. R. Gilmore and D.H. Feng, Nucl. Phys. A301 (1978) 189.
6. A. Klein, Phys. Letters (to be published).
7. A. Klein, H. Rafelski, and J. Rafelski, U. of Pennsylvania report UPR-0159T (1980).
8. T. Otsuka, A. Arima, and F. Iachello, Nucl. Phys. A309 (1978) 1.

RENORMALIZATION OF THE IBA HAMILTONIAN FOR THE EFFECTS OF THE *g* BOSON

Keith A. Sage and Bruce R. Barrett

Department of Physics
University of Arizona
Tucson, Arizona 85721

I. ABSTRACT

The IBA Hamiltonian is renormalized in lowest order for the effects of the *g* boson. The renormalization of the single *d* boson energy ε is calculated explicitly, with an application of this result to the barium isotopes.

II. PERTURBATION THEORY FOR THE *g* BOSON

The IBM proton-neutron Hamiltonian is given by[1-2]

$$H_{IBA2} = \varepsilon(n_{d_\pi} + n_{d_\nu}) + \kappa Q_\pi \cdot Q_\nu + M_{\pi\nu} + V_{\pi\pi} + V_{\nu\nu} \qquad (1)$$

where n_{d_π} (n_{d_ν}) is the number of proton (neutron) bosons with L = 2, and the other interaction terms are defined in Ref.3. This Hamiltonian has been used by many investigators[1-4] to do phenomenological analyses of nuclei in the mass region Z = 50-82 and N = 50 to N $\geq$ 126, in which the parameters of the Hamiltonian are determined by fitting the low-lying energy spectra for an individual nucleus. A question of current theoretical interest is how these parameters are affected by terms left out of the model Hamiltonian, such as the *g* boson, i.e., proton or neutron pairs coupled to J = 4.

Rather than expand the boson model space of *s* and *d* (L = 0 and 2) bosons by putting in *g* bosons, we use second-order Rayleigh-Schrodinger perturbation theory to examine the effect of a *g* boson on the IBA Hamiltonian. We work in a model space constructed from correlated pairs of fermions because the boson model space is spann by s and d boson states only, and calculate the second-order perturbative correction to Eq.(1) for an interaction between correlated fermion pairs of the form $\kappa Q_\pi \cdot Q_\nu$, $\kappa < 0$. Here we assume that the quadrupole operator $Q_{\pi(\nu)}$ is a two-body operator for fermions, as given in Ref.5. The completely general algebraic expression for v_2 in second quantized notation, is given by

$$v_2 = (1/4)^4 \kappa^2 <(\alpha\beta)_\pi(\gamma\delta)_\nu|Q_\pi \cdot Q_\nu|(\rho\sigma)_\pi(\tau\xi)_\nu>_A \frac{P}{\Delta E}$$

$$<(\alpha'\beta')_\pi(\gamma'\delta')_\nu|Q_\pi \cdot Q_\nu|(\rho'\sigma')_\pi(\tau'\xi')_\nu>_A \qquad (2)$$

$$\{a^\dagger_\alpha a^\dagger_\beta a^\dagger_\gamma a^\dagger_\delta a_\xi a_\tau a_\sigma a_\rho\} \quad \{a^\dagger_{\alpha'} a^\dagger_{\beta'} a^\dagger_{\gamma'} a^\dagger_{\delta'} a_{\xi'} a_{\tau'} a_{\sigma'} a_{\rho'}\}$$

where the first two state labels in the bras and kets refer to protons and the last two refer to neutrons. The subscript "A" on the kets means that the kets are independently antisymmetrized for protons and neutrons. The sum is taken over all state labels of the fermion creation and annihilation operators, except that the intermediate states $|(\rho\sigma)_\pi(\tau\xi)_\nu>$ and $<(\alpha'\beta')_\pi(\pi'\delta')_\nu|$ are restricted by the projection operator P to a specific set of states lying outside the subspace of S and D paired fermion states (corresponding to s and d boson states, respectively). The grouping of single fermion labels as in $<(\alpha\beta)_\pi(\gamma\delta)_\nu|$ indicates the correlation between fermions making up a pair. The states in Eq.(2) are then the unperturbed, paired-fermion states in the complete model space.

After contracting the operators corresponding to the intermediate state and performing considerable Racah algebra, we obtain for the perturbative correction, assuming single j shells for protons and for neutrons,

$$v_2 = \kappa^2 \sum_{J\, K_\pi K_\nu I_\pi I_\nu J_\pi J_\nu} <(K_\pi K_\nu)J|| Q_\pi \cdot Q_\nu || (I_\pi I_\nu)J>_A$$

$$\frac{P}{\Delta E} <(I_\pi I_\nu)J|| Q_\pi \cdot Q_\nu || (J_\pi J_\nu)J>_A (-1)^J(2J+1)$$

$$\sum_{L\quad J_\nu J_\pi L}^{K_\pi K_\nu J} T_L(K_\pi, J_\pi) \cdot T_L(K_\nu, J_\nu$$

where

$$T_{LM}(K,J) = (A_K^\dagger \tilde{A}_J)^L_M = \sum_{M'M''} (KJM'M''|LM)\, A^\dagger_{KM'}\, \tilde{A}_{JM''} \tag{4}$$

and $A^\dagger_{JM}$, A_{JM} are the usual paired-fermion creation and annihilation operators, with $\tilde{A}_{JM} = (-)^{J-M+1} A_{J-M}$. The action of the projection operator P in this multipole-multipole form of a proton-neutron interaction is to require that there be one and only one I = 4 fermion pair in the intermediate states. With the initial and final states made up only of $\mathcal{S}$ and $\mathcal{D}$ paired fermions, L can be 0, 1, 2, 3, or 4. We examine only the L = 0 part of Eq.(3) in detail. Defining the quantities

$$\hat{N}_{J_\pi} = \sum_{M_\pi} A^\dagger_{J_\pi M_\pi} A_{J_\pi M_\pi}, \quad \hat{N}_{J_\nu} = \sum_{M_\nu} A^\dagger_{J_\nu M_\nu} A_{J_\nu M_\nu},$$

$$g^J_{J_\pi J_\nu} = \frac{(2J+1)}{(2J_\pi+1)(2J_\nu+1)} \tag{5}$$

we have the simple form of the monopole-monopole part of the perturbation,

$$v_2(L=0) = \kappa^2\, |<(J_\pi J_\nu)J||Q_\pi \cdot Q_\nu||(I_\pi I_\nu)J>_A|^2 \; \frac{P}{\Delta E}\, g^J_{J_\pi J_\nu}\, \hat{N}_{J_\pi} \hat{N}_{J_\nu} \tag{6}$$

Comparing the $\hat{N}_J$ operator defined above to the fermion number operator, we see that the two operators are identical in form, but the creation and annihilation operators in Eq.(5) relate to pairs of fermions; hence, while the fermion number operator is a one-body operator, $\hat{N}_J$ is a one-pair operator. Another difference is that the fermion operators obey anti-commutation rules, while the pair operators have more complex commutation rules.[5] However, if we make the "quasi-boson" approximation that the paired-fermion operators do commute, then the operator $\hat{N}_J$ can be interpreted as a number operator counting the number of fermion pairs coupled to angular momentum J. We also apply the "simple correspondence" argument of Ginocchio and Talmi[6] to map the resulting correction v_2 into a boson space spanned by s and d bosons

$$v_{2,B}(L = 0) = \kappa^2 \, |<(J_\pi J_\nu)J|| \, Q_\pi \cdot Q_\nu || \, (I_\pi I_\nu)J> _A|^2 \quad \frac{P}{\Delta E} g^J_{J_\pi J_\nu} \hat{n}_{J_\pi} \hat{n}_{J_\nu} \tag{7}$$

That is, we simply replace the paired-fermion operators with analogous boson operators $\hat{n}_J$ and consider v_2 to be a correction to the boson Hamiltonian. The energy denominator in Eq.(7) is determined in terms of the unperturbed boson states. The matrix elements of $Q_\pi \cdot Q_\nu$ have been given elsewhere.[5]

III. CALCULATIONS FOR THE BARIUM ISOTOPES

We have evaluated Eq.(7) for the Ba isotopes, using experimental 2^+ and 4^+ energies for the *d* and *g* boson energies (the boson energy is set to zero) in E. The correction to the IBA Hamiltonian (in MeV) is

$$v_{2,B}(L = 0) = C_1 \hat{n}_{d_\pi} \hat{n}_\nu + C_2 \hat{n}_\pi \hat{n}_{d_\nu} + C_3 \hat{n}_{d_\pi} \hat{n}_{d_\nu} \tag{8}$$

where the coefficients C_1, C_2, C_3 are (in units of barns^4/MeV

$$C_1 \cong -\kappa^2 A^{4/3} \; 8 \times 10^{-6}$$

$$C_2 \cong -\kappa^2 A^{4/3} \; 6 \times 10^{-6}$$

$$C_3 \cong \kappa^2 A^{4/3} \; 5.5 \times 10^{-5}$$

and where A is the mass number and the $\hat{n}$'s are number operators for the total number of proton or neutron bosons or the number of proton or neutron *d* bosons. Combining Eq.(8) with the unperturbed boson Hamiltonian in Eq.(1), we may clearly see the renormalization of the single boson energy ε,

$$H_{IBA2} + v_{2,B}(L = 0) = (\varepsilon + C_1\hat{n}_\nu)\hat{n}_{d_\pi} + (\varepsilon + C_2\hat{n}_\pi)\hat{n}_{d_\nu} + C_3\hat{n}_{d_\pi}\hat{n}_{d_\nu} + \kappa Q_\pi \cdot Q_\nu + M_{\pi\nu} + V_{\pi\pi} + V_{\nu\nu} \quad (9)$$

Using the parametrization of Kisslinger and Sorensen[7] for the strength of the interaction $Q_\pi \cdot Q_\nu$, where their quadrupole operator is a one-body operator in a single j-shell, we find that $\kappa \cong -116\ A^{-2/3}$ MeV/barns^2 for A = 130-140 in a single j shell with $j = 31/2$.

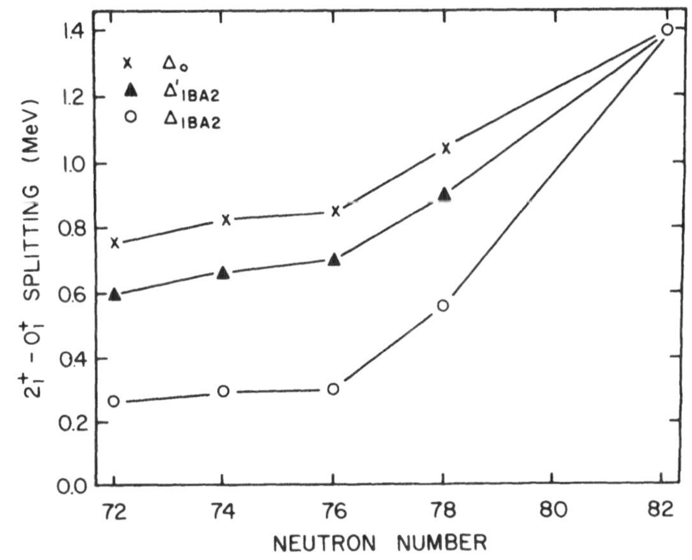

Fig.1. Calculated splittings of the 2_1^+ state from the ground state for several Ba isotopes.

Otsuka et al.[1] have reported that it is necessary to renormalize the single boson energy ε as a function of the number of valence bosons (or in this case, boson holes), when fitting data from the Ba isotopes, as reflected in the empirical curve in Fig.1. Because we seek to provide a microscopic explanation of this renormalization, we have calculated the splitting of the first excited 2^+ state from the ground state for the isotopes of barium with A = 128, 130, 132, 134, and 138, which we compare to the phonomenological IBA splittings rather than experimental data. In this comparison we ignore the effect of all but the first two terms in H_{IBA2} in determining the empirical curve (for which we have used the parametrization of Otsuka et al.[1]). The perturbed Hamiltonian obtained from Eq.(9) is then

$$H'_{IBA2} = (\varepsilon_0 + C_1\hat{n}_\nu)n_{d_\pi} + (\varepsilon_0 + C_2\hat{n}_\pi)n_{d_\nu} + C_3\hat{n}_{d_\pi}\hat{n}_{d_\nu} \tag{10}$$

where ε_0 = 1.4 MeV, the value at the closed shell. The corresponding splittings are given by

$$\Delta_{IBA2} = \langle 2_1^+ | \varepsilon(\hat{n}_{d_\pi} + \hat{n}_{d_\nu}) | 2_1^+ \rangle - \langle 0_1^+ | \varepsilon(\hat{n}_{d_\pi} + \hat{n}_{d_\nu}) | 0_1^+ \rangle \tag{11}$$

$$\Delta'_{IBA2} = \langle 2_1^+ | H'_{IBA2} | 2_1^+ \rangle - \langle 0_1^+ | H'_{IBA2} | 0_1 \rangle \tag{12}$$

We have also plotted in the figure the unrenormalized, unperturbed splittings Δ_0, obtained by setting $\varepsilon = \varepsilon_0$ in Eq.(11), which show that a significant part of the renormalization seen by Otsuka et al. is accomplished merely by changes with neutron number in the matrix elements of the number operators for neutrons and protons. The figure also shows that the perturbation $v_{2,B}(L = 0)$ has the proper trend with neutron number, i.e., it decreases the splitting as n_ν increases toward mid-shell, in agreement with the phenomenological results.

Finally, we have made the same comparisons using a perturbation constructed from single-particle quadrupole operators; the change in the matrix elements of $Q_\pi \cdot Q_\nu$ is less than 1% of the value of the matrix element for two-body quadrupole operators.

IV. CONCLUSIONS

We have calculated the effects of the *g* boson on the IBA Hamiltonian using second-order perturbation theory in a model space of paired-fermion states. The calculations were simplified by assuming degenerate j shells for protons and neutrons and by approximating the operators of Eq.(5) as number operators. In a specific application of the resulting correction term, the agreement between the predicted $2_1^+ - 0_1^+$ splittings and the phenomenological IBA values for several Ba isotopes suggests that the perturbative approach to including the g boson in the IBA Hamiltonian accounts for part of the observed renormalization of the phenomenological single boson energies in the IBA.

REFERENCES

1. T. Otsuka, A. Arima, F. Iachello and I. Talmi, Phys. Lett. 76B, 139 (1978).
2. A. Arima and F. Iachello, Ann. Phys. (N.Y.) 99, 253 (1976); 111, 201 (1978).
3. F. Iachello in Interacting Bosons in Nuclear Physics, ed. F. Iachello (Plenum Press, New York, 1979), p.1.
4. O. Scholten, F. Iachello and A. Arima, Ann. Phys. (N.Y.) 115, 325 (1978).
5. K. Sage and B. R. Barrett, to be published in Physical Review C (1980).
6. J. Ginocchio and I. Talmi, Nucl. Phys. A337, 431 (1980).
7. L. Kisslinger and R. A. Sorensen, Rev. Mod. Phys. 35, 853 (1963).

THE INTERACTING-BOSON-FERMION APPROXIMATION

O. Scholten

Cyclotron Laboratory
Michigan State University
East Lansing, Michigan 48824

I. INTRODUCTION

In this contribution the extension will be discussed of the Interacting Boson Approximation (IBA) model to odd-A nuclei where the fermion degrees of freedom need to be considered explicitly. The IBA model describes even-even nuclei as a system of interacting s- and d-boson. By coupling the degrees of freedom of a single nucleon to these bosons one can describe also odd-A nuclei. The general Hamiltonian for this coupled system can be written as

$$H = H_B + H_F + V_{BF} \tag{1}$$

The IBA Hamiltonian, H_B, describes the system of s- and d-bosons. H_F is the Hamiltonian for the single odd particle and thus contains only a one-body term

$$H_F = \sum_{j,m} \varepsilon_j a^{\dagger}_{jm} a_{jm} \tag{2}$$

In our convention the summation index j denotes the shell model orbits in the valence shell. The quasi-particle energies are denoted by ε_j.

The most general two-body boson-fermion interaction is given in Ref.1. In this formulation the physical interpretation of the various terms is not very transparent and it is impractical to give a phenomenological description of odd-A nuclei. In the

Interacting Boson Fermion Approximation (IBFA) model only a few terms are taken into account, selected on the basis of microscopic arguments. It will be shown that with this simple Hamiltonian it is possible to reproduce the wide variety of observed collective spectra in odd-A nuclei.

II. THE MODEL

In the IBA model the bosons are considered a collective state of nucleon pairs. The pairing property of the interaction between like nucleons plays an important role in the formation of the L=0 (S) pairs, while the strong neutron-proton quadrupole interaction gives rise to the collective properties. The latter interaction is also dominating in the coupling of the odd nucleon to the bosons.

Because of the correspondence between bosons and fermion pair states it is possible to make an equivalence between states in the IBA model and those in the shell-model. Using this correspondence, the boson quadrupole operator, for example, can be constructed by equating the matrix elements of the operator in the boson space to the equivalent ones in the shell-model space, as is described extensively in Refs. 2 and 3. One can apply the same procedure to the IBFA model. One can divide the matrix elements of the quadrupole operator in the fermion space into two classes, the seniority conserving ones and those connecting states which differ in seniority by two units. By retaining only the terms which are in lowest order in the number of d-boson operators for each of these classes, one obtains the IBFA image of, for example, the proton quadrupole operator[4].

$$Q^{(2)} = Q_B^{(2)} + Q_F^{(2)} \quad , \tag{3}$$

$$Q_B^{(2)} = \alpha_\pi[(s^\dagger\tilde{d} + d^\dagger s)^{(2)} + \chi_\pi(d^\dagger\tilde{d})^{(2)}] \tag{4a}$$

$$Q_F^{(2)} = \sum_{jj'} Q_{jj'}(u_j u_{j'} - v_j v_{j'})(a_j^\dagger \tilde{a}_{j'})^{(2)} - \frac{\sqrt{10}}{N_\pi} \sum_{jj'j''} n_\beta^{-1} Q_{jj'} \tag{4b}$$

$$\times (u_j v_{j'} + v_j u_{j'})\beta_{j''j} \frac{1}{\hat{j}}\left[(d^\dagger\tilde{a}_{j''})^{(j)}(s a_{j'}^{+})^{(j)'}\right]^2 + \text{h.c.}$$

where $\tilde{a}_{jm} = (-1)^{j-m} a_{j-m}$ and for convenience of writing the index π has been dropped at most places. In eq.(3), $Q_B^{(2)}$ is the normal boson quadrupole operator. The coefficients $Q_{jj'}$ are the single particle matrix elements of the quadrupole operator in the shell model basis.

The particle structure of the d-boson[2] is determined by the coefficients β_{jj}. The normalization constant n_β is defined as

$$\sum_{jj'} (\beta_{jj'})^2 = n_\beta^2 \tag{5}$$

Under the assumption that the $|D>$ state [2,3] (the fermion equivalent of the d-boson) exhausts the full E2 strength ($|D> = Q^{(2)}|S>$), $\beta_{jj'}$ is given by

$$\beta_{jj'} = (u_j v_{j'} + v_j u_{j'}) Q_{jj'} \tag{6}$$

The coefficients v_j, ($u_j = \sqrt{1-v_j^2}$, can be regarded as the occupation probabilities for the different spherical shell-model states $|j>$. They are related to the coefficients α_j, appearing in the definition of the $|S>$ state[3], $|S> = \sum_j \alpha_j \sqrt{\Omega_j/\Omega_e}\; 1\; (j^2)^{(o)}>$ as

$$v_j = \alpha_j \sqrt{N_\pi/\Omega_e} \tag{7}$$

where $\Omega_e = \sum_j \alpha_j^2 (j + \frac{1}{2})$ is the effective pair degeneracy of the major shell.

Note that $Q_F^{(2)}$ in eq.(4b) has two terms. The first term in Q_F^2 can be regarded as the direct term while the second is the exchange term. This latter solely comes from the fact that the nucleon degrees of freedom are contained in the d-boson. In our case this gives rise to an additional term in the quadrupole operator since by definition a_j and $a_j^\dagger$ commute with the boson creation and annihilation operators. In the shell model these effects are taken into account by a nonvanishing commutator. It must be noted that the Pauli principle is also responsible for the $(u^2 - v^2)$ factor in the first term in $Q_F^{(2)}$.

The boson-fermion interaction can now be written as

$$V_{BF} = \bar{\kappa} Q_{\nu B}^{(2)} \cdot Q_{\pi F}^{(2)} \tag{8}$$

for the case that an odd proton is coupled to the system of bosons. However to do the calculations in a basis in which the neutron and proton degrees of freedom are treated explicitly will give rise to practical problems ; large matrices would have to be diagonalized.

The boson space can be truncated by considering only the states which are symmetric in the neutron and proton degrees of freedom[5]. Phenomenologically, this truncation has proven to work well for even-even nuclei[8]. The projection onto the symmetric states can also be applied to eq.(8). If we furthermore keep only the terms which are at most quadratic in the boson operators, the IBFA boson-fermion interaction becomes

$$\begin{aligned} V_{BF} = & \sum_j A_j [(d^\dagger \tilde{d})^{(0)} (a_j^\dagger \tilde{a}_j)^{(0)}]^{(0)} \\ & + \sum_{jj'} \Gamma_{jj'} [Q^{(2)} (a_j^\dagger \tilde{a}_{j'})^{(2)}]^{(0)} \\ & + \sum_{jj'j''} \Lambda_{jj'}^{j''} :[(d^\dagger \tilde{a}_j)^{(j'')} (\tilde{d} a_{j'}^\dagger)^{(j'')}]^{(0)}: \end{aligned} \tag{9}$$

where a monopole force has been added, which could originate from the interaction between like nucleons and

$$Q^{(2)} = (s^\dagger \tilde{d} + d^\dagger s)^{(2)} + \chi_\nu (d^\dagger \tilde{d})^{(2)} \tag{10}$$

On the basis of the microscopic theory as outlined before, the j-dependence of the coefficients A, Γ and Λ can be derived, giving

$$A_j = \sqrt{5}\,\sqrt{2j+1}\, A_0, \tag{11a}$$

$$\Gamma_{jj'} = \sqrt{5}\,\Gamma_0\,(u_j u_{j'} - v_j v_{j'}) Q_{jj'} \tag{11b}$$

$$\Lambda_{jj'}^{j''} = -2\sqrt{5}\,\Lambda_0 \beta_{jj''} \beta_{j'j''} / \sqrt{2j''+1}, \tag{11c}$$

$$\Gamma_0 = \Lambda_0 n_\beta / \sqrt{N_\pi} = \bar{\kappa} N_\nu / (N_\nu + N_\pi) \tag{11d}$$

where β and $\beta_{jj'}$ are given by eqs.(5) and (6). It should be noted that in deriving these formulas, explicitly the approximation given in eq.(6), has been used. In the phenomenologic calculations as presented later in this contribution, A_0, Γ_0, Λ_0, and v_j will be treated as free parameters. The parameter χ is determined by the even-even core nucleus.

In a first approximation the single particle matrix elements of the quadrupole operator can be taken proportional to the reduced matrix elements of $Y^{(2)}$,

$$Q_{jj'} = \langle \ell \tfrac{1}{2} j || Y^{(2)} || \ell' \tfrac{1}{2} j' \rangle \qquad (12)$$

thus neglecting the radial integrals.

III. LIMITING CASES

The spectra calculated in the IBFA model are in general very complex since they depend both on the structure of the core and on the boson-fermion interaction. However in the cases that the core Hamiltonian exhibits a dynamical symmetry, also in the spectra for odd-A nuclei some simple structures emerge. I will discuss here only two cases, those in which a $j = 9/2$ particle is coupled to an SU(5) and to an SU(3) core. In the contribution of J. Wood to this workshop the third limit, O(6), is discussed extensively, while further examples can be found in Refs.4, 6, and 7.

In the case that the odd particle occupies a unique shell model orbit $|j\rangle$, eqs.(11) can be simplified,

$$\Gamma = \Gamma_{jj} = \sqrt{5}\,\Gamma_0\,(u^2 - v^2)Q \qquad (13)$$

$$\Lambda = \Lambda^j_{jj} = -\,8\sqrt{5}\,\Lambda_0\,u^2v^2Q^2/\sqrt{2j+1}$$

where $Q = Q_{jj} = -\,0.982$ for a $j = 9/2$ shell. Since the monopole force is less important for the structure of the odd-A spectrum we put $A_0 = 0$ in the schematic calculations presented here. In the numerical calculations the program ODDA[8] has been used.

IIIA. THE SU(5) LIMIT

Since the SU(5) limit of the IBA model corresponds to the anharmonic vibrator, we expect that by coupling an odd-particle we recover the features of the particle-vibration coupling model.

The most convenient scheme to label the states in this limit is the weak-coupling scheme,

$$|\alpha,L,j;\,JM\rangle = [\,|\alpha L\rangle_c \otimes |j\rangle\,]^{(J)}_M \qquad (14)$$

where the square bracket denotes angular momentum coupling and $|\alpha L\rangle_c$ is the wavefunction of a state in the even-even core. The ground state is obtained by coupling the particle to the ground state of the core. The lowest excited states form a quintuplet, based on the 2^+ first excited state in the core. The splitting of the members of this quintuplet is given by

$$\langle n_d = 1,\ L = 2,\ j;\ JM|V_{BF}|n_d = 1,\ L = 2,\ j;\ JM\rangle$$
$$= \sqrt{5}\,\Gamma\chi(-1)^{J-j}\begin{Bmatrix}2 & 2 & 2\\ j & j & J\end{Bmatrix} + \Lambda\sqrt{2j+1}\begin{Bmatrix}2 & j & j\\ 2 & j & J\end{Bmatrix} \qquad (15)$$

A convenient way to study the different types of spectra that can be obtained is to solve the Hamiltonian as a function of v^2 for constant values of Γ_0, χ and Λ_0. In Fig.1 the calculated[4] energies

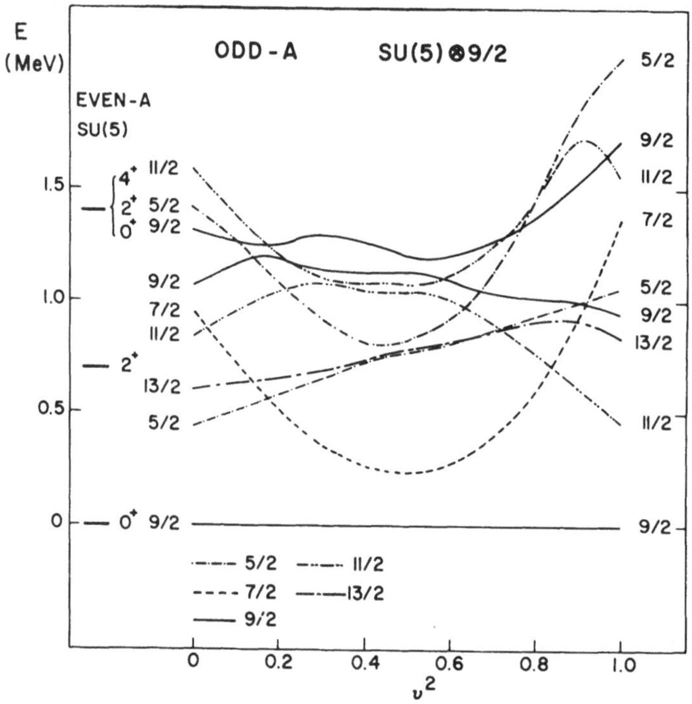

Fig.1 Calculated low-lying states for the system of a j = 9/2 particle coupled to an SU(5) core (N = 5, ε = 0.7 MeV) as a function of v^2, using Γ_0 = 0.567 Mev, Λ_0 = 1.83 MeV and $\chi = -\frac{1}{2}\sqrt{7}$

are given for Γ_0 = 0.567 MeV, $\chi = -\frac{1}{2}\sqrt{7}$ and Λ_0 = 1.83 MeV, while the core is given by N = 5 and ε = 0.7 MeV to simulate a realistic vibrational nucleus. For v^2 = 0 the J = j-2 level is the lowest member of the quintuplet, a well known feature that occurs for the coupling of an empty unique parity level to a vibrator. For $v^2 = u^2 = 0.5$ the spectrum is characterized by a low lying J = j-1 level. It is known that this is entirely due to the effect of the Pauli principle. In the IBFA model this is the result of the action of the exchange force.

A clear example of this behavior can be found in the spectrum of ${}^{101}_{45}Rh_{56}$. The positive parity states are described by the coupling of a $g_{9/2}$ particle to the ${}^{100}_{44}Ru_{56}$ core. At Z = 45 one expects the $g_{9/2}$ level to be half filled, v^2 = 0.5 thus giving rise to a pure exchange coupling. In the spectrum there should therefore be a low lying $7/2^+$ = j-1 level. As is shown in Fig.2, this is indeed the case. The calculation presented in this figure, using only an exchange force further reproduces also the positions of other states quite well.

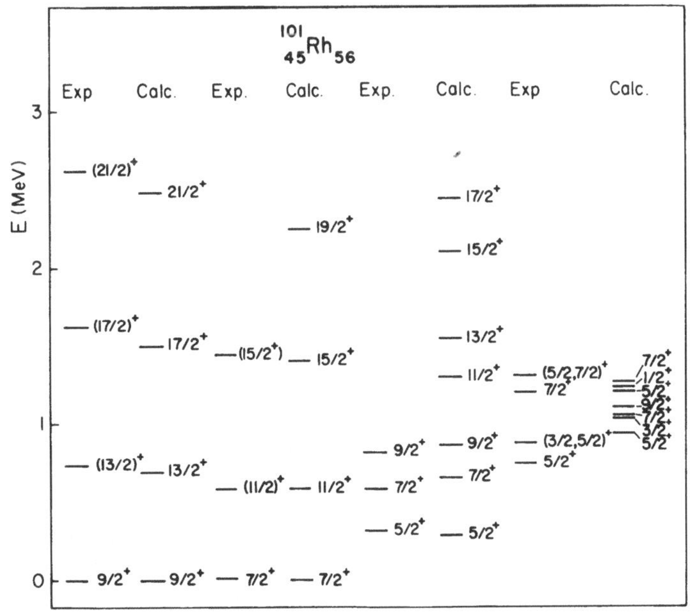

Fig.2 Calculated and experimental low lying positive parity states in ${}^{101}_{45}Rh$. The core parameters were taken from a best fit ${}^{100}_{44}Ru$, while for the coupling of the $g_{9/2}$ proton a pure exchange force has been taken, v^2=0.5 and Λ_0=2.32 MeV.

IIIB. THE SU(3) LIMIT

By coupling a particle to an SU(3) core one obtains a similar spectrum as in the geometrical model by coupling a particle to a rigid axially symmetric rotor. As is seen from Fig.3, the spectrum can be ordered into bands in which the energies are proportional to J(J + 1). The bands can be labeled with a quantum number K_0 = j, j - 1, ..., 1/2. The effect of the Coriolis face can be seen in the distorted structure of especially the $K_0 = 1/2$ band. To each K_0 band there belongs the equivalent of the β and γ bands of the even-even core.

In the SU(3) scheme the wavefunctions are most easily described in the strong coupling scheme.

$$|[N],(\lambda,\mu),K_c;\, jK_0;\, KJM> \;=\; \sum_R \sqrt{1+\delta_{K_c 0}}\,\sqrt{2R+1}\begin{pmatrix} j & R & J \\ K_0 & K_c & -K \end{pmatrix} \times \qquad (16)$$

$$|[N](\lambda,\mu),K_c,R,j,JM>$$

where the wavefunction on the r.h.s. is the weak coupling wavefunction as is defined in eq.(14). In eq.(16) the quantum number

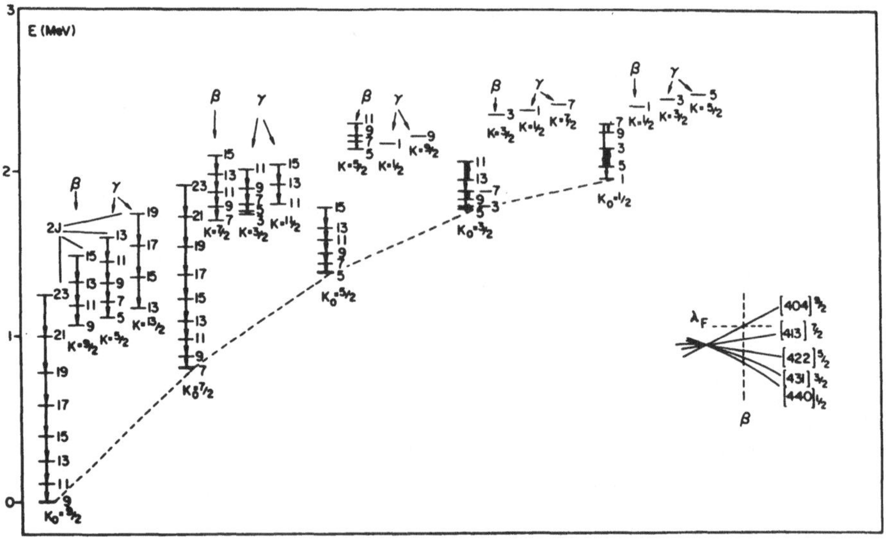

Fig.3 A typical spectrum, obtained by coupling a j = 9/2 particle to an SU(3) core (N = 6, K = 12.5 keV). In this case a pure quadrupole coupling has been taken, $v^2 = 1$, $\Gamma_0 = 0.23$ MeV and $\chi = -\frac{1}{2}\,7$. In the inset the corresponding situation in the Nilsson model is shown.

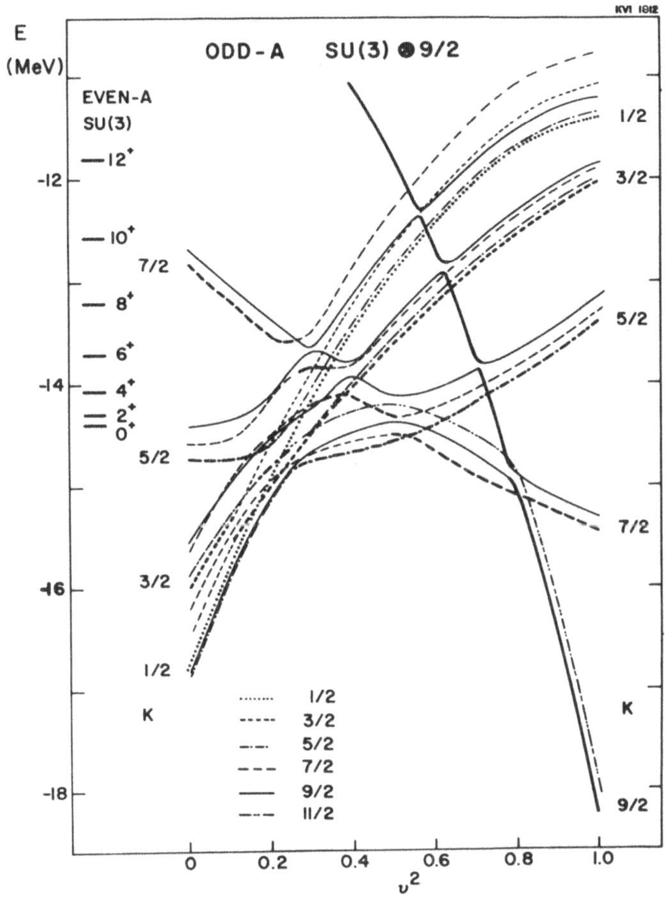

Fig.4 Calculated levels for the system of a j=9/2 particle coupled to an SU(3) core (N=9, K=50 keV, K'=20.8 keV) as a function of v^2, using Γ_0=0.567 MeV, Λ=1.83 MeV and $\chi= -\frac{1}{2}\sqrt{7}$. The bandheads are indicated by the heavy line.

K have been introduced in analogy with the Nilsson model. In the limit of large N ($N \to \infty$) the wavefunctions given in eq.(16) are eigenstates and the energy of the bandheads, based on the $(\lambda, \mu) = (2n, 0)$ ground state band in the even-even core, can be calculated in leading order in N as

$$\langle [N],(2N,0),0;\ jK_0;\ K_0JM|V_{BF}|[N],(2N,0),0;\ jK_0;\ K_0JM\rangle =$$
$$= -2N[\Gamma B_j(3K^2 - j(j+1))/\sqrt{2} + \Lambda\sqrt{2j+1}\, B_j^2(3K^2 - j(j+1))^2/3] \tag{17}$$

where

$$B_j = [(2j-1)(2j+1)(2j+3)j(j+1)]^{-\frac{1}{2}}$$

From eq.(17) clearly the relative importance of the quadrupole and the exchange force can be seen. For the case of an empty shell ($v^2 = 0$, $\Gamma_0 > 0$, $Q < 0$ in eq.13) we obtain $\Gamma < 0$, $\Lambda = 0$. Substituting these values in eq.(17) we obtain a $K_0 = 1/2$ band as ground state band. For a filled shell ($v^2 = 1$) the picture is reversed in the sense that now a $K = j$ band is lowest. When going from one to the other the K value of the ground state band changes gradually due to the presence of the exchange force. In the case of a half filled shell ($v^2 = u^2 = 0.5$, $\Lambda_0 > 0$ in eq.(13)) for example, we have $\Gamma = 0$, $\Lambda < 0$. The K value (an integer number) will be close to $\sqrt{j(j+1)/3}$.

As is seen from Fig.4 these simple features persist also for finite N(N = 9). To clarify the picture, the β and γ bands have not been drawn in Fig.4. It also shows the similarity between v^2 in IBFA and λ_F, the position of the Fermi surface, in the Nilsson model. Properties like single particle transfer amplitudes confirm the correspondence between the two models. The main difference is that in the present case all energies are calculated with a single Hamiltonian and that therefore the positions of the bandheads, the moments of inertia of the bands, the amount of "Coriolis" mixing, the positions of the β and γ bands, etc., are all interrelated.

IV. THE TRANSITIONAL REGIONS

Up to this point we have only shown the resemblances and equivalences of the limiting cases of the IBFA model with the geometrical model. An important point in favor of the IBA model is that it works equally well for the limiting cases as for the transitional regions in between and that it does not involve complicated calculations. In this section we will show that this

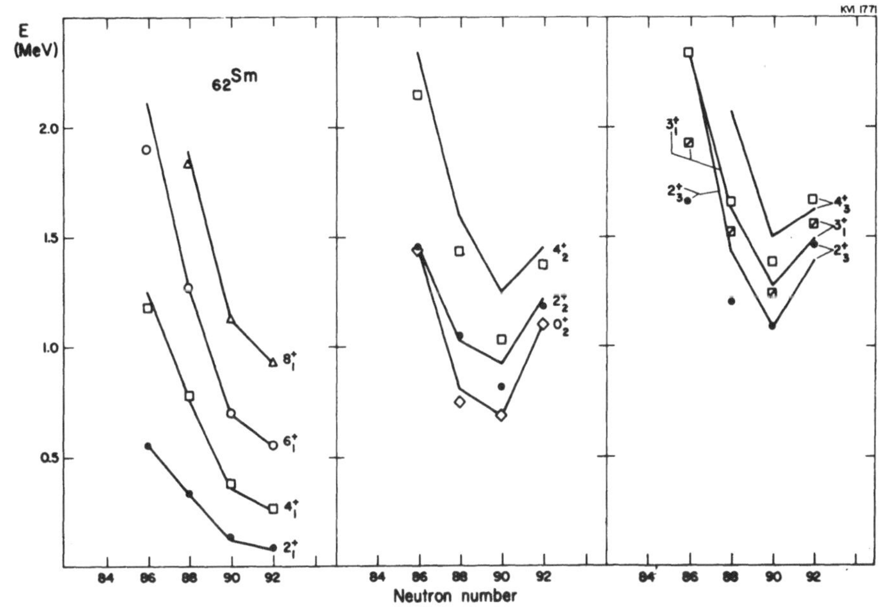

Fig.5 Calculated[4] and experimental low lying levels in the even mass Samarium isotopes.

point is also valid for the IBFA model by applying it to an SU(5) - SU(3) transitional region. As an example we have chosen the Europium isotopes. The calculations presented here were performed by N. Balsi and myself at the KVI, Groningen, The Netherlands. The O(6) - SU(3) transitional region is discussed by J. Wood.

In the IBFA model the odd mass ${}^{A}_{63}$Eu-isotopes are described by coupling a proton to the even ${}^{A-1}_{62}$Sm core. In the Sm isotopes one observes a clear transition from spherical to deformed[8] (see Fig.5). In the lighter isotopes one recognizes the SU(5) symmetry by the occurence of a triplet of states, 0^+_2, 2^+_2 and 4^+_1 at about twice the energy of the 2^+_1 state. The SU(3) limit is characterized by a clear ground state band with energies proportional to J(J + 1) while the 0^+_2 and $2^+_2(2^+_3)$ states, the band heads of the β and γ bands lie relatively high in the spectrum.

In Fig.6 the approximate positions of the proton single particle energies for the mass region A = 150 are given. The 13 valence protons of Eu will primarily occupy the $g_{7/2}$, $d_{5/2}$ and $h_{11/2}$ levels. The negative parity states can thus be calculated by considering only the odd particle in the $h_{11/2}$ shell. For the positive parity states both the $g_{7/2}$ and the $d_{5/2}$ levels have to be taken into account. Test calculations have shown that $s_{1/2}$ and $d_{3/2}$ levels which lie much higher, can be omitted from the calculations.

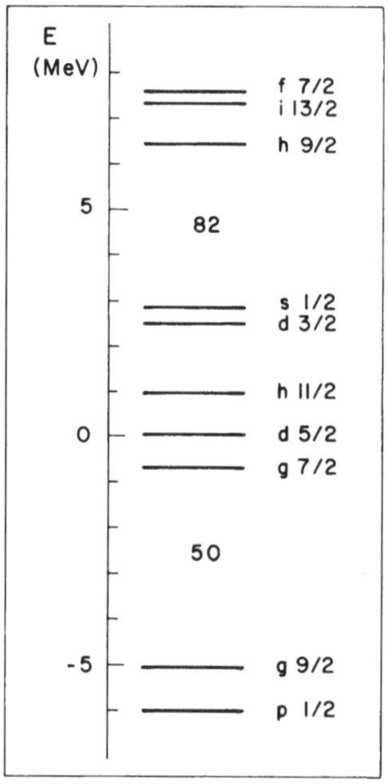

Fig.6 Relative proton single particle energies for mass number A = 150.

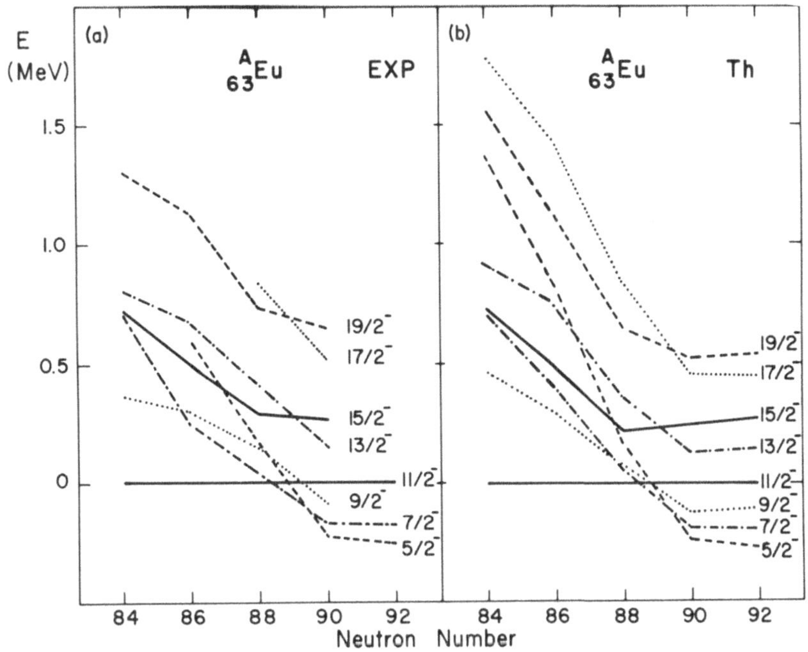

Fig.7. Experimental and calculated relative energies of negative parity states in the Eu isotopes.

The calculated energies of positive and negative parity states, using the parameters given in Table I, are compared with experiment in Figs.7 and 8. The transition from spherical to deformed is characterized by the behavior of the $5/2^-$ level. In the lighter isotopes it lies very high in the spectrum where it belongs to the n_d=2 multiplet. With increasing neutron number it comes down rapidly in energy to form the bandheads of the K=$5/2^-$ lowest negative parity band in ^{153,155}Eu. The fact that the K=$5/2^-$ band is lowest in the heavier isotopes depends only on the occupancy of the $h_{11/2}$ level (see Fig.4). This occupancy also determines in the lighter isotopes the splitting of the levels in the quintuplet (see Fig.1), which means that in the IBFA model the two are related. For the positive parity levels (Fig.8) the changes that occur in the spectra look much less dramatic. In the lighter isotopes however, the $5/2^+_1$ and $7/2^+_1$ levels are rather pure single particle states while at an excitation energy of about ε (the energy of the 2^+_1 state in the core) one finds the two n_d=1 quintuplets built on these single particle configurations. For the heavier isotopes the situation is completely different. The $5/2^+_1$ level is still the ground state, but it is now the bandhead of a K=$5/2^+$ rotational band in which the $d_{5/2}$ and $g_{7/2}$ single particle components are strongly mixed. To

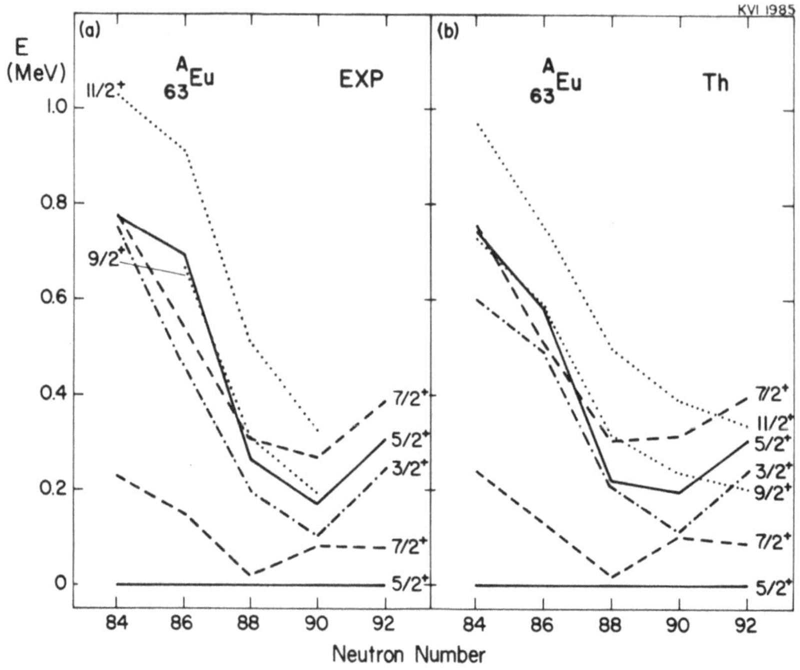

Fig.8. Experimental and calculated excitation energies of positive parity states in the Eu isotopes. For some levels the spin assignment is tentative.

show the change in structure more clearly, in Fig.9 the spectra of ^{149}Eu (typical particle-vitration) and ^{155}Eu (typical particle-rotor) have been plotted next to each other. The levels in a band are connected by strong B(E2) transitions. In the IBFA model both spectra are calculated within the same space.

The values of the parameters used in the boson-fermion interaction are given in Table I. The description of the Sm-cores is given in Ref.4. Since the $d_{5/2}$ and $g_{7/2}$ single particle energies lie very close to each other the occupation probabilities have been taken equal for these levels. As a consequence the j-dependence given in eq.(11) can be rewritten as,

$$A_j = \sqrt{5}\,\sqrt{2j+1}\,A_0$$

$$\Gamma_{jj'} = \sqrt{5}\,\Gamma_0\,(u^2 - v^2)Q_{jj'}, \qquad (18)$$

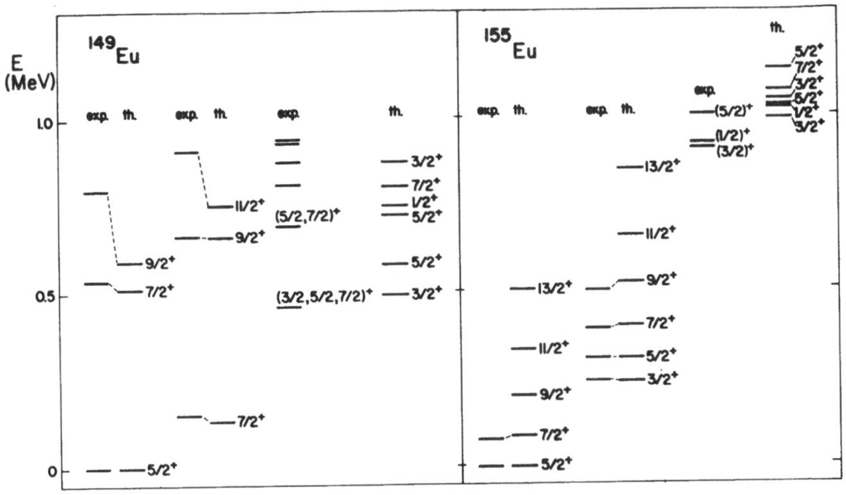

Fig.9 Experimental and calculated positive parity spectra of ^{149}Eu and ^{155}Eu. For some levels the spin assignment is tentative.

$$\Lambda_{jj'}^{j''} = -8\sqrt{5}\,\Lambda_0 u^2 v^2 Q_{jj''} Q_{j'j''}/\sqrt{2j''+1}.$$

As a consequence only two of the three parameters Γ_0, Λ_0 and v are linearly independent. In Table I therefore the values of the parameters γ and λ, defined as

$$\gamma = \Gamma_0 (u^2 - v^2) \tag{19a}$$

and

$$\lambda = \Lambda_0 u^2 v^2 \tag{19b}$$

have been quoted.

From Table I it can be seen that the parameters change smoothly from isotope to isotope. There are however, large differences between the values used for positive and negative parity states. For example γ has changed sign. In order to see the physical significance of the parameters we have to deduce the values of Γ_0 and Λ_0 from them. For ^{149}Eu this was sone, using Eqs.(19) while imposing the further conditions that $\Sigma v^2(2j + 1) = 13$, the number of protons outside the Z = 50 closed shell, and that $(\Gamma_0^+/\Gamma_0^-)^2 = \Lambda_0^+/\Lambda_0^-$,

TABLE 1

Parameters used in the calculation of the Eu-isotopes (in MeV). The strength of the monopole force was kept constant, $A_o = +0.1$ MeV.

A	$d_{5/2} - g_{7/2}$			$h_{11/2}$	
	γ	λ	$\varepsilon_{g_{7/2}} - \varepsilon_{d_{5/2}}$	γ	λ
147	-0.14	0.12	0.25	0.25	0.34
149	-0.17	0.14	0.145	0.33	0.40
151	-0.14	0.20	0.02	0.41	0.57
153	-0.12	0.26	-0.20	0.44	0.77
155	-0.065	0.27	-0.50	0.45	0.80

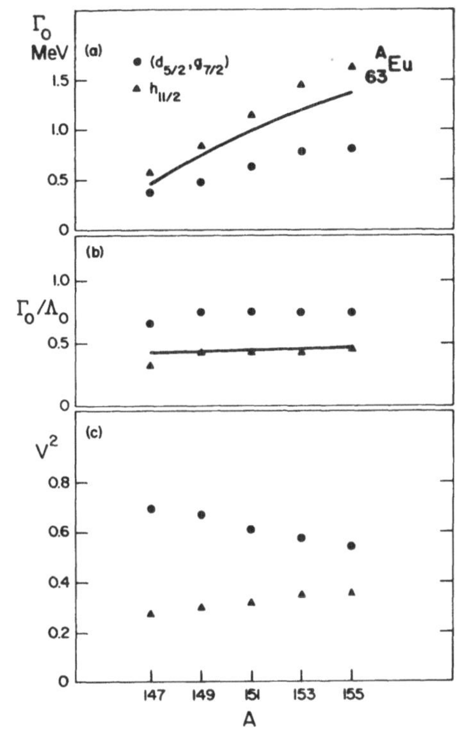

Fig.10 Deduced values of Γ_0, Λ_0 and v^2 from the parameters given in Table I. The drawn curves give the microscopic estimates as given by Eq.(11) for $\bar{\kappa}$ = 3 MeV.

where the superscript +(-) stands for the values as extracted for positive (negative) parity states. This latter condition can be shown to be equivalent with the fact that the relation between $Q_{jj'}$ for positive and negative parity states might not be given by Eq.(6) since the odd parity levels originate from a different harmonic oscillator shell. Otherwise Γ_0 and Λ_0 should be equal for positive and negative parity states. These six equations suffice to determine the six parameters for ^{149}Eu. For the other isotopes the parameters were determined by imposing the conditions that Γ_0/Λ_0 stays (approximately) constant and that v^2 changes smoothly from nucleus to nucleus cleus. The values extracted in this way are plotted in Fig.10.

The first thing one notices is that the sign difference in γ is explained by the fact that the $h_{11/2}$ level is less than half occupied while the positive parity levels have $v^2 > 0.5$. It can even be shown that by increasing Q_{jj} by a factor 1.7 for the $h_{11/2}$ level[10], both the curves for Γ_0^+ and Γ_0^- and those for Γ_0^+/Λ_0^+ and Γ_0^-/Λ_0^- fall on top of each other. On the basis of microscopic arguments[10], as is noted shortly above, one expects an increase of Q_{jj} for the unique parity level.

The drawn curves in Fig.10 give the N_ν dependence of Γ_0 and Γ_0/Λ_0 as calculated from Eq.(11d), using the obtained values of v^2. The assumed constancy of Γ_0/Λ_0 and its value appears to be consistent with the determined values of v^2. Furthermore, the strong increase in Γ_0 with increasing N_ν is in excellent agreement with Eq.(11d). The value of $\bar{\kappa}$ = 3 MeV, however, seems to be nearly an order of magnitude larger than what is used in the boson-boson interaction. This last point requires further investigation.

V. ELECTROMAGNETIC TRANSITIONS

In the IBFA model the E2 operator can be defined as

$$T^{(E2)} = e_B Q^{(2)} + e_F Q_F^{(2)} \tag{20}$$

where

$$Q_F^{(2)} = (s^\dagger \tilde{d} + d^\dagger s)^{(2)} + \chi (d^\dagger \tilde{d})^{(2)}$$

$$Q_F^{(2)} = \sum_{jj'} (u_j u_{j'} - v_j v_{j'}) Q_{jj'} (a_j^\dagger \tilde{a}_{j'})^{(2)}$$

In Eq.(20) $e_B Q_B^{(2)}$ is the same operator as used for the core and $e_F Q_F^{(2)}$ gives the contribution of the odd-particle. In the latter only the one body part has been retained (see also Eq.(4b)). There exist only very little experimental information on E2 matrix elements in the Eu isotopes. We have therefore compared our results only to the measured quadrupole moments, Fig.11. The boson effective operator is determined from the $2_1^+ \rightarrow 0_1^+$ transition in ^{154}Sm, e_B = 0.13 eb, using $\chi = -\frac{1}{2}\sqrt{7}$ as in previous calculations[8]. The fermion effective charge e_f has been set to zero since for the allowed transitions the contribution of the odd nucleon $(Q_F^{(2)})$ is roughly a factor 1/N smaller than that of the bosons $(Q_B^{(2)})$. The introduction of the fermion contribution, which can be done easily, would therefore not change appreciably the results, given in Fig.11. The calculated values change considerably across the transitional region. The few experimental points agree well with the calculation, especially in view of the fact that there were no parameters adjusted. However, because of the scarcity of data, no firm statement can be made on whether or not the IBFA model is indeed able to describe the electromagnetic properties correctly. More measurements are needed in order to clarify this point.

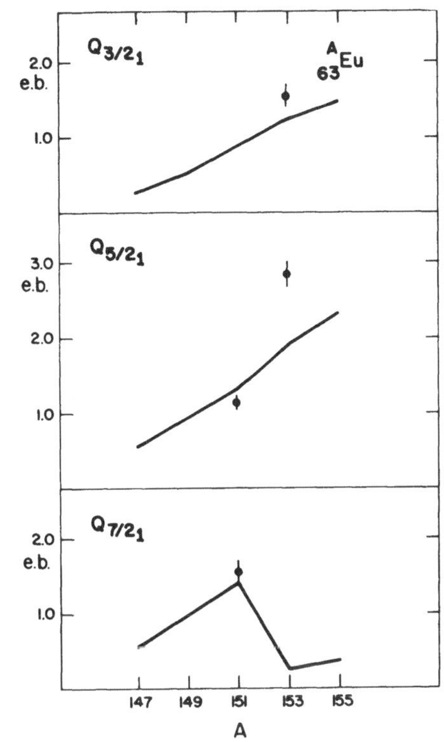

Fig.11 Experimental and calculated values of the quadrupole moments of the $3/2_1^+$, $5/2_1^+$ and $7/2_1^+$ states.

VI. SUMMARY AND CONCLUSION

In this contribution a short outline of the microscopic basis of the IBFA model has been given. Calculations in the limiting cases, SU(5), SU(3) and O(6), show the correspondence of this model with the particle-vibration coupling model, the Nilsson model and the gamma-unstable rotor plus particle model respectively. Application to the Eu isotopes, in which one observes a transition from spherical to deformed, shows that within the IBFA model it is possible to calculate also transitional odd-A nuclei, without doing complicated calculations.

REFERENCES

1. A. Arima and F. Iachello, Phys. Rev. C14 (1976) 761.
2. T. Otsuka, A. Arima and F. Iachello, Nucl. Phys. A309 (1978) 1.
3. T. Otsuka, Ph.D. Thesis, University of Tokyo, Japan (1979).
4. O. Scholten, Ph.D. Thesis, University of Groningen, The Netherlands (1980).

5. A. Arima, T. Otsuka, F. Iachello and I. Talmi, Phys. Lett. 66B (1977) 205.
6. F. Iachello, Phys. Rev. Lett. 44 (1980) 772.
7. F. Iachello and O. Scholten, Phys. Lett. 91B (1980) 189.
8. O. Scholten, Computer code ODDA, KVI int. rep. 253 (1980) Groningen, The Netherlands.
9. O. Scholten, F. Iachello and A. Arima, Ann. Phys. (NY) 115 (1978) 325.
10. O. Scholten and A.E.L. Dieperink, "Interacting Bose-Fermi Systems in Nuclei," F. Iachello, ed., Plenum Press, New York (1981).

QUANTIZED BOGOLYUBOV TRANSFORMATION AND

MICROSCOPIC FOUNDATION OF IBM

Li-Ming Yang

Institute of Theoretical Physics
Beijing University
Beijing, People's Republic of China

ABSTRACT

A new transformation similar in form to the Bogolyubov transformation, but with the coefficients U_j, V_j replaced by operators is deviced, in which the fermion operators (C_α^+, C_α) are transformed into mutually commuting quasi-fermions (a_α^+, a_α) and s-pairs, with the nucleon number always conserved. Among other possible applications it is particularly suited for the construction of a microscopic theory of the Interacting Boson Model (IBM).

If the success of IBM (1-4) is not fortuitous, it would mean that the correlated L = 0 and L = 2 pairs play a particularly important role in the ground state and low excited collective states. Of these two types of pairs, the L = 0 pair is even more important. Therefore in a proper formulation of a microscopic theory of nuclear structure, the presence of L = 0 pairs as fundamental constituent should be taken into account seriously. Usually the Bogolyubov transformation is utilized to incorporate the pairing effect into the one-body part of the Hamiltonian at the cost of particle number non-conservation. The coefficients U_j and V_j determined for the ground state are used for all excited states. As a result, the blocking effect is not properly treated. Besides, in this kind of treatment, the effect of L = 0 pair is only implicitly and partially contained in the quasi-particles. In a microscopic theory of IBM the L = 0 pair ought to appear explicitly.

A quantized Bogolyubov transformation is derived here in which a modified L = 0 pair behaving like a real boson appears explicitly after the transformation.

Suzuki and Matsuyanagi[5] in 1976 proposed a "quantized Bogolyubov transformation", but theirs is not a transformation but a mapping, the original fermion space is mapped into a new space, a product space of an ideal boson space and an ideal quasi-particle space. There is a one-to-one correspondence between states in the fermion space and in the new space, but the ideal boson space has no direct connection with the original fermion space.

In the following, a genuine transformation similar in form to the Bogolyubov transformation is given. Consider first the case of a single j-shell. Using the quasi-spin formalism one can express a general normalized state vector as $|S_j, S_{jo}, \Gamma_j\rangle$ where S_j, S_{jo} are the quasi-spin quantum number and its projection respectively, and Γ_j consists of angular momentum and other necessary quantum numbers. The index j will be omitted in the following. By definition, we have

$$2S = \Omega - \nu \quad , \qquad 2S_o = n - \Omega$$

where $\Omega = j+1/2$, ν is the seniority number, n is the nucleon number, and $2S = 0, 1, 2, \ldots \Omega$, $S_o = -S \ldots +S$. Besides one can write

$$|S, S_o, \Gamma\rangle = |S, -S+p, \Gamma\rangle$$

since $S_o = -\frac{1}{2}(\Omega - \nu) + \frac{1}{2}(n - \nu) = -S + \frac{1}{2}(n - \nu) = -S + p$.
p is the number of pairs coupled to L = 0. With the use of the quasi spin formalism, it can be shown that

$$|S, -S+p, \Gamma\rangle = 0^{\dagger}(p)|S, -S, \Gamma\rangle \tag{1}$$

where

$$0^{+}(p) = \frac{1}{\sqrt{p!}} (\mathscr{S}^{+}(j))^{p} \tag{2}$$

$$\mathscr{S}^{+}(j) = \hat{S}_{+}(j)(\hat{S}(j) - \hat{S}_{o}(j))^{-1/2} \tag{3}$$

S_+ S_o are the usual quasi-spin operators. S(j) is defined as follows

$$\hat{S}(j)|S, \Gamma\rangle = S|S, \Gamma\rangle \tag{4}$$

where $|S, \Gamma\rangle = |S, -S, \Gamma\rangle$. The operator $\mathscr{S}^+(j)$ and its hermitian conjugate $\mathscr{S}(j) = (\hat{S}(j) - \hat{S}_o(j))^{-1/2}\hat{S}_-(j)$ can be easily shown to satisfy the boson commutation rule

$$[\mathscr{S}(j), \mathscr{S}^+(j)] = 1 \qquad (5)$$

The operators $\mathscr{S}^+(j)$ and $\mathscr{S}(j)$ are the modified L = 0 pair operators mentioned above. In order to separate an operator that operates on states with no L = 0 pairs from the rest, a transformation operator T is defined as follows:

$$T = \sum_{\Gamma} \sum_{S=0}^{\Omega/2} \sum_{S_o=-S}^{S} 0^+(p)|S, \Gamma\rangle\langle S, S_o, \Gamma| \qquad (6)$$

Since $p = S+S_o$, summation over S_o for a given s implies summation over p as well. T is in fact equal to 1 since the bra and ket are in one-to-one correspondence and the summations run over the complete set. Nevertheless, the transformation induced by T is non-trivial. With the help of Wigner-Eckart theorem and the tensor property of the nuclear creation (annihilation) operator $C^+_{jm}(C_{jm})$ with respect to the quasi-spin group,

$$C^+_{jm} \sim T_{\frac{1}{2}\frac{1}{2}} \qquad C_{j\tilde{m}} \text{ (time reverse of } C_{jm}) \sim T_{\frac{1}{2}, -\frac{1}{2}}$$

one can show that

$$\left.\begin{aligned} TC^+_{jm}T &= C^+_{jm} \\ &= a^+_{jm}\hat{u}(j) - \hat{v}^+(j)a_{j\tilde{m}} \\ TC_{j\tilde{m}}T &= C_{j\tilde{m}} \\ &= \hat{u}(j)a_{j\tilde{m}} + a^+_{jm}\hat{v}(j) \end{aligned}\right\} \qquad (7)$$

where

$$\hat{u}(j) = \sqrt{1 - \frac{\mathscr{S}^+(j)\,\mathscr{S}(j)}{2\,\hat{S}(j)}} \qquad (8)$$

$$\hat{v}^{+}(j) = \frac{\mathscr{S}^{+}(j)}{\sqrt{2\hat{S}(j)}} \tag{9}$$

$$\left.\begin{aligned} a^{+}_{jm} &= \sum_{p} O^{+}(p) a^{+}_{jm} O(p) \\ a_{j\tilde{m}} &= \sum_{p} O^{+}(p) a_{j\tilde{m}} O(p) \end{aligned}\right\} \tag{10}$$

$$\left.\begin{aligned} a^{+}_{jm} &= \sum_{S\Gamma} \sum_{S'\Gamma'} |S\Gamma\rangle\langle S\Gamma|C^{+}_{jm}|S'\Gamma'\rangle\langle S'\Gamma'| \\ a_{j\tilde{m}} &= \sum_{S\Gamma} \sum_{S'\Gamma'} |S\Gamma\rangle\langle S\Gamma|C_{j\tilde{m}}|S'\Gamma'\rangle\langle S'\Gamma'| \end{aligned}\right\} \tag{11}$$

where a_{jm}, $a_{j\tilde{m}}$ are the operators that act only on states with no L = 0 pair. These operators have the following properties:

$$[a^{+}_{nm}, \mathscr{S}^{+}(j)] = 0 , \qquad [a^{+}_{jm}, \mathscr{S}(j)] = 0$$

$$a^{+}_{jm}|S, \Gamma\rangle = a^{+}_{jm}|S, \Gamma\rangle$$

Starting from the commutation relations of C^{+}_{jm}, $C_{j\tilde{m}}$ in the fermion space and performing the transformation as above, one obtains

$$\left.\begin{aligned} \left\{a^{+}_{jm}, a^{+}_{jm'}\right\} &= 0 \\ \left\{a_{j\tilde{m}}, a^{+}_{jm'}\right\} &= \delta_{j\tilde{m},jm'} + a^{+}_{jm} \frac{1}{2\hat{S}(j)} a_{j\tilde{m}} \end{aligned}\right. \tag{14}$$

where

$$\delta_{j\tilde{m},jm'} = \delta_{-m,m'}(-1)^{j+m}$$

When Eq.14 is to act on states with no L = 0 pair, one obtains

$$\left\{a^{+}_{jm},\ a^{+}_{jm'}\right\} = 0$$

$$\left\{a_{j\tilde{m}},\ a^{+}_{jm'}\right\} = \delta_{j\tilde{m},jm'} + a^{+}_{jm}\frac{1}{2\hat{S}(j)}a_{j\tilde{m}} \quad . \tag{14'}$$

Since $\alpha_{j\tilde{m}}(a_{j\tilde{m}})$ and $\alpha^{+}_{jm'}(a^{+}_{jm'})$ satisfy somewhat different commutation relations from that of fermions, we shall call them quasi-fermions.

Note that in the transformation Eq.7, the nucleon number is conserved. The operator $\hat{S}(j)$ as defined in Eq.4 becomes

$$\hat{S}(j) = \Omega(j) - \hat{n}_a(j)$$

$$\hat{n}_a(j) = \sum_m {}^{+}_{jm}\,{}_{jm} \tag{15}$$

The eigenvalue of $\hat{n}_a(j)$ is just the seniority ν. The space spanned by states with no L = 0 pair is called the intrinsic space, any state in this space can be expressed as

$$|S,\ \Gamma> = (\alpha^{+}_{j})^{\nu}_{\Gamma}|0> = (a^{+}_{j})^{\nu}_{\Gamma}|0>$$

where $|0> = |S = -\Omega/2,\ -S = -\Omega/2,\ \Gamma>$ is the vacuum state. Since $\mathscr{S}^{+}(j)$ does not alter S, and $S > S_o$ holds for any state $|S\ S_o\Gamma>$ that is not annihilated by $\mathscr{S}^{+}(j)$, i.e., $\mathscr{S}^{+}(j)|S,\ S_o,\ \Gamma> \neq 0$, one can expand Eq.3 for a given eigenvalue S of $\hat{S}(j)$, using $\hat{S} = (\hat{n}(j) - \Omega(j))/2$

$$\mathscr{S}^{+}(j) = \frac{\hat{S}_{+}}{\sqrt{S}}\sum_{p=0}^{\infty} C_p(\hat{n}(j))^p/p!$$

$$C_p = \sum_{k=0}^{\infty}\frac{[2(k+p)-1]!!}{(4S)^{k+p}}(-\Omega(j))^k/k! \tag{16}$$

One sees from Eq.16 that $\mathscr{S}^{+}(j)$ is a modified L = 0 pair, i.e., the addition of an L = 0 pair is accompanied by a series of $C^{+}_{jm}C_{jm}$, or by an attenuation factor depending on the number of nucleons already present. Further, the operators α^{+}_{jm} and a^{+}_{jm} have the properties

$$(a_j^+, a_j^+)_o = 0 \qquad (a_j^+ a_j^+) = 0$$

$$\left.\begin{aligned} (a_j^+)^n &= 0 \\ (a_j^+)^n &= 0 \end{aligned}\right\} \quad \text{for} \quad n > \Omega$$

Thus the operators a_{jm}^+ have the same properties as the ideal quasi-particle introduced by Suzuki et al[5].

For the many j-shell case, say $j = j_1, j_2, \ldots j_f$, the same transformation is carried out for each j-shell, since fermion operators belonging to different j-shell anti-commute with each other. Interaction among the j-shells tends to mix $\mathscr{S}^+(j)$ for different j, i.e.,

$$\mathscr{S}^+(\sigma) = \sum_j \phi_j^{(\sigma)} \mathscr{S}^+(j) \tag{17}$$

where $\phi_j^{(\sigma)}$ is the mixing coefficient ($\sigma = 1, 2, \ldots f$). For sufficient strong pairing interaction, coherent superposition of the $\mathscr{S}^+(j)$'s is expected to lead to a state $\mathscr{S}^+(\sigma = 1)$ of lowest energy and highest collectivity. In other cases, one may have a few such collective states and the rest non-collective. Since $[\mathscr{S}(j), \mathscr{S}^+(j')] = \delta_{jj'}$ one has

$$[\mathscr{S}(\sigma), \mathscr{S}^+(\sigma)] = \delta_{\sigma\sigma'} \tag{18}$$

provided

$$\sum_j \phi_j^{(\sigma)} \phi_j^{(\sigma')} = \delta_{\sigma\sigma'}$$

To determine $\phi_j(\sigma)$ one needs to start from the hamiltonian[6]. Thus for the many j-shell, the operators to be dealt with are $\mathscr{S}(\sigma)$, $\mathscr{S}^+(\sigma)$, a_{jm}^+ and a_{jm}^+($\sigma = 1, 2, 4 \ldots f$; $j = j_1, j_2, \ldots j_f$), their commutation relations besides Eq.18 are

$$[\mathscr{S}(\sigma), a_{jm}^+] = 0 \qquad [\mathscr{S}_\sigma^+, a_{jm}^+] = 0 \tag{19}$$

$$\left\{a_{jm}, a^+_{j'm'}\right\} = \delta_{j\tilde{m},j'm'} + a_{jm} \frac{\delta_{jj'}}{2\hat{S}(j)} a_{j'\tilde{m}'}$$
$$\delta_{j\tilde{m},j'm'} = \delta_{-m,m'}\delta_{j,j'}(-1)^{j+m} \tag{20}$$

The transformation here introduced is particularly suited for the construction of a microscopic theory of IBM, for we have here the L = 0 pair appears explicitly in the formulation and in its modified form $\mathcal{S}^+(\sigma)$ behaves like a real S-boson. Besides one can construct in the single j-shell case the L = 2 pair from the quasi-fermion operators $(a^+_j a^+_j)_{2\mu}$ which commutes with $\mathcal{S}^+(j)$ and $\mathcal{S}(j)$. and after suitable modification, it can be made to behave approximately like a d-boson[6]. In the many j-shell case, the L = 2 pair can be formed by coupling two quasi-fermions from the same j-shell $(a^+_j a^+_j)_{2\mu}$ or from two different j-shells $(a^+_j a^+_{j'})_{2\mu}$, if angular momentum and parity are allowed. One must then superpose all these L = 2 pairs to form a new set of L = 2 pairs,

$$X^+_{2\mu}(\rho) = \frac{1}{\sqrt{2}} \sum_{jj'} \psi^{(\rho)}_{jj'} \left(a^+_j a^+_{j'}\right)_{2\mu} ,$$

from which only the most collective ones are retained, which after properly modified can be made to behave approximately like a d-boson boson[6]. The modified new pairs so constructed commute by Eq.19 with $\mathcal{S}^+(\sigma)$ and $\mathcal{S}(\sigma)$. Thus a fully microscopic constructions of basic building blocks with L = 0 and 2 in the case of non-degenerate many j-shells in terms of the fermions and having just the right properties as required by IBM can be realized[6].

Since the transformation introduced in Eq.7 is exact, it may have much wider application than the construction of a microscopic theory of IBM. As long as pairing effect is strong enough to make correlated L = 0 pair a more-or-less stable unit in the low energy region, then it would be profitable to make use of the transformation Eq.7 together with Eq.17. The usual many-fermion problem may be highly simplified when some of the high-lying states are left out. This transformation may further be applied to the case of deformed field.

REFERENCES

1. A. Arima and F. Iachello, Phy. Rev. Lett. 35 1069 (1975).
2. A. Arima and F. Iachello, Ann. Phys. (N.Y.) 99, 253 (1976).
3. A. Arima and F. Iachello, Phy. Rev. Lett. 111, 201 (1978).
4. A. Arima and F. Iachello, Phy. Rev. Lett. 123, 468 (1979).
5. T. Susuku and U. Matsuyanagi Prof. Theo. Phys. 56, 1156 (1976).
6. L.M. Yang, D.H. Lu, H.P. Li, H.Y. Pan (to be published).

THEORY OF NUCLEAR SINGLE PARTICLE POTENTIAL

Shi-Shu Wu

Department of Physics
Jilin University of China
Changchung, Jilin, China

I. ABSTRACT

The single particle (sp) potential $u_{\alpha\beta}=M_{\alpha\beta}(\varepsilon_\beta)$ [or $M_{\alpha\beta}(\varepsilon_\alpha)$] defined in terms of the mass operator $M_{\alpha\beta}(\omega)$ is investigated in detail. First it is proven that although $u_{\alpha\beta}$ is non-hermitian, the energy eigen values of the single particle Schrödinger equation

$$\underset{\sim}{h}|\gamma\rangle = (\underset{\sim}{t} + \underset{\sim}{u})|\gamma\rangle = \varepsilon_\gamma|\gamma\rangle$$

are all real and satisfy rigorously the relation

$$\varepsilon_\gamma = \pm [E_\gamma(A \pm 1) - E_0(A)]$$

where $E_0(A)$ denotes the exact ground state energy of a closed shell nucleus (A), $E_\gamma(A\pm1)$ are exact energy eigenvalues of its neighboring nuclei of mass number $A\pm1$, and the upper (lower) sign holds if γ refers to a particle (hole) state. Then the principle of maximal cancellation of perturbation diagrams is analyzed by a new and non-perturbative method, which not only tells what terms can be cancelled by $\underset{\sim}{u}$ [referred to as the cancellation property of $\underset{\sim}{u}$], but also yields an analytic expression for the remaining terms. It is proved that the cancellation property of u can be expressed analytically in the form

$$M_{\alpha\beta}(\varepsilon_\beta) = 0,$$

where $M_{\alpha\beta}(\omega)$ is the reducible mass operator. The consequence of the above relation is studied. It is shown that the terms which can be cancelled by $\underset{\sim}{u}$ are more plentiful than what were known previously. Further, it is discussed under what conditions the following approximation for $G_{\alpha\beta}(t)$ holds

$$G_{p\beta}(t > 0) \simeq A_{p\beta} e^{-i\varepsilon_p t}$$

$$G_{h\beta}(t < 0) \simeq -A_{h\beta} e^{-i\varepsilon_h t}$$

$$G_{p\beta}(t < 0) \simeq G_{h\beta}(t > 0) \simeq 0$$

where $A_{\alpha\beta}$ [referred to as amplitude renormalization factor] is rigorously given by

$$A_{\alpha\alpha} = 1 - [\frac{d}{d\omega} M_{\alpha\alpha}(\omega)]_{\omega=\varepsilon_\alpha} = 1 - M'_{\alpha\alpha}(\varepsilon_\alpha)$$

$$A_{\alpha\beta} = M_{\alpha\beta}(\varepsilon_\alpha) G^o_\beta(\varepsilon_\alpha), \qquad (\alpha \neq \beta)$$

and α, β may denote either a particle (p) or a hole (h) state. Finally a method for the calculation of $A_{\alpha\beta}$ is suggested. One obtains, for instance, $A_{\alpha\alpha} = 1 - M'_{\alpha\alpha}(\varepsilon_\alpha) = [1 - M'_{\alpha\alpha}(\varepsilon_\alpha) - \sum_{\gamma(\neq\alpha)} x_{\alpha\gamma} G^o_\gamma (\varepsilon_\alpha) M'_{\gamma\alpha}(\varepsilon_\alpha)]^{-1}$, which is reduced to the well known Brandow formula, if one neglects the non-diagonal elements and introduced the following approximation for $M_{\alpha\alpha}(\omega)$: $M_{\alpha\alpha}(\omega) \simeq \Sigma_h\, G_B(\alpha h, \alpha h; \omega + \varepsilon_h)\, A_{hh}$, where G_B is Brueckner's G-matrix.

II. INTRODUCTION

The single particle (sp) potential plays an important role in the theory of structure of fermion systems. Although we are interested mainly in problems of nuclear structure, it is evident that the general results obtained apply to other fermion systems as well. A long standing question in the theory of sp potential is: Does it exist a sp potential $\underset{\sim}{u}$ such that the energy eigenvalues ε_γ of the sp Hamiltonian $\underset{\sim}{h} = \underset{\sim}{t} + \underset{\sim}{u}$ satisfy exactly the following relation

$$\varepsilon_\gamma = \pm [E_\gamma(A \pm 1) - E_o(A)] \tag{1}$$

In the above equation, $E_o(A)$ denotes the exact ground state energy of a closed shell nucleus (A), $E_\gamma(A\pm 1)$ are exact energy eigenvalues of its neighbouring nuclei of mass number $A\pm 1$, and the upper (lower) sign holds if γ refers to a particle (hole) state.

According to the principle of maximal cancellation of perturbation diagrams [PMCPD], Jones, Mohling and Becker[1] (JMB) have shown by means of the time dependent perturbation theory that the sp potential $u_{\alpha\beta}$ defined in terms of the mass operator $M_{\alpha\beta}(\omega)$ by

$$u_{\alpha\beta} = M_{\alpha\beta}(\varepsilon_\beta) \qquad [\text{or} \quad M_{\alpha\beta}(\varepsilon_\beta)] \tag{2}$$

achieves greater cancellation of terms [or diagrams if folded diagrams are included[2]] than the potential derived previously from the generalized time-ordering procedure by Brandow and Kirson . However, they argued that the choice given by Eq.(2) has the following drawbacks: (1) since $\underset{\sim}{u}$ is non-hermitian, the eigenvectors of the Schrödinger equation

$$\underset{\sim}{h}|\gamma\rangle = (\underset{\sim}{\iota} + \underset{\sim}{u})|\gamma\rangle = \varepsilon_\gamma|\gamma\rangle \tag{3}$$

do not form an orthogonal set; (2) the discrete eigenvalues of Eq.(3) are complex.

We have pointed out that the above defect (1) really causes no great trouble, because one may introduce a bi-orthonormal set $\{|\gamma\rangle, |\bar{\gamma}\rangle$ with $\langle\bar{\alpha}|\gamma\rangle=\delta_{\alpha\gamma}$ and define, for instance, the sp Green function as follows:

$$G_{\alpha\beta}(t = t_1 - t_2) = \langle\Psi_o|T\{\underset{\sim}{\bar{\xi}}_\alpha(t_1)\underset{\sim}{\xi}^+_\beta(t_2)\}|\Psi_o\rangle \tag{4}$$

where $\bar{\xi}_\gamma$ and $\bar{\xi}^+_\gamma$ are the creation operators of the sp states $|\gamma\rangle$ and $|\bar{\gamma}\rangle$ respectively. They satisfy the anticommutation relation

$$\left\{\xi^+_\alpha, \xi^+_\beta\right\} = \left\{\bar{\xi}_\alpha, \bar{\xi}_\beta\right\} = 0, \qquad \left\{\xi^+_\alpha, \bar{\xi}_\beta\right\} = \delta_{\alpha\beta}$$

This implies that the contractions are given by

$$\overbrace{\xi_\alpha{}^+(t_1)\xi^+_\beta(t_2)} = \overbrace{\bar{\xi}_\alpha(t_1)\bar{\xi}_\beta(t_1)} = 0,$$

$$\bar{\xi}_\alpha(t_1)\xi^+_\beta(t_2) = \langle\bar{\Phi}_o|T\left\{\bar{\xi}_\alpha(t_1)\xi^+_\beta(t_2)\right\}|\phi_o\rangle$$

$$= G^o_{\alpha\beta}(t_1 - t_2) = \delta_{\alpha\beta}G^o_\alpha(t_1 - t_2)$$

$$= \delta_{\alpha\beta}[\theta(t_1 - t_2) - n_\alpha]\, e^{-i\varepsilon_\alpha(t_1-t_2)}$$

where $\theta(t)$ is the step function and n_α takes the value 0 or 1 accord ing as α refers to a particle (p) or hole (h) state, while the zeroth order approximation ot $|\Psi_o\rangle$, and its dual state $|\Phi_o\rangle$ are given by

$$|\phi_o\rangle = \prod_{\alpha=1}^{A} \xi^+_\alpha|0\rangle, \qquad |\bar{\Phi}_o\rangle = \prod_{\alpha=1}^{A} \bar{\xi}^+_\alpha|0\rangle$$

Thus, one may apply the Wick theorem to calculate $G_{\alpha\beta}(t)$ just as if it were defined with respect to an orthonormal set of sp basis functions. However, defect (2) would be a real one, if it were true. But we observe that JMB's argument is not conclusive, since it is based on a finite order perturbation theory[2]. One of our purposes is to give an exact proof of the following conclusion (referred to as conclusion A):

Although the sp potential defined by Eq.(2) is non-hermitian, the energy eigenvalues of Eq.(3) are all real and satisfy Eq.(1) exactly.

Thus, the above result not only gives an affirmative answer to the question raised at the non-hermitian potential $u_{\dot{\alpha}\hat{\beta}} = M_{\alpha\beta}(\alpha_\beta)$. We do believe that Eq.(2) is a better choice than the symmetrized herimtian potential which JMB suggested in order to cure the above two drawbacks. The proof of conclusion A will be given in Sec.III, where some applications of Eq.(1) will also be discussed.

It is clear that there is no loss of generality if we restrict our discussion to the calculation of Green's functions. Lee and Yang pointed out a long time ago that Green's functions can be calculated more adequately and conveniently by considering propagator-renormalized skeleton diagrams, if proper care has been taken for over-counting. Analytically, propagator renormalization simply means that we substitute the exact sp Green function $G_{\alpha\beta}(t)$ for its zeroth order approximation $\delta_{\alpha\beta}G^o_\alpha(t)$. This says that we must first know how to calculate $G_{\alpha\beta}(t)$. Can the sp potential be so chosen that $G_{\alpha\beta}(t)$ may be calculated easily and accurately? We note that what PMCPD asks or can achieve is really not more than this. It is known that in the time-dependent formulation each renormalized sp propagator can be calculated independently of the rest of a diagram. Thus in order to realize PMCPD, one only needs to ask what diagrams contained in $G_{\alpha\beta}(t)$ can be cancelled by $\underset{\sim}{u}^1$. Obviously, cancellation alone does not ensure that $\underset{\sim}{u}$ will be a good choice, because the remaining terms still form an infinite series. It is therefore important to know how to handle these terms and to make sure that the terms neglected are truly small. To our knowledge, the latter question has not yet been studied seriously. In Sec.IV, PMCPD will be analysed by a new and non-perturbative method, which not only tells what terms can be cancelled by $\underset{\sim}{u}$, but also lets us gain a deeper insight into the property of the remaining terms. It is proved that the cancellation property of $\underset{\sim}{u}$ can be expressed analytically in the form

$$M_{\alpha\beta}(\varepsilon_\beta) = 0 \tag{5}$$

where $M_{\alpha\beta}(\omega)$ is the reducible mass operator. By means of the above relation we shall show that the terms which can be cancelled by $\underset{\sim}{u}$ are more plentiful than what were known previously. Further, it will be discussed under what conditions the following approximation for $G_{\alpha\beta}(t)$ holds:

$$\begin{aligned} G_{p\beta}(t > 0) &\simeq A_{p\beta} e^{-i\varepsilon_p t} \\ G_{h\beta}(t < 0) &\simeq -A_{h\beta} e^{-i\varepsilon_h t} \\ G_{p\beta}(t < 0) &\simeq G_{h\beta}(t > 0) \simeq 0 \end{aligned} \tag{6}$$

where the amplitude renormalization factor $A_{\alpha\beta}$ is rigorously given by

$$\begin{aligned} A_{\alpha\alpha} &= 1 - [\frac{d}{d\omega} M_{\alpha\alpha}(\omega)]_{\omega=\varepsilon_\alpha} = 1 - M'_{\alpha\alpha}(\varepsilon_\alpha) \\ A_{\alpha\beta} &= M_{\alpha\beta}(\varepsilon_\alpha) G^o_\beta(\varepsilon_\alpha) \qquad (\alpha \neq \beta), \end{aligned} \tag{7}$$

and α, β may be either a particle (p) or a hole (h) state.

With the help of Eq.(7), a method for calculating $A_{\alpha\beta}$ is suggested in Sec.V. Finally, a short summary and some comments will be made in Sec.VI.

III. PROOF OF CONCLUSION A

Let $\underset{\sim}{u}$ be an arbitrary sp potential. It may be non-hermitian and the eigenvalues determined by Eq.(3) may be complex. It will only be assumed that the eigenvectors of Eq.(3) form a complete set. Since this set is in general non-orthogonal, we shall introduce the biorthonormal basic set $\{|\gamma>, |\bar{\gamma}>\}$, which is reduced to the simple set $\{|\gamma>\}$ if the latter is orthonormal. The sp Green function will be defined according to Eq.(4). We shall first show that regardless whether the eigenvalues ε_γ are real or complex, the following equation

$$M_{\alpha\beta}(\omega) = (\omega - \varepsilon_\alpha)\delta_{\alpha\beta} + u_{\alpha\beta} + G^{-1}_{\alpha\beta}(\omega) \tag{8}$$

holds. This is just the Dyson equation if ε_γ are real. The reason why we have to generalize it to complex ε_γ will become clear later. The proof of Eq.(8) is almost trivial. However, we shall still write it down because it is vital to our argument. The total Hamiltonian of a many-fermion system may be written generally in the following form:

$$\underset{\sim}{H} = \underset{\sim}{H}_o + \underset{\sim}{V} - \underset{\sim}{U} \tag{9}$$

where $\underset{\sim}{H}_o$ and $\underset{\sim}{U}$ are given by

$$\underset{\sim}{H}_o = \sum_\alpha \varepsilon_\alpha \xi^+_\alpha \bar{\xi}_\alpha , \qquad \underset{\sim}{U} = \sum_{\alpha\beta} u_{\alpha\beta} \xi^+_\alpha \bar{\xi}_\beta . \tag{10}$$

Of course, $\underset{\sim}{H}$ and $\underset{\sim}{V}$ are hermitian, though $\underset{\sim}{U}$ and thus $\underset{\sim}{H}_o$ may not. In the following, $\underset{\sim}{V}$ will not be restricted to be a two-body interaction. We shall only assume that for the $\underset{\sim}{V}$ considered, the mass operator $M_{\alpha\beta}(\omega)$ exists or can be made to exist by a proper convergence procedure. According to the definition of the chronological T operator, we have

$$\frac{\partial}{\partial t_1} T\left\{\bar{\underset{\sim}{\xi}}_\alpha(t_1)\underset{\sim}{\xi}^+_\beta(t_2)\right\} = \delta_{\alpha\beta}\,\delta(t_1 - t_2) + T\left\{\frac{\partial}{\partial t_1}\bar{\underset{\sim}{\xi}}_\alpha(t_1)\underset{\sim}{\xi}^+_\beta(t_2)\right\} . \tag{11}$$

Since

$$i \frac{\partial}{\partial t} \bar{\xi}_{\alpha}(t) = - [H, \bar{\xi}_{\alpha}(t)]$$

$$= \varepsilon_{\alpha}\bar{\xi}_{\alpha}(t) - \sum_{\beta} u_{\alpha\beta}\bar{\xi}_{\beta}(t) + V_{\alpha}(t) \tag{12}$$

where

$$V_{\alpha}(t) = e^{iHt}[\bar{\xi}_{\alpha}, V]e^{-iHt} \tag{13}$$

we obtain from Eq.(4) that the following equation

$$(i \frac{\partial}{\partial t_1} - \varepsilon_{\alpha})G_{\alpha\beta}(t_1 - t_2) = i\delta_{\alpha\beta}\delta(t_1 - t_2)$$

$$- \sum_{\gamma} u_{\alpha\gamma}G_{\gamma\beta}(t_1 - t_2)$$

$$+ G(V_{\alpha}, \beta; t_1 - t_2) \tag{14}$$

holds regardless whether ε_{α} are real or complex. In Eq.(14), $G(V_{\alpha}, \beta; t_1-t_2)$ is given by

$$G(V_{\alpha}, \beta; t_1 - t_2) = \langle\Psi_o|T\left\{V_{\alpha}(t_1)\xi_{\beta}^{+}(t_2)\right\}|\Psi_o\rangle \tag{15}$$

As is wellknown, the mass operator $M_{\alpha\beta}(t_1-t_2)$ may be defined by

$$i\int_{-\infty}^{+\infty} d\sigma \sum_{\gamma} M_{\alpha\gamma}(t_1 - \sigma)G_{\gamma\beta}(\sigma - t_2) = G(V_{\alpha}, \beta; t_1 - t_2), \tag{16a}$$

or[1]

$$\sum_{\gamma} M_{\alpha\beta}(\omega)G_{\alpha\beta}(\omega) = G(V_{\alpha}, \beta; \omega) \tag{16b}$$

Thus, the Fourier transform of Eq.(14) yields immediately

$$\sum_{\gamma} [M_{\alpha\gamma}(\omega) - u_{\alpha\gamma} - (\omega - \varepsilon_{\alpha})\delta_{\alpha\gamma}]G_{\gamma\beta}(\omega) = \delta_{\alpha\beta} \tag{17}$$

[1]The Fourier transform is defined as $f(\omega) = i\int_{-\infty}^{+\infty} dt\, e^{i\omega t} f(t)$

which implies that $G^{-1}_{\alpha\beta}(\omega)$ exists and we have Eq.(8), Q.E.D. We note that we may also first convert Eq.(14) to an integral equation and then find its Fourier transform, which will again give us Eq.(17). However, here care must be taken, because the zeroth order sp Green function $G^{o}_{\alpha\beta}(t)$ is meaningless if Im $\varepsilon_p>0$ and Im $\varepsilon_h<0$. But it is easily seen that the solution of the differential equation

$$(i \frac{\partial}{\partial t_1} - \varepsilon_\alpha)K_{\alpha\beta}(t_1 - t_2) = i\delta_{\alpha\beta}\delta(t_1 - t_2) \tag{18}$$

may be written not only as $K_{\alpha\beta}(t) = G^{o}_{\alpha\beta}(t)$, but also as

$$\begin{aligned} K_{\alpha\beta}(t) &= \tilde{G}^{o}_{\alpha\beta}(t) = - \langle\bar{\phi}_o|\tilde{T}\left\{\bar{\xi}_\alpha(t_1)\xi^+_\beta(t_2)\right\}|\phi_o\rangle \\ &= \delta_{\alpha\beta}[n_\alpha - \theta(t_2 - t_1)]\, e^{-i\varepsilon_\alpha(t_1-t_2)} \end{aligned} \tag{19}$$

where $\tilde{T}$ denotes an inverse time ordering operator and is defined as follows:

$$\tilde{T}\left\{\bar{\xi}_\alpha(t_1)\xi^+_\beta(t_2)\right\} = \begin{cases} -\xi^+_\beta(t_2)\bar{\xi}_\alpha(t_1) \quad , & \text{if } t_1 > t_2 \, ; \\ \bar{\xi}_\alpha(t_1)\xi^+_\beta(t_2) \quad , & \text{if } t_1 < t_2 . \end{cases}$$

Obviously, $\tilde{G}^{o}_{\alpha\beta}(t)$ is well defined if Im $\varepsilon_p>0$ and Im $\varepsilon_h<0$. Thus we may introduce $\hat{G}^{o}_{\alpha\beta}(t)$ by

$$\hat{G}^{o}_{\alpha\beta}(t) = \begin{cases} G^{o}_{\alpha\beta}(t) \; , & \text{if Im } \varepsilon_p \leq 0 \text{ and Im } \varepsilon_h \geq 0; \\ \tilde{G}^{o}_{\alpha\beta}(t) \; , & \text{if Im } \varepsilon_p > 0 \text{ and Im } \varepsilon_h < 0. \end{cases}$$

It is clear the $\hat{G}^{o}_{\alpha\beta}(t)$ is now meaningful for either real or complex ε_α. Since $\hat{G}^{o}_{\alpha\beta}(t)$ satisfies Eq.(18), Eq.(14) may be converted straightforwardly into the following integral form:

$$\begin{aligned} G_{\alpha\beta}(t_1 - t_2) &= \delta_{\alpha\beta}\hat{G}^{o}_{\alpha}(t_1 - t_2) + \iint_{-\infty}^{\infty} d\sigma_1 d\sigma_2 \hat{G}^{o}_{\alpha}(t_1 - \sigma_1) \\ &\quad \times \sum_{\gamma} [M_{\alpha\gamma}(\sigma_1 - \sigma_2) + iu_{\alpha\gamma}\delta(\sigma_1 - \sigma_2)] \\ &\quad \times G_{\gamma\beta}(\sigma_2 - t_2) \; , \end{aligned} \tag{20}$$

which now holds not only for real, but also for complex ε_α. Thus we have again established Eq.(17), because it is just the Fourier transform of Eq.(20). According to Eq.(19) we observe that if Im $\varepsilon_p>0$ (Im $\varepsilon_h<0$), the particle (hole) line should run backward (forward) in time. Since Eq.(2) is defined in terms of $M_{\alpha\beta}(\omega)$, it may be desirable to derive Eq.(8) directly in the ω-representation. Such a proof has also been given in ref.7.

We shall now use Eq.(8) to prove conclusion A. Since we do not know beforehand whether the eigenvalues ε_γ are real or not, it is necessary that Eq.(8) should hold for either real or complex ε_γ. For definiteness we shall only discuss $u_{\alpha\beta}=M_{\alpha\beta}(\varepsilon_\beta)$, because the other choice $u_{\alpha\beta}=M_{\alpha\beta}(\varepsilon_\alpha)$ can be considered similarly. Set $\omega=\varepsilon_\beta$. From Eqs.(2) and (8) we find

$$G^{-1}_{\alpha\beta}(\varepsilon_\beta) = 0. \tag{21}$$

This says that the eigenvalue determined by the non-hermitian potential of Eq.(2) must also be a root of $G^{-1}_{\alpha\beta}(\omega)$. Now, the definition we have

$$\sum_\alpha G_{\beta\alpha}(\varepsilon_\beta)G^{-1}_{\alpha\beta}(\varepsilon_\beta) = 1 \tag{22}$$

Since Eq.(21) holds for all α, from Eq.(22) we conclude that ε_β must satisfy

$$G_{\beta\alpha}(\varepsilon_\beta) = \infty \tag{23}$$

for at least one α.

As is wellknown, the Lehmann representation of $G_{\alpha\beta}(\omega)$ is given by

$$G_{\alpha\beta}(\omega) = -\sum_n \left\{ \frac{g^{(+)}_{\alpha\beta}(n)}{\omega - {}^{+}_{n} + i\eta} + \frac{g^{(-)}_{\alpha\beta}(n)}{\omega + {}^{-}_{n} - i\eta} \right\}_{\eta\to 0^+} \tag{24}$$

where

$${}^{\pm}_{n} = E_n(A \pm 1) - E_o(A) \tag{25}$$

$$g^{(+)}_{\alpha\beta}(n) = \langle\Psi_o|\bar{\xi}_\alpha|\Psi_n(A+1)\rangle \langle\Psi_n(A+1)|\xi^+_\beta|\Psi_o\rangle$$
$$g^{(-)}_{\alpha\beta}(n) = \langle\Psi_o|\xi^+_\beta|\Psi_n(A-1)\rangle \langle\Psi_n(A-1)|\bar{\xi}_\alpha|\Psi_o\rangle \tag{26}$$

From Eqs.(23), (24) and (25) we obtain immediately Eq.(1), which implies at the same time that ε_γ are all real. Thus, conclusion A is proved. Although the above argument is rigorous and sufficient to prove conclusion A, there still remain some questions which one would like to ask further. In order to make our notation clear, let us express Eq.(24) in more detail. Clearly, Σ should not only sum over the discrete, but also integrate over the continuous spectra of nuclei $A\pm1$. Correspondingly, $G_{\alpha\beta}$ may be separated into two parts $G^d_{\alpha\beta}$ and $G^c_{\alpha\beta}$

$$G_{\alpha\beta}(\omega) = G^d_{\alpha\beta}(\omega) + G^c_{\alpha\beta}(\omega) \tag{27}$$

where $G^d_{\alpha\beta}$ is a meromorphic function, while $G^c_{\alpha\beta}$ contains branch cuts. We shall use ($\mp$, Λ) to label eigenstates belonging to the continuous spectrum, where $\pm$=E(A$\pm$1)-E (A) and Λ denotes all the other quantum numbers. Obviously, each of the nuclei $A\pm1$ can be split into different kinds of fragments. Every kind of such a split will be called a channel as usual. Let us set

$$\Delta^{\pm}_k = E^{(k)}_{th}(A \pm 1) - E_o(A) \tag{28}$$

where $E^{(k)}_{th}(A\pm1) - E_o(A\pm1)$ is the threshold energy of the k-th channel. Further, $\Delta^{\pm}_k$ will be ordered as

$$\Delta^{\pm}_k \leq \Delta^{\pm}_{k+1} \qquad (k = 1, 2, \ldots)$$

$G^c_{\alpha\beta}$ can then be written explicitly in the form

$$G^c_{\alpha\beta}(\omega) = F^{(+)}_{\alpha\beta}(\omega) + F^{(-)}_{\alpha\beta}(\omega) \tag{29a}$$

$$F^{(\pm)}_{\alpha\beta}(\omega) = \pm \int_{\Delta^{\pm}_1}^{\infty} d \quad \frac{g^{(\pm)}_{\alpha\beta}(\ )}{\pm\ \omega - i\eta}$$

where

$$g^{(\pm)}_{\alpha\beta}(\ ) \quad = \quad \sum_k g^{(\pm)}_{\alpha\beta}(\ ) \tag{30a}$$

and corresponding to the k-th channel of nucleus A+1 and A-1, we have respectively

$$g^{(+)k}_{\alpha\beta}(\ ) \quad = \quad \langle\Psi_o|\bar{\xi}_\alpha|\Psi_{\Lambda_k}(A+1)\rangle\langle\Psi_{\Lambda_k}(A+1)|\xi^+_\beta|\Psi_o\rangle$$

$$g^{(-)k}_{\alpha\beta}(\ ) \quad = \quad \langle\Psi_o|\xi^+_\beta|\Psi_{\Lambda_k}(A-1)\rangle\langle\Psi_{\Lambda_k}(A-1)|\bar{\xi}_\alpha|\Psi_o\rangle \tag{30b}$$

Thus $F^{(\pm)}(\omega)$ may be written as

$$F^{(\pm)}_{\alpha\beta}(\omega) \quad = \quad \sum_k F^{(\pm)k}_{\alpha\beta}(\omega) \tag{31a}$$

where

$$F^{(+)k}_{\alpha\beta}(\omega) \quad = \quad \int_{\Delta_k^+}^{\infty} d\ \frac{g^{(+)k}_{\alpha\beta}(\ )}{\ -\omega - i\eta}$$

$$F^{(-)k}_{\alpha\beta}(\omega) \quad = \quad -\int_{\Delta_k^-}^{\infty} d\ \frac{g^{(-)k}_{\alpha\beta}(\ )}{\ +\omega - i\eta}$$

$$= \quad \int_{-\infty}^{-\Delta_k^-} dZ\ \frac{g^{(-)k}_{\alpha\beta}(-Z)}{Z - \omega + i\eta} \tag{31b}$$

The analytic properties of the above Cauchy integrals are wellknown[8]. The line of integration L or L (Fig.1) is a branch cut, and

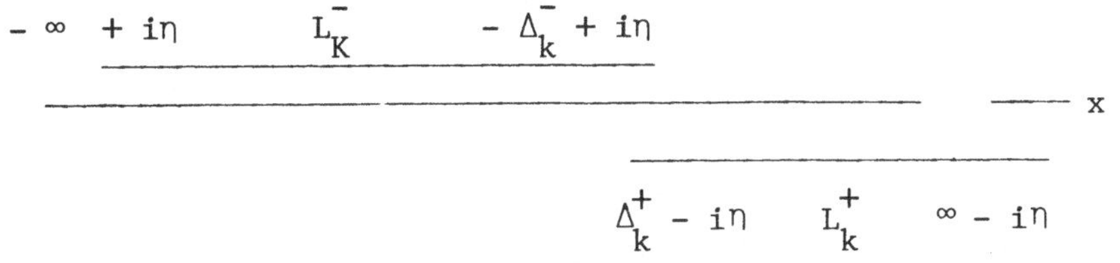

Fig.1 Line of integration $L_k^{\pm}$

$F_{\alpha\beta}^{(+)k}$ ($F_{\alpha\beta}^{(-)k}$) is holomorphic in the entire complex plane excluding L_k^+ (L_k^-). Its end point $\Delta_k^+ - i\eta(-\Delta_k^- + i\eta)$ plays the part of a branch point, at which $F_{\alpha\beta}^{(+)k}$ ($F_{\alpha\beta}^{(-)k}$) has a logarithmic singularity (this singularity will not occur, if $g_{\alpha\beta}^{(\pm)k}(\Delta_k^{\pm}) = 0$). Therefore, Eq.(23) implies that ε_β must coincide with either a pole of $G_{\beta\alpha}^d$ or a point lying on one of $L_k^{\pm}$. Any one of these cases gives us Eq.(1) and so conclusion A. However, Eq.(23) alone cannot tell which one of the above two cases does occur. That is to say, the following questions have not yet been answered: How is the discrete spectrum of $\underset{\sim}{h}$ [with $\underset{\sim}{u}$ given by Eq.(2)] related to the poles of G^d, may a part of it lie on some of $L_k^{\pm}$, and how is its continuous spectrum related to $L_k^{\pm}$, may some of the poles of G^d embedded in it?

Certainly we cannot expect that all the discrete energy eigenvalues of nuclei (A$\pm$1) are given by $\{\varepsilon_\gamma\}$ according to Eq.(1). Those eigenstates whose eigenvalues $E_\gamma(A+1)[E_\gamma(A-1)]$ satisfy Eq.(1) will be referred to as the single particle (hole) states of nucleus A+1 (A-1), while all the other eigenstates as their collective states. Naturally, this does not exclude the possibility that some eigenvalues of the collective states may also satisfy Eq.(1) accidentally. However, as is wellknown, in the case of discrete spectrums accidental degeneracy can always be lifted by an appropriate limiting procedure. Therefore, the above definitions specifies the discrete sp (sh) states of nucleus A+1 (A-1) uniquely [cf. Section IV]. For the case of continuous spectrum, the channel consisting of the closed shell nucleus (A) at its ground state and a single nucleon will be defined as the sp channel. We shall call it the s-th channel. Let ε_o^c denote the starting point of the continuous spectrum of h. It will be shown in the next section that $\varepsilon_o^c = \Delta_s^+$, and there is a unique correspondence between the eigenstate belonging to the continuous spectrum of h and the sp channel state of nucleus A+1. That is to say, we have the following conclusion B: The sh states of nucleus A-1 and the sp states (including the sp channel states) of nucleus A+1 are completely and uniquely specified by the eigensolutions of Eq.(3) if $\underset{\sim}{u}$ is chosen according to Eq.(2).

Though in general s = 1, i.e. $\Delta_s^+ = \Delta_1^+$, we may also have s > 1. For instance, if the ground state of nucleus A+1 is unstable against α-decay of spontaneous fission, then Δ_1^+ will be determined by the α-decay channel or fission channel. It is not difficult to see that Eqs.(24) and (27) still hold, even if some or all of the discrete eigenvalues are embedded in the continuous spectrum. Thus, if $\Delta_s^+ > \Delta_1^+$ and Eq.(3) possesses discrete eigensolutions, then nucleus (A+1) may have metastable sp states which are stable against decay into the sp channel, but unstable against decay into some other channels. Similarly, nucleus A-1 may also possess metastable sh states. It is evident that the number of hole states of nucleus A-1 is finite. For example, if the closed shell nucleus is chosen to be ^{16}O, we obtain from Eq.(1) the following hole-state eigenvalues of ^{15}O or ^{15}N:

$$E_h(15) = E_o(16) - \varepsilon_h,$$

where $h = 0s_{1/2}$, $0p_{3/2}$, and $0p_{1/2}$ and it may refer to neutron or proton hole.

We note that in our argument no special form of $\tilde{V}$ has been assumed. This shows that conclusions A and B will be valid for a variety of choices of V, including the cases where $\tilde{V}$ contains many-body forces, or $\tilde{V}$ is considered as mediated by mesons[9]

If $\tilde{V}$ is known, the non-hermitian potential $\tilde{u}$ defined by Eq.(2) is determined completely. Since Eq.(1) is theoretically rigorous and $E_\gamma(A\underline{+}1) - E_o(A)$ are experimentally measurable, conclusions A may serve as one of the means to check the foundation of the present nuclear many-body theory (FNMBT). However, as the form of $\tilde{V}$ is yet uncertain, we shall assume in the following that FNMBT is reliable. This implies that the results predicted by conclusion A should be exactly the same as those measured experimentally. Let ε_γ(cal) denote the energy eigenvalues calculated by means of Eqs.(2) and (3). According to Eq.(1) we see that ε_γ(cal) are useful at least in the following two aspects:

1. For regions where experimental dat of $E_\gamma(A\underline{+}1) - E\ (A)$ are available, we may use ε_γ(cal) as a criterion to discriminate between different choices of $\tilde{V}$.

2. If an adequate $\tilde{V}$ has been found, ε_γ(cal) will then give us reliable information on $E_\gamma(A\underline{+}1) - E_o(A)$ for regions where experimental data are yet unavailable. It is clear that this information will be useful for the prediction of stable islands of superheavy nuclei.

IV. SINGLE PARTICLE GREEN'S FUNCTION AND CHOICE OF SINGLE PARTICLE PARTICLE POTENTIAL

The purpose of this section is twofold:

1. We shall give an answer to the question: "How to choose u so that the sp Green function $G_{\alpha\beta}(t)$ can be calculated easily and accurately?"

2. It will be proven that conclusion B is true.

Since the zeroth order sp Green function has the property: $G^o_{\alpha\beta}(t) = \delta_{\alpha\beta}G^o_\alpha(t)$ where $G^o_p(t < 0) = 0$ and $G^o_h(t > 0) = 0$, correspondingly $G_{p'p}(t > 0)$ and $G_{h'h}(t < 0)$ will be referred to as the principal components of $G_{\alpha\beta}(t)$. As a preliminary step towards our goal, we shall first ask how to choose $\tilde{u}$ such that these principal components can be calculated easily and accurately (Proposition A).

We observe that if we know the principal components, we can already carry out Lee-Yang's procedure mentioned in the introduction. Besides, they also facilitate the calculation of the non-principal components of $G_{\alpha\beta}(t)$. Such an example for calculating $G_{p'p}(t < 0)$ is illustrated in Fig.2. Furthermore, we shall show that Proposition A determines $\underset{\sim}{u}$ completely and the $\underset{\sim}{u}$ so determined, in fact, simplifies the expressions of the non-principal components at the same time. Before proceeding, we would like to make a trivial remark concerning the calculation of the Fourier integral

$$G^{c}_{\alpha\beta}(t) = \frac{1}{2\pi i}\int_{-\infty}^{\infty} d\omega e^{-i\omega t}\, G^{c}_{\alpha\beta}(\omega) \tag{32}$$

Though $G^{c}_{\alpha\beta}(\omega)$ contains branch cuts, it is known that the above integral can still be calculated by the residue theorem as if the singularities of $G^{c}_{\alpha\beta}(\omega)$ are just continuous series of poles distributed on the lines of integration $L^{\mp}_{k}$. This is because the double integral $\int d\omega \int d\mathscr{E}$ resulted by substituting Eq.(29) in (32) may be calculated by reversing the order of integration. Thus for the calculation of the Fourier integral, no special distinction between the discrete and continuously distributed poles is necessary.

It is evident that Brandow-Kirson's potential is only an approximation to Eq.(2), while they both contain the HF potential as their lowest order approximation. In order to clarify the relationship between JMB's result and ours, we shall give two methods of derivation. One is also based on the perturbation theory. However, we shall use an expansion where each term of the series is itself a highly summed partial infinite series. Then it will be shown that there exists a very simple non-perturbative argument, by means of which we can not only derive easily all the results obtained by the perturbation theory, but also gain a deeper insight into the property of the remaining terms. With the help of this

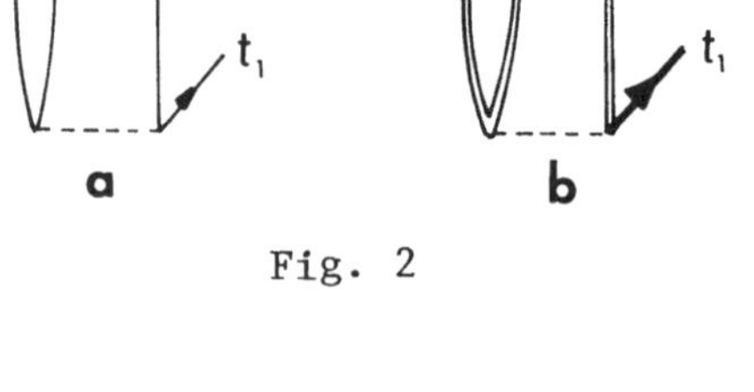

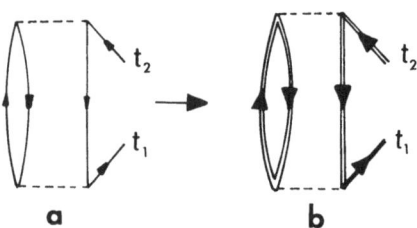

Fig. 2

argument we shall prove that the terms which can be cancelled by $\underset{\sim}{u}$ are more plentiful than what were known previously. This together with conclusion B justifies that $\underset{\sim}{u}$ may be regarded as the best choice among sp potentials.

According to Dyson's equation we have

$$\begin{aligned} G_{\alpha\beta}(\omega) &= \delta_{\alpha\beta}G^{o}_{\alpha}(\omega) + \sum_{\gamma} G^{o}_{\alpha}(\omega)[u_{\alpha\gamma} - M_{\alpha\gamma}(\omega)]G_{\alpha\gamma}(\omega) \\ &= \delta_{\alpha\beta}G^{o}_{\alpha}(\omega) + G^{o}_{\alpha}(\omega)[u_{\alpha\beta} - M_{\alpha\beta}(\omega)]G^{o}_{\beta}(\omega) \\ &\quad + \sum_{\gamma} G^{o}_{\alpha}(\omega)[u_{\alpha\gamma} - M_{\alpha\gamma}(\omega)]G^{o}_{\gamma}(\omega)[u_{\gamma\beta} \\ &\quad - M_{\gamma\beta}(\omega)]G^{o}_{\beta}(\omega) + \ldots \\ &= G^{o}_{\alpha\beta}(\omega) + G^{(1)}_{\alpha\beta}(\omega) + G^{(2)}_{\alpha\beta}(\omega) + \ldots \end{aligned} \tag{33}$$

As is well known, $M_{\alpha\beta}(\omega)$ may be written as

$$M_{\alpha\beta}(\omega) = M^{o}_{\alpha\beta} + M^{+}_{\alpha\beta}(\omega) + M^{-}_{\alpha\beta}(\omega) \tag{34}$$

where M^{o} is independent of ω and the superscripts $(\pm)$ indicate that the poles of M^{+} and M^{-} lie in the lower and upper half complex plane respectively. Let us introduce

$$\hat{\rho}_{\delta\gamma} = \langle\Psi_{o}|\xi_{\gamma}\xi_{\delta}|\Psi_{o}\rangle = \delta_{\gamma\delta}{}_{\ \gamma} \quad \hat{\rho}^{(1)}_{\delta\gamma} + \ldots \tag{35}$$

For example, if $\underset{\sim}{V}$ is a two-body interaction, the perturbation expansion of $M_{\alpha\beta}(\omega)$ has the form

$$\begin{aligned} M_{\alpha\beta}(\omega) &= {}_{\gamma\delta} \quad {}_{\alpha\gamma\ \beta\delta}[\delta_{\gamma\delta}n_{\gamma} + \hat{\rho}^{(1)}_{\delta\gamma} + \ldots] + \frac{1}{2}\sum_{\mu\nu\rho} \quad {}_{\alpha\rho\ \mu\nu} \\ &\quad \times \left(\frac{(1-n_{\mu})(1-n_{\nu})n_{\rho}}{\omega-\varepsilon_{\mu}-\varepsilon_{\nu}+\varepsilon_{\rho}+i_{\eta}} + \frac{n_{\mu}n_{\nu}(1-n_{\rho})}{\omega-\varepsilon_{\mu}-\varepsilon_{\nu}+\varepsilon_{\rho}-i\eta} \right) {}_{\mu\nu,\beta\rho} + \ldots \end{aligned} \tag{36}$$

We must remark that the poles expressed by the perturbation expansion (36) are in general not the true poles $M^{\pm}(\omega)$, because they may change drastically after the perturbation series has been summed, except if the latter converges quickly. In order not to be misled by Eq.(36), only the exact expression (34) will be used in our argument.

In the following we shall first consider the discrete spectrum of Eq.(3). Let us now discuss $G_{p'p}(t > 0)$. From Eq.(33) we have

$$G^{(1)}_{p'p}(t > 0) = \frac{1}{2\pi i}\int d\omega e^{-i\omega t} G^{o}_{p'}(\omega)[u_{p'p} - M_{p'p}(\omega)]G^{o}_{p}(\omega). \quad (37)$$

Since

$$G^{o}_{\gamma}(\omega) = -\left(\frac{1 - n_{\gamma}}{\omega - \varepsilon_{\gamma} + i\eta} + \frac{n_{\gamma}}{\omega - \varepsilon_{\gamma} - i\eta}\right) \quad (38)$$

by means of the residue theorem we obtain immediately for $p' \neq p$

$$\begin{aligned} G^{(1)}_{p'p}(t > 0) = {} & [u_{p'p} - M_{p'p}(\varepsilon_{p'})]G^{o}_{p}(\varepsilon_{p'})e^{-i\varepsilon_{p'}t} \\ & + [u_{p'p} - M_{p'p}(\varepsilon_{p})]G^{o}_{p'}(\varepsilon_{p})e^{i\varepsilon_{p}t} \\ & + R^{(1)}_{p'p} \end{aligned} \quad (39)$$

where $R^{(1)}_{p'p}$ denotes the contribution due to the poles of $M^{\pm}_{p'p}(\omega)$. Hence, if we choose

$$u_{p'p} = M_{p'p}(\varepsilon_{p}), \quad (40a)$$

Eq.(39) is simplified to

$$G^{(1)}_{p'p}(t > 0) = [u_{p'p} - M_{p'p}(\varepsilon_{p'})]G^{o}_{p}(\varepsilon_{p'})e^{-i\varepsilon_{p'}t} + R^{(1)}_{p'p} . \quad (41a)$$

Similarly, if we choose

$$u_{p'p} = M_{p'p}(\varepsilon_{p'}) \quad (40a')$$

we obtain

$$G^{(1)}_{p'p}(t > 0) = [u_{p'p} - M_{p'p}(\varepsilon_p)]G^o_{p'}(\varepsilon_p)e^{-i\varepsilon_p t} + R^{(1)}_{p'p} \qquad (41a')$$

Clearly, the two choices (40a) and (40a') are equivalent. Henceforth forth, we shall only consider the former choice. For the case $p' = p$, $\omega = \varepsilon_p - i\eta$ is a double pole. From Eq. (37) we then have

$$G^{(1)}_{pp}(t > 0) = \left\{M'_{pp}(\varepsilon_p) + it[u_{pp} - M_{pp}(\varepsilon_p)]\right\} e^{-i\varepsilon_p t} + R^{(1)}_{pp}$$

where $M'_{pp}(\varepsilon_p) = [dM_{pp}(\omega)/d\omega]_{\omega=\varepsilon_p}$. Thus, if we assume that Eq. (40a) also holds for $p' = p$, the term proportaional to t in the curved bracket will be zero and $G^{(1)}_{pp}(t > 0)$ is reduced to

$$G^{(1)}_{pp}(t > 0) = M'_{pp}(\varepsilon_p)e^{-i\varepsilon_p t} + R^{(1)}_{pp} \qquad (41b)$$

Let us further consider $G^{(2)}_{p'p}(t > 0)$. From Eq. (33) we have

$$G^{(2)}_{p'p}(t > 0) = \frac{1}{2\pi i}\int d\omega e^{-i\omega t} \sum_\gamma G^o_{p'}(\omega)[u_{p'\gamma} - M_{p'\gamma}(\omega)] \times G^o_\gamma(\omega)[u_{\gamma p} - M_{\gamma p}(\omega)]G^o_p(\omega) \qquad (42)$$

Clearly, only if γ refers to a particle state, will the pole of $G^o_\gamma(\omega)$ contribute to the above integral. In case $\underset{\sim}{u}$ is chosen according to Eq. (40a), this contribution is obviously zero when $\gamma \neq p'$ or p. It is not difficult to see that for $p' \neq p$, the contribution from the double pole ε_p contained in $[G^o_p(\omega)]^2$ due to $\gamma = p$ is also zero. However, the term contributed by the simple pole of $G^o_p(\omega)$ in case $\gamma \neq p$ will only then become zero, if we further require

$$u_{hp} = M_{hp}(\varepsilon_p) \qquad (40b)$$

Eqs. (40a) and (40b) are just the part of Eq. (2) when we set $\beta = p$. This says if $\underset{\sim}{u}$ is defined by Eq. (2), the only non-zero contributions to Eq. (42) in case $p' = p$ are those contributed by (i) the poles of $M^+(\omega)$, and (ii) the poles of order n contained in $[G^o_{p'}(\omega)]^n$, where n = 1 and 2 according as $\gamma \neq p'$ and $\gamma = p'$. Thus from Eq. (42) we obtain

$$G^{(2)}_{p'p}(t > 0) = \Big\{M'_{p'p'}(\varepsilon_{p'})[u_{p'p} - M_{p'p}(\varepsilon_{p'})]G^{o}_{p}(\varepsilon_{p'}) + \sum_{\gamma(\neq p')}[u_{p'\gamma} - M_{p'\gamma}(\varepsilon_{p'})]G^{o}_{\gamma}(\varepsilon_{p'})[u_{\gamma p} - M_{\gamma p}(\varepsilon_{p'})]G^{o}_{p}(\varepsilon_{p'})\Big\}e^{-i\varepsilon_{p'}t} + R^{(2)}_{p'p} \tag{43}$$

where $R^{(2)}_{p'p}$ again denotes the terms contributed by the poles of $M^{+}(\varepsilon)$. For $p' = p$, the sources of non-zero contribution are again the above two, except that n of the second source should now be changed to $n = 2$ and 3 according to as $\gamma \neq p$ and $\gamma = p$, respectively. The unpleasant terms proportional to t and t^2 coming from these poles of order 2 and 3 vanish too. One sees easily that the above argument also applies to all the higher order terms i.e., one has the following result: if $\underset{\sim}{u}$ is given by Eq.(2), $G_{p'p}(t > 0)$ can be written exactly in the form

$$G_{p'p}(t > 0) = A_{p'p}e^{-i\varepsilon_{p'}t} + R^{(+)}_{p'p} \tag{44}$$

where $A_{p'p}$ is independent of t and $R^{(+)}_{p'p}$ contains only those terms which are contributed by the poles of $M^{+}(\omega)$. $A_{p'p}$ is still an infinite series. We have shown[7] that this series can be summed and expressed as

$$A_{pp} = 1 - M'_{pp}(\varepsilon_p),$$

$$A_{p'p} = M_{p'p}(\varepsilon_{p'})G^{o}_{p}(\varepsilon_{p'}) \qquad (p' \neq p) \quad . \tag{45}$$

However, we shall not repeat this argument here, as a simpler derivation will be given below.

Since for $G_{h'h}(t < 0)$ we have

$$G_{h'h}(t < 0) = \frac{1}{2\pi i}\int d\omega e^{-i\omega t}G_{h'h}(\omega) \tag{46}$$

substituting Eq.(33) into the above equation and following ths same argument leading to Eq.(44), we obtain easily the following result: if $\underset{\sim}{u}$ is giben by

$$u_{\alpha h} = M_{\alpha h}(\varepsilon_h) \tag{40c}$$

then $G_{h'h}(t < 0)$ is reduced to

$$G_{h'h}(t < 0) = - \left\{ A_{h'h} e^{-i\varepsilon_{h'} t} + R^{(-)}_{h'h} \right\} \quad (47)$$

where

$$A_{hh} = 1 - M'_{hh}(\varepsilon_h) ,$$

$$A_{h'h} = M_{h'h}(\varepsilon_{h'}) G^o_h(\varepsilon_{h'}) \qquad (h' \neq h) \quad (48)$$

and $R^{(-)}_{h'h}$ only contains terms contributed by the poles of $M^-(\omega)$. From Eq.(40) one sees that $\underset{\sim}{u}$ is completely determined by proposition A.

In order to gain a deeper insight into the property of the rest terms $R_{\alpha\beta}$, let us now consider a non-perturbative derivation of the above results. As is wellknown, between $G_{\alpha\beta}(\omega)$ and the reducible mass operator $M_{\alpha\beta}(\omega)$ there exists the relation

$$G_{\alpha\beta}(\omega) = \delta_{\alpha\beta} G^o_\alpha(\omega) + G^o_\alpha(\omega) M_{\alpha\beta}(\omega) G^o_\beta(\omega) \quad (49a)$$

or

$$M_{\alpha\beta}(\omega) = G^o_\alpha(\omega)^{-1} G_{\alpha\beta}(\omega) G^o_\beta(\omega)^{-1} - \delta_{\alpha\beta} G^o_\beta(\omega)^{-1} . \quad (49b)$$

From Eqs.(24) and (38) we note that the above equation implies

$$M_{\alpha\alpha}(\varepsilon_\alpha) = 0 \quad (50)$$

If ε_α does not satisfy Eq.(1), i.e. if it is not equal to a pole of $G_{\alpha\beta}(\omega)$, Eq.(49) further gives us

$$1 - M'_{\alpha\alpha}(\varepsilon_\alpha) = M_{\alpha\beta}(\varepsilon_\alpha) = M_{\beta\alpha}(\varepsilon_\alpha) = 0 \quad (51)$$

This shows that for any well defined sp potential we have the following criterion: if $1 - M'_{\alpha\alpha}(\varepsilon_\alpha) \neq 0$, or if there exists at least a $\beta(\neq \alpha)$ such that either $M_{\alpha\beta}(\varepsilon_\alpha) \neq 0$ or $M_{\beta\alpha}(\varepsilon_\alpha) \neq 0$, i.e. if any one of the above conditions holds, then ε_α satisfies Eq.(1)[5].

In case ε_p or ε_h does satisfy Eq.(1), from Eqs.(24), (38) and (49) one easily finds

$$M'_{pp}(\varepsilon_p) = 1 - g^{(+)}(n_p),\quad M'_{hh}(\varepsilon_h) = 1 - g^{(-)}_{hh}(n_h) \;; \tag{52a}$$

$$\left.\begin{aligned} M_{p\alpha}(\varepsilon_p) &= g^{(+)}_{p\alpha}(n_p)G^o_\alpha(\varepsilon_p)^{-1} \\ M_{\alpha p}(\varepsilon_p) &= G^o_\alpha(\varepsilon_p)^{-1}g^{(+)}_{\alpha p}(n_p) \end{aligned}\right\}\ (\alpha \neq p)$$

$$\left.\begin{aligned} M_{h\alpha}(\varepsilon_h) &= g^{(-)}_{h\alpha}(n_h)G^o_\alpha(\varepsilon_h)^{-1} \\ M_{\alpha h}(\varepsilon_h) &= G^o_\alpha(\varepsilon_h)^{-1}g^{(-)}_{\alpha h}(n_h) \end{aligned}\right\}\ \begin{matrix}(\alpha \neq h)\\ (\ \neq h)\end{matrix}$$

It is evident that Eq.(52) can also be derived from the following wellknown relations

$$G_{\alpha\beta}(t > 0) = \frac{1}{2\pi i}\int d\omega e^{-i\omega t}\left[\delta_{\alpha\beta}G^o_\alpha(\omega) + G^o_\alpha(\omega)M_{\alpha\beta}(\omega)G^o_\beta(\omega)\right]$$

$$= \sum_n g^{(+)}_{\alpha\beta}(n)\ \exp(-i\mathcal{E}^+_n t) \tag{53a}$$

$$G_{\alpha\beta}(t < 0) = \frac{1}{2\pi i}\int d\omega e^{-i\omega t}\left[\delta_{\alpha\beta}G^o_\alpha(\omega) + G^o_\alpha(\omega)M_{\alpha\beta}(\omega)G^o_\beta(\omega)\right]$$

$$= -\sum_n g^{(-)}_{\alpha\beta}(n)\ \exp(i\mathcal{E}^-_n t) \tag{53b}$$

Now let us consider $G_{p'p}(t > 0)$ again. For $\underset{\sim}{u}$ given by Eq.(2) we have already proved that ε_γ satisfies Eq.(1). Therefore, according to Eq.(53) and our definition of sp and collective states of nucleus A+1 we may write

$$G_{pp}(t > 0) = A_{pp}e^{-i\varepsilon_p t} + R^{(+)}_{pp}$$

$$R^{(+)}_{pp} = \sum_{s(\neq p)} g^{(+)}_{pp}(n_s)e^{-i\varepsilon_s t} + \sum_c g^{(+)}_{pp}(n_c)e^{-i\mathcal{E}^+_c t} \tag{54a}$$

and for $p' \neq p$

$$G_{p'p}(t>0) = A_{p'p}e^{-i\varepsilon_{p'}t} + B_{p'p}e^{-i\varepsilon_p t} + R^{(+)}_{p'p}$$
$$R^{(+)}_{p'p} = \sum_{s(\neq p',p)} g^{(+)}_{p'p}(n_s)e^{-i\varepsilon_s t} + \sum_c g^{(+)}_{p'p}(n_c)e^{-i\mathcal{E}^+_c t} \tag{54b}$$

where the subscripts s and c refer to sp and collective states, respectively. From Eq.(52) we obtain that $A_{p'p}$ is indeed given by Eq.(45) and

$$B_{p'p} = g^{(+)}_{p'p}(n_p) = G^o_{p'}(\varepsilon_p)M_{p'p}(\varepsilon_p) , \quad (p' \neq p) \tag{55}$$

According to Eqs.(33) and (49a) we have

$$M_{\alpha\beta}(\omega) = [u - M(\omega)]_{\alpha\beta} + \sum_\gamma [u - M(\omega)]_{\alpha\gamma}G^o_\gamma(\omega)[u - M(\omega)]_{\gamma\beta}$$
$$+ \sum_{\gamma\delta} [u - M(\omega)]_{\alpha\gamma}G^o_\gamma(\omega)[u - M(\omega)]_{\gamma\delta}G^o_\delta(\omega) \tag{56}$$
$$\times [u - M(\omega)]_{\delta\beta} + \ldots$$

Since $[u - M(\varepsilon_\beta)]_{\alpha\beta} = 0$ and

$$\lim_{\omega\to\varepsilon_\beta} [u - M(\omega)]_{\alpha\beta}G^o_\beta(\omega) = M'_{\alpha\beta}(\varepsilon_\beta) \tag{57}$$

Substituting $\omega = \varepsilon_\beta$ into Eq.(56) we observe that each term on its right hand side is equal to zero. Thus

$$M_{\alpha\beta}(\varepsilon_\beta) = 0 \tag{58}$$

which together with Eq.(55) gives us

$$B_{p'p}(\varepsilon_p) = 0 , \quad (p' \neq p) \tag{59}$$

Furthermore we would like to point out that it is worthwhile to derive the above result from a somewhat different argument. It is wellknown that $M_{\alpha\beta}(\omega)$ satisfies

$$M_{\alpha\beta}(\omega) = [u - M(\omega)]_{\alpha\beta} + \sum_{\gamma} [u - M(\omega)]_{\alpha\gamma} G^{o}_{\gamma}(\omega) M_{\gamma\beta}(\omega) \tag{60a}$$

$$= [u - M(\omega)]_{\alpha\beta} + \sum_{\gamma} M_{\alpha\gamma}(\omega) G^{o}_{\gamma}(\omega) [u - M(\omega)]_{\gamma\beta} \tag{60b}$$

From Eq. (8) one easily finds that $M_{\alpha\beta}(\omega)$ cannot have discrete poles at

$$\omega = \pm\mathcal{E}^{\pm}_{n} = \pm [E_{n}(A \pm 1) - E_{o}(A)]$$

Since ε_{γ} satisfies Eq. (1), this says that all $M_{\alpha\beta}(\varepsilon_{\gamma})$ are well-defined. According to Eq. (49) ε_{α} and ε_{β} are regular points of $M_{\alpha\beta}(\omega)$, however ε_{γ} $(\gamma \neq \alpha, \beta)$ may still be one of its poles. Therefore, it is not advantageous to use Eq. (60b) to prove Eq. (58) before we can make sure that $M_{\alpha\beta}(\varepsilon_{\beta})$ are all finite -- this is indeed a conclusion following from Eq. (58). Substituting $\omega = \varepsilon_{\beta}$ into Eq. (60a), we obtain

$$\sum_{\gamma(\neq\beta)} \left\{\delta_{\alpha\gamma} - [u - M(\varepsilon_{\beta})]_{\alpha\gamma} G^{o}_{\gamma}(\varepsilon_{\beta})\right\} M_{\gamma\beta}(\varepsilon_{\beta}) = 0, \quad (\alpha \neq \beta), \tag{61}$$

where use has been made of Eqs. (50) and (58). Clearly, Eq. (58) is a solution to Eq. (61). Let the order of its coefficient matrix $D(\varepsilon_{\beta})$ be denoted by N - 1, N may tend to ∞. If the rank of $D(\varepsilon_{\beta})$ is N - 1, i.e., if its determinant $|D(\varepsilon_{\beta})| \neq 0$, then Eq. (58) is the only solution to Eq. (61). However, if the rank is N - f, then Eq. (61) will possess f linearly independent solutions, among which one is again Eq. (58) while the other f - 1 are non-zero. According to Eqs. (24), (49) and (52) this means that ε_{β} is f-fold degenerate, i.e., there are f linearly independent eigenstates $\Psi_{\beta i}(A \pm 1)$ $(i = 1, 2, \ldots, f)$ which correspond to the same eigenvalue: $E_{\beta}(A \pm 1) = E_{o}(A) \pm \varepsilon_{\beta}$. Therefore even if ε_{β} is degenerate, among the linear combinations of $\Psi_{\beta i}(A \pm 1)$ there is one and only one eigenstate

$$\Psi_{\beta}(A \pm 1) = \sum_{i} c_{i} \Psi_{\beta i}(A \pm 1)$$

which will make Eq. (58) hold. Let $\beta = p(h)$ refer to a particle (hole) state. From Eqs. (26), (52) and (58) we obtain

$$\left.\begin{aligned} g^{(+)}_{\alpha p}(n_{p}) &= \langle\Psi_{o}|\bar{\xi}_{\alpha}|\Psi_{p}(A+1)\rangle\langle\Psi_{p}(A+1)|\xi^{+}_{p}|\Psi_{o}\rangle = 0, \quad (\alpha\neq p) \\ g^{(-)}_{\alpha h}(n_{h}) &= \langle\Psi_{o}|\xi^{+}_{h}|\Psi_{h}(A-1)\rangle\langle\Psi_{h}(A-1)|\bar{\xi}_{\alpha}|\Psi_{o}\rangle = 0, \quad (\alpha\neq h) \end{aligned}\right\} \tag{62}$$

Cleary, $<\Psi_p(A+1)|\xi_p^+|\Psi_o>$ and $<\Psi_o|\xi_h^+|\Psi_h(A-1)>$ cannot be equal to zero (cf. next section). Thus, Eq.(62) tells us

$$\left.\begin{aligned} <\Psi_o|\bar{\xi}_\alpha|\Psi_p(A+1)> &= \delta_{\alpha p}<\Psi_o|\bar{\xi}_p|\Psi_p(A+1)> \\ <\Psi_h(A-1)|\bar{\xi}_\alpha|\Psi_o> &= \delta_{\alpha h}<\Psi_h(A-1)|\bar{\xi}_h|\Psi_o> \end{aligned}\right\} \quad (63)$$

Accroding to our definition, $\Psi_p(A+1)[\Psi_h(A-1)]$ have been referred to as the sp (sh) states of nucleus A + 1 [A - 1]. We see that Eq.(63) specifies them unambiguously. By means of Eqs.(26) and (63) we conclude

$$g_{p'p}^{(+)}(n_s) = 0 \quad (64)$$

if $s \neq p'$. Eq.(64) holds for both $p' = p$ and $p' \neq p$. In the following the double indices α, β will always mean that α may $=\beta$ and $\neq\beta$, if not stated otherwise. Comparing Eq.(64) with (54), one sees that the choice of $\tilde{u}$ given in Eq.(2) makes not only $B_{p'p}$, but also every term in the summation over s contained in the rest term $R_{p'p}^{(+)}$ equal to zero. That is, if $\tilde{u}$ is chosen according to Eq.(2), $G_{p'p}(t > 0)$ will be reduced to the simple form

$$G_{p'p}(t > 0) = A_{p'p}e^{-i\varepsilon_{p'}t} + R_{p'p}^{(+)} \quad (65a)$$

$$R_{p'p}^{(+)} = \sum_c g_{p'p}^{(+)}(n_c)\ \exp(-i\,\mathcal{E}_c^+\ t) \quad (65b)$$

where $\sum_c$ only sums over the collective states of nucleus A + 1. Obviously, the above argument also applies to the other components of $G_{\alpha\beta}(t)$. From Eqs.(52), (53) and (63) one derives easily the following result:

$$\left.\begin{aligned} G_{p\beta}(t > 0) &= A_{p\beta}e^{-i\varepsilon_p t} + R_{p\beta}^{(+)} \\ G_{p\beta}(t < 0) &= -R_{p\beta}^{(-)}\ ; \end{aligned}\right. \quad (66a)$$

$$\left.\begin{aligned} G_{h\beta}(t > 0) &= R_{h\beta}^{(+)} \\ G_{h\beta}(t < 0) &= -[A_{h\beta}e^{-i\varepsilon_h t} + R_{h\beta}^{(-)}]\ . \end{aligned}\right. \quad (66b)$$

where β may refer to either a particle or a hole state, the expressions for A_{pp} and A_{hh} have already been given in Eqs.(45) and (48) respectively, while

$$A_{p\beta} = g^{(+)}_{p\beta}(n_p) = M_{p\beta}(\varepsilon_p)G^o_\beta(\varepsilon_p), \quad (p \neq \beta) \tag{67a}$$

$$A_{h\beta} = g^{(-)}_{h\beta}(n_p) = M_{h\beta}(\varepsilon_h)G^o_\beta(\varepsilon_h), \quad (h \neq \beta) \tag{67b}$$

and the rest terms $R^{(\pm)}_{\alpha\beta}$ can be written as

$$R^{(+)}_{\alpha\beta} = \sum_c g^{(+)}_{\alpha\beta}(n_c) \exp(-i \mathscr{E}^+_c t) , \tag{68a}$$

$$R^{(-)}_{\alpha\beta} = \sum_c g^{(-)}_{\alpha\beta}(n_c) \exp(i \mathscr{E}^-_c t) . \tag{68b}$$

Al the above results are exact. Furthermore, Eq.(63) suggests that the sp Green function will have an especially simple form, if it is defined as

$$\bar{G}_{\alpha\beta}(t = t_1-t_2) = <\Psi_o|T\left\{\bar{\xi}_{\alpha}(t_1)\bar{\xi}^+_{\beta}(t_2)\right\}|\Psi_o> . \tag{69}$$

Since in this case

$$\bar{g}^{(+)}_{\alpha\beta}(n) = <\Psi_o|\bar{\xi}_\alpha|\Psi_n(A + 1)><\Psi_n(A + 1)|\bar{\xi}^+_\beta|\Psi_o>$$

$$\bar{g}^{(-)}_{\alpha\beta}(n) = <\Psi_o|\bar{\xi}^+_\beta|\Psi_n(A - 1)><\Psi_n(A - 1)|\bar{\xi}_\alpha|\Psi_o>$$

and obviously

$$<\Psi_n(A + 1)|\bar{\xi}^+_\beta|\Psi_o>^* = <\Psi_o|\bar{\xi}_\beta|\Psi_n(A + 1)>$$

$$<\Psi_o|\bar{\xi}^+_\beta|\Psi_n(A - 1)>^* = <\Psi_n(A - 1)|\bar{\xi}_\beta|\Psi_o>$$

Eq.(63) implies

$$\bar{A}_{p\beta} = \bar{g}^{(+)}_{p\beta}(n_p) = 0 , \quad (p \neq \beta)$$

$$\bar{A}_{h\beta} = \bar{g}^{(+)}_{h\beta}(n_h) = 0, \quad (h \neq \beta).$$

Therefore, $\bar{G}_{\alpha\beta}(t)$ will have the same form as that given by Eq.(66), but with the additional simplification, namely, $\bar{A}_{p\beta} = \bar{A}_{pp}\delta_{p\beta}$ and $\bar{A}_{h\beta} = \bar{A}_{hh}\delta_{h\beta}$. The relation between $G_{\alpha\beta}(t)$ and $\bar{G}_{\alpha\beta}(t)$ is clear, because

$$\bar{\xi}^+_\beta = \sum_\gamma \xi^+_\gamma <\bar{\gamma}|\bar{\beta}> , \qquad \xi^+_\beta = \sum_\gamma \bar{\xi}^+_\gamma <\gamma|\beta>$$

Since the anticommutation relation between $\bar{\xi}_\alpha$ and $\bar{\xi}^+_\beta$ is more complicated than that between $\bar{\xi}_\alpha$ and $\bar{\xi}_\beta$, we see that although the form of $\bar{G}_{\alpha\beta}(t)$ is simpler, its calculation is more complicated. This says that $G_{\alpha\beta}(t)$ and $\bar{G}_{\alpha\beta}(t)$ are in essence equivalent.

Now let us consider whether the rest terms $R_{\alpha\beta}$ can be neglected or not. Since

$$g^{(+)}_{\alpha\beta}(n_c) = <\Psi_o|\bar{\xi}_\alpha|\Psi_c(A+1)><\Psi_c(A+1)|\xi^+_\beta|\Psi_o>$$

$$g^{(-)}_{\alpha\beta}(n_c) = <\Psi_o|\xi^+_\beta|\Psi_c(A-1)><\Psi_c(A-1)|\xi^-_\alpha|\Psi_o>$$

and $\Psi_c(A+1)[\Psi_c(A-1)]$ is a collective state of nucleus A+1[A-1], we expect that $|g^{(+)}_{\alpha\beta}(n_c)|\{|g^{(+)}_{\alpha\beta}(n_c)|\}$ compared with $|g^{(+)}_{pp}(n_p)|\{|g^{(-)}_{hh}(n_h)|\}$ are all small quantities of second order. Clearly, even if this estimation is correct, it still does not imply that $R^{(+)}_{\alpha\beta}$ $(R^{(-)}_{\alpha\beta})$ can be neglected as it contains infinitely many such second order terms. According to the anticommutation relation we have

$$\delta_{\alpha\beta} = <\Psi_o|\bar{\xi}_\alpha\xi^+_\beta + \xi^+_\beta\bar{\xi}_\alpha|\Psi_o>$$

From the above relation and Eq.(63) we find

$$\delta_{p\beta} - A_{p\beta} = \delta_{p\beta} - g^{(+)}_{p\beta}(n_p) = \sum_c \left\{g^{(+)}_{p\beta}(n_c) + g^{(-)}_{p\beta}(n_c)\right\}, \quad (70a)$$

$$\delta_{h\beta} - A_{h\beta} = \delta_{h\beta} - g^{(-)}_{h\beta}(n_h) = \sum_c \left\{g^{(+)}_{h\beta}(n_c) + g^{(+)}_{h\beta}(n_c)\right\}. \quad (70b)$$

From the above relation and Eq.(63) we find

$$|\delta_{p\beta} - A_{p\beta}| << |g^{(+)}_{pp}(n_p)| = |A_{pp}| ,$$

$$|\delta_{h} - A_{h\beta}| << |g^{(-)}_{hh}(n_h)| = |A_{hh}| . \tag{71}$$

The above inequality requires that the non-diagonal elements $|A_{\alpha\beta}|$ $(\alpha \neq \beta)$ should be much smaller than $|A_{\alpha\alpha}|$. This implies at the same time that we should have

$$|<\alpha|\beta>| (\alpha \neq \beta) << |<\alpha|\alpha>| ,$$

i.e., the eigenvectors $|\alpha>$ obtained from Eq.(3) should form an approximately orthogonal set. If condition (71) is realized, we conjecture that the following approximation holds:

$$G_{\alpha\beta}(t) \simeq \delta_{\alpha\beta}[\theta(t) - n_\alpha]A_{\alpha\alpha}e^{-i\varepsilon_\alpha t} . \tag{72}$$

In order to prove conclusion B, we still have to generalize Eq.(63) to the continuous spectrum. This can be done by making use of the artifice of box normalization. Let us imagine that the whole nucleus A+1 is included in a large sphere of radius R centred at its center of mass*. For finite R we have Eq.(63a). As $R \to \infty$, we obtair

$$<\Psi|\bar{\xi}_\alpha|\Psi \mathscr{E}_{\Lambda_s}(A + 1)> = \delta(\varepsilon_\alpha - \mathscr{E})\delta_{\Lambda_\alpha,\Lambda_s} C_\alpha(A + 1) \tag{73}$$

where we have used the notation $\alpha \equiv (\varepsilon_\alpha, \Lambda_\alpha)$. Clearly, concludion B follows from Eqs.(63) and (73), Q.E.D. Let us substitute Eq.(73) into Eqs.(30) and (31). For the sp channel we further have

$$g^{(+)s}_{\alpha\beta}(\mathscr{E}) = \delta(\varepsilon_\alpha - \mathscr{E})\hat{g}^{(+)s}_{\alpha\beta}(\varepsilon_\alpha)$$

$$F^{(+)s}_{\alpha\beta}(\omega) = - \frac{g^{(+)s}_{\alpha\beta}(\varepsilon_\alpha)}{\omega - \varepsilon_\alpha + i\eta}$$

V. AMPLITUDE RENORMALIZATION FACTOR OF SP GREEN'S FUNCTION

In this section we shall discuss how to calculate $A_{\alpha\beta}$. According to Eq.(67), it is equivalent to ask how to calculate the following quantities:

*Appropriate boundary conditions are assumed.

V. AMPLITUDE RENORMALIZATION FACTOR OF SP GREEN'S FUNCTION

In this section we shall discuss how to calculate A . According to Eq.(67), it is equivalent to ask how to calculate the following quantities:

$$1 - M'_{\alpha\alpha}(\varepsilon_\alpha)\ , \quad M_{\alpha\beta}(\varepsilon_\alpha) \quad \text{and} \quad M_{\alpha\beta}(\varepsilon_\beta) \tag{76}$$

This fact makes the calculation of $A_{\alpha\beta}$ quite simple, because we can now make use of Eq.(60). First we would like to point out that Eq.(60) can be reduced to a simpler form by an appropriate transformation. Let us define

$$\hat{M}^{\boldsymbol{\cdot}}_{\alpha\beta}(\omega) = [1 - \hat{M}_{\alpha\alpha}(\omega)G^{o}_{\alpha}(\omega)]M_{\alpha\beta}(\omega) \tag{77}$$

Multiplying Eq.(60b) from the left by $1 - \hat{M}_{\alpha\alpha}(\omega)G^{o}_{\alpha}(\omega)$, we obtain

$$\hat{M}_{\alpha\beta}(\omega) = [u - M(\omega)]_{\alpha\beta} + \sum_{\gamma(\neq\alpha)} \hat{M}_{\alpha\gamma}(\omega)G^{o}_{\gamma}(\omega)[u - M(\omega)]_{\gamma\beta}\ , \quad (\alpha \neq \beta) \tag{78a}$$

and

$$\hat{M}_{\alpha\alpha}(\omega) = [u - M(\omega)]_{\alpha\alpha} + \sum_{\gamma(\neq\alpha)} \hat{M}_{\alpha\gamma}(\omega)G^{o}_{\gamma}(\omega)[u - M(\omega)]_{\gamma\alpha} \tag{78b}$$

We see that Eq.(78a) determines $\hat{M}_{\alpha\beta}(\omega)$ $(\alpha \neq \beta)$ completely. After we have solved Eq.(78a) for $\hat{M}_{\alpha\beta}(\omega)$, $\hat{M}_{\alpha\alpha}(\omega)$ can be calculated directly by means of Eq.(78b). Similarly, if we define

$$\check{M}_{\alpha\beta}(\omega) = M_{\alpha\beta}(\omega)[1 - G^{o}_{\beta}(\omega)\check{M}_{\beta\beta}(\omega)] \tag{79}$$

we get from Eq.(60a)

$$M_{\alpha\beta}(\omega) = [u - M(\omega)]_{\alpha\beta} + \sum_{\gamma(\neq\beta)} [u - M(\omega)]_{\alpha\gamma}G^{o}_{\gamma}(\omega)\check{M}_{\gamma\beta}(\omega)\ , \quad (\alpha \neq \beta) \tag{80a}$$

$$\check{M}_{\alpha\alpha}(\omega) = [u - M(\omega)]_{\alpha\alpha} + \sum_{\gamma(\neq\alpha)} [u -$$

$$+ \sum_{\gamma(\neq\alpha)} [u - M(\omega)]_{\alpha\gamma} G^{o}_{\gamma}(\omega) \check{M}_{\gamma\alpha}(\omega) \quad (80b)$$

Eqs.(78) and (80) are clearly equivalent. In the following we shall only use Eq.(78). Since a general discussion which also implies to other sp potentials has already been given elsewhere , here only the case where u is given by Eq.(2) will be considered. In this case we know that $M_{\alpha\beta}(\varepsilon_\beta) = 0$. Further, for simplicity, we shall only consider the case of discrete spectrum. Substituting $\omega = \varepsilon_\alpha$ into Eq.(78a), we obtain

$$\sum_{\gamma(\neq\alpha)} x_{\alpha\gamma} \Big\{ \delta_{\alpha\beta} - G^{o}_{\gamma}(\varepsilon_\alpha) \times [u - M(\varepsilon_\alpha)]_{\alpha\beta} \Big\} = [u - M(\varepsilon_\alpha)]_{\alpha\beta} , \quad (\alpha \neq \beta) \quad (81)$$

where we have set

$$x_{\alpha\gamma} = M_{\alpha\gamma}(\varepsilon_\alpha) \quad (82)$$

It is not difficult to see that the rank Ω of the coefficient matrix of Eq.(81) is the same as that of Eq.(61), i.e., we also have $\Omega = N - f$. Moreover, one can show by means of the eigenvalue equation derived from Eqs.(17) and (24) that the rank of the augmented matrix of Eq.(81) by including its non-homogeneous terms is again Ω. This says that Eq.(81) always possesses solutions. In case f = 1, its solution is unique. For the case f > 1, we may introduce $\tilde{V} = \tilde{V} + \lambda\tilde{W}$ and choose $\tilde{W}$ so that the accidental degeneracy of $\varepsilon_{\tilde{\alpha}}$ is lifted. Thus even in the latter case, the solution of Eq.(81) can be determined uniquely by

$$x_{\alpha\gamma} = \lim_{\lambda\to 0} x_{\alpha\gamma}(\lambda)$$

From Eqs.(2) and (78b) we have

$$M_{\alpha\alpha}(\varepsilon_\alpha) = 0 \quad (83)$$

which gives us

$$\lim_{\omega \to \varepsilon_\alpha} \hat{M}_{\alpha\alpha}(\omega) G^{o}_{\alpha}(\omega) = - \hat{M}'_{\alpha\alpha}(\varepsilon_\alpha) \tag{85}$$

Multiplying Eq.(78b) by $G^o_\alpha(\omega)$ and using Eqs.(57) and (84), we obtain

$$1 + \hat{M}'_{\alpha\alpha}(\varepsilon_\alpha) = 1 - M'_{\alpha\alpha}(\varepsilon_\alpha) - \sum_{\gamma(\neq\alpha)} x_{\alpha\gamma} G^o_\gamma(\varepsilon_\alpha) M'_{\gamma\alpha}(\varepsilon_\alpha) \tag{85}$$

Clearly, according to Eq.(50) we also have

$$\lim_{\omega \to \varepsilon_\alpha} M_{\alpha\alpha}(\omega) G^o_\alpha(\omega) = - M'_{\alpha\alpha}(\varepsilon_\alpha) \tag{86}$$

Thus, Eq.(77) implies

$$1 - M'_{\alpha\alpha}(\varepsilon_\alpha) = [1 + \hat{M}'_{\alpha\alpha}(\varepsilon_\alpha)]^{-1} \tag{87a}$$

$$M_{\alpha\beta}(\varepsilon_\alpha) = \hat{M}_{\alpha\beta}(\varepsilon_\alpha)[1 + \hat{M}'_{\alpha\alpha}(\varepsilon_\alpha)]^{-1} \tag{87b}$$

From Eqs.(7) and (85) we obtain

$$\begin{aligned} A_{\alpha\alpha} &= 1 - M'_{\alpha\alpha}(\varepsilon_\alpha) \\ &= \left\{1 - M'_{\alpha\alpha}(\varepsilon_\alpha) - \sum_{\gamma(\neq\alpha)} x_{\alpha\gamma} G^o_\gamma(\varepsilon_\alpha) M'_{\gamma\alpha}(\varepsilon_\alpha)\right\}^{-1} \end{aligned} \tag{88}$$

wehich reduced to the wellknown Brandow formula

$$A_{\alpha\alpha} \simeq [1 - M'_{\alpha\alpha}(\varepsilon_\alpha)]^{-1} \tag{89}$$

if one neglects the non-diagonal elements and introduces the following approximation for $M_{\alpha\alpha}(\omega)$:

$$M_{\alpha\alpha}(\omega) \simeq \sum_h G_B(\alpha h, \alpha h; \omega + \varepsilon_h) A_{hh} ,$$

where G_B is Brueckner's G-matrix.

Finally we would like to show that $A_{\alpha\alpha} \neq 0$. From Eqs.(49)

and (77) one easily finds

$$M_{\alpha\alpha}(\omega) = G_{\alpha}^{o}(\omega)^{-1}G_{\alpha\alpha}(\omega)G_{\alpha}^{o}(\omega)^{-1} - G_{\alpha}^{o}(\omega)^{-1}$$

$$= \hat{M}_{\alpha\alpha}(\omega)G_{\alpha\alpha}(\omega)G_{\alpha}^{o}(\omega)^{-1} \tag{90}$$

or

$$\hat{M}_{\alpha\alpha}(\omega)G_{\alpha\alpha}(\omega) = G_{\alpha}^{o}(\omega)^{-1}G_{\alpha\alpha}(\omega) - 1 \tag{91}$$

We note $G_{\alpha}(\varepsilon_{\alpha})^{-1} = 0$. Therefore, if $G_{\alpha\alpha}(\varepsilon_{\alpha}) \neq \infty$, we shall have

$$\lim_{\omega \to \varepsilon_{\alpha}} \hat{M}_{\alpha\alpha}(\omega)G_{\alpha\alpha}(\omega) = -1 \tag{92}$$

According to Eq.(83), this is impossible. We conclude that $G_{\alpha\alpha}(\varepsilon_{\alpha})$ $G_{\alpha\alpha}(\varepsilon_{\alpha}) = \infty$. This implies $A_{\alpha\alpha} \neq 0$. Hence, according to Eq.(26) we necessarily have Eq.(63).

VI. CONCLUSION

We have investigated the non-hermitian sp potential $\underset{\sim}{u}$ defined by Eq.(2) in some detail. Conclusions A and B as well as Eqs.(7), (63), (66) and (68) are all general and exact. They show that with respect to the energy spectrum, the cancellation property and the relation to the sp (sh) states of nucleus A+1 (A-1), $\underset{\sim}{u}$ may be regarded as the best choice among sp potentials. However, in order to understand the consequence of $\underset{\sim}{u}$ more thoroughly, further works, especially numerical calculations, still have to be done. We note that there are also results which will depend on the explicit form of $\underset{\sim}{V}$ and the ground state property of the relevant closed shell nucleus A (which may also be referred to as the shell dependence and will be designated simply by Ψ_o). One of these is the question concerning whether the rest terms $R_{\alpha\beta}$ can be neglected or not*. In order to compare the energy spectra calculated from Eq.(3) for different $\underset{\sim}{V}$ and to understand the dependence of $A_{\alpha\beta}$ on $\underset{\sim}{V}$ and Ψ_o, numerical calculations are underway. Finally we would like to remark that though $\underset{\sim}{u}$ simplifies the expression of $G_{\alpha\beta}(t)$ and may, therefore, facilitate the calculation of propagator renormalization, this clearly does not imply that the series of propagator-renormalized

*Clearly, if $R_{\alpha\beta}$ are not negligible according to Eq.(68) this means that at least the overall effect of the overlapping between $\xi_{\alpha}^{+}(\bar{\xi}_{\alpha}^{+})|\Psi_o\rangle$ $[\xi_{\alpha}(\xi_{\alpha})\Psi_o\rangle]$ and the collective states $\Psi_c(A+1)[\Psi_c(A-1)]$ is significant.

skeleton diagrams of a mp-nh Green's function will be made convergent or its speed of convergence will be enhanced in this way. The latter is still a question which should be studied further and is also being studied.

REFERENCES

1. R.W. Jones and F. Mohling, Nucl. Phys. A151 (1970) 420; R.W. Jones, F. Mohling and R.L. Becker, Nucl. Phys. A220 (1974) 45.
2. R.L. Becker and R.W. Jones, Nucl. Phys. A174 (1971) 449.
3. B.H. Brandow, Phys. Rev. 152 (1966) 863; Ann. of Phys. 57 (1970) 214.
4. M.W. Kirson, Nucl. Phys. A115 (1968) 49; A139 (1969) 57.
5. S.S. Wu, Acta Physica Sinica 25 (1976) 433.
6. T.D. Lee and C.N. Yang, Phys. Rev. 117 (1960) 22
7. S.S. Wu, Physica Energiae Forties et Physica Nuclearis 2 (1978) 10; 3 (1979) 469.
8. N.I. Muskhelishvili, Singular Integral Equations (P. Noordhoff N.V., Groningen, 1953).
9. C.B. Dover and R.H. Lemmer, Phys. Rev. 165 (1968) 1105; F. Catara et al., Nucl. Phys. A276 (1977) 433; also ref.7.

APPLICATION OF THE BOSON-FERMION HYBRID REPRESENTATION:

WHY IS SPIN-POLARIZED ATOMIC HYDROGEN A BOSONIC SYSTEM?†

Cheng-Li Wu*, Jian-Min Yuan and Da Hsuan Feng

Department of Physics and Atmospheric Science
Drexel University
Philadelphia, PA 19104

In the previous talk by Wu[1], a formalism known as the boson-fermion hybrid representation (BFHR) was proposed. The BFHR was utilized to form a theoretical foundation of the nuclear field theory[2] (NFT). However, one must bear in mind that the interplay of bosons and fermions is by no means a mere nuclear phenomena, it is displayed in many other disciplines of physics as well. Therefore, if the BFHR is a general quantum mechanical representation, it can be utilized to study other physical systems which display this kind of intricate and fundamental phenomena. In this paper, we shall present the use of the BFHR in the case of the spin-polarized atomic hydrogen system.

The study of spin-polarized atomic hydrogen (denoted commonly in the literature as H↑), viz., atomic hydrogen with the electronic spin polarized antiparallelly to an external magnetic field of about 10 tesla, at about 0.1°K, has stimulated enormous experimental and theoretical interests recently[3-12]. It is believed[5] that since H↑ has a shallower potential well in the $b^3\Sigma_u^+$ pair potential and with lighter mass than ^{4}He, namely a higher quantum parameter $\eta=\hbar^2/(m\sigma^2\varepsilon)$ (where σ and ε are the Lennard-Jones potential parameters), it should exhibit even more pronounced quantum properties than ^{4}He. A successful recent attempt to prepare a "stable" H↑ gas with a density of about 10^{17}atoms/cm^3 lends support to the assertion that H↑ is indeed a Bose gas, even down to the absolute zero temperature[1,3]. Furthermore, it is believed, that one is within the realm

†Work partially supported by NSF under grant # Phy-7908402 and Research Corporation.

*Permanent Address: Department of Physics, Jilin University
Changchun, Jilin, The People's Republic of China.

of observing the Bose-Einstein Condensation soon. With such excitement in the field, it is all the more reason to know if there exist a (more) fundamental reason, besides the plausible and qualitative quantum parameter argeument[5], as to why H↑ could behave as a bosonic system.

It is well known that the constituents of a hydrogen atom are fermions (i.e. electrons and protons). This is very much analogous to a pion which is now believed to consist of a quark and antiquark pair. The quark (antiquark) is, of course, a fermion. Furthermore, in nuclear structure physics, the SU(6) boson model of Arima and Iachello (which was extensively discussed in this workshop) has met with exceptional success in predicting and explaining many behavior of the nucleus. Thus, this common denominator of the H↑ system, the pions, and the nucleons must stimulate one to ask the important question: Under what physical conditions can fermions pairs (or clusters) behave as a boson?

Freed[12] tried to understand, from a molecular physics viewpoint, as to why H↑ is a Bose gas. Unfortunately, his arguement is somewhat restrictive and does not provide an answer to the above raised question. On the other hand, the BFHR is a tailor-made formalism for this purpose. Briefly, in the BFHR, one begins by defining precisely the exact meaning of a boson with the context of an interacting two fermion system and then proceed to determine the probability of the boson structure in the fermion system in a well-defined interacting boson-fermion and boson-boson system. For detailed mathematical formalism, the reader is referred to ref.1.

In this report, we shall present some of our preliminary investigation of bosonic properties of the H↑ system.

Let us consider a system of two hydrogen atoms. Analogous to the nuclear case, we shall define a "frozen" hydrogen atom (in either the ground or the excited state) as a Boson. However, one must bear in mind that the H↑ system must be inherently a multi-centered system. This is not true, of course, in the nuclear structure problem where all the fermions are attracted by a common (Hartree-Fock or Shell model) potential. In this sense, the application of BFHR for the H↑ system may be regarded as an additional test for the theory. Another important difference between the H↑ and the nuclear structure problem is that the interaction between an electron and a proton is basically known, i.e., a coulomb force. This is clearly not the case in the nuclear problem where the interaction is, at best, vaguely known. In our treatment of the H↑ system, we have imposed the usual Born-Oppenheimer approximation (BOA) thereby fixing the nuclear separation in calculating the electronic properties.

To write down the BFHR Hamiltonian in the second quantized form we need to define a chosen space for the model. For simplicity, but without loosing the essential physics, we decided to take only the simplest basis states so that the problem is analytically solu able in both the conventional (i.e., fermion space) and the BFHR. The basis states which we have chosen are molecular orbitals which can be constructed from the linear combination of atomic orbitals (LCAO-MO). If we restrict ourselves to only the 1s hydrogen orbital, the basis states are

$$|\psi_1^{el}> = |e_1> = \frac{1}{\sqrt{2(1+S)}} (|H_a> + |H_b>) \tag{1}$$

$$|\psi_2^{el}> = |e_2> = \frac{1}{\sqrt{2(1+S)}} (|H_a> - |H_b>) \tag{2}$$

where $H_{a(b)}$ is the hydrogen 1s orbital located at proton a(b), S is an overlapping integral between the two 1s orbitals located each at the respective protons. In this model space, the ground state and the excited state wavefunction of the H atom (located at proton a a) is given as

$$|H_a> = \sqrt{\frac{1+S}{2}} |e_1> + \sqrt{\frac{1-S}{2}} |e_2> \tag{3}$$

$$|H_a^*> = \sqrt{\frac{1-S}{2}} |e_1> + \sqrt{\frac{1+S}{2}} |e_2> \tag{4}$$

It is very clear that $|H_a>$ and $|H_a^*>$ are orthogonal to one another. The wave function $|H_b>$ and $|H_b^*>$ ca be similarly expressed.

The fermionic Hamiltonian for this system can be written as

$$H_F = H_e^o + H_p^o + V_e + V_p + V_{ep} \tag{5}$$

where

$$H_e^o = \sum_k \varepsilon_k^e a_{e_k}^\dagger a_{e_k} \tag{6}$$

$$H_p^o = \sum_i \varepsilon_i^p a_{p_i}^\dagger a_{p_i} \tag{7}$$

where $a^\dagger(a)$ are the usual fermion creation (annihilation) operators and k(i) denotes a set of quantum numbers to fully specify the single particle state, ε_k^e is the field-free energy of the electronic state k plus its Zeeman energy in a magnetic field B and ε_i^p is the Zeeman energy associated with the proton spin state i. The terms V_e and V_p in eq.5 are, respectively, the electron-electron and proton-proton interaction terms while V_{ep} is the electron-proton interaction. In the fermion space, the electronic part of the Schrödinger equation is given by

$$H_F \Psi(\underset{\sim}{x}, R) = U(R)\Psi(\underset{\sim}{x}, R) \tag{8}$$

where $\underset{\sim}{x}$ is the electronic coordinate and R is the interproton distance which, under the BOA, is just a parameter.

In the BFHR, the Hamiltonian is written as

$$H_{BFH} = H_B^o + H_F + V_{BF} \tag{9}$$

where

$$H_B^o = \sum_\alpha \omega_\alpha B_\alpha^\dagger B_\alpha \tag{10}$$

and

$$V_{BF} = \sum_\alpha Z_\alpha (B_\alpha^\dagger A_\alpha + A_\alpha^\dagger B_\alpha) \tag{11}$$

In eq.10 and 11, $B_\alpha^\dagger(B_\alpha)$ is the creation (annihilation) operator for the "real" boson and $A_\alpha^\dagger(A_\alpha)$ are the fermion pair operators. Notice that in eqs.10 and 11, the sum is carried out for all possible kinds of bosons (ee, pp and ep). Furthermore e denotes either the $|e_1\rangle$ or $|e_2\rangle$ state. In eq.10, ω_α is the single boson energy and Z_α the boson-fermion vertices, both can be defined according to the way that similar quantities are defined in ref.1. We want to solve the eqn.

$$H_{BFH}\, \Psi_{BFH} = U(R)\Psi_{BFH} \tag{12}$$

As in ref.1, Eq.12 is solved via the standard Brillouin-Wigner perturbation scheme. Let us rewrite Eq.9 as

$$H_{BFH} = H_B^o + H_F^o + V \tag{13}$$

and

$$V = V_F + V_{BF} \tag{14}$$

where we have used the fact that $H_F = H_F^o + V_F$. V_F is the interaction between a pair of fermions and V_{BF} the boson-fermion interaction. Following the usual P and Q spaces separation and denoting the P projected wave function as ψ_B, we thus have

$$(H_B^o + V_{eff})\psi_B = U(R)\psi_B \tag{15}$$

where

$$V_{eff} = V + VDV + \ldots \tag{16}$$

$$D = (U(R) - H_B^o - H_F^o)^{-1}\,\hat{Q} \tag{17}$$

In eq.17, $\hat{Q}$ is the usual Q-space projection operator. According to the Brillouin-Wigner Scheme, the total wave function is given by

$$\Psi_{BFH} = \Psi_B + D\,V_{eff}\,\Psi_B \tag{18}$$

For H_2^+ system, we set

$$\psi_B = |B_1^{(a)}\,B_1^{(b)}> \tag{19}$$

where $B_1^{(a)}$ denotes a ground state boson corresponding to a ground state H atom located at proton a as given by eq.3. If we multiply both sides of eq.15 by ψ_B^* and integrate over the electronic coordinates, we obtain

$$U(R) = V_{eff}\,(R) \tag{20}$$

after setting the energy of the ground state boson to zero. As shown in ref.1 $V_{eff}(R)$ can be put into a closed form by summing up all the diagrams having no bubbles. On the other hand U(R) can be calculated by applying a variational method to eq.8 in the fermion space spanned by the same basis set. We have carried out both calculations and obtained exactly the same results for the ground state ($X^1\Sigma g$) of the field-free H_2 molecule. Thus we believe that BFHR can be applied to molecular problems as well. The closed form of $V_{eff}(R)$ obtained is also very useful in calculating the coefficient of a certain bosonic state in the total wave function according to eq.18. With the choice of eq.19, eq.18 becomes

$$\psi = |B_1^{(a)} B_1^{(b)}> + \sum_q C_q \phi_q \tag{21}$$

where

$$C_q = <\phi_q \left\{ D \, V_{eff} | B_1^{(a)} B_2^{(b)} > \right. \tag{22}$$

The appearance of left curly bracket in eq.22 is to signify the that an overcomplete, nonorthogonal basis set is used. Through a proper normalization procedure1, we can then obtain the bosonic probability distribution, P(R), of the spin-polarized state ($b^3\Sigma_u^+$) as a function of R. Preliminary results that we have obtained show that P(R) is a monotonically increasing function of R and becomes essentially 1 for R $\geq$ 6 bohrs.

One last comment about the H system is that the effect of a magnetic field with a strength around 10 tesla is only to shift U(R) downward in energy by a constant amount proportional to the field strength B. Since U(R) in the field has the same shape as that in the field-free case within the Born-Oppenheimer approximation the bosonic probability distribution remains the same for a magnetic-field strength up to at least 10 tesla.

ACKNOWLEDGEMENT

One of us (CLW) would like to thank the Physics Department of Drexel University for the warm hospitality extended to him during the summer of 1980. He is grateful to the National Science Foundation as well as Drexel and Vanderbilt Physics Departments for financial assistance.

REFERENCES

1. C.L. Wu, M.W. Guidry, J.Q. Chen and D.H. Feng, Preceding paper; C.L. Wu and D.H. Feng, Phys. Lett. 96B, 243 (1980); ibid., Ann. of Physics (in press).

2. P.F. Bortignon, R.A. Broglia, D.R. Bes and R. Liotta, Phys. Rev. 30C, 305 (1977) and references therein.
3. I.F. Silvera and J.T.M. Walraven, Phys. Rev. Lett. 44, 164 (1980); ibid, 45, 1268 (1980); J.T.M. Walraven, I.F. Silvera and A.P.M. Matthey, Phys. Rev. Lett. 45, 449 (1980); A.P.M. Matthey, J.T.M. Walraven and I.F. Silvera, Phys. Rev. Lett. 46, 668 (1981).
4. M. Morrow, R. Jochemsen, A.J. Berlinsky and W.N. Hardy, Phys. Rev. Lett. 46, 195 (1981).
5. S.B. Crampton, I. Greytak, D. Kleppner, W. Phillips, D.A. Smith and A. Weinrib, Phys. Rev. Lett. 42, 1039 (1980).
6. W.C. Stwalley and L.H. Nosanow, Phys. Rev. Lett. 36, 910 (1976).
7. J.M. Yuan, T.K. Lim and L.H. Nosanow, Phys. Lett. 81A, 61 (1981).
8. J.T. Jones, M.H. Johnson, H.L. Mayer, S. Katz and R.S. Wright, Aeronautronics Systems, Inc. Report No.6-216, 1958, unpublished.
9. W.J. Mullin, Phys. Rev. Lett. 44, 1420 (1980).
10. E.D. Siggia and A.E. Ruckenstein, Phys. Rev. Lett. 44, 1423 (1980).
11. R.W. Cline, D.A. Smith, T.J. Greytak and D. Kleppner, Phys. Rev. Lett. 45, 2117 (1980).
12. J.H. Freed, J. Chem. Phys. 72, 1414 (1980).

THE MICROSCOPIC MECHANISM OF THE (p, π^+) REACTION

Liu Bo and Zhang Zong-Ye

Institute of High Energy Physics
Academia Sinica
Beijing, The People's Republic of China

I. ABSTRACT

The microscopic description of the π nucleus scattering is generalized to study the (p, π) reactions. The differential cross section of the ^{3}Hc(p, π^+)^{4}He at T_p^{lab} = 415 Mev is calculated by this method. The shape of the theoretical angular distribution coincides with the experimental data, but the absolute value is larger than the experimental value by a factor 2.

In recent years, there are several reports about experiments on (p, π^+) reactions in the (3,3) resonance region[1]. The differential cross section of ^{3}He(p, π^+)^{4}He has been calculated using the one nucleon mechanism model[2], in which the neutron wave function is taken to be the experimental charge form factor. The results did not coincide with the experimental data. (See Fig.1). On the other-hand, the isobar configuration model for pion production has proved so successful in explaining the p(p, π^+)d reaction. But in the process ^{3}He(p, π^+)^{4}He, the theoretical cross section of the quasi-deuteron approach was not in good agreement with the experimental data, especially the angular distributions. (See Fig.1). We think that in (p, π^+) reactions both the one nucleon mechanism and the isobar configuration model of two nucleon mechanism are important, and the coherence between them is also considerable. So we have generalized the isobar door-way state model[4] of π-nucleus scattering to study the (p, π^+) reaction. Its differential cross section in C.M. system is written as

$$\left(\frac{d\sigma}{d\Omega}\right)_{C.M.} = \frac{E_p E_\pi}{(2\pi)^2} \frac{k_\pi}{k_p} \sum_{i,f} \left| T_{fi}^{p\pi^+} \right|^2 \qquad (1)$$

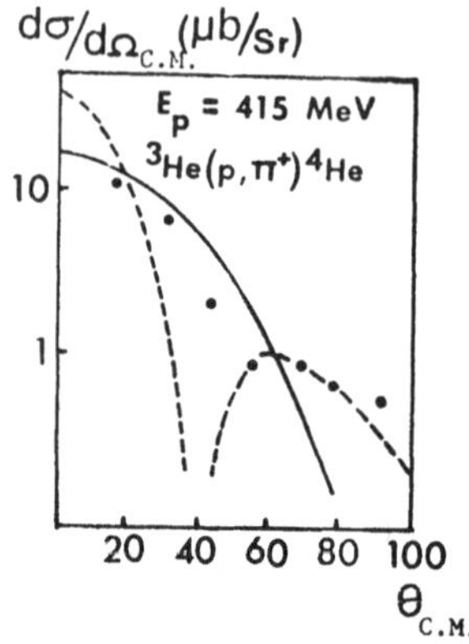

Fig.1 The differential cross section[2] of $^3He(p, \pi^+)^4He$.
----- One nucleon mechanism. (x 0.2)
——— Quasi-deuteron approach. (x 1.5)

and the transition matrix element can be expressed as

$$2E_\pi T_{fi}^{p\pi^+} = H_{NN\pi}^+ + H_{\Delta N\pi}^+ GV \qquad (2)$$

Here E_π, E_p are the energies and k_π, k_p are the momenta of π^+ and proton in the C.M. System respectively.

$$H_{NN\pi} = i \frac{f_{NN\pi}}{m_\pi} (\vec{\sigma}\cdot\vec{q})\vec{\tau}\cdot\vec{\phi}_\pi \qquad (3)$$

$$H_{\Delta N\pi} = i \frac{f_{\Delta n\pi}}{m_\pi} (\vec{\sigma}_{\Delta N}\cdot\vec{q})\vec{\tau}_{\Delta N}\cdot\vec{\phi}_\pi \qquad (4)$$

$$<\tfrac{3}{2}||\sigma_{\Delta N}||\tfrac{1}{2}> = <\tfrac{3}{2}||\tau_{\Delta N}||\tfrac{1}{2}> = 2 \qquad (5)$$

$$G = \sum_i \frac{|\Psi_i><\Psi_i|}{E_\pi - E_i} \tag{6}$$

is the Green function of the eigenstates for the Δ-h configurations, Ψ_i and E_i are the corresponding eigenstate and the eigen-energy respectively.

$$V = H^{\pi}_{NN\to N\Delta} + H^{\rho}_{NN\to N\Delta} \tag{7}$$

is a transition potential[4].

$$H^{\pi}_{NN\to N\Delta} = \frac{f_{NN\pi}\, f_{\Delta N\pi}}{m_\pi^2} (\vec{\tau}\cdot\vec{\tau}_{\Delta N}) \left\{ \frac{(\vec{\sigma}\cdot\vec{q})(\vec{\sigma}_{\Delta N}\cdot\vec{q}) - \frac{1}{3}(\vec{\sigma}\cdot\vec{\sigma}_{\Delta N})q^2}{\omega^2 - m_\pi^2 - \vec{q}^{\,2}} - \frac{1}{3}(\vec{\sigma}\cdot\vec{\sigma}_{\Delta N}) + \frac{1}{3}\frac{\omega^2 - m_\pi^2}{\omega^2 - m_\pi^2 - \vec{q}^{\,2}}(\vec{\sigma}\cdot\vec{\sigma}_{\Delta N}) \right\} \tag{8}$$

$$H^{\rho}_{NN\to N\Delta} = \frac{f_{NN\rho}\, f_{\Delta N\rho}}{m_\rho^2} (\vec{\tau}\cdot\vec{\tau}_{\Delta N}) \left\{ \frac{-(\vec{\sigma}\cdot\vec{q})(\vec{\sigma}_{\Delta N}\cdot\vec{q}) + \frac{1}{3}(\vec{\sigma}\cdot\vec{\sigma}_{\Delta N})q^2}{\omega^2 - m_\rho^2 - \vec{q}^{\,2}} - \frac{2}{3}(\vec{\sigma}\cdot\vec{\sigma}_{\Delta N}) + \frac{2}{3}\frac{\omega^2 - m_\rho^2}{\omega^2 - m_\rho^2 - \vec{q}^{\,2}}(\vec{\sigma}\cdot\vec{\sigma}_{\Delta N}) \right\} \tag{9}$$

In Eq.8 and 9, the first term gives rise to a tensor-like interaction; the second term, when Fourier transformed, will behave like $(\vec{\sigma}\cdot\vec{\sigma}_{\Delta N})(\vec{\tau}\cdot\vec{\tau}_{\Delta N})\delta(\vec{r}_1-\vec{r}_2)$ in configuration space, and the third term is Yikawa in nature.

In Eq.2 the one nucleon mechanism and the coherence between them are all included. It can be shown graphically in Fig.2.

We have calculated the differential cross section of $^3He(p, \pi^+)$ $^4He(T_p^{lab}=415$ MeV) by using Eq.2 and compared it with the experimental data[5].

In our calculation, the plane wave function is taken for the incident proton and the harmonic oscillator wave function ($\alpha=0.71$ fm^{-1}) for the Δ-isobar states. The neutron wave function

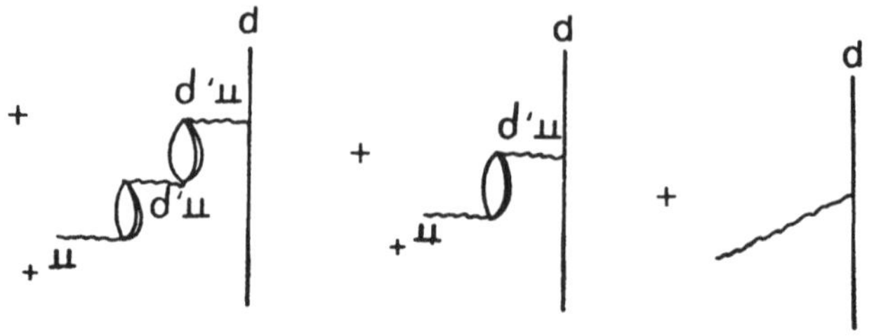

Fig.2 The mechanism for (p, π^+) reaction

in ^{4}He is given by the experimental charge form factor[6]. E_i and ψ_i are adopted as follows: ψ_i can be taken as the quasi-doorway states:

$$\psi_i = \psi_{NLJ} = \frac{1}{N_{nL}} \sum_{n_\Delta \ell_\Delta n_N \ell_N} (-1)^{\ell_\Delta} A_{\ell_\Delta \ell_N L} F_{n_\Delta \ell_\Delta n_N \ell_N L}(k) \quad (2n_\Delta + \ell_\Delta - 2n_N - \ell_N = n) \quad (a^+_{n_\Delta \ell_\Delta} b^+_{n_N \ell_N})_{LSJT} \psi_o \tag{10}$$

Here

$$A_{\ell_\Delta \ell_N L} = \frac{\hat{\ell}_\Delta \hat{\ell}_N}{L} C^{LO}_{\ell_\Delta O \ell_N O} \tag{11}$$

and

$$F_{n_\Delta \ell_\Delta n_N \ell_N L}(k) = \int R_{n_\Delta \ell_\Delta}(r) R_{n_N \ell_N}(r) j_L(kr) r^2 dr \tag{12}$$

$$E_i = E_{nLJ} = M_\Delta - M_N + n\hbar\omega + \mu_\Delta - i\Gamma_\Delta + 3 \sum_{j_\Delta} (\Delta\varepsilon_{j_\Delta}) \hat{j}^2_\Delta W^2 (1\tfrac{3}{2} J j_\Delta; \tfrac{1}{2} \ell_\Delta) \tag{13}$$

Here

$$\Delta\varepsilon_{j_\Delta} = \frac{1}{2} V_{\Delta\vec{\ell}\cdot\vec{s}} (j_\Delta(j_\Delta + 1) - \ell_\Delta(\ell_\Delta + 1) - \frac{15}{4}) \tag{14}$$

In eq.13, there are three parameters: μ_Δ, Γ_Δ and $V_{\Delta\vec{\ell}\cdot\vec{s}}$; they are determined from the cross section of the π-^{4}He elastic scattering[7].

$$\mu_\Delta = -43 \text{ MeV}$$

$$\frac{1}{2}\Gamma_\Delta = 63 \text{ Mev} \tag{15}$$

$$V_{\Delta\vec{\ell}\cdot\vec{s}} = 6 \text{ Mev}$$

(The harmonic oscillator energy $\hbar\omega$ is taken to be 22.5 Mev). The Δ-h configurations in eq.2 are included up to $5\hbar w$:

n	ℓ_Δ	ℓ_N	L	S	J
0	1s	1s	0	1	1
1	1p	1s	1	1	0
					2
2	2s	1s	0	1	1
	1d	1s	2	1	1
					3
3	2p	1s	1	1	0
					2
	1f	1s	3	1	2
					4
4	3s	1s	0	1	1
	2d	1s	2	1	1
					3
	1g	1s	4	1	3
					5
5	3p	1s	1	1	0
					2
	2f	1s	3	1	2
					4
	1h	1s	5	1	4
					6

To calculate the matrix elements of V, we neglect the tensor forces in the transition potential $H^{\pi}_{NN\to N\Delta}$ and $H^{\pi}_{NN\to N\Delta}$. For $H^{\pi}_{NN\to N\Delta}$, we drop the δ function piece, this is equivalent to making the Ericson-Lorentz-Lorenz correction. For $H^{\rho}_{NN\to N\Delta}$, the δ-function term has been discarded also. Since the modifications due to ω-exchange are considerable, a two-body correlation function is considered in $H^{\rho}_{NN\to N\Delta}$. Thus we have got an effective interaction:

$$\hat{H}^{\pi}_{NN\to N\Delta} = \frac{1}{3}\frac{f_{NN\pi}\, f_{\Delta N\pi}}{m_\pi^2}(\vec{\tau}\cdot\vec{\tau}_{\Delta N})\frac{\omega^2 - m_\pi^2}{\omega^2 - m_\pi^2 - \vec{q}^{\,2}}(\vec{\sigma}\cdot\vec{\sigma}_{\Delta N}) \tag{16}$$

$$\hat{H}^{\rho}_{NN\to N\Delta} = \frac{2}{3}\frac{f_{NN\rho}\, f_{\Delta N\rho}}{m_\rho^2}(\vec{\tau}\cdot\vec{\tau}_{\Delta N})\left\{\frac{\omega^2 - m_\rho^2}{\omega^2 - m_\rho^2 - \vec{q}^{\,2}} + \frac{\omega^2 - m_\rho^2}{K_c\,|\vec{q}|}\frac{1}{4}\,\ell n\,\frac{-(\omega^2-m_\rho^2)+(K_c+|\vec{q}|)^2}{-(\omega^2-m_\rho^2)+(K_c-|\vec{q}|)^2}\right\}(\vec{\sigma}\cdot\vec{\sigma}_{\Delta N}) \tag{17}$$

($K_c = 3.93\ \text{fm}^{-1}$ for ^{4}He)

For ^{3}He(p, π^+)^{4}He, the initial state is

$$|i\rangle = a^+_{k_p}\, b^+_{\ell_n}\,|0\rangle \tag{18}$$

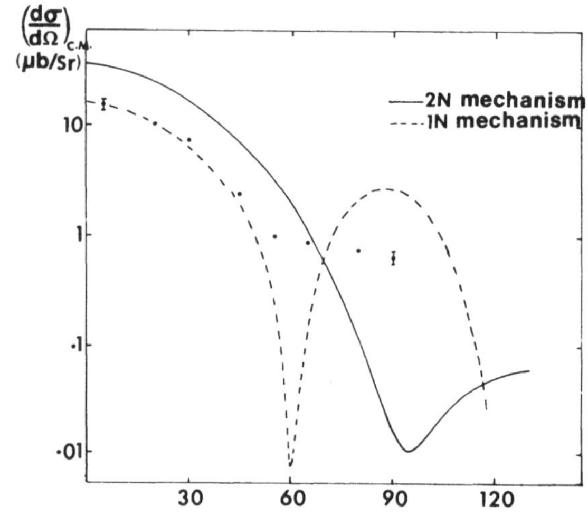

Fig.3a

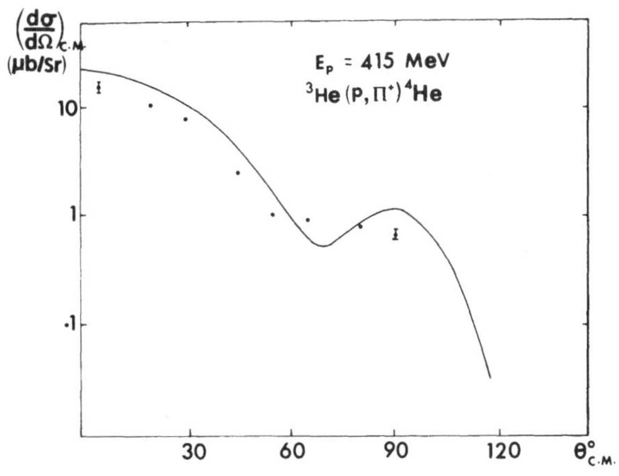

Fig.3b

and the final state is

$$|f> \; = \; \alpha^+_{k_\pi} |0> \tag{19}$$

The results of the differential cross section of $^3He(p, \pi^+)^4He$ are shown in Fig.3

In Fig.3a, the resulting differential cross sections of the one-nucleon mechanism are shown by dash line, and the isobar configurations model of two nucleon mechanism shown by solid line. In Fig.3b is shown the coherent result of the two. One can see that: (1) For one nucleon mechanism, there is a dip near $\theta=60°$, and for two-nucleon mechanism, there is a dip near $\theta=90°$. But experimentally there is no obvious dip from $\theta=0°$ to $\theta=100°$. (2) When the one-nucleon mechanism, the two nucleon mechanism and the coherence between them are all considered, a good agreement with the shape of the experimental angular distribution is obtained. The absolute values are about 2 times larger than the experimental values. We think it may be refined by taking the distrotion of the proton wave function.

This work shows that for (p, π^+) reaction in (3,3) resonance region not only the isobar configuration model of two nucleon mechanism is important but also the one nucleon mechanism and its coherence with the two nucleon mechanism are also important. The coherence between them has the property of helping to fill in the type of the dip.

REFERENCES

1. T. Baller, Phys. Lett. 69B, (1977) 433.
E. Aslanides, et al., Phys. Rev. Lett. 39, (1977) 1654.
J. Carroll, et al., Nucl. Phys. A305, (1978) 502.
2. B. Tatischeff, et al., Meson Nuclear Physics 1976, 0.470.
3. W. Weise, Nucl. Phys. A278, (1977) 402.
E. Oset and W. Weise, Nucl. Phys. A319, (1979) 477.
4. G.E. Brown, et al., Phys. Reports 50C, (1979) No.4.
5. B. Tatischeff, Phys. Lett. 63B, (1976) 158.
6. R.F. Frosch, Phys. Rev. 160, (1967) 874.
7. Gao Qin, et al., Physica Energiae Fortis et Physica Nuclearis (to be published).

SUPERMULTIPLET CLASSIFICATION AND COLLECTIVE SPECTRA OF TWO-ELECTRON ATOMS

Michael E. Kellman

Department of Chemistry
Columbia University
New York, N.Y. 10027

I. ABSTRACT

We review recent work on excited states of two-electron atoms. Starting with the O(4) symmetry of the hydrogen atom, we describe approximate two-electron O(4) "multiplets" which have spectra resembling that of a cutoff rigid rotor. We then expand the classification scheme beyond O(4) by grouping the multiplets into "supermultiplets". These reveal a spectral pattern similar to that expected if the two-electron atom had a structure like that of a highly nonrigid linear molecule undergoing collective rotation and vibration.

II. INTRODUCTION

In this paper we discuss recent work on highly excited levels of two-electron atoms aimed at providing a coherent picture of the electronic motion and spectra. With coworkers we have succeeded in obtaining a group theoretically based classification of these levels with a physical interpretation in terms of collective rotation-vibration motion. We feel that this development is important for two reasons:

1. It offers the prospect of a fruitful re-examination of atomic and molecular electronic structure at a more fundamental level using powerful new ideas and techniques.

2. It may serve as a bridge linking problems of atomic and molecular physics with those of nuclear and elementary particle physics.

It has generally been thought that collective behavior was not of relevance in atomic and molecular electronic structure problems. However, the situation is changing as experimental and computational advances open to scrutiny previously inaccessible phenomena. Generally, these involve highly excited states where correlation effects would be expected to be important.

Let us consider the simplest example, the two-electron atom. In a zeroth approximation, one would describe its excited states with a term symbol $^{2S+1}L^{\pi}$ and a pair of numbers $(n_1 n_2)$ specifying single particle atomic shell numbers. We shall deal here with "intrashell" levels $(n_1 n_2)$ where both the electrons reside in the same principal shell.

III. COLLECTIVE ROTATIONAL SPECTRA

Why should one consider these levels as possibly exhibiting a collective spectrum? First, because of the Coulomb repulsion, in energetically favorable states, the electrons should tend to reside on opposite sides of the nucleus. Second, with both electrons in the same shell, they should both be about equidistant from the nucleus. Analysis[1] of model doubly-excited wave functions[2] confirms these expectations. Consequently, we were led to propose[3] a linear, symmetric XYX "molecular" structure for the atom, with X's as electron and Y the nucleus.

We then considered which intrashell levels might bave a collective rotational spectrum corresponding to this structure. It turns out that among each intrashell manifold there is a series of levels $^1S^e$, $^3P^o$, $^1D^e$, ..., $^1L^e_{max}$ with $L_{max} = 2(n-1)$ such that each level is the lowest energy term within the (nn) manifold having its value of orbital angular momentum. These low energy levels are the ones most likely to correspond to our energetically favorable XYX structure.

Figs.1 and 2 compare for these levels energies obtained from extensive configuration interaction calculations with those of the rigid rotor, for various values of the nuclear charge Z and shell number n. We note that for low Z and high n, where correlation effects are expected to be greatest, there is marked resemblance to the rotor spectrum, a resemblance which fades as Z increases and n decreases.

A salient feature of our atomic rotor-like spectra is that they are cut off at $L_{max} = 2(n-1)$. This is a direct consequence of the limited number of states within a shell, and is thus a point of similarity to collective spectra in nuclei[4].

Having found rotor-like behavior for some of the intrashell

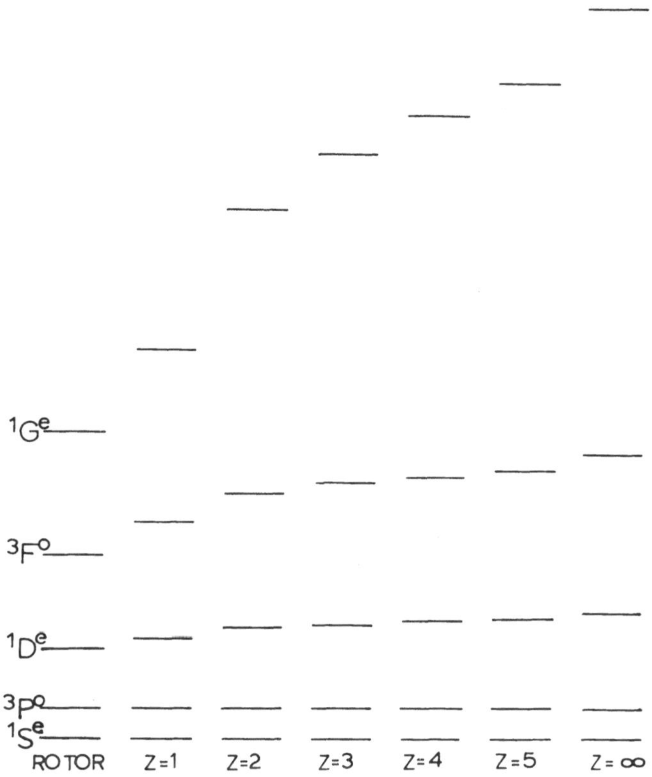

Fig.1 Comparison of relative energies of rigid rotor levels with low energy two-electron series $^1S^e$, $^3P^o$, $^1D^e$, $^3F^o$, $^1G^e$ for n = 3 and Z = 1-5, ∞. (From Ref.3).

levels, we sought a way to classify the entire intrashell manifold in terms of collective behavior which presumably would include both rotations and vibrations. To accomplish this we exploited and extended recent work on dynamical group theoretic classifications of two-electron atoms.

IV. GROUP THEORETIC CLASSIFICATION

It is well-known that the hydrogen atom is invariant under transformations of the four-dimensional rotation group O(4)[5]. The generators of this group are the ordinary angular momentum operators $\vec{l} = \vec{r} \times \vec{p}$ and the energy-weighted Runge-Lenz operators $\vec{b} = N(\vec{p}(\vec{r}.\vec{p}) - \vec{r}(\vec{p}^2 - 1/r))$. These operators satisfy O(4) commutation relations:

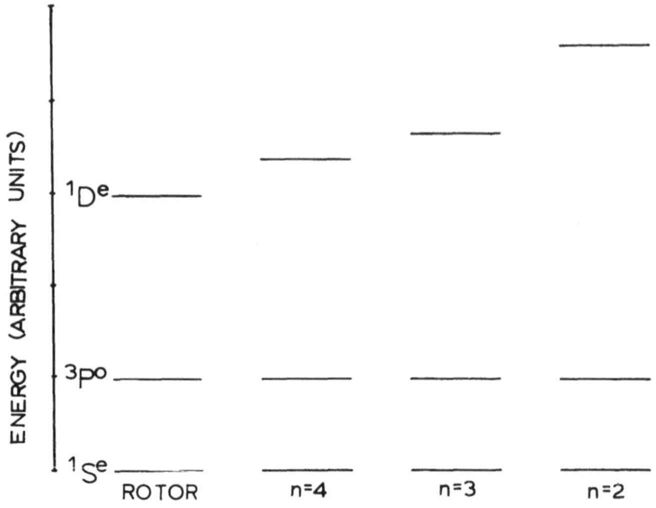

Fig.2 Comparison of relative energies of rigid rotor levels with low energy $^1S^e$, $^3P^o$, $^1D^e$ levels for Z = 2, n = 2-4. (From Ref.3).

$$
\begin{aligned}
[\vec{l}_i, \vec{l}_j] &= i\varepsilon_{ijk}\vec{l}_k \\
[\vec{l}_i, \vec{b}_j] &= i\varepsilon_{ijk}\vec{b}_k \\
[\vec{b}_i, \vec{b}_j] &= i\varepsilon_{ijk}\vec{b}_k
\end{aligned}
\tag{1}
$$

From the one-particle operators, one may form the following two-particle operators which also satisfy O(4) commutation relations:

$$
\begin{aligned}
\vec{L}_i &= \vec{l}_{i1} + \vec{l}_{i2} \\
\vec{B}_i &= \vec{b}_{i1} - \vec{b}_{i2}
\end{aligned}
\tag{2}
$$

$$
\begin{aligned}
[\vec{L}_i, \vec{L}_j] &= i\varepsilon_{ijk}\vec{L}_k \\
[\vec{L}_i, \vec{B}_j] &= i\varepsilon_{ijk}\vec{L}_k \\
[\vec{B}_i, \vec{B}_j] &= i\varepsilon_{ijk}\vec{B}_k
\end{aligned}
\tag{3}
$$

Note the unconventional choice of a minus sign in the definition of $\vec{B}_i$ in Eq.(2). It has been found[2] that diagonalization of $\vec{B}^2 = (\vec{b}_1 - \vec{b}_2)^2$ approximately diagonalizes the two-electron Hamiltonian in a model basis set. We now consider how this helps in classifying the two-electron states in the desired manner.

The irreducible representations of O(4) are labelled by two quantum numbers [P, T]. For instance, the degenerate manifold of hydrogenic orbitals $l = 0,1, \ldots, n-1$ is contained within a single irreducible representation of O(4) labelled [n-1, 0]. Two-electron representations of O(4) satisfy the Clebsch-Gordan series

$$[n-1, 0]_1 \times [n-1, 0]_2 = \sum [P, T]$$

with $T = 0,1, \ldots, n-1$

$$P = T, T+2, \ldots, 2n-2-T$$

These quantum numbers give a set of labels for classifying two-electron states over and above the usual L, S, and π. For instance, the low energy rotor series $^1S^e$, $^3P^o$, ..., $^1L^e$ with $L = 2(n-1)$ discussed above turns out to correspond to the O(4) multiplet with [P, T] = [2n, 0]. This is an indication of the physical usefulness of O(4) labels for classification purposes, even though O(4) is by no means an exact symmetry of the two-electron system.

V. SUPERMULTIPLET CLASSIFICATION

We now turn to the problem of classifying the entire intrashell manifold in terms of rotational and vibrational collective motion. We saw that the rotor series could be classified in terms of a single O(4) multiplet. It therefore is reasonable to hope that a classification of the O(4) multiplets themselves in terms of (as yet unspecified) "supermultiplets" should be the key to unraveling the problem. A procedure found by Herrick and Kellman[6] is to collect O(4) multiplets into supermultiplets with the quantum number $d \equiv \frac{1}{2}(P + T)$. A supermultiplet will then be denoted by the symbol {d}. In terms of this quantum number, the Clebsch-Gordan series of Eq.(4) may be written

$$[n-1, 0]_1 \times [n-2, 0]_2 = \{0\} + \{1\} + \ldots + \{n-1\} \quad (5)$$

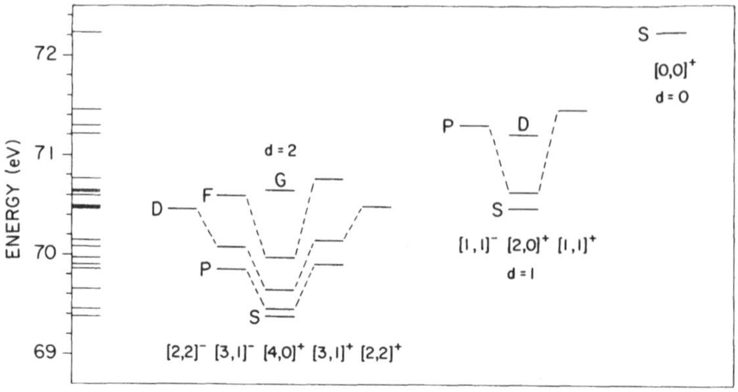

Fig.3 Supermultiplet classification of helium levels for n = 3. Note unresolved spectrum at left. (From Ref.6).

Each supermultiplet {d} in turn has the following O(4) multiplet reduction:

$$\{d\} = [2d, 0] + [2d - 1, 1] + \ldots + [d, d] \qquad (6)$$

Fig.3 shows the energy levels of the n = 3 manifold for helium as classified by the supermultiplet prescription. It is evident that this scheme resolves the spectrum into a very regular structure. Let us examine this from the point of view of collective motion[7].

VI. ROTATION-VIBRATION INTERPRETATION

First consider the largest supermultiplet in Fig.3, d = 2. Each vertical series of levels within the supermultiplet is labeled by $[P, T]^{\eta}$ where $\eta = (-1)^L$. Note the series $[4, 0]^+$. This is just the n = 3 rotor series considered previously (Fig.1, Z = 2). It is evident that the remaining O(4) multiplets also have an approximate rotor-like energy pattern. It turns out that each O(4) multiplet can be thought of as a rotational series built upon a vibrational level. Thus $[4, 0]^+$ has zero vibrational quanta; $[3, 1]^{\pm}$ have one; and $[2, 2]^{\pm}$ have two. (That $[P, T]^{\pm}$ are paired in this way we call "T-doubling", concerning which we shall have more to say shortly).

What is the nature of the vibrational excitation? In a linear XYX model, three kinds of vibration are possible[8]: a symmetric stretch, an asymmetric stretch, and (doubly degenerate) bending motions. The streches both change the distance of the X particles (electrons) from the nucleus; while in first approximation the

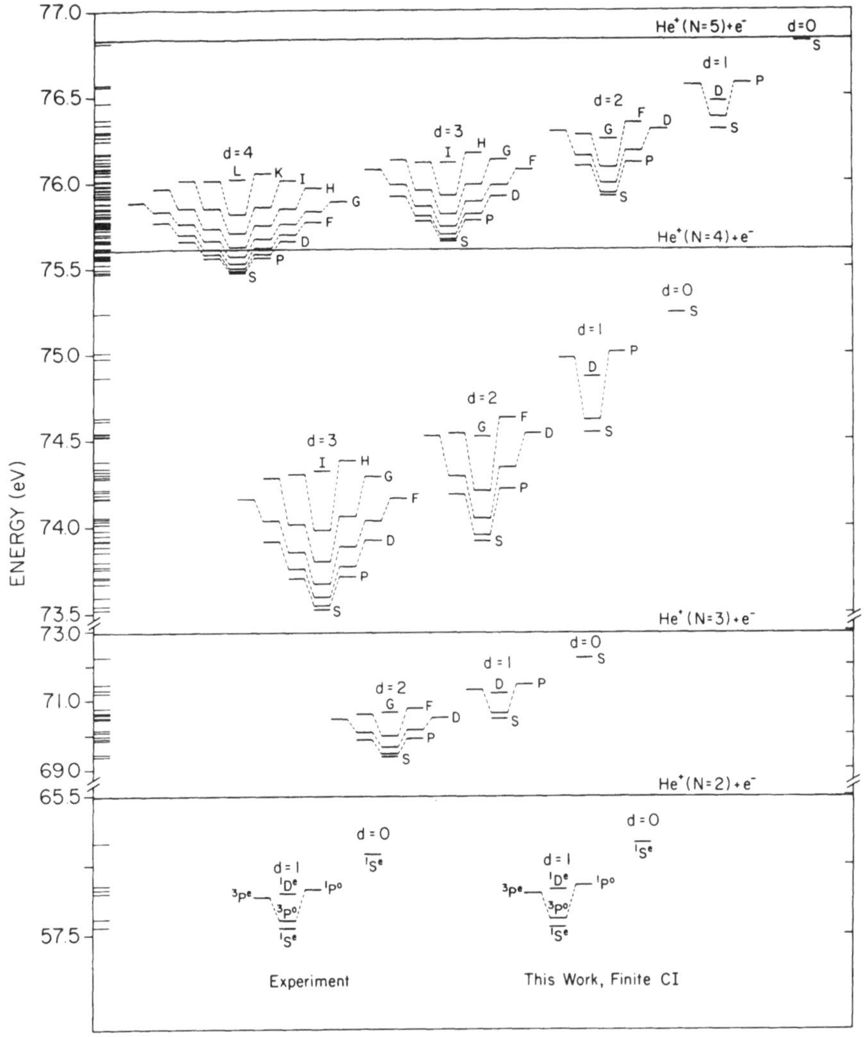

Fig.4 Supermultiplet classification of helium levels for n = 2-5. Note unresolved spectrum at left. (From Ref.9).

bending mode does not. Since the excitations we are concerned with keep both electrons in the same shell n, and therefore do not change radial distances appreciably, we might expect the vibrations we are looking for to be the bending modes.

The bending vibrations have two further favorable properties:

1. They combine to give a projection K of angular momentum along the axis of the XYX structure[8], much as a prolate nucleus can have

"intrinsic" angular momentum along its symmetry axis. We thus postulate identification of the quantum number K with the O(4) quantum number T. Note, for example, how the angular momentum of the lowest level of each multiplet fits this prescription. Thus, the lowest level of the $[3,1]^{11}$ multiplet is a P state with L = T = K = 1

2. For each value of K, there are two "intrinsic" states corresponding to the two degrees of freedom of the bending mode. This accounts for the "T-doubling" pattern noted earlier.

We have thus interpreted the d = 2 supermultiplet in terms of rotational and vibrational excitations. What is the interpretation to be applied to the other supermultiplets? It turns out[7] that one can make the identification $n-d-1 \leftrightarrow n_2$ where n_2 is the number of nodes in the wave function along a radial coordinate perpendicular to the symmetry axis of the XYX structure. Thus, the lowest level of the d = 1 supermultiplet has $n_2 = n-d-1 = 1$; this corresponds to two quanta of bending vibration coupled to give intrinsic angular momentum K = 0. The higher levels of this supermultiplet may be interpreted in terms of rotational excitations and further bending excitations, analogous to d = 2. However, note that there are fewer levels in the d = 1 supermultiplet than in the d = 2. This means the rotation-vibration progressions are truncated at fewer quanta of excitation. Finally, the d = 0 supermultiplet has $n_2 = 2$, corresponding to four bending quanta. There are no other levels in this supermultiplet -- it has no further rotation-vibration excitations.

We have thus been able to classify the entire n = 3 manifold in terms of supermultiplets which can be interpreted as rotation-vibration spectra. We now consider higher shells[9].

Fig.4 shows the helium intrashell spectra for n = 2, 3, 4, 5 classified according to the supermultiplet scheme. Note that in going from one shell to the next higher one, the same general structure is reproduced with successively greater complexity, with the rotation-vibration pattern clearly evident in each case.

VII. DISCUSSION

The work described here clearly demonstrates a collective spectral pattern for the doubly excited two-electron states, revealed by the supermultiplet classification scheme. Many future developments of this work immediately suggest themselves:

1. Extension to levels outside the intrashell manifolds, i.e., with values of the single particle shell numbers $n_1 \neq n_2$.

2. Explicit construction of an internal coordinate system for the

"intrinsic" XYX molecular structure of the atom.

3. Derivation of rotation-vibration parameters[7] (such as moment of inertia) from first principles.

4. Application of similar ideas to electronic structure of larger atoms[10] and molecules.

5. Elucidation of the relationship between this work and work on nonrigid molecular three-body systems[11].

REFERENCES

1. P. Rehmus, M.E. Kellman and R.S. Berry, Chem. Phys. 31, 239 (1978). See also: P. Rehmus, C.C.J. Roothaan and R.S. Berry, Chem. Phys. Lett. 58, 321 (1978); P. Rehmus and R.S. Berry, Chem. Phys. 38, 257 (1978).

2a. O. Sinanoglu and D.R. Herrick, J. Chem. Phys. 62, 886 (1975).

b. C. Wulfman, Chem. Phys. Lett. 23, 370 (1973).

3. M.E. Kellman and D.R. Herrick, J. Phys. B11, L755 (1978).

4. A. Bohr, Rev. Mod. Phys. 48, 365 (1976).

5a. V.A. Fock, Z. Phys. 98, 145 (1935).

b. V. Bargmann, Z. Phys. 99, 576 (1936).

6. D.R. Herrick and M.E. Kellman, Phys. Rev. A21, 418 (1980).

7. M.E. Kellman and D.R. Herrick, Phys. Rev. A22, 1536 (1980).

8. G. Herzberg, Molecular Spectra and Molecular Structure, vol.II (Van Nostrand Reinhold, New York, 1945).

9. D.R. Herrick, M.E. Kellman and R.D. Poliak, Phys. Rev. A22, 1517 (1980).

10. D.R. Herrick and M.E. Kellman, Phys. Rev. A18, 1770 (1978).

11a. M.E. Kellman and R.S. Berry, Chem. Phys. Lett. 42, 327 (1976).

b. M.E. Kellman, Chem. Phys. 48, 89 (1980).

c. M.E. Kellman, F. Amar and R.S. Berry, J. Chem. Phys. 73, 2387 (1980).

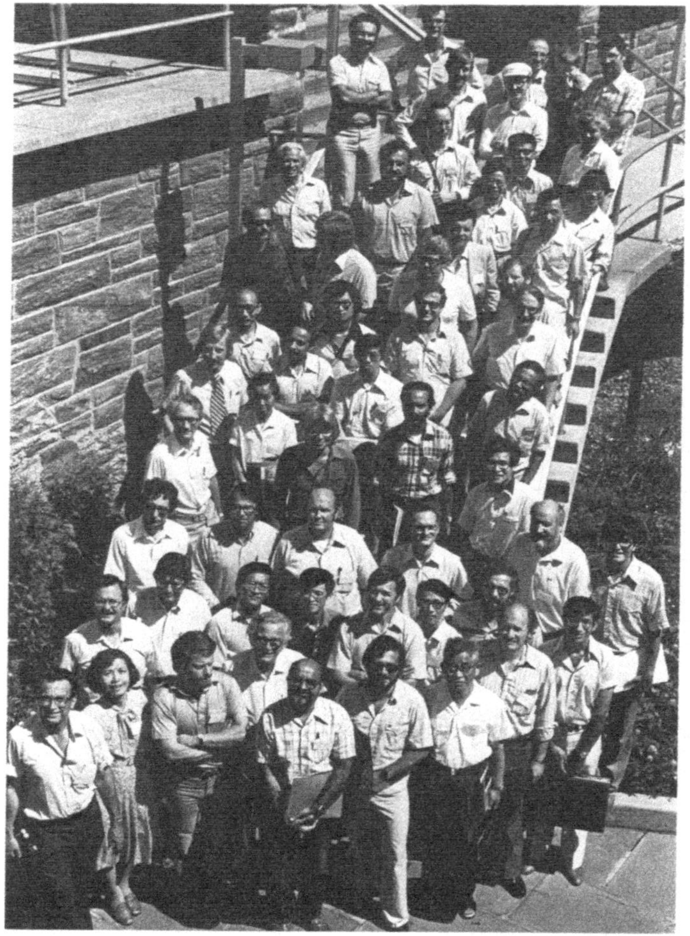

1
2
3
4
5
6
7
8
9
10
11
12
13
14
15
16
17
18
19
20
21
22
23
24
25
26
27
28
29
30
31
32
33
34
35
36
37
38
39
40
41
42
43
44
45
46
47
48
49
50
51
52
53
54
55
56

WORKSHOP IN NUCLEAR PHYSICS, DREXEL UNIVERSITY, September 1 - 3, 1980

1. Igal Talmi, Weizmann Institute
2. Ching-Liang Lin, National Taiwan University
3. Eckart Grosse, GSI Darmstadt
4. Rubby Sherr, Princeton University
5. Krishna Kumar, Tennessee Tech. University
6. Stu Pittel, Bartol Research Foundation
7. Sing-Nan King, Institute of Atomic Energy, Academia Sinica
8. Michel Vallieres, Drexel University
9. A. Moalem, University of Pennsylvania
10. Mike W. Guidry, University of Tennessee
11. Cheng-Li Wu, Jilin University
12. John S. Lilley, Daresbury Laboratory
13. Jian-Min Yuan, Drexel University
14. Teck-Kah Lim, Drexel University
15. Joe Hamilton, Vanderbilt University
16. William W. Eidson Drexel University
17. Keith Sage, University of Arizona
18. Rick Casten, Brookhaven National Laboratory
19. Jerry D. Garrett, Niels Bohr Institute
29. Jean Barrette, Brookhaven Nationa Laboratory
21. Sven Hjorth, Oak Ridge National Laboratory
22. Da Hsuan Feng, Drexel University
23. Osamu Hashimoto, LBL
24. S. Chakravarti, University of Minnesota
25. John Wood, Georgia Tech.
26. Zong-Ye Zhang, Institute of High Energy Physics, Academia Sinica
27. James Griffin, University of Maryland
28. Lee L. Riedinger, University of Tennessee
29. Afsar Abbas, Rutgers University
30. C.T. Li, University of Pennsylvania

31. R. Bhaduri, McMaster University
32. D.R. Bes, Buenos Aires
33. Pedro Federman, UNAM
34. C.K. Lin, Rutgers University
ty 35. Gen-Ming Jin, Institute of Modern Physics Academia Sinica
36. S. Frauendorf, Rossendorf
37. Jan Vaagen, Bergen University
38. G.G. Dussel, Buenos Aires
39. Robert Gilmore, Insti. for Defense Analyses
40. Zhu-Xia Li, SUNY at Stony Brook
41. Li-Ming Yang, Peking University
42. Jin-Quan Chen, Nanjing University
43. Peter von Brentano, Koln
44. W. Andrejtschess, Rutgers University
45. Peter Hodgson, Oxford
46. O. Scholten, MSU
47. Shi-Shu Wu, Jilin University
48. Hui Ye, Fudan University
49. Uzi Smilansky, Weizmann Institute
50. Son of Lutz Cleemann
51. Mrs. Lutz Cleemann
52. Lutz Cleemann, Koln
53. E.R. Marshalek, Notre Dame
54. John Rasmussen, LBL
55. M.E. Kellman, Columbia
56. C. Theodosiou, Drexel University

MISSING A. Klein, University of Pennsylvania
L. Zamick, Rutgers University
Gerold J. Borse, Lehigh University
Robert T. Folk, Legigh University
George Temmer, Rutgers University

INDEX